USEFUL PROCEDURES
OF INQUIRY

BY

ROLLO HANDY and E. C. HARWOOD

INCLUDING

KNOWING AND THE KNOWN

John Dewey and Arthur F. Bentley

AND

INTRODUCTION TO JOHN DEWEY'S PHILOSOPHY

Joseph Ratner

BEHAVIORAL RESEARCH COUNCIL
Great Barrington, Massachusetts

Library of Congress Catalog Card No. 72-93865
International Standard Book Number 0-913610-00-3

Manufactured in the United States of America

PREFACE

Many workers have devoted innumerable man-hours to the problem of what "knowledge" is and how we can obtain it, often without useful results. We are indebted to a great many other people in our attempt to describe more useful procedures of inquiry. Although we have endeavored to give credit for our borrowings by way of footnotes, the field is so large that we surely have not acknowledged all who have contributed to our efforts. In addition to those mentioned in the text, we wish to thank a succession of graduate students who pursued their research at the American Institute for Economic Research. Those students, along with some Institute staff members, participated in a series of seminars on procedures of inquiry and aided materially in the development of the contents of this volume.

We also wish to record our appreciation to the following:

Random House, Inc., for permission to reprint the whole of Joseph Ratner's "Introduction to John Dewey's Philosophy," which originally appeared in Joseph Ratner, editor, *Intelligence in the Modern World: John Dewey's Philosophy,* New York, The Modern Library, Random House, 1939;

Mr. Julius Altman, for permission to reprint the whole of *Knowing and the Known,* by John Dewey and Arthur F. Bentley, originally published by Beacon Press, Boston, in 1949;

Warren H. Green, Inc., for permission to reprint Chapters 1, 2, and 3 of Section D;

American Institute for Economic Research, for permission to reprint Chapters 4, 5, and 6 of Section D;

The American Journal of Economics and Sociology, for permission to reprint Chapter 7 of Section D.

The laborious work on the Index was done by Martha Lane and Ann Bloch to whom we are deeply grateful.

R. H.
E. C. H.

TABLE OF CONTENTS

INTRODUCTION

(Intended To Be Read)

For at least 6,000,000 years men presumably have been inquiring into their problem situations. Survival of the human race evidences some degree of success in those inquiries, but our current problem situations provide equally good evidence that much more effective inquiry could be used to advantage.

The available records inadequately describe the procedures of man's early inquiries. Only a hint as to the procedures of inquiry that were applied prior to a few thousand years ago are provided by a few artifacts that survived the millennia. After written communications became generally used, however, many additional and more adequate descriptions of inquiry into problem situations were preserved for subsequent generations. Approximately 2,000 years ago, concern with the procedures of inquiry had become a primary occupation of the Greek philosophers. They were the scientists of their day, and their procedures of inquiry became so well established that even today those same procedures are still used by some inquirers.

Not surprisingly, the procedures carried over from earlier times involved language uses and habitual attitudes that were firmly established in human culture, as is illustrated in the functioning of witch doctors, tribal medicine men, oracles, and other purveyors of word magic, as well as community historians whose oral communications included ancient proverbs and much mythology.

Less than four centuries ago, however, a revolution in conducting inquiries began. Galileo ordinarily is regarded as the leader of that revolution. The new procedures initially were applied in the physical sciences, and the consequences were startling in that they superseded much that had been considered irrefutable "knowledge." But this was only the beginning of the revolution. In the physical sciences it has continued to date, facilitating not only wider applications of new "knowledge," but also the evolutionary development of the procedures of inquiry themselves. Thus new "knowledge" becomes established and widely applied, only to be superseded by even more useful "knowledge," as when Newton's work in part superseded Galileo's and Einstein's work to some extent superseded Newton's.[1]

Several decades after Galileo, aspects of the new procedures were applied in the physiological sciences with similarly startling and continuing consequences. As in the physical sciences, the cultural lag in applying the new procedures was great. The eminent physicians who hounded Dr. Semmelweiss out of his profession, after he had ascertained the circumstances in which puerperal fever could be expected, continued to endanger the lives of mothers in childbirth, not because of evil intent but because of their failure to grasp the new ideas; they were

[1] Many others also were concerned both in the application of the new procedures and in their further development, but for present purposes the detailed record need not be described.

too confident that they knew what they claimed to know. Even the students of those who denounced Semmelweiss were continuing the unpremeditated killing of mothers in childbirth as long as a half century after Semmelweiss' great discovery.

More recently, primarily in the present century although earlier traces can be found, a beginning has been made in applying the new procedures of inquiry to the problems of men in society. In several of the behavioral sciences progress is discernible, albeit miniscule in relation to the job to be done.

To date the outstanding success appears to have been in inquiry into inquiry in order to describe those procedures that have proven to be most useful and therefore perhaps offer the greatest promise for the future. That epistemology, as the philosophers designated inquiry into inquiry, had become the most dismal semantic swamp in which human inquirers ever had been lost became apparent to many observers decades ago. Not until the middle of this century had men applied the new procedures of inquiry with marked success. The results seem comparable in their potential revolutionary effects to the initial efforts of Galileo. If the new procedures are applied in all the behavioral sciences, the methods may be as successful as they already have been when applied to inquiry into inquiry. Some beginnings have been made, and the results are encouraging.

No one should suppose, however, that success will be rapid and unhindered. Again the cultural lag will be encountered. Thousands of tenured professors and their textbooks will not soon be superseded by better "knowledge," even if hundreds instead of a few dozen inquirers soon apply the new procedures with increasing and cumulative success.

We venture one prediction: Those who are confident that they already know what they claim to know will hinder progress and will continue to waste their lives in substantial part for the privilege of retaining old linguistic and other procedural habits; but other inquirers more receptive to new ideas in the decades to come will lead the way out of the dismal swamp toward better solutions to the problems of men in society.

We are aware that our prediction, our guess, is not as yet a scientifically warranted assertion. Only the successful outcome of future applications of the new methods of inquiry can provide the warrant that our views are sound. Therefore, we should welcome reports from those who apply the new procedures. If the results are unsatisfactory, we should like to be among the first to know; and we should especially appreciate being informed about the further development of the new procedures, which we expect in the decades ahead as a matter of course. We suspect that the modest claims for the new procedures (not purporting to achieve certainty), coupled with the provisions for further development embodied in them, are among their more significant and useful characteristics.

Section A

SCIENTIFIC INQUIRY

I.

ORIENTATION FOR THE READER

The extensive literature on scientific inquiry provides many conflicting recommendations as to the most useful procedures of inquiry to be followed. The tentative conclusions we have reached in our inquiry into inquiry are different in many respects from the descriptions of scientific method usually found. The following summary of some of our major, but tentative, conclusions may therefore be useful to readers. Our immediate objective in this short chapter is not to satisfy the reader that our views are useful, for that is the task undertaken in the entire volume, but rather to provide orientation for the reader before he grapples with detailed consideration of the problem.

We have attempted to make this summary report immediately useful to the reader by avoiding specialized terminology and by stating our conclusions with a minimum of elaboration. The full account of what we should both include and exclude in the suggested procedures of inquiry will require detailed consideration of the material provided in the rest of the volume; the orientation summary below should indicate the general direction of the advance we believe has been made and thus facilitate the reader's task.

How Inquiry Proceeds

1. The inquirer becomes aware of a problem situation.

2. He observes some facts that appear to be pertinent. Various aspects and phases of the situation are differentiated, some changes among them are measured, and a tentative partial description of what is happening is begun.

3. In noting connections among some of the things observed and measured, other connections may be imagined. The inquirer focuses on what seem to be the pertinent aspects and phases of the situation, and develops a conjecture as to what may happen under specified circumstances.

4. That conjecture may involve other facts to be observed, perhaps including some facts originally not believed to be pertinent. As the inquirer proceeds, he may find that the original problem situation is quite different than it first had seemed.

5. The tentative description of what happens is supplemented and perhaps revised. Transformations via verbal or mathematical logic may be used. What were earlier taken as facts may be revised or rejected.

6. Perhaps another conjecture occurs to the inquirer about possible connections among facts, including measured changes.

7. Investigation of the new conjecture requires further observation and perhaps results in the development of a more adequate description. These procedures of observation, reconsideration, renewed observation; i.e., the interweaving or reciprocal stimulation of what are sometimes called empirical observations and the formulation of hypotheses, may be repeated many times in succession.

8. Finally, if the inquirer is successful, a description adequate for resolving the immediate problem situation is developed.

9. Future inquiries may further supplement the description of what happens; in some instances new inquiries may reveal aspects or phases that force drastic amendment of the best earlier description.

10. Inquiry has no necessary end. A *complete* description of even a simple problem situation apparently never has been achieved and may never be, but an adequately useful description is the goal of modern scientific inquiry.

The actual order of successful inquiry seems to be: awareness, observation, partial description, conjectures leading to further observation, etc., until an adequate description has been formulated. Descriptions of a small part of the full sequence may be mistaken as the key to the whole process; e.g., when inquiry is understood as beginning with a "well-formulated" hypothesis and then searching for evidence, or when mathematical transformations are assumed to be the essence of scientific inquiry, or when logical deductions are emphasized.

If the interweaving of observation and tentative descriptions that has been so productive in past inquiries is departed from, the verbal and mathematical transformations used by the inquirer may not be applicable to the data involved. Many inquirers have endeavored to achieve useful descriptions by means of extended verbal logic, or by mathematical transformations not closely connected with observable data, or by the computer models that recently have become popular (by means of which so-called "theoretical constructions" and mathematical transformations have been mechanized or automated). Some of the displays of technical proficiency may be impressive, especially to those less skilled in mathematics, but there is little evidence that useful scientific inquiry has been advanced thereby, and there is much evidence that such elaborate theorizing lends a false appearance of authenticity to assertions that by no means are scientifically warranted.

II.
PRELIMINARY COMMENTS

Inquiry usually begins with awareness of a problem situation; i.e., when human behaving encounters some kind of obstacle, hindrance, or "hitch." The problem may be relatively simple, such as the necessity of choosing between two directions at a fork in the road, or may be far more complex, such as a cholera epidemic with the attendant risk of health or even life.

In the course of man's long experience, now believed to have been many millions of years,[1] many different procedures of inquiry have been applied in attempting to solve such problems. What is accepted as a solution has varied considerably; sometimes emphasis is placed on what is intellectually or subjectively satisfying, or what is in accordance with deeply held beliefs. Compared to inquirers in other fields, scientists often have been remarkably successful in developing useful answers to the problems they have encountered. More accurate description of those successful procedures of inquiry may lead to further improvement in scientific work and facilitate the application of those procedures to problems not yet solved, in some instances because they have been regarded as beyond the range of scientific inquiry.

Many disagreements about the application of the name "science" are found in the accounts of scientific inquiry by scientists, philosophers of science, and historians of science, particularly concerning the appropriate procedures of inquiry and the criteria by which work is to be judged. Some recent writers, for example, emphasize "superb taste," or individual intuition, or rational insight transcending empirical observation, as necessary in the development of major scientific theories.[2] Other writers, probably far more numerous, emphasize public confirmation based on observations and the logical/mathematical connections involved.

Such arguments typically occur within the context of conventional epistemological categories, as illustrated by debates about the relative contributions of the "rational" and the "empirical" in scientific "knowledge." The following quotation from Polanyi, a physical chemist, illustrates the type of epistemological framework within which the arguments between the "rationalists" and the "empiricists" sometimes occur:

"...the discovery of objective truth in science consists in the apprehension of a reality which commands our respect and arouses our contemplative admiration; ...while using the experience of our senses as clues, [it] transcends this experience by embracing the vision of a reality beyond the impressions of our senses.... Twentieth-century physics...[shows] the power of science to make contact with reality in nature by recognizing what is rational in nature."[3]

Compared to the work of philosophers, theologians, literary scholars, etc., scientists have achieved remarkable results; yet discussions of the procedures of inquiry used by scientists typically are carried on within the same framework of assumptions about the objectives and criteria of "knowledge" that is accepted in nonscientific and prescientific inquiries. That state of affairs is not surprising in view of man's cultural history, but it also suggests another possibility: attempting to describe as adequately as possible the procedures used in successful scientific inquiry without initial reliance on the traditional epistemological procedures.

A detailed and thorough alternative to traditional epistemologies was developed by a group of American thinkers. John Dewey, making use of important work by Charles Sanders Peirce and William James, provided a detailed report on inquiry in his *Logic: The Theory of Inquiry* (1938). Later Dewey collaborated with Arthur F. Bentley over a period of many years. They published *Knowing and the Known* in 1949, in which substantial improvements were made in the *Logic*. The Dewey-Bentley *Correspondence* (1964) continues a record of many of those improvements and further criticism of traditional views.[4]

Unfortunately, the Dewey-Bentley description of useful procedures of inquiry frequently is misunderstood, even by those who seem to be sympathetic to it. A major reason for such misunderstandings apparently is that many readers fail to grasp the extent to which Dewey and Bentley have rejected the traditional epistemological framework for inquiry. Anyone who wishes to judge the merits of Dewey and Bentley's work should first acquire an understanding of what they are saying.

Joseph Ratner has reported in considerable detail on the historical background of leading contemporary notions about "knowledge," how inquiry is most usefully conducted, and related matters. His account, written at the same time as Dewey's *Logic* was being completed, is so useful that we are reprinting his entire account as Section B of the present volume.

Knowing and the Known, which is the culmination of decades of work by Dewey and Bentley, may present considerable difficulties for readers not already conversant with the chief materials found in the book. In order to facilitate understanding of that book, which is reprinted as Section C of the present volume, and to update Ratner's Introduction, the next two chapters explore additional background material. First a survey is made of some of the principal procedures of inquiry that currently are advocated, but that we regard as being outmoded, and then a summary restatement of the principal aspects of Dewey and Bentley's procedures is given.

---

[1] This time estimate includes the "man-apes" believed to be the progenitors of modern man. New discoveries have consistently pushed back such estimates. *Ramapithecus*, the oldest form most authorities regard as belonging to the hominid line of development, is now estimated to have lived about 14 million years ago. *Australopithecus* is now estimated to go back approximately 6 million years. And the anthropoids in general date back some 40 million years. See David Pilbeam, "The Earliest Hominids," *Nature*, Vol. 219, Sept. 28, 1968. Interesting discussions of recent findings are given in *Science News*, Feb. 27, 1971 and Feb. 26, 1972.

[2] See for example, Gerald Holton, "Einstein, Michelson, and the 'Crucial' Experiment," *Isis*, Vol. 60, 1969; Giorgio de Santillana, *Reflections on Men and Ideas*, Cambridge, M.I.T. Press, 1968; Michael Polanyi, *Personal Knowledge*, Chicago, University of Chicago Press, 1958.

[3] Polanyi, *op. cit.*, pp. 5-6.

[4] John Dewey, *Logic: The Theory of Inquiry*, New York, Holt, Rinehart and Winston, 1938; John Dewey and Arthur F. Bentley, *Knowing and the Known*, Boston, Beacon Press, 1949, paperback edition, 1960; Sidney Ratner, Jules Altman, and James Wheeler, eds., *John Dewey and Arthur F. Bentley: A Philosophical Correspondence, 1932-1951*, New Brunswick, Rutgers University Press, 1964. Also pertinent to the later phases of this line of development is Arthur F. Bentley, *Inquiry Into Inquiries* (Sidney Ratner, ed.), Boston, Beacon Press, 1954.

III.
SOME OUTMODED PROCEDURES OF INQUIRY

Many alleged "ways of knowing" have been tried in the course of man's cultural history; diverse views are found, either explicitly stated or implicitly assumed, concerning what knowing is, what the successful outcomes of knowing are, and what the relation is between knowings and knowns. Many useful ways of grouping these methods for purposes of analysis and discussion could probably be found. For present purposes three emphases (Mentalistic/Rationalistic, Formal Model Building, and Subjectivism) will be considered separately in order to facilitate discussion, although in a given inquiry aspects of all three, and of other procedures also, may be found.

A preliminary point of some consequence will be discussed first. Many, perhaps most, of the alleged "ways of knowing" currently found involve what Dewey called the "quest for certainty."[1] The difficulties, hazards, and uncertainties of life are frequently so frustrating that humans long for an absolute certainty that will give the type of security we fail to achieve in grappling with problems in the here and now. That absolute certainty may be said to reside in knowledge of a Platonic heaven of ideas, of Aristotelian essences, of supernatural eternities, or of epistemological "incorrigibles," but some absolutely certain way is allegedly available for apprehending truths that are absolutely certain.

For the Greeks, the procedures of inquiry that purportedly led to absolutely certain knowledge were coherently related to many other major aspects of their cultural system. Aristotle's logic, for example, was "of a piece" with the scientific inquiry of his day and the prevailing notions of what reality was like. As scientific inquiry progressed, and shifted from taking immutable objects as subject matter to a focus on correspondences of changes, many of the major procedures of inquiry associated with the old immutable-object framework were still retained, with consequent inconsistency or incoherence.

To illustrate, Aristotle said:

"It is absurd to make the fact that the things of this earth change and never remain the same the basis of our judgments about the truth. For in pursuing the truth one must start from things that are always in the same state and never change. Such are the heavenly bodies; for they do not appear to be now of one nature and now of another, but are always manifestly the same and do not change."[2]

Almost no one today would make the fundamental contrast Aristotle did between the earth and the heavenly bodies, yet a great many do accept the notion that knowledge must be based on some indubitable or incorrigible starting point, on which we build other knowledge, perhaps of a transcendent reality. While surely Polanyi would not accept the Aristotelian distinction between earth and heavens, the statement quoted on the preceding page suggests that he retains much of the characteristic Greek view about inquiry. Aristotle, of course, could be mistaken about the relation of the earth and the heavens, and still say useful things

about "knowledge." But the more we depart from the old Greek views about scientific subject matter, the more questionable seems the retention of associated notions about procedures of inquiry.

A. *Mentalistic/Rationalistic Method.* The procedures of inquiry in this method assume a basic split between the mental and the nonmental, which are viewed as fundamentally different types or levels of reality. Knowledge about nonmental reality is said to be possible for the mind, and attainable by means of propositions or some other intervening entities that purportedly represent nonmental reality to or for the mind. Knowledge about much human behavior is alleged to be attainable through direct inspection of the mind by the mind, introspection, or some other type of action of the mind upon its own processes. Ratiocination is emphasized; important knowledge is believed to be attainable independently of observation or experience.

Illustrations can be found in many fields of inquiry. Albert Einstein once wrote: "Nature is the realization of the simplest conceivable mathematical ideas. . . .In a certain sense, therefore, I hold it true that pure thought can grasp reality, as the ancients dreamed." Ludwig von Mises asserted that to understand human behavior "there is but one scheme of interpretation and analysis available, namely, that provided by the cognition and analysis of our own purposeful behavior," and that the "ultimate yardstick of an economic theorem's correctness or incorrectness is solely reason unaided by experience." In recent years Noam Chomsky's work has had a strong influence not only on linguists but on inquirers in many other fields. Chomsky has revived a Cartesian dualism, maintains that the mind possesses innate knowledge, and claims that psychology is the science of mind, not of behavior (he argues that to view psychology as a behavioral science would be like viewing physics as a science of meter readings). Chomsky goes so far as to describe a child as being born "with a perfect knowledge of universal grammar, that is, with a fixed schematism" that is used in acquiring language.[3]

Probably the number of inquirers working in scientific fields who espouse rationalistic mentalism as boldly as the authors just quoted is not large, but milder or more disguised views (and as such, at least potentially more dangerous) frequently are found. The belief that the mind exists as some sort of entity, and that it can provide important knowledge in advance or independently of observation, is widespread.

Historically the assumption of the existence of a substance with fixed properties has frequently impeded scientific progress, and in instance after instance such assumptions have been replaced by inquiry into processes. For example, heat was regarded as a thing, rather than a

1 John Dewey, *The Quest for Certainty*, New York, Minton, Balch, 1929. See also Ratner's account, pp. 27-28 of the present volume.

2 Aristotle, *Metaphysics*, 1063ª, Ross translation.

3 The statement from Einstein is quoted in Gerald Holton, "Mach, Einstein, and the Search for Reality," *Daedalus*, Spring, 1968, p. 650. Ludwig von Mises, *Human Action: A Treatise on Economics*, New Haven, Yale University Press, 1949, p. 26, p. 858. For Chomsky's views, see his *Language and Mind*, New York, Harcourt, Brace & World, 1968; *Cartesian Linguistics*, New York, Harper and Row, 1966; and his article, "Linguistics and Philosophy," in Sidney Hook, ed., *Language and Philosophy*, New York, NYU Press, 1969. The direct quotation is from p. 88 of the Hook volume.

doing, for a long period; the substance framework was so cherished that when it was discovered that heat had no weight, many argued that it was a special kind of weightless substance, rather than concluding that it was not a substance.

In the specific instance of mind, not only do we have all the difficulties associated with trying to observe mind as an entity, but also the "internal" problems generated by the theory, such as the relation between mind and body. All those difficulties might be overlooked (i.e., regarded as relatively insignificant) if the conclusions reached were generally warranted and useful, but in fact the presumed universal or normal operations of the mind notoriously give rise to conflicting results that vary from person to person and from cultural setting to cultural setting.

B.F. Skinner has argued that "mentalistic or psychic explanations of human behavior almost certainly originated in primitive animism."[1] The attractiveness of mentalism for many should not be underestimated, despite its apparent origin in animism. Perhaps the recurrence in the field of linguistics is at least partially attributable to failures of many allegedly scientific inquirers to describe adequately what we call here "sign-behavior." Such behavior is so distinctively human that failures to inquire successfully into it may tend to reinforce strongly the belief that man's language abilities make him ontologically different from the "natural" world.

B. *Formal Model Building.* In recent years there has been a marked emphasis on formal model building across the behavioral science fields. Some models are mathematical, some are cybernetic, some are from game theory, some are from information theory, etc. As Alphonse Chapanis notes: "It is almost as though there was a special form of magic attached to the word 'model'." He goes on to say that what previously would have been described with "words like hypothesis, theory, hunch, and empirical equation, are now very often called models because it is the thing to do."[2]

In view of all the things that are called models, and the complexities of scientific inquiry, our criticisms in this section should not be taken as a general opposition to models in inquiry, for they may be highly useful. Rather, we are objecting to misleading or uncritical uses of models. Often in the history of scientific inquiry a technique that leads to progress is taken as a kind of master-key to inquiry; the success of axiomatization in some areas of physical inquiry has apparently convinced some commentators that all scientific subject matters should be axiomatized.[3] Similarly, the success of model building in some areas of inquiry has led some workers to believe that great progress generally is to be expected through the use of models. But, as Dewey emphasized, often we are tempted to convert a useful function or procedure in inquiry into an "independent structure," and to overdo the "ritual of scientific practice at the expense of its substance" by insisting on some procedure not required by the consequences to be effected. He also notes that premature formalization can "perpetuate existing mistakes while strengthening them by seeming to give them scientific standing."[4]

Often the justification given for the construction of models is that a more direct attack on the problem at hand is impossible, difficult, or ineffective. At other times, however, the notion seems to be that a formal model can provide insights into "reality" that direct inquiry into the data cannot give. Whatever the reasons for building a model, the following stages are often found:

1. Various notions about human behavior are taken as sound (as axiomatic, or as truisms, or as otherwise assured).

2. These notions are translated into a formal model, sometimes by traditional logic and sometimes by mathematics.

3. Numerous transformations, perhaps requiring marked ingenuity and technical proficiency, are performed within the model.

4. The results of those transformations, when translated out of the model, hopefully describe important aspects of human behavior.

Various difficulties arise often enough to warrant specific mention. The initial notions taken as sound may be dubious. For example, George Homans developed a model of "elementary social behavior" (viewed as "face-to-face contact between individuals, in which the reward each gets from the behavior of others is relatively direct and immediate") based primarily on economic theory and behaviorist psychology. In the light of the results of other inquiries, one would suspect that differing cultural settings and social structures would make for marked differences even in "elementary" social behavior. But Homans assumes that in such behavior there is "neither Jew nor Gentile, Greek nor barbarian, but only man" and goes on to say that "though I believe that the general features of elementary social behavior are shared by all mankind, I believe it as a matter of faith only, and the evidence that I shall in fact adduce is almost wholly American."[5] Even if Homans' faith turned out to be justified, should not evidential confirmation of such a basic point precede the elaboration of a model of universal elementary behavior?

Frequently—perhaps even typically—the greatest effort and concern goes into the transformations possible within the model. Such work, as noted, may require a high degree of competence and ingenuity, but it neither demonstrates the correctness of the original notions nor the usefulness of the results for describing the range of behavior the model was designed to represent (or to "illuminate," for some inquirers maintain that models can provide insight into what the models do not represent). In short, a warranted assertion about processes within a model is not the same as a warranted assertion about the behavior the model purports to describe. Dewey has made a parallel point that, although elementary, often is overlooked in the context of formalized procedures. After emphasizing the usefulness of deductively developed mathematical formulae in physics, he notes that "the value of the deduced result for physical science is not determined by the correctness of the deduction."[6]

The confidence model builders have in the soundness of their assumptions, combined with their interest in working out the details of the model, seems often to lead to poor workmanship when it comes to testing the results against observed behavior. Chapanis, for example, notes that: "Even when we find model builders attempting to make

1 B.F. Skinner, "Behaviorism at Fifty," in T.W. Wann, ed., *Behaviorism and Phenomenology*, Chicago, University of Chicago Press, 1964, p. 79.

2 Alphonse Chapanis, "Men, Machines, and Models," *American Psychologist*, Vol. 16, 1961, p. 114.

3 See, for example, David Greenwood, *The Nature of Science*, New York, Philosophical Library, 1959.

4 John Dewey, *Logic: The Theory of Inquiry*, New York, Holt, Rinehart and Winston, 1938, pp. iv, 149, 205.

5 George Homans, *Social Behavior: Its Elementary Forms*, New York, Harcourt, Brace & World, 1961, pp. 6-7.

6 John Dewey, *op. cit.*, p. 11.

some validation of their models we sometimes find them using as scientific evidence the crudest form of observations collected under completely uncontrolled conditions."[1]

Moreover, the "delights" of model building can lead to a retention of the models as somehow significant, even though one may be forced to change radically the type of significance the model is believed to have, or to use within the model assumptions that are inconsistent with the relevant behavior that the model is supposedly modelling. Such points will be illustrated by developments in game theory.

The original hope of John von Neumann and Oskar Morgenstern was to show that "the typical problems of economic behavior *became strictly identical* with the mathematical notions of suitable games of strategy."[2] Even initially, some doubts intrude. For example, the assumption was made that each player has "complete knowledge" of what the other players are attempting to maximize, but such is often not the case in behavioral situations. Also, much of the work is on "zero sum" games, in which the winnings of some players must be equal to losses by other players; yet in many economic transactions there is a general improvement or net gain for all involved, rather than winners and losers as in gambling. (If I buy a highly useful $10.00 tool from a merchant, both he and I may be better off than before; a situation markedly different from a $10.00 loss at poker.)

Even when non-zero sum games or other extensions of the original work are developed, other problems of the type von Neumann and Morgenstern encountered may arise. In order to develop solutions for zero sum *n*-person games, the technical limitations of the devices they had available forced them to assume that "utilities" were substitutable and unrestrictedly transferable among the various players; characteristics that may fail to apply in many economic transactions.[3] To generalize, in various models the mathematical transformations required to make the model "work" may require assumptions that are inconsistent with observed behavior.

Although the original hope was that game theory would help describe actual human behavior, and the theory was sometimes hailed as an outstanding achievement comparable to Newton's celestial mechanics, many predictions of behavior resulting from game theory were not confirmed, and some who initially were enthusiastic became disillusioned. Others began to view game theory as "normative" or "prescriptive"; Anatol Rapoport, for example, saw as the goal a prescription of how a rational player should play, given certain types of situations. In a later paper, he concluded that in some situations game theory "ceases to be normative"; it then is "neither prescriptive nor descriptive," but is rather a structural theory that mathematically describes "the *logical* structure of a great variety of conflict situations."[4] Whatever the advantages accruing from a mathematical description of the logical structure of many conflict situations that neither describes the actual behavior of humans in conflict situations nor serves as a guide as to how humans might behave, such an interpretation of game theory is a far cry from the original advantages claimed for the theory.

C. *Subjectivism.* The material discussed in this section overlaps with that discussed under the Mentalistic/Rationalistic method, but the subjective methods considered here do not necessarily involve the assumption of a separate mind. Subjectivisms emphasize the "inner" world of the individual as contrasted to what is said to occur outside the individual. At present there is considerable interest in many behavioral science fields in existentialist, phenomenological, and humanistic approaches, all of which focus on "inner" processes in human behavior. A representative statement is from the psychologist, Carl Rogers: "The inner world of the individual appears to have more influence on his behavior than does the external environment stimulus."[5]

The appeal to some inner source of certainty or other warrant for knowledge takes many forms. Sometimes an inquiry is based on an initial "truth" or "truths" guaranteed as such by "intuition," and other knowledge is based on those intuitions. Sometimes an inner sense of certainty is claimed to operate, not at the beginning of inquiry, but at its end. The physicist, P.W. Bridgman, is well known not only for his emphasis on operationalism, but for his preoccupation with the notion that ultimately even scientific knowledge is private; i.e., that public confirmation is not vital to scientific inquiry. The final test in scientific inquiry is when a conclusion "clicks," as it were, for the individual scientist.[6]

Perhaps the most common form of subjectivism today is found in an insistence that an understanding of inner states, such as feelings, is necessary in inquiry into human behavior, but that natural science procedures are restricted to external happenings. *Verstehen* theories have been urged for a long time in many behavioral science fields, despite decades of criticism.[7] In general, an introspective understanding is said to be necessary to explain the behavior being inquired into; "the actor's point of view" is understood from the "inside" through a sympathetic grasp of his motives, intentions, feelings, etc.

"Put yourself in the other person's shoes" is, of course, often excellent advice, but how do we know when we have achieved that feat? Not infrequently we go wrong in projecting our tastes, preferences, and proclivities on others who turn out not to share them; even the most useful insights into our own behavior may not carry over into the behavior of others. Ascertaining whether or not the carry-over is successful seems to require the kind of observation the subjectivists wanted to rule out in the beginning. Moreover, alleged explanations in terms of motives, etc., often turn out to be tautological. To be told that a mother who made an unusually great sacrifice for her child did so because of her great love for the child hardly marks progress.

More generally, all the subjectivisms described above appeal in a major way to some sort of inner knowledge that is self-validating. In view of the human record, one might well conclude that strong skepticism is called for toward what-

1 Chapanis, *op. cit.*, p. 130.

2 John von Neumann and Oskar Morgenstern, *Theory of Games and Economic Behavior*, 3rd ed., Princeton, Princeton University Press, 1953, pp. 1-2, emphasis added.

3 *Ibid.*, p. 604.

4 Anatol Rapoport, *Fights, Games, and Debates*, Ann Arbor, University of Michigan Press, 1960, pp. 226-227; Anatol Rapoport, "Game Theory and Human Conflict," in Elton B. McNeil, ed., *The Nature of Human Conflict*, Englewood Cliffs, Prentice-Hall, 1965, p. 196.

5 Carl R. Rogers, "Toward a Science of the Person," in T.W. Wann, *op. cit.*, p. 125.

6 Bridgman emphasized such "privacy" many times; for example, in his books *The Nature of Physical Theory*, Princeton, Princeton University Press, 1936, pp. 13-15, pp. 135-136; *The Intelligent Individual and Society*, New York, Macmillan, 1938, pp. 157-159; and *Reflections of a Physicist*, 2nd ed., New York, Philosophical Library, 1950, pp. 36-61.

7 For a sampling of *Verstehen* theories see William Dray, *Laws and Explanation in History*, London, Oxford University Press, 1957; Peter Winch, *The Idea of a Social Science and Its Relation to Philosophy*, London, Routledge & Kegan Paul, 1958; F.A. Hayek, *The Counter-Revolution of Science*, Glencoe, Free Press, 1952; and Leonard S. Krimerman, ed., *The Nature and Scope of Social Science*, New York, Appleton-Century-Crofts, 1969.

ever seems subjectively certain, unquestionable, or immediately known.[1] Hunches, intuitions, and senses of what is certain or self-evident have been wrong so often and have so frequently impeded inquiry that the persistence of defenses of subjective methods is surprising. Two possible clues to that persistence will conclude the present discussion.

First, subjectivisms tend to assume the separation in some fundamental way of man and the rest of the cosmos. Once human thinking, feeling, etc., is regarded, not as biosocial adjustive behavior, but as psychic products or processes (contrasted in some fundamental way to the nonhuman), one faces a gap that cannot be bridged or can be bridged only with difficulty. Despite those difficulties, the dualistic tradition in Western civilization is so entrenched that challenging it is nearly unthinkable for some individuals. *Within their framework*, to do so is tantamount to denying the humanity of human beings.

Second, some purported scientific procedures that reject mentalism, consciousness, etc., still retain important aspects of dualistic assumptions. Some forms of American behaviorism, for example, reject the mental half of the traditional mind-body dualism, but fail to reject the assumptions underlying that dualism. Either the brain (or the brain plus other parts of the body) is construed in a "mentaloid" way and is given many of the old psychic functions of the mind, or much that is distinctively human is overlooked by an insistence on using physical or physiological techniques of inquiry.[2] The failure of those techniques to come to grips with distinctive human behavior does not, of course, argue in favor of subjectivist methods, but that failure seems to have led some to turn to subjectivism.

Many procedures of inquiry can be found in our cultural history in addition to those we have just considered. We conclude our discussion in this chapter by presenting and discussing a tabular summary for comparing descriptions of inquiry. Our presentation emphasizes a) what the inquirer begins with, b) what he then proceeds to do, and c) what he terminates with:

	Procedure I	Procedure II	Procedure III
Begins with:	Alleged Truth (axiomatic or other)	General Laws	Tentatively selected facts of observation and partial description of them
Proceeds by means of:	Logical transformations, including use of mathematics as shorthand logic	Hypothesis development, followed by observation of facts for the purpose of "testing" the hypothesis	When initial phase is blocked, conjectures emerge about connections among facts, Which merge into observations of new or modified facts, permitting further description, When blocked, improved conjectures emerge, which merge into new observations, Etc.
Terminates with:	Proven theories (certain; true)	New general laws	Assertions warranted by the prior procedures

Procedure I reflects some of the leading notions of the Greek philosophers, and represents what many Rationalists assume is the means of obtaining new "knowledge." The movement is from truth known with certainty to new truths known with equal certainty, via transformations that cannot be doubted. Within such a general pattern, many variations can be found. The initial truths may be said to be innate ideas, or self-evident, or to result from revelation, intuition, the intellectual apprehension of the real, etc. The logical transformations may be those of the Aristotelian syllogism or some other form of deduction, they may include mathematical transformations, the "clear and distinct" ideas of Descartes, etc. The general pattern is that of the early geometers: begin with certain truths and by the transformation of those truths according to rules designed to insure consistency, arrive at new theorems that also are certainly true.

Procedure I is followed not only by Rationalists, but also to some extent by many Empiricists who vigorously oppose the Rationalists. Such Empiricists substitute for the initial *a priori* truths of the Rationalists various inductive generalizations. However, once the basic generalizations are obtained, they are viewed as being unquestionable and as axiomatic for further inquiry. Classical political economists, such as both Mills, proceeded in this way, and many traces can be found in the work of contemporary economists, game theorists, and others emphasizing formal model construction.

Procedure II has some family resemblances to Procedure I, but reflects the language often used after the Galilean revolution in physics. Many discussions of the hypothetical-deductive method of inquiry fall under Procedure II. General laws of the type found in physics form the background of inquiry. Hypotheses are constructed, consistent with those general laws, in order to "explain" data (actual or to be obtained) that are not as yet explained; i.e., are not deducible from the general laws. The hypothesis is "tested" by observational materials via deductions from the hypothesis; if the hypothesis is true, what else must follow logically? On the basis of such "testing," the hypothesis is rejected or accepted. Accepted hypotheses (perhaps several linked together logically) then are said to be new general laws.

Procedure II involves a more significant role for empirical observation than does Procedure I, but still calls for considerable theorizing in advance of observational tests. In some versions, emphasis is placed on developing hypotheses through insight into the structure of reality. In practice the direction of the work is first to develop the hypothesis, and then to look for confirming evidence.

Procedure III is discussed more adequately in the next two chapters. The inquirer begins with what seem to be the pertinent facts in a problem situation. When his advance toward a solution of the problem is blocked, he imaginatively then develops conjectures about possible connections among the facts; which lead to investigation of new facts (including improved descriptions of earlier facts); when progress again is blocked new or improved conjectures are developed; and the process may be repeated many times in succession. The outcome, when inquiry is successful, is a warranted assertion; i.e., a useful description of what happens under specified circumstances. But even the best available warranted assertion is not "certain" or an embodiment of "truth"; later inquiry may lead to its modification or abandonment. A warranted assertion developed in one inquiry may become the problem situation of a new inquiry.

---

1 In his critique of "immediate knowledge," John Dewey not only analyzes in detail the defects in many views relying on such a doctrine, but he also gives numerous historical examples of the blockage of inquiry that often follows from the acceptance of something as immediately known. See *Logic: The Theory of Inquiry*, pp. 139-158.
2 For trenchant comments on the retention of dualistic assumptions by some materialists, see J.R. Kantor, *The Logic of Modern Science*, Bloomington, Principia Press, 1953, pp. 258-259, and Dewey and Bentley, *Knowing and the Known*, p. 129.

THE DEWEY-BENTLEY VIEW OF SCIENTIFIC METHOD

In this chapter we consider the main aspects of scientific inquiry as discussed by Dewey and Bentley. The quest for certainty has already been considered. Dewey and Bentley reject certainty as the objective of inquiry and emphasize forcefully that all data, all facts, and all interpretations of facts are subject to modification and possible rejection as inquiry proceeds. What is known is not a terminus outside or beyond inquiry, but is a goal *within* inquiry. In contrast with conventional views, their procedures take human knowings as behaviors that can be observed; man's most complex inquiries are themselves to be inquired into in the same general way that any scientific subject matter is investigated.

Much of *Knowing and the Known* is devoted to analyzing the inconsistencies, incoherencies, and lack of progress characteristic of inquiries that are based on traditional theories of how we attain knowledge about various subject matters; those difficulties are especially great when knowledge itself is taken as the subject matter of inquiry, as in epistemology. Hence, Dewey and Bentley conclude, developing "firm names" in inquiry into knowings and knowns is of crucial importance. For Dewey and Bentley, however, naming is quite a different form of behavior than it is taken to be by many other authors.

Naming is a behavior of organism-in-environment, rather than a third something separate from, and intermediate between, the organism and the environment. In other words, names do not have an intervening status between the organism and external reality, but are behavioral processes. Dewey and Bentley also reject the commonly found view that a *word*, taken as physical, can be split or detached from its *meaning*, taken as mental. More generally, they do not find useful any procedure that is based on a mind-body dualism or that involves a splitting of a thing from its function. Naming, according to Dewey and Bentley, is itself directly knowing, and knowing itself is behaving; they suggest that better progress can be made by so treating naming than by proceeding along traditional lines. They say:

"Naming [not as noun but as behavior] does things. It states. To state, it must both conjoin and disjoin, identify as distinct and identify as connected. If the animal drinks, there must be liquid to drink. To name the drinking without providing for the drinker and the liquid drunk is unprofitable except as a tentative preliminary stage in search. Naming [again using 'naming' as the name for knowing behavior] selects, discriminates, identifies, locates, orders, arranges, systematizes."[1]

Unlike many writers who hope to provide final or definitive clarifications of meanings, Dewey and Bentley emphatically reject that version of the quest for certainty and urge the importance of continuing to improve naming. They seek *firm* names of the type achieved in scientific specification, but not *final* names:

"In seeking firm names, we do not assume that any name may be wholly right, nor any wholly wrong. We introduce into language no melodrama of villains all black, nor of heroes all white. We take names always as namings: as living behaviors in an evolving world of men and things. Thus taken, the poorest and feeblest name has its place in living and its work to do, whether we can today trace backward or forecast

ahead its capabilities; and the best and strongest name gains nowhere over us completed dominance."[2]

As indicated above, Dewey and Bentley reject many of the dualisms or bifurcations commonly found, such as that between mind and body, man and nature, thinking and doing. Of particular importance in the present setting is their procedure of always regarding knowings and knowns as inseparable and as twin aspects of a common fact. They name such proceeding *transactional*, as contrasted to *self-actional* and *interactional* procedures.

In self-actional proceeding, things are viewed as acting under their own powers. A primitive version is illustrated by the belief that rain is caused by Jupiter Pluvius; the use of *substance*, *essence*, *actor*, *creator*, etc., may reveal the use of a self-actional framework. The interactional procedure was typified by Newtonian mechanics. Thing is balanced against thing in causal interconnection and the separate "reals" can be detached or isolated from each other. In the transactional proceeding, however, the "reality" of the components is dependent upon the field; a borrower cannot borrow without a lender to lend. Just as the severing of borrower from lender is an impediment to inquiry, and raises the question of how to bring the severed parts back together, so too the severing or isolation of borrowing from the other components in the transaction impedes inquiry. The process as a whole is what Dewey and Bentley emphasize, rather than the components, connections, reciprocal relations, etc., taken as separate "reals." (This mode of proceeding does not preclude focusing attention on aspects or phases of a transaction whenever that may be useful, but it does preclude proceeding as though, for example, borrowers could exist independently of lenders.)

Transaction, then, names the full ongoing process in a field (a cluster of connected things and events) in which the inquirer himself may be in various connections with many aspects and phases of that field. In knowing transactions, there is nothing like the ancient notion of a pure intellect that neither affects, nor is affected by, what it knows. Rather, the inquirer is influenced by what he is inquiring into, and vice-versa. Dewey and Bentley decisively reject the notion that the human investigator can somehow know that which goes beyond what is found in inquiry, for such a feat would involve knowing what we cannot know.[3]

Frequently a sharp differentiation is made between the physical and the behavioral sciences on the alleged ground that inquiries into physical subject matters do not affect those subject matters in the way that inquiries into behavioral matters do, and that special methods, such as the participant-observer, are called for in behavioral fields. However, analysis of knowing transactions in general reveals that all knowings are alike in that mutually influencing relations occur between the inquirer and other portions of the transactional field. As Bentley pointed out nearly forty years ago:

"We must face the condition that we, the investigators, are participants in what we investigate; that our participation is 'local' within it, not as a simple attachment to animal

---

[1] *Knowing and the Known*, p. 133.

[2] *Ibid.*, p. 90.

[3] Some philosophers find no great difficulty in going beyond all that is. For example, note the title of Paul Weiss' paper, "On What There Is Beyond the Things There Are," in J.E. Smith, ed., *Contemporary American Philosophy*, 2nd series, New York, Humanities Press, 1970.

bodies in a mechanistic world, but in full behavioral presentation; that the definite determination of such localizations, however difficult, is essential to the interpretation of what we, thus localized, observe; . . .and that the two-fold construction of the observation, in terms, on the one side, of what is observed, and, on the other, of the position from which the observation is made, is essential to any dependable knowledge of the kind we call scientific."[1]

Dewey and Bentley do differentiate the physical, the physiological, and the behavioral subject matters of inquiry, but the differences they find do not imply hierarchical levels. Rather, the differentiation concerns the *techniques* of inquiry (not the general procedures) that have been found useful. Just as the techniques used by physicists are not appropriate for many physiological inquiries, so physiological techniques are not appropriate for inquiry into human behavior. Although human behavior requires research that is as "natural" (scientific) as the other two large subject matter areas, the immediate procedures are quite different. Knowing, then, is to be investigated as scientifically as anything else, but inquiry into inquiries is not restricted to the technology that has been used so successfully in the physical and physiological areas.

As noted earlier, there are disputes not only concerning which areas of inquiry can be investigated scientifically, but also disputes concerning the adequate description of scientific inquiry. Dewey and Bentley inquired into past successful inquiries in order to find helpful guides for future inquiries. The word "successful" should not be interpreted as what is psychologically satisfying to an individual, as what leads to certainty, or as what conforms to formal tests such as deducibility from an axiom or axiom set, but rather as what leads to warranted assertions providing useful solutions to problems. The work of Semmelweiss and others in insuring sterile techniques in childbirth, for example, led to effective control over puerperal fever, at one time the leading cause of maternal deaths during childbirth.

Dewey and Bentley conclude that in successful inquiry theoretical work[2] (making conjectures about connections among facts) and laboratory work (measurement of changes) do not exist in isolation from each other and that neither has methodological primacy over the other. Existing data are useful in formulating conjectures; new data may be required for testing those conjectures; the original conjecture may be appropriately modified in the light of available data or be discarded; what seemed to be reliable data may turn out to be faulty; etc. Warranted assertions are the outcome of successful inquiries and may be modified, corrected, or even discarded as further evidence becomes available.

The relation of the roles of what are sometimes called the theoretician and the laboratorian is so important that we shall consider the matter in detail. Some commentators separate these roles in a basic way, others give theoretical constructions the major place in inquiry, and others defend an antitheoretical position. Each will be considered in turn.

Views Separating the Roles of Theoretician and Laboratorian. In some accounts of the hypothetical-deductive method, hypotheses are taken as existing in a realm of their own, as it were, and data as existing in some other realm. Hypotheses, according to this view, can be graded or judged to some extent solely as hypotheses (e.g., rules saying that, other things being equal, always choose the logically simplest hypothesis), and data can also be judged on their own, so to speak. Hypotheses and data are brought together eventually, but in the fashion called interactional by Dewey and Bentley.

One may find, for example, accounts saying that after a hypothesis is formulated, logical consequences of that hypothesis are derived and then are empirically tested, and that if even one consequence is disconfirmed by the empirical test, the hypothesis must be modified or rejected. Such an account, although apparently clear and straightforward, runs into difficulty when actual inquiries are investigated. For example, some "certainty" may be attributed to the data that goes far beyond any warranted assurance; sometimes the conjecture being tested may itself help indicate which data are in need of correction. There are many ways in which data that seem firm can be misleading, even in physical inquiry.

To illustrate, the first discussion of Einstein's work in relativity theory printed in *Annalen der Physik* (1906) was by a respected experimenter, W. Kaufmann. Kaufmann announced that the experimental measurements he had performed were not compatible with Einstein's theory. Einstein admitted that Kaufmann's work seemed sound and that some other theories of electron motion yielded predictions closer to Kaufmann's results. Einstein did not give up his theory, however, among other reasons because he believed that the issue could be settled only when "a great variety of observational material is at hand." After several years, physicists concluded that Kaufmann's equipment was not adequate for the purposes of his inquiry; what seemed to be "hard" data requiring the modification of a hypothesis turned out to be mistaken.[3] Such occurrences do not show the primacy of theory over observation, as some have argued, but rather illustrate once again the pitfalls of the quest for certainty. To assume at any given moment that a candidate hypothesis is confronted by unchallengeable data can be just as misleading as assuming that a hypothesis meeting certain formal criteria must therefore be superior to the data in hand. In order to emphasize this point, we have found the name "conjecture," to be more useful in many instances (as we hope the reader has noted) than the word "hypothesis," which all too frequently is misunderstood to be somehow less conjectural.

Some writers putting emphasis on prediction argue that "unreal" assumptions may be useful scientifically and that a conjecture should not be rejected simply because its assumptions run counter to well-substantiated findings. In a well-known work, Milton Friedman says:

"Consider the density of leaves around a tree. I suggest the hypothesis that the leaves are positioned as if each leaf deliberately sought to maximize the amount of sunlight it receives, given the position of its neighbors, as if it knew

---

[1] Arthur F. Bentley, *Behavior, Knowledge, Fact*, Bloomington, Principia Press, 1935, p. 381.

[2] Both "theory" and "hypothesis" have been used to designate various things. Among other applications, "theory" sometimes is applied to an inquirer's conjectures, working hypotheses, or notions about possible connections; sometimes is applied to a system of related hypotheses (or even "principles") in a field of inquiry; and sometimes is applied to well-established warranted assertions, as in "theory of evolution." "Hypothesis" sometimes is applied to any tentative notion about possible connections, but sometimes is restricted to relatively exact formulations that may emerge in an advanced stage of inquiry. Sometimes "hypothesis" is embedded in the terminology of traditional logic and epistemology, as when a hypothesis is said to be a proposition not known to be true or false initially, but from which consequences are deduced; if sufficient deductions are confirmed by the facts, the "hypothesis" is said to become a "truth."

We believe that using "hypothesis" to designate any notion or conjecture about possible connections, and "theory" to designate related warranted assertions as in "theory of evolution," could be useful. In the present volume, however, we use both names infrequently and only casually, in order to avoid confusion with other uses found in the literature on methodology.

[3] The materials here are taken from Gerald Holton, "Mach, Einstein, and the Search for Reality," *Daedalus*, Spring, 1968.

the physical laws determining the amount of sunlight that would be received in various positions and could move rapidly or instantaneously from any one position to any other desired and unoccupied position. Now some of the more obvious implications of this hypothesis are clearly consistent with experience: for example, leaves are in general denser on the south than on the north side of trees but, as the hypothesis implies, less so or not at all on the northern slope of a hill or when the south side of the trees is shaded in some other way. Is the hypothesis rendered unacceptable or invalid because, so far as we know, leaves do not 'deliberate' or consciously 'seek,' have not been to school and learned the relevant laws of science or the mathematics required to calculate the 'optimum' position, and cannot move from position to position? Clearly none of these contradictions of the hypothesis is vitally relevant; the phenomena involved are not within the 'class of phenomena the hypothesis is designed to explain'; the hypothesis does not assert that leaves do these things but only that their density is the same *as if* they did."[1]

Several comments seem appropriate: (1) From the point of view of actual scientific inquiry, is it not misleading to consider only the density of leaves as what is to be "explained"? As Friedman himself indicates later on, certain alternatives to the conscious leaf hypothesis may be preferred because they are connected with other scientific findings. If we consider *only* what Friedman takes as relevant, we can imagine many conjectures that conform as well as his conscious leaf hypothesis does to observed leaf density. For example, we might propose that God, who "obviously" knows the relevant physics and mathematics and can move leaves as He wishes, always arranges leaves so they will receive maximum sunlight. Or we could stipulate a god wanting to maximize leaf exposure to sunlight who is locked in combat with a devil wanting to minimize sunlight exposure, and further that the devil always loses the battle. Both conjectures, presumably, would have the same implications about leaf density as does Friedman's hypothesis. (The "as if" qualification introduced in the last sentence of the quotation from Friedman allows for any number of bizarre notions.) (2) The conscious leaf hypothesis reeks of the *ad hoc* hypotheses often encountered in the history of science that typically lead to a "dead-end." Even worse, explaining change through the workings of an unobserved and perhaps unobservable-in-principle conscious mind, which time and again has hindered scientific inquiry, is here paraded as if there were no cumulative evidence as to its difficulties. (3) Friedman's hypothesis may have some implications not mentioned by him that would not be confirmed by observation. The stipulated complete mathematical and physical information of the leaves combined with their freedom to move rapidly or instantaneously to any desired, unoccupied location, might well result in a different leaf density than the observed density; i.e., actual density may only poorly approximate such an idealized distribution.

Creative and imaginative efforts to develop notions about possible connections among things and events should not be inhibited merely by an inconsistency between those notions and traditional beliefs, for traditional beliefs often turn out to be wrong. But to encourage conjectures that not only run counter to what has been scientifically warranted, but also are similar in principle to conjectures that again and

again have impeded progress, seems scientifically irresponsible. An enormous number of silly or worse notions can be dreamed up that will yield predictions that can be confirmed for a narrow range of data. In general, then, to separate conjectures from testing, or predicting from other parts of the scientific transaction, may create more problems than it solves.

Views Giving Theoretical Constructions the Primary Role in Scientific Inquiry. Some contemporary authors uphold what they call "scientific rationalism."[2] As support for that position, often Einstein's later methodological views are mentioned. Those later views are in marked contrast with Einstein's earlier empiricism, which is expressed in a letter he wrote to a physicist friend in 1918:

"In your last letter I find, on re-reading, something which makes me angry: That speculation has proved itself to be superior to empiricism. You are thinking here about the development of relativity theory. However, I find that this development teaches something else, that it is practically the opposite, namely that a theory which wishes to deserve trust must be built upon generalizable facts."
And in 1921, Einstein said:

". . .I am anxious to draw attention to the fact that this theory [relativity theory] is not speculative in origin; it owes its invention entirely to the desire to make physical theory fit observed facts as well as possible. We have here no revolutionary act, but the natural continuation of a line that can be traced through centuries."

Einstein later changed his views considerably, and in 1933 held that Nature was the realization of "the simplest conceivable mathematical ideas" and that our knowledge of physics shows a way in which "pure thought can grasp reality." In 1938, in a letter to a friend, Einstein wrote:

"Coming from sceptical empiricism. . ., I was made, by the problem of gravitation, into a believing rationalist, that is, one who seeks the only trustworthy source of truth in mathematical simplicity. The logically simple does not, of course, have to be physically true; but the physically true is logically simple, that is, it has unity at the foundation."

Possibly the most interesting brief statement comes from a 1929 essay. Einstein says that physical theory "desires. . . to help us not only to know *how* Nature is and *how* her transactions are carried through but also to reach as far as possible the perhaps utopian and seemingly arrogant aim of knowing *why* Nature is *thus and not otherwise*." He goes on to say that when making deductions from a "fundamental hypothesis," such as the connections among pressure, volume, and temperature deduced from the kinetic-molecular theory, "one experiences, so to speak, that God Himself could not have arranged those connections in any other way than that which factually exists, any more than it would be in His power to make the number 4 into a prime number."[3]

Even within his own frame of reference, Einstein's later statements seem to yield difficulties. To say, for example, that one "seeks the only trustworthy source of truth in mathematical simplicity" and yet that the "logically simple does not, of course have to be physically true" seems inconsistent, or at least embarrassing in that the "only trustworthy source" apparently has to be supplemented by something not so trustworthy. Moreover, to hope to learn why ultimately Nature is as it is and not otherwise seems doomed to frustration. Even if we had proof that physical

---

1 Milton Friedman, "The Methodology of Positive Economics," expanded version printed in William Breit and Harold M. Hochman, eds., *Readings in Microeconomics*, New York, Holt, Rinehart and Winston, 1968, p. 33. The original version is from Friedman's book, *Essays in Positive Economics*, Chicago, University of Chicago Press, 1953.

2 For a recent defense of scientific rationalism, see the book by the eminent historian of science, Giorgio de Santillana, *Reflections on Men and Ideas*, Cambridge, The M.I.T. Press, 1968.

3 The quotations from Einstein are taken from Holton, *op. cit.*, pp. 645-646, 649-650, 657-658, and 658-659.

connections are as they are because God wanted them that way, we have no "ultimate explanation," for we can always inquire why God should be the way He is. If mathematical limitations exist on God's freedom, as Einstein says, we can ask why those limitations—why not none, or some other ones? If mathematical or logical simplicity is urged as the final arbiter, we can ask why Reality should necessarily be simple or logical. Rather than grasping some ultimate cosmic necessity, we suggest that Einstein was simply exemplifying some of the views deeply enmeshed in Western culture about certainty and perfect knowledge.

Anti-Theoretical Views. From time to time writers can be found who put far more emphasis on the role of the laboratorian than on the role of the theoretician. Sometimes such emphases may tend in the direction of maintaining that scientific inquiry is only data-collection, but more often those emphases are protests against premature theorizing or insufficient attention being given to testing. Thus Alvin G. Goldstein says: "In psychology, most theories are stated before enough solid information has been collected, and as a result there is a ridiculous profusion of theories." A statement made in 1937 by Goerge P. Murdock would find acceptance among many workers today, and not only for sociology: "Sociology . . .has a plethora of hypotheses. What it most needs. . .is more factual studies to test them."[1]

B.F. Skinner is an eminent behavioral scientist often cited as anti-theoretical. Attention here will be focused on an article in which he describes his early research methods. Skinner says:

"This account of my scientific behavior up to the point at which I published my results in a book called *The Behavior of Organisms* is as exact in letter and spirit as I can now make it. The notes, data, and publications which I have examined do not show that I ever behaved in the manner of Man Thinking as described by John Stuart Mill or John Dewey or in reconstructions of scientific behavior by other philosophers of science. I never faced a Problem which was more than the eternal problem of finding order. *I never attacked a problem by constructing a Hypothesis.* I never deduced Theorems or submitted them to Experimental Check. So far as I can see, I had no preconceived Model of behavior—certainly not a physiological or mentalistic one, and, I believe, not a conceptual one. The 'reflex reserve' was an abortive, though operational concept which was retracted a year or so after publication in a paper at the Philadelphia meeting of the APA. *It lived up to my opinion of theories in general by proving utterly worthless in suggesting further experiments.* Of course, I was working on a basic Assumption—that there was order in behavior if I could only discover it—but such an assumption is not to be confused with the hypotheses of deductive theory. It is also true that I exercised a certain Selection of Facts but not because of relevance to theory but because one fact was more orderly than another. If I engaged in Experimental Design at all, it was simply to complete or extend some evidence of order already observed."[2]

Despite Skinner's specific disclaimer that he did not behave as Dewey suggests problem-solvers behave, Dewey described just the kind of procedures Skinner tells us about in the case history. Dewey's emphasis on the whole transaction in which a scientist carries out his inquiry, including the role of the equipment and technology available to the scientist at work in a given setting, is also what Skinner emphasizes. Indeed, the "earthiness" Skinner stresses is what Dewey, unlike many writers on scientific inquiry, also stresses. Again and again Skinner mentions problems he encountered in studying rat behavior, or in devising machinery to aid his observations, or in improving that machinery to make the experimenter's work easier, or resulting from his limited time for experimentation under given life conditions. The detail Skinner gives about his own case history fits with Dewey's methodological advice: look at the full transaction, in which one organism (the scientist) is involved with other things and events (in this instance, rats and the equipment used to observe them), rather than to adopt procedures in which the various aspects and phases of a transaction are taken as separate "reals" existing independently of each other.

According to Skinner's own words, he does make use of hypotheses, in the sense of conjectures or notions about possible connections among facts. To illustrate, let us consider one of the incidents Skinner cites to show that luck sometimes plays a part in scientific inquiry. He had constructed an apparatus for measuring the delay between a rat's eating the food it gets after going down a runway and the rat's returning "home." A wood disc, taken from a store of discarded apparatus, had been fashioned into a food magazine for releasing the rat's food. The disc happened to have a spindle that Skinner had not bothered to remove. He says that one day it occurred to him that "if I wound a string around the spindle and allowed it to unwind as the magazine was emptied," then he could record the delays as a curve rather than as the polygraph-like pips his earlier apparatus had produced. Skinner notes that although the differences between the old type of record and the new one may not appear great, "as it turned out the curve revealed things in the rate of responding, and in changes in that rate, which would certainly otherwise have been missed."[3] Here we clearly have a conjecture being formed about the apparatus and then tested, with useful scientific results, which is just what Dewey was talking about.

To take one more example, when Skinner was working half-time at the Medical School and hence had difficulty maintaining the desired work schedule with his rats, he attempted to overcome that problem by devising a way in which the rats could be kept at a constant level of food deprivation. He conjectured that if "you reinforce the rat, not at the end of a given period, but when it has completed the number of responses ordinarily emitted in that period," the rat "should operate the lever at a constant rate around the clock" except when sleeping, and if the reinforcement were set "at a given number of responses it should even be possible to hold the rat at any given level of deprivation." However, Skinner goes on to say, "nothing of the sort happens"; what he actually got was "fixed-ratio" rather than "fixed-interval" reinforcement.[4] So here we have an instance of a set of conjectures being formulated that did not have the consequences it was first expected to have, but which did lead to useful results.

Summarizing Skinner's early work, then, although he did not usually formulate deductive hypotheses about rat behavior before he carried out his observations, but rather observed rats under carefully controlled circumstances to

1 Alvin G. Goldstein, Review of G.M. Solley and G. Murphy, *Development of the Perceptual World,* in *Philosophy of Science,* Vol. 29, 1962, p. 326. George P. Murdock, ed., *Studies in the Science of Society,* New Haven, Yale University Press, 1937, p. x.
2 B.F. Skinner, "A Case History in Scientific Method," *The American Psychologist,* Vol. 11, May, 1956, p. 227, emphasis added.
3 *Ibid.,* pp. 224-225.
4 *Ibid.,* pp. 226-227.

see what order could be discerned in their behavior, he did use conjectures and paid great attention to seeing what warranted generalizations ("laws" in his terminology) could be arrived at on the basis of careful observation. His discussion of technological and instrumental problems, often de-emphasized in accounts of scientific inquiry, along with the constant theme of the inquirer running into difficulties and solving them, seems to us to illustrate neatly the account of inquiry given by Dewey and Bentley.

Many of the points emphasized in this Section concerning the interweaving (or reciprocal movement and stimulation) of notions about things and measurement of changes can be illustrated in the following example from the history of science. In the late years of the 18th century, Lazzarro Spallanzani conducted an extensive inquiry into the question of how bats avoid obstacles when they fly at night.[1] He took as his initial basic notion that bats fly at night by means of some sense organ.

The first test concerned vision; possibly bats have such keen eyesight that they can see in almost total darkness. Spallanzani blinded several bats, but found that they could all fly as well as before. His second conjecture concerned the sense of touch, but the bats in flight do not actually touch the walls of the caves. Spallanzani then speculated that perhaps bats have such an acute sense of touch that they are responsive to slight disturbances in the air near walls or other obstacles. To test that notion, Spallanzani covered some bats with a thick varnish, but they flew as well as unvarnished bats. Next he tested the sense of taste, although it seemed incomprehensible that bats can guide themselves through taste. Excising the tongues of the bats, however, did not diminish their flying ability.

Next he tested smell by plugging up the noses of the bats. Some bats flew as well as ever, but others lost the ability or had it diminished. Further inquiry led Spallanzani to believe his experimental procedures had interfered with the bats' breathing and that accounted for the decrease in flying ability. He then turned to hearing, and plugged the ears of the 11 bats. One of the 11 flew only with difficulty, but the other 10 flew as well as before. Spallanzani concluded that the single exception was an accident that did not weaken the force of the positive evidence.

Possibly two or more sense organs are required for guided flight. Another scientist, Rossi, tested that conjecture by covering all the head organs with a hood. When that was done, the bats did lose their ability to avoid obstacles. However, Spallanzani objected to the combination-of-organs hypothesis, arguing that if the senses taken separately cannot account for a phenomenon, their combination cannot do so either. This led Spallanzani finally to conclude that bats guide themselves by "some new organ or sense which we do not have and of which, consequently, we can never have any idea"; he was driven to explaining a mystery by another mystery.

For about a century and a half the matter rested

until Hartridge suggested, by analogy with echo-location devices of World War I, that bats locate obstacles through the reflection of their high-pitched cries. That conjecture was consistent with Rossi's findings, and suggests that the one instance of plugged-ear bats that Spallanzani ignored was the important instance. Griffin and Galambos performed several experiments that led to a highly warranted assertion. When a bat's snout is carefully tied, so that it can emit no sound, it cannot avoid obstacles. When a bat's ears are *very carefully* plugged, it also cannot avoid obstacles. When an instrument that will convert supersonic sounds into audible sounds was used, it was found that bats emitted such inaudible cries while flying. It was also shown that bats respond to such sounds, as well as emit them. Thus the combination-of-sense-organs hypothesis that Spallanzani ruled out was confirmed, and what he viewed as an insignificant "renegade" instance was a key finding.

Data that seem unquestionable, then, may not be so. Technological limitations may keep a conjecture from being adequately tested. Apparently logical arguments may be mistaken or inapplicable. Yet when inquirers get "on the right track," the interweaving between conjectures and observation may lead rapidly to a very high degree of corroboration of a description of the connections among things and events. Although data are not self-interpreting, interpretations made far in advance of the data almost invariably have been misleading.[2]

To summarize some of Dewey and Bentley's main points: When we observe humans we see organisms living in-and-by-means-of their environment (not separated from, or merely placed in, an environment), struggling to adjust behaviorally to a multitude of things and events. Among the many transactional processes men are engaged in are transactions that we call scientific inquiry. The inquirer's behavior as a scientist is continuous with other of his behaviors; his knowings are behavioral adjustments, not god-like apprehensions of the Real. Successful inquiries exhibit an interweaving of notions about things and measurements of changes. Knowing transactions are best investigated as they occur, rather than by postulating actors, or actions, or other aspects and phases of the transactions as separate reals that somehow come together on some occasions.

---

[1] The materials that follow are taken from a useful summary by Lewis W. Beck, *Philosophical Inquiry*, New York, Prentice-Hall, 1952, pp. 109-115. Beck's summary is based on: Robert Galambos, "The Avoidance of Obstacles by Flying Bats," *Isis*, Vol. XXXIV, 1942, pp. 132-140; Robert Galambos, "Flight in the Dark," *Scientific Monthly*, Vol. LVI, 1943, pp. 155-162; Robert Galambos and Donald R. Griffin, "Obstacle Avoidance by Flying Bats," *Journal of Experimental Zoology*, Vol. LXXXVIII, 1942, pp. 475-490.

[2] Throughout this volume we emphasize the dangers of proceeding by means of an elaborate extension of hypotheses in the absence of observations of facts. Solving the vexing problems of men-in-society typically requires many stages of inquiry, and at each stage almost always more than one conjecture is initially plausible. The crucial importance of using *each successive* conjecture as a guide for further observation and measurement readily can be understood. Each time that progress in inquiry is blocked and a tentative description of what happens under specified circumstances remains inadequate, the inquirer in imagination conjectures (develops hypotheses) about the possibilities. If the inquirer selects the possibility that to him seems most plausible and proceeds on to the next blockage, without returning to observation and measurement, he is confronted with rapidly increasing odds against the success of his inquiry. To illustrate, if the number of possible conjectures at each stage is 10, his chances of selecting the correct conjecture 10 times in succession would be only 1 in 10 billion. Even if there were only two alternative possibilities at each successive point, and 10 stages, his chances of selecting all the correct or more useful conjectures in succession would be only 1 in 1,024.

Yet often we find the "free creation" of hypotheses being advocated to help solve the complex problem situations of men-in-society. The formidable amount of work required in making the measurements of changes from which warranted assertions could be developed apparently constitutes such a frustrating barrier to those who want immediate solutions that they prefer a "short-cut" via elaborate theorizing, despite the overwhelming odds against achieving useful results.

V.
THE COURSE OF INQUIRY

In the course of inquiry, as the inquirer initially observes and measures he may note connections among the things measured and imagine other possible connections. Such notions encourage him to differentiate and focus attention on other aspects and phases of the problem situation. When changes among them are measured, additional connections may be noted and still more may be imagined, which encourages further differentiation and measuring. Thus the inquirer progressively develops increasingly useful descriptions of what happens under specified circumstances until, if he succeeds, he achieves description adequate for coping with the immediate problem situation.

Partial tentative descriptions are developed from initial observations. The inquirer, temporarily baffled in his effort to achieve usefully adequate description, imagines various possibilities or notions among which he selects the seemingly more promising as the guide to further observation. If lucky or skillful or both, his additional observations develop further the original tentative descriptions. If not adequate for the problem confronted by the inquirer, he again is baffled and develops notions or imagines other possible connections among things, from which he chooses a guide for further observations and so on, until satisfactory description is achieved.

In the course of these procedures of inquiry, the inquirer may explore many blind alleys, discard many of his observations, and begin over again at various stages, perhaps many times. The succession of notions about possible additional things to be investigated may be labeled in technical jargon "hypotheses," but the sequence of proceeding is not from elaborately formulated hypotheses to testing of them by subsequent observation of facts. Rather the sequence in successful inquiry seems always to be from observation and measurement of initially selected aspects and phases of the problem situation to partial, tentative, inadequate descriptions, followed by conjectures about possible but as yet unobserved connections, which in turn require new observations, etc. When further development of description is blocked at any stage the inquirer imagines what could have happened or might happen under the given circumstances. These notions or conjectures, whether elaborately stated or merely fleeting, enable the inquirer to find additional steps toward his goal. By further observation and measurement, often with the aid of elaborate instruments and especially arranged experiments, but not necessarily so, the observer investigates the various possibilities.

From observation and measurement emerge tentative description, and from the blocked procedures of describing emerge one or more conjectures to be investigated; the investigating of which may result in the emergence of new or modified conjectures to be investigated in turn, etc., until adequate useful scientific description is achieved.[1]

Two related objections to our description of the course of inquiry are often made by other workers.

(1) Some maintain that the movement from observation to a useful theoretical formulation represents a break, perhaps an "illogical" step. The linguist, Martin Joos, says:

"Hypothesis comes about in this way: you collect a mass of data which have been interesting to you. . .and you try to make sense out of your collection. You worry about it anywhere from three days to three weeks. By conscious thinking, you can't get a satisfactory accounting for what you have collected. Then, in my case, it wakes me up at night about 3 a.m. A hunch suddenly comes to you, seemingly out of nowhere. . . . If, in the waking hours, the hunch does not seem utterly ridiculous, you adopt it as a working hypothesis and hope to improve it or disprove it, in which case you will have to try again. . . . All experience shows that you can't take the step from data to theory logically; it has to be an illogical or irresponsible notion—a hunch, as I call it."[2]

(2) Others sharply separate the "generation" and "verification" of theories and stress the importance of creative theory construction in scientific inquiry. For example, Arthur Stinchcombe argues that at present it is more important for sociologists to invent than to test theories; Barney Glaser and Anselm Strauss believe there is an excessive emphasis on testing in current sociology and regret that young inquirers are often taught they are not functioning as sociologists unless they are involved in verifying their theories.[3]

Both types of objection, we suggest, misinterpret certain aspects of successful inquiry. What Joos describes seems to happen quite often; an "inspiration" comes, apparently "out of the blue," that may be quite unlike the earlier notions the inquirer had considered. But such occurrences do not represent a break with the data. As Joos points out, and indeed emphasizes, the hunch is taken as a working hypothesis that the inquirer tries to improve or disprove on the basis of further evidence. A marked break in the type of notions entertained during the course of an inquiry is not necessarily "illogical or irresponsible"; it may not be a step in traditional Aristotelian logic, but it may well be both responsible and logical in the sense that logic is a name for the procedures of successful inquiry.

Perhaps what underlies the conclusions of Stinchcombe and others is that in retrospect sometimes the single most important part of a particular successful inquiry appears to be the innovative, unusual, or daring formulation of a possible relation among facts. Inquiry into the ability of bats to fly at night made very little progress until it was suggested that perhaps they do so through echo-location; once that idea occurred, a marked convergence of old and new evidence led to rapid progress. On many occasions the statement of the possible relation among facts that turns out to be sound may conflict with established

---

1 This account of inquiry is similar to some points made by Karl R. Popper, who emphasized conjectures as proposed solutions to problems. If a conjecture survives all attempts to refute it, the conjecture is provisionally accepted. See his *The Logic of Scientific Discovery*, New York, Basic Books, 1959, and his *Conjectures and Refutations: The Growth of Scientific Knowledge*, New York, Basic Books, 1962. However, in a great many important respects Popper's "critical rationalism" diverges from our work; he sees as useful many procedures in inquiry that we regard as outmoded epistemology.

2 Martin Joos, "Discussion," in Paul L. Garvin, ed., *Method and Theory in Linguistics*, The Hague, Mouton, 1970, p. 21.
3 Arthur L. Stinchcombe, *Constructing Social Theories*, New York, Harcourt, Brace and World, 1968; Barney Glaser and Anselm Strauss, *The Discovery of Grounded Theory*, Chicago, Aldine, 1967.

views, with what the experts are convinced of, or with what is sanctioned by the centuries. But "distance" from accepted notions, traditions, or habitual ways of thinking is *not* necessarily distance from the evidence; indeed the evidence is precisely what encourages us to give up the accepted notions and to explore innovative conjectures.

The emphasis often placed on the development of creative and imaginative hypotheses, however, can be misleading in the extreme if taken out of the context of the course of inquiry, wherein observations and conjectures are continuously interwoven in the sense of emerging from and merging into each other. In field after field what seemed to be brilliant theories when first proposed have been quietly forgotten after a few years. What Garvin says about historical linguistics is applicable to many other areas:

"A look at the development of historical linguistics will show that there has been no scarcity of explanatory theories about linguistic history. Note, however, how most of those explanations, no matter how attractive they may have seemed at the time they were proposed, have since been relegated to oblivion. Whatever deep insights into linguistic history they may have suggested turned out to be unacceptable to succeeding generations of linguists."[1]

At times important progress in inquiry occurs when someone is bold enough to formulate a statement of possible connections among facts that is daring, or even shocking to some people, but that significantly helps to describe what happens under specified circumstances. To conclude that therefore the key to successful inquiry is the free, innovative, or creative formation of hypotheses in advance of, or despite, the evidence, is quite another matter.

To generalize, we suggest that often mistakes in describing the course of inquiry occur because one or another aspect of inquiry is taken as the "essence" of all inquiry. For example, quite often commentators suggest that a hypothesis (or a set of competing hypotheses) is relatively fully developed and relatively fully tested against a set of data that is firm beyond reasonable doubt. Although such an account may describe a *particular phase of a particular inquiry* when the full process of inquiry is "frozen" for the sake of analysis, it may also obscure the changes characteristically occurring in the tentative descriptions of connections among facts, the inquirer's estimate of the "hardness" of facts, and their pertinence to the problem at hand. Those changes are highly significant for the development and the self-correcting features of successful inquiry.

When a specific successful inquiry is analyzed in retrospect, some particular step in that inquiry may emerge as crucial to the success of that particular inquiry. Unfortunately, sometimes that crucial aspect is mistakenly generalized as being crucial for *all* inquiry; what performed a vital function in one inquiry (or in a set of similar inquiries) is assumed to be generally or universally necessary.

Illustrations may be helpful. In recent physical inquiries unusually complex mathematical transformations sometimes were what advanced the inquiry; some observers have been misled into taking such mathematicization as necessary or desirable for all inquiry. In some instances deductions from axioms have proved useful in moving an inquiry ahead, and this has led some to maintain that such deductions are the mark of any "advanced" inquiry. Sometimes a hypothesis that apparently was conclusively refuted by data later is found to be useful because the data were erroneous; such occurrences sometimes have

encouraged a high degree of confidence in hypotheses unsupported by evidence. Sometimes the key to success in an inquiry was to depart radically from widely accepted notions (as in the development and successful application of non-Euclidean geometry), and this has led some to regard the creation of innovative hypotheses as the single most important feature of all inquiry.

Putting our general point in another way, in surveying inquiries one finds that a principal obstacle may be encountered anywhere throughout the course of an inquiry. At times the main difficulty may be that the problem itself is badly formulated. At times the observations made may be so inadequate that most of the effort must go into improving observation, even to the extent of inventing new or improved instruments. At times the development of new tools of analysis may be required, as when Newton's invention of integral calculus facilitated the solution with ease of problems that previously had been insoluble. At times the difficulty may consist in obtaining the relevant data for choosing among conflicting conjectures; the gravitational deflection of light predicted by Einstein was twice that predicted by Newton's theory, but the measurements required for testing were not obtained until a solar eclipse occurred. In the past progress was sometimes slow because massive quantities of data had to be transformed and analyzed; some problems simply were too cumbersome to be attacked until modern devices such as computers became available. And even when difficulties of the type just mentioned do not occur, some inquiries may be blocked because the only workable procedures presently available are too costly, or violate cultural taboos, or require cooperation from others that is not forthcoming, etc.

In short, any step within a particular inquiry may be the most difficult part of that inquiry; but the procedures that lead to success in overcoming that difficulty should not be assumed to be essential for all inquiry. Moreover, even within one area of inquiry, procedures that prove successful may meet with failure later on, as is illustrated by the discovery of Neptune and the non-discovery of Vulcan. According to the Newtonian account of gravitation, the planet Uranus should move in a way that was not in accord with its actually observed motion. Rather than giving up Newton's "law," Adams and Leverrier suggested that an undiscovered planet existed that would account for the deviations. Leverrier then calculated exactly where such a planet should be on a particular date, and through the use of a powerful telescope the new planet (Neptune) was observed just as predicted. Newton's "law" was thus given additional confirmation and the problem of deviations in Uranus' motion was solved. Leverrier also discovered irregularities in Mercury's motion. He suggested the existence of another undiscovered planet, Vulcan, moving between Mercury's orbit and the sun. Although strong confidence in Vulcan's existence was exhibited by some workers, astronomers could find no trace of the predicted planet.

That situation led to doubt about the accuracy of Newton's "law" of gravitation. Later Einstein developed a theory of gravitation in which planetary motions were not exactly elliptical and should show a slight precession of the perihelion. His predictions were in accord with the observations of Mercury's motion and thus helped to disestablish the Newtonian theory.[2] Two similar prob-

---

[1] Paul L. Garvin, "Introduction," in Garvin, *op. cit.*, p. 11.

[2] A detailed and yet relatively non-technical discussion of the development of this inquiry can be found in A. d'Abro, *The Evolution of Scientific Thought from Newton to Einstein*, 2nd ed., New York, Dover, 1950, pp. xiii-xiv, 276-278.

lems, then, were resolved in two different ways, one involving the use of Newton's "law," and one involving the abandonment of that "law." In both instances inquiry progressed, even though eventually the conjecture about gravitation that was supported by the discovery of Neptune was replaced by a much different conjecture. New data resulting from successful predictions may give impressive support to a tentative description of connections among facts, but later that description may have to be abandoned because of additional data, and the process may go on indefinitely. What seems to be a triumph of human intellectual ability (e.g., Newton's "law") may later function as a serious impediment to progress. The constant interweaving between observations and conjectures provides a useful way of avoiding such impediments to inquiry or of overcoming them when encountered.

Discussing in some detail an interesting inquiry that was not successful may help to illustrate further our account of the course of inquiry. In addition to his many political activities, Thomas Paine maintained strong scientific and technological interests. In 1806 he published "The Cause of the Yellow Fever."[1] The ports of Philadelphia and New York had recently become the scenes of serious outbreaks of yellow fever; in severe forms of the disease the mortality rate approximated 60%. Although Paine believed that he had found a means of preventing yellow fever, he was mistaken, and nearly a century passed before Walter Reed and others solved the problem Paine had been working on. Reed discovered that yellow fever is transmitted from person to person by a particular form of mosquito prevalent in the West Indies, Central America, and West Africa.

In the period just before Paine wrote his article, extensive trade had developed with the West Indies. The water tanks of the sailing vessels coming from there provided a good environment for the mosquitos, and when the ships docked in New York and Philadelphia many people were infected with the disease. The mosquitos were able to survive for only a short time in the climate of those cities; consequently, the disease did not spread far beyond the docks.

Paine observed many aspects and phases of the problem situation. Some of the main facts he believed were interconnected and pertinent to the problem, along with our parenthetical comments based on later work, follow:

1. In earlier times yellow fever was not known in the U.S.A., and the occurrence there dated only from about 12 years prior to Paine's work. He therefore looked for other events occurring only within the past 12 years. (Extensive trade with the West Indies had developed only during the 12-year period.)

2. Yellow fever begins in the lowest part of populous mercantile towns, near the water, but doesn't spread to the higher parts of the town. This suggests the importance of a geographic, perhaps even an altitudinal, factor. (The restricted geographic occurrence was a result of the mosquitos' inability to survive long in our climate.)

3. Yellow fever is most widespread where new, solid earth wharves had been built out of soil dredged from the "muddy and filthy bottom" of the river. Such soil is

1 Printed in Philip S. Foner, ed., *The Complete Writings of Thomas Paine*, Vol. II, New York, Citadel, 1945, pp. 1060-1066.

quite different from the natural condition of soil in higher parts of the city. This led Paine to conjecture that the character of the soil was a primary factor in the cause of yellow fever. (As it happened, the ships carrying the mosquitos docked almost exclusively at the new wharves, which were constructed as a result of the increasing West Indies trade. The type of soil was not pertinent to the problem at hand.)

4. Earlier, during the Revolutionary War, Paine had done some work with gases trapped in muddy river bottoms. He, along with George Washington and members of Washington's military staff, investigated a creek that could be set on fire. They found that when the bottom was disturbed inflammable marsh gas was released. River bottoms, then, sometimes contain "impure" gases that can be injurious to life. Paine concluded that different types of soil produce different types of what he called "effluvia or vapor." (The release of injurious gases from river bottoms, unfortunately, was not pertinent to the yellow fever problem.)

5. The failure of yellow fever to spread from the wharves to the other parts of the city led Paine to conclude that the disease couldn't be imported from the West Indies, as some had conjectured. Paine argued that if the disease could not travel more than a short distance within a city, it surely could not travel from the West Indies, a distance of more than a thousand miles. Hence a local cause for the disease must exist. (Quite often a logical argument turns out not to be applicable to a particular problem situation. Although it may appear paradoxical to argue in the abstract that what cannot be transmitted beyond a mile or so can be transmitted over a much longer distance, under the specific circumstances encountered that did happen. The distances *per se* turned out not to be pertinent, but climatic differences that happened to be correlated with the observed differences in distance were important.)

On the basis of his inquiry, Paine concluded that yellow fever is caused by the "pernicious vapor" given off from the mud used in constructing the new wharves, and he had so much confidence in his conclusion that he enthusiastically recommended a new method of constructing wharves that would allow the river bottoms to remain in their natural state. He did *not* suggest using a differently constructed wharf as a test of his conjecture about yellow fever; instead he urged construction of new wharves as a sure way of preventing the disease. The great confidence he had in his conclusion seems to have been based on the following: His conjecture about vapors causing yellow fever was consistent with the evidence he had at the time; he had refuted some other conjectures; there were no plausible competing conjectures available; hence his conjecture must be correct. But as later inquiry showed, the problem he began with was not solved, and his confidence proved to be badly misplaced.

We suggest that a similar pattern often is found in inquiry into problems of men-in-society. A conjecture consistent with some evidence is prematurely regarded as warranted, and then remedies for the original problem situation are confidently advanced. The problem, however, is not resolved (and sometimes may be worsened), considerable resources may be wasted in the process of applying the "remedy," and useful inquiry is halted.

VI.

DESCRIPTION OF PROCEDURES TENTATIVELY SUGGESTED FOR TRIAL IN SCIENTIFIC INQUIRY

The procedures suggested below have been selected as those believed to be most useful in facilitating progress toward the objective of scientific inquiry. Those procedures are based primarily on the work of Dewey and Bentley, but we have not hesitated to make changes when that seemed desirable.

Human beings have found that many problems can be solved by ascertaining what happens under specified circumstances. Therefore the first suggestion is that the objective or goal of scientific inquiry is a description of what happens under specified circumstances. Ascertaining what happens is part of the scientific inquirer's job, but his task is not completed until he has provided a scientifically useful description of his findings. "Scientifically useful" as here applied is a name or short-hand designation for a description that can be used by others as well as the inquirer concerned for rechecking the inquiry, or as a basis for further inquiry, or as a means of modifying either external events or internal adjustive behavior or for any combination of such purposes.

The objective of scientific inquiry here suggested does not include achievement of "knowledge" in any absolute or final form, does not purport to establish "certainty," and does not offer its findings as unalterable indestructible Truth (whatever that may be). The goal is assertions warranted by the procedures of inquiry but not guaranteed to be fixed and immutable. The reports of scientific inquiry are invariably provisional, always subject to revision if and when better means of observation and measurement or other improvements in procedures of inquiry make possible more useful descriptions of what happens under specified circumstances.

At this stage of the scientific inquiry into scientific inquiry itself, no one inquirer or any group of inquirers has offered a comprehensive and systematically organized description of the procedures of inquiry that have proven to be most useful in solving problems. Nevertheless, much work has been done, and the procedures suggested for trial have been described in some detail by certain observers, including Dewey and Bentley. For the purposes of this report, rewording and rearranging has been undertaken in the hope of facilitating application of the procedures; but neither categorical verbal form nor apparent finality of expression should be misunderstood as altering the provisional status of the suggested procedures. The inquirer who attempts to use these procedures is asked to regard his use of them as an experiment in the conduct of inquiry.

Involved in any inquiry is the bundle of habits the inquirer has acquired. All humans are subjected to enculturation from the day they are born, perhaps earlier. Family living, formal education, and other aspects of experience combine to influence the habits of observation, of talking, of reading and writing, and of responding in various circumstances. Much human behavior reflects the gradual acquiring of such habits with their tendencies to dominate action. For example, since the publication of Ames' experiments at Dartmouth, who can doubt that much of what people observe is determined by what they have observed in the past, by habits formed in the course of repeated "seeing" under certain circumstances.[1] And in attempting to report their observations, to communicate, men are greatly influenced by their habitual attitudes toward words, by the ways of talking to which they have become accustomed, however primitive from the scientific inquirer's viewpoint those ways of talking may be.

Consequently, readers should not be surprised to find that applying the procedures of inquiry suggested here requires concentrated effort, at least in the beginning. Acquired habits can be changed, but precisely because they are habits they usually are not easily changed.

We begin our description of useful procedures of inquiry by noting the vast universe of the world, sun, stars, and all that we can see, smell, taste, hear, and feel. We wish to discuss the sum total of such things without repeatedly having to describe them in detail. For that purpose we need a short name, and we select "cosmos." This name is applied to the universe as a whole system, including the speaking-naming thing who uses the name.

Next we differentiate (or note differences) among the vast number of things in the cosmos and select for naming the living things; for these we choose the name "organism." Note that selecting for naming does not imply detaching the physical thing from the cosmos. Everything named remains a part of cosmos with innumerable relations to other parts.

Among the organisms, we further differentiate and select for naming ourselves, our ancestors, and our progeny; these we name "man."

We then observe the transactions of man with the remainder of cosmos and note the transactions named "eating," "breathing," etc. Among the numerous transactions, we differentiate further and select for naming those transactions typical of man but rarely characteristic of other organisms.

Human behavior involves transactions wherein something is regarded as standing for or referring to something else. This process we name "sign behavior," or simply "sign." Note that "sign" is not the name of the thing that is regarded as standing for something else; "sign" is the name of the transaction as a whole (i.e., is the short name for "sign process"). Sign or sign process is the type of organism-environmental transaction that distinguishes a behavioral from a physiological process, a behavioral transaction from a transaction such as eating, digesting, seeing, etc.

Sign process has evolved through the following still-existing stages:

a. The signaling or perceptive-manipulative stage of

[1] For accounts of these experiments, see William H. Ittelson, *The Ames Demonstrations in Perception*, Princeton, Princeton University Press, 1952; Hadley Cantril, ed., *The Morning Notes of Adelbert Ames, Jr.*, New Brunswick, Rutgers University Press, 1960; Franklin P. Kilpatrick, ed., *Explorations in Transactional Psychology*, New York, New York University Press, 1961. For an account of the role of habits, see John Dewey, *Human Nature and Conduct*, New York, Henry Holt, 1922.

sign in transactions such as beckoning, whistling, etc.

b. The naming stage used generally in speaking and writing.

c. The symboling stage as used in mathematics. (Border regions remain to be explored and characterized; i.e., tentatively named).

Focusing our attention now on the naming stage of sign process, we choose to name it "designating." Designating always is behavior, or organism-environmental transaction typical primarily of man in cosmos. Designating includes:

1. The earliest stage of designating or naming in the evolutionary scale, which we shall name "cueing." Cueing, as primitive naming, is so close to the situation of its origin that at times it is not readily differentiated from signal. Face-to-face perceptive situations are characteristic of cueing. It may include cry, expletive, or other single-word sentences; in fully developed language it may appear as an interjection, exclamation, abbreviated utterance, or other casually practical communicative convenience.

2. A more advanced type of designating or naming in the evolutionary scale, which we shall name "characterizing." This name applies to the everyday use of words, usage reasonably adequate for many practical purposes of life.

3. The, at present, farthest advanced type of designating, which we shall name "specifying." This name applies to the highly developed naming behavior found in modern scientific inquiry.

For the purpose of economizing words in discourse, we need a general name for the aspects and phases of cosmos differentiated and named. For this general name we choose "fact." Fact is the name for aspects and phases of cosmos differentiated and named by man (and man himself being among the aspects of cosmos) in descriptions sufficiently developed to include definite time and space aspects. Fact includes all namings—named durationally and extensionally spread; it is not limited to what is differentiated and named by any one man at any moment or in his lifetime.

Frequently, we have need to discuss a limited range of fact where our attention is focused for the time being. For this we choose the name "situation." This is the blanket name for those facts localized in time and space for our immediate attention.

Within a situation we frequently have occasion to refer to durational changes among facts. For these we choose the name "events."

Finally, in discussing events we frequently have occasion to refer to aspects of the fact involved that are least vague or more firmly determined and more accurately specified. For those we choose the name "object." Object is differentiated from event in that it is relatively stable subject matter of inquiry, at least for the time being.

Further tentative comments on sign process may be helpful. The transition from sign process at the perceptive-manipulative stage (here designated signaling) to the initial naming stage (designated cueing) is a change from the simplest attention-getting procedures, by evolutionary stages, to a somewhat more complex sign process used to describe things and events. No clear lines of demarcation are found. Some perceptive-manipulative signalings as well as primitive word cues are more than simple alerting behavior; they are also descriptive.

The transition from cueing to characterizing also reflects evolutionary development with increasing com-plexity of process including formal grammar, etc. The further transition from characterizing to specifying in the manner of modern science reflects the further evolutionary development of sign process, a still more complicated procedure. Moreover, all designating, even the technical naming used in modern scientific inquiry and here classified as specifying, names only *some* aspects or phases of any object or event. All naming involves abstracting; that is, involves focusing attention on some (not all) aspects or phases. All naming also is incomplete in that the possibility always exists that new and heretofore unknown aspects and phases of any object or event may at some future time be discovered (differentiated).

At first thought the stage we have here designated "symboling" may seem to be a marked departure from, or to reflect a break in, the evolutionary development of sign process. However, mathematical symboling may be considered a shorthand means of specifying. Each symbol replaces one or more words. A single mathematical equation may replace a long and involved sentence, even a paragraph, or a longer description in words.

Thus, sign process in its evolutionary progress to date may be described as the efforts of man to communicate and to record: first by simple perceptive-manipulative processes; then by verbal processes of increasing complexity, until this increasing complexity of verbal procedure became so much of a barrier to further progress that a shorthand system was devised in order to facilitate further communicating. This shorthand system has been most extensively developed in mathematical symboling.

At this point, readers are reminded that no part of the foregoing is asserted to be *the* correct procedure for scientific inquiry and reporting. Only suggestions for trial in the procedures of inquiry have been offered. Whether or not the suggested procedures prove to be useful will be determined not by anyone's preconceptions about "knowledge," nor by any formal logic (Aristotelian or other), nor by any revelation, secular or otherwise, but simply by the results obtained through application of the suggested procedures.

Anyone attempting to apply the indicated procedures may find helpful the additional tentative suggestions that follow:

Knowings—namings are organic—environmental be-havings of men in and as parts of cosmos.

All the subject matters of scientific inquiry are aspects and phases of cosmos; all are natural in that modern scientific inquirers do not purport to provide warranted assertions (useful descriptions) about the allegedly super-natural. Nor do modern scientific inquirers assert that nothing ever will be found beyond the scope of present means of observing by sight, smell, feeling, tasting, hearing and such extensions of sense perception as telescopes, microscopes, and other instruments at present provide.

Various subject matters of inquiry may be classified into groups from time to time in accordance with the various techniques of inquiry that may be applicable. The most recent widely recognized major classifications are: physical, physiological, and behavioral. None of these fields of inquiry is subject to the domination of one over another, yet in each an inquirer may make use of some findings in another, and all remain in the general system of cosmos becoming known by means of man's knowing behavior.

Within much of the realm of knowing-behavior, wherever sign process is involved, knowing is naming.

Naming is application of verbal or other signs to things differentiated in cosmos. Things are differentiated by observing, hearing, touching, or otherwise noticing that this differs from that in some aspect or phase. Differences are ascertained by comparison, one thing with another, one aspect with another, one phase with another, etc.

Differences may be: in size, shape, or color, etc.; hot-cold, smooth-rough, hard-soft, elastic-inelastic, few-many, early-later, etc.

For some solutions to many problems encountered, crude comparisons (observance of differences) may be sufficient. For modern scientific inquiry more precise measurements of differences frequently are required. Such more precise measurements may be reported in numbers as digits, on some developed scale or instrument, etc.

Differences sometimes labeled "qualitative" simply are differences noted. Differences sometimes labeled "quantitative" are differences reported more accurately by measurements, recorded usually in numbers.

As natural events in cosmos, knowings and knowns are observable and are enduring and extensive within enduring and extensive situations. A knowing is an event just as is an eclipse, a fossil, an earthquake, or any other subject matter of inquiry. Knowings and knowns are to be investigated by methods similar in principle, albeit sometimes varying in technical details (such as the instruments used, perhaps), to those that have been successful in the physical and biological sciences.

Space and time (extensional and durational) aspects of inquiry developed in one of the three principal subject matters of inquiry may be useful aids when investigating other principal subject matters, but should not be made limiting controls over inquiry beyond their usefulness as established in the course of inquiry.

Objects designated as such in practical, everyday experience prior to the application of modern scientific methods of inquiry have no permanent place or priority in relation to the objects of scientific inquiry. All objects, even those established with some degree of assurance by scientific inquiry, always are provisional, are subject to re-examination, in whole or in part, and may be superseded by other objects as found in the processes of improving observation. In short, objects as fixed, final, eternal and absolute things are neither known through modern inquiry nor assumed to exist.

Durationally and extensionally observable events are sufficient for inquiry. Nothing more real than the observable is established by using the word "real" or by attempting to peer behind or beyond the observable for something to which the name can be applied. Abandoned is the notion that "reals" exist as matter, or that "minds" exist as manifestations of organically specialized "reals" or that the "certainty" of matter somehow survives all the "uncertainties" of increasing knowledge about it. Finally, nothing is accepted or assumed in modern scientific inquiry that is alleged to be inherently nonobservable or as requiring some type of supernatural observation.

Namings and named, or knowings and known, (each phrase being a different name for the same behavioral event) are aspects of one event, not combinations of separate or separable events. Namings and named develop and fade away together; one does not leave the other behind like the grin of Alice's Cheshire cat. Although either principal aspect of the naming and named (or knowing and known) may be examined for some purposes as though it were separate, full scientific report requires transactional (as contrasted with interactional) observa-tion. All facts or purported facts have aspects of the knowing as well as the known, with the knowings among the facts known.

The observable extensions of knowings and knowns include all of cosmos observable by man; the observable durations extend across cultures, backward into the historical-geological record, and forward into indefinite futures as subject matters of inquiry. Knowings and knowns tend to persist as habitual behavior, but are not assured of permanence.

Namings may be segregated for special investigation within knowings much as any special region within scientific subject matter may be segregated for special consideration. The namings thus segregated are taken as themselves the knowings to be investigated. The namings are directly observable in full behavioral durations and extensions. No instances of naming are observed that are not themselves directly knowings; and no instances of knowings are observed within the range of naming-behaviors that are not themselves namings. The namings and the named are one transaction. No instance of either is observable without the other.

The scientist's descriptive report of what happens under specified circumstances may include assertions that are warranted or established by the course of inquiry in varying degrees. The most strongly warranted assertion is the hardest of hard fact, but that does not enthrone it beyond the reach of future inquiry nor guarantee its permanence. What is "hard fact" at one time may or may not be "hard" forever.

The study of written texts (or their spoken equivalents) in provisional severance from the particular organisms who wrote them, but nevertheless as durational and extensional behaviors under cultural description, is legitimate and useful. Such examination is comparable to that of species in life, of a slide under a microscope, or of a cadaver on the dissection table—directed strictly at what is present to observation, and not in search for nonobservables presumed to underlie observation, but always in search for more and more pertinent observables. Behavioral investigation of namings is to be correlated with the physiology of organism-in-environment rather than with the intradermal formulations that physiologists initially employed in reporting their earlier inquiries.

Subject matters of inquiry are to be taken in full durational spread as present through durations of time, comparable to that direct extensional observation they receive across extensions of space.

Namings of subject matters are to be taken as durational, both as names and with respect to all that they name. Neither instantaneities nor infinitesimalities, if taken as lacking durational or extensional spread, are to be set forth as within the range of named fact.

Secondary namings falling short of these requirements are imperfections, often useful, but to be employed safely only under express recognition at all critical stages of report that they do not designate subject matters in full.

Rejected are:

1. All "reals" beyond knowledge.

2. All "minds" as bearers of knowledge.

3. All assignments of behaviors to locations "within" an organism in disregard of the transactional phases of "outside" participation (and, of course, all similar assignments to "outsides" in similar disruption of transactional event).

4. All forcible applications of Newtonian space and time forms (or of the practical forms underlying, and

antedating, the Newtonian) to behavioral events as frameworks or checkerboard type grills, which are either (1) insisted upon as adequate for behavioral description, or (2) considered as so repugnant that behavior is divorced from them and expelled into some separate realm of its own.

5. Any notion that "reals" exist in a realm of their own; that "minds" exist in another realm; and that some kind of magic is required for the mind in its separate realm to achieve its knowing of the "real."

A highly significant characteristic of the suggested methods of inquiry is that they are self-corrective; that is, included among them are the procedures for correcting them. Men have used various methods of inquiry including those of common sense, of revealed religion, of secular revelations, of seeking the aid of spooks and fairies, of consulting the oracles, of Aristotelian logic, of the philosophers' quest for certainty, and of Newtonian mechanics, to name a few. By one or more of these or other means men have claimed to find what von Mises calls "apodictic certainty,"[1] that is to say, absolutely certain certainty, as though a sufficient application of earnestly offered verbiage could embalm their findings in a copper-riveted, indestructible, and forever established form never to require amendment, updating, or reconsideration. Some men have been so sure that the methods of inquiry satisfactory to them had yielded absolute certainty that they have tied their fellow men to stakes and burned them; and, although these are glaring examples, they may well have been the least harmful actions, all things considered, that men have done while laboring under the delusion that the methods of inquiry satisfactory to them have yielded ultimate truth. As will have been realized by any reader, the procedures of inquiry suggested herein are radically different. They apparently are the only ones that provide for continuing development, including revision, as may be advisable for themselves as methods as well as for the findings of inquiry.

1 Ludwig von Mises, *Human Action: A Treatise on Economics*, New Haven, Yale University Press, 1949, p. 39.

Section B

INTRODUCTION TO
JOHN DEWEY'S PHILOSOPHY

JOSEPH RATNER

Editorial note:

We reprint here, in full and unaltered, Ratner's *Introduction to Dewey,* with the exception of certain footnote citations to other specific pages of the book in which Ratner's *Introduction* was first published. For the convenience of the reader, we have changed those citations either to the page numbers of the present volume, or to the page numbers of the original books by Dewey, as appropriate.

R. H.
E. C. H.

INTRODUCTION TO

JOHN DEWEY'S PHILOSOPHY

by

Joseph Ratner

I

For those who believe it is the philosopher's task to juggle the universe on the point of an argument, Dewey is a complete disappointment. The world he starts out with and also ends with is the common world we all live in and experience every day of our lives. To start out with the familiar world of common experience is not altogether a philosophic novelty. Some philosophers have consciously done that before and the others, despite their more exalted intentions, have had to do the same thing to some extent; for they too are human beings and to hoist themselves into another world by their intellectual bootstraps, they must first at least take hold of those common things. But for a philosophy which emcompasses every important intellectual and cultural activity to end, as well as begin, with the world of everyday life is altogether novel, an achievement unique in the history of thought.

There are, of course, arguments in Dewey's philosophy. It could not be otherwise, for philosophy is just one long argument. But the world Dewey argues about is not a world his arguments have created. His arguments rest on, refer to and are controlled by experience of the common world. Control of philosophic arguments about the world by experience of the world is what Dewey fundamentally means by empiricism in philosophy, by scientific or experimental method. There are arguments in science and plenty of them. But the last word in science always rests not with the arguments, but with the facts, with the observations and experiments, with the laboratory experience of the scientist, be that laboratory one which he has artificially constructed for himself, or be it the laboratory into which he has converted the natural world of stones and stars. This relationship between arguments and experience is so firmly established, so integral a part of scientific technique and practice, that it is now taken as a matter of course and no scientist, no matter how mathematical or theoretical he may be, would even dream of disputing it. A scientist who refused to submit his arguments or theory to the test of observation and experiment on the ground that theory was higher than practice, or on any other ground he could imaginably concoct, would be laughed out of scientific court. And the same attitude would be taken toward such an imaginary scientist by every philosopher today.

But there are many philosophers still extant who, with regard to their own arguments or theories, disdain to recognize similar obligation. Certainly, scientific theories must submit to the test of practical experience, but philosophic theories, ah! they are different! In the realm of philosophy, theory is superior to practice, theory is completely independent of practice, theory is entirely separated from practice, theory has its own infallible ways and means of establishing its own irrefutable Truth, and practice and experience have, with respect to these philosophic ways and means and this philosophic Truth, no authority whatsoever. They are immaterial, incompetent and irrelevant.

Now Dewey's basic position, his basic argument about philosophic method, is that theory in philosophy is no more privileged than theory in science. If theory in science must submit to the test of practice and experience, theory in philosophy must do likewise. Philosophers are the same breed of men as scientists, the brains of both are alike, the product of the same earthly evolution. For philosophers to believe they are endowed with unique powers giving them access to special realms of Being and revealing to them knowledge of special Truths is a gross piece of self-delusion. Philosophers are gifted with no supernal powers of insight denied other mortals. There are no exclusive regions of Being or Reality into which a philosopher alone can enter because he carries a philosophic passport—made out by himself. The only genuine passport, the only passport commanding entrance into Being, Reality, Nature or whatever else you care to call it, by capital letter or small, is the passport that is filled out, signed, countersigned, stamped and sealed by public experience. And until philosophers recognize this, until they accept their common humanity with good grace and without mental reservations, they cannot hope to perform any intelligent function and make philosophy a living thing, a progessive force in our common human life.

That it should still be necessary to argue for experimentalism in philosophy is anomalous, as Dewey has tirelessly driven home. In science, the practical issue over experimentalism was fought in the sixteenth and seventeenth centuries and to the practical victory of experimentalism in science all the marvelous scientific achievements of the past two hundred and fifty years are to be ascribed. That this issue should still be of primary debate in philosophy is as bad a case of cultural lag as one could ever hope to come across, especially when one takes into account the fact that it has been the boast of philosophy that she is by history and by nature the intellectual leader, the one that is always found at the very head of the line of march.

What is the reason for the backwardness of philosophy? Why do philosophers continue to oppose the demand that their theories be based on experimental grounds and undergo experimental tests? Such opposition smacks of antediluvian kicking against the pricks, true enough, but it would be silly to think it is due merely to unregenerate antediluvianism. Nor can the opposition with any show of justice be chalked up as a result of obtuseness, natural or acquired, or as the result of ignorance of scientific history. The reasons, as Dewey has shown with voluminous clarity, are not ascribable to any personal shortcomings of philosophers as a class, but lie embedded in the heart of our culture, in the traditional forces operative in the social, political, religious, educational, philosophic and even scientific spheres.

Modern culture stems from heterogeneous roots. It is more a cultural compendium than a cultural complex. There is everywhere a medley of forces at work and the scene, wherever one looks, is full of strife. In some very few and very restricted areas, there has already been achieved some measure of outer harmony and integration, but it is outer, superficial and not thorough. The conflict between modern methods, understanding and ideals, and traditional attitudes, beliefs and objectives where it has disappeared from the surface has disappeared only to persist below. The least probing discloses intensified discord and widened division. And this is true whether one considers together and in relation two or more areas of modern culture or considers each area separately and alone. The spectacular conflicts rage across the open fields, while confusions smoulder and agitate underground. The backwardness of philosophy is an expression and reflection of the widespread and varying cultural lags. It is both a symptom and a symbol of the outer clashes and inner confusions, of the essentially discordant, unintegrated character of modern culture. Philosophy, aspiring to a secret vision of eternal existence, has fallen heir to the social ills of mortal experience.

The general state of modern culture explains the situation in philosophy but, for Dewey, explanation is not excuse. Explanation of any trouble is, for him, the starting point for intelligent and thorough re-examination; it defines the problem to be faced and the task to be done. Philosophy is omnipotent and philosophers exercise absolute sway only in their Platonic dreams—dreams that have never beguiled Dewey. However, it cannot be significantly denied by any one that philosophy has had and still has some social power and whatever the measure of that power may be, that is also, for Dewey, the measure of philosophy's social responsibility for the future of human culture as well as for the present and past. To what precise extent philosophy has helped bring about the existing divisions, the multifarious splits in modern cultural life is a question that can never be accurately answered. It will also forever be impossible to estimate to what precise extent the discords and confusions have been perpetuated by the theoretical sanctions they have received from the great systems of modern philosophy. But quantitatively exact answers to these questions are not at all necessary for reaching the sound conclusion that philosophy is in fact and in honor bound to shoulder some of the blame in both instances. Answers of quantitative exactitude are even less necessary, if that is possible, for reaching an intelligent judgment as to what should be the function and purpose of philosophy in the present juncture in our social life.

In so far as philosophy has wielded social influence it is responsible for the existing state of affairs; and in so far as it does and will continue to wield such influence its real task in the present cultural epoch is mapped out by the indisputable nature of the epoch itself. It is not to help perpetuate and justify the existing state and disorder of things, but, to the reach of its ability, to help find a way into a better order, an order in which there will be social unity of mind as a consequence of achieving civilized integration of intelligent life. This is, for Dewey, the supremely important task confronting philosophy, its all-comprehensive task, the only one that genuinely brings

philosophy into commerce with the universal. If it can be justly said of philosophy that it is uniquely equipped to undertake any task, then it is this one. And yet this is just the task modern philosophy has either approached obliquely or else outrightly shirked on the pretext that philosophy had more vital concerns, more universal objectives to attain, that its elected destiny was to circumnavigate the great ocean of Being.

Philosophic pretensions to superhuman universality inevitably generate theories that degrade human life and experience to subhuman estate. While pretending jurisdiction over all time and existence, such theories actually function to support and justify the practices of intolerance and the barbarities of fanaticism. To rid philosophy of pretentiousness—the prolific mother of evil—is the all-controlling, all-permeating purpose of Dewey's lifework. By the example of his own work, a work eloquent with the fire of his conviction that philosophy has a real and useful, a vitalizing and humanizing function to perform, modest though it may be, Dewey has, for close on half a century, continued to call upon his fellow philosophers to have done with their building of sandpiles on the shores of human life and to come inland and help build habitations fit for men.

II

Properly to understand Dewey's conception of the relation between philosophy and culture, it is essential to keep focused in mind that Dewey conceives philosophy to be one part of culture, interacting with all other parts with varying degrees of sensitivity and effectiveness. Put strictly, Dewey always thinks of philosophy *in* culture, not of philosophy *and* culture. That he does not always write as he thinks, on this matter as on others, is not something for which he is solely to blame. Dewey, after all, did not inherit his own mature philosophy, nor was he taught it in school, college or university. His mind, like the mind of everyone, was first informed with the issues, ideas and language produced before his time. And, like every original thinker, it was only by working with the material he acquired that he was able to work through it. That there should be signs and evidences in his writings of the uphill intellectual road he has travelled is a natural consequence, and something every intelligent person not only does find but expects to find in the work of every creative thinker, no matter of what period or place. Faultless lucidity and articulation in writing, like faultless execution in painting, are possible only for those who are superficial in treating their subjects, or superficial in accepting what is current, or superficial in both.

Because philosophy is *in* culture, one part interacting with other parts, it is basically to falsify matters to interpret philosophy's cultural role as being either all cause or all effect. In any interacting system there is inevitable a cross-weaving or intersection of cause and effect. In fact, such cross-weaving or intersection is precisely what interaction means.

Only by doing intellectual violence to the actual condition and state of affairs can philosophy be torn out of its cultural environment and be set up as something isolated and apart. The violence of this act is not mitigated but compounded when it is made for and followed up by bringing philosophy into relation again with the remainder of culture in a way that makes philosophy exclusively either the cause or the effect of that remainder. The net amount of logical falsity and empirical distortion stands unaltered whether the act of violence and its sequel are done in the name of a theological, idealistic, materialistic or dialectic theory or any combination and permutation of these.

Of course it would also be false to assert that philosophy has maintained equally effective interactive relations with all the cultural forces constituting its environment, that its interactive integration with its environment has been complete and perfect. The actual history of philosophy—like the actual history of all affairs human and divine—is a mixed record of failures and successes.

In so far as modern philosophy has been in interactive relations with the social, political, economic and scientific forces and movements, it has developed in fruitful and distinctive ways so that modern philosophy is actually and recognizably different from the philosophy of any other epoch in human history. And by virtue of the same interactive relations, and in the measure that they have been effectively sustained, modern philosophy has undoubtedly contributed to the distinctive and fruitful development of other members in the cultural system. But none of the areas of modern culture has been in full interaction with any other area, let alone with all the others. This is true, with especial emphasis, of philosophy. In consequence, it has also been stagnant and uniform, repetitious of its modern self and its pre-modern history. To be sure, modern philosophy has successfully escaped being outwardly marked by stagnation, uniformity and repetition. It presents a continually changing face and seems to be always going in at least ascending and widening spirals. This pleasing, even flattering outer appearance is, however, mostly deceptive, for it is maintained by the momentum derived from modern philosophy's few genuine interactive relations and which could not be entirely wiped out.* Inwardly, and for the most part, modern philosophy has been going in narrowing and flattening circles.

III

It can be argued—as many have actually argued—that since there is internal division in every area of modern culture and loosejointedness and conflict between all, modern philosophy, by exhibiting like features, shows it is really in the modern step and that it would be badly out of step if it exhibited contrary features. There is virtue in this argument, but it is the virtue of its content of fact, not of argument. True enough, only a fanciful philosophy can be fully integrated in a culture which is mostly otherwise. Those modern systems wherein all things are neatly disposed of and settled down in permanent wedlock, as in Hegel, or in a dual state of permanent marriage and irremediable divorce, as in Kant, are intellectual fantasias rather than philosophies. In the one case as in the other they do not fulfill but betray the cause of intelligence which is the supreme cause of philosophy—the cause philosophy cannot forsake without losing her mind and soul. This is the virtue of fact in the argument.

But the argument, if it is an argument at all, implies more than the facts. All arguments worthy of the name carry the mind forward by presenting possibilities that lead to the discovery and help in the making of further fact. Thus understood, how stands it with the argument in question? What possibilities does it present? What conception of philosophy does it imply or assume? How does it see and define the function of philosophy in the changing course of cultural history and development? The answers to these questions are not difficult to find. Clearly, the argument fundamentally implies or assumes that philosophy is inherently merely an effect produced by the remainder of the cultural forces which are alone really operative as cause. And in necessary line with this basic principle or assumption, the only possibility it presents is that philosophy must forever continue in this ineffectual role. This point of view was advanced by the Hegelian argument, though it runs counter to Hegel's cardinal tenet that Mind is the one and absolute cause and philosophy (in fact, his own philosophy) is the highest realization of Mind in the empirical world. The Marxian argument advances the same doctrine and also with the self-same contradictoriness, though its contradiction runs in the opposite direction. It is consistent with the ground-plan of the materialist interpretation of past cultural history and inconsistent with the revolutionary program for realizing future cultural history. This is but one instance of the net identity in logical falsity of idealist and materialist theories pointed to before.

Determinism is a magical word and, when supplemented by the adjective "rigid," its magical effectiveness is beyond all hindrance and recall. In the mouth of Idealism, rigid determinism—called by the more pleasant sounding names Destiny, Divine Will, etc.—instantaneously converts the confused and conglomerate history of man into the inevitable unrolling of the pellucid Divine Idea; in the mouth of Materialism it performs a no less magical act of conversion, but instead of the unrolling being the fulfillment of an all-necessitating Idea, designedly leading us by the nose, the unrolling becomes the fulfillment of an all-necessitating congress of material forces pushing us blindly from behind. If choice were absolutely restricted to either one or the other of these two the latter is, beyond doubt, on human and practical not cosmic and theoretical grounds preferable.* For it keeps us unremittingly conscious of the fact that we have behinds, something those who are led by the nose will only rarely and hesitantly admit. However, and fortunately, our choice is not restricted to the either-or of these two. Actually, as is empirically verified and verifiable, our anatomy faces both ways and there is nothing inherent in the nature of the mind, no constitutional, ineradicable defect which forever prevents it from displaying a like vituosity.

On the contrary! To look before and after and think of what is not but may become through our efforts controlled and directed by what we see before and after is the very essence of mind and what it naturally does when not blocked in the exercise of its function. This, too, is the essence and function of philosophy as a phase of cultural mind or intelligence.

The development of philosophy can be aided or hindered by the cultural forces with which it contemporaneously interacts. About this there can be no sensible

---

*In a pioneering essay, *"The Significance of the Problem of Knowledge,"* Dewey traced in illuminating outline the interactive relations between, on the one hand, the leading social, economic, political and scientific movements of modern times, and on the other, the development of the two dominant issues in modern philosophy—the sensationalist-empiricist and the rationalist. This essay, reprinted in *The Influence of Darwin on Philosophy* (1910) was first published in 1897 and so antedates by some years the socio-economico-politico-culturo-historical wave of interpretation which has recently swept over many current writers and swept away so many more.

*As cosmic doctrines, there are no intelligible, let alone intelligent reasons for preferring one over the other. In this respect they are more than alike: they are identical.

question. But the aid or hindrance is in every such case partial, not total. God helps those who help themselves because nothing exists which cannot in some way help itself. And what is capable of self-help is capable also of self-hindrance. The blockage philosophy may suffer or the freedom it may derive from the operation upon it of other forces is never an automatic effect of a one-way operating cause or series of causes; it is always the consequence of interaction. The order and connection of ideas in philosophy are not the same as the order and connection of events in society. If there were this one-to-one correspondence or parallelism, philosophy would always be exactly abreast of its times. It could never possibly fall behind or get ahead. Actually, however, as history empirically verifies, philosophy has done both. Therefore, whenever philosophy is frustrated or liberated, the causes for that must also be partly ingredient in philosophy itself, in its own complement of ideas, in its history and development up to the time under examination. For philosophy does not merely interact with contemporaneous social forces and events; more than any other human intellectual enterprise it interacts with its own past. In this case, "interact" is too generous a word because, as is the great burden of one of Dewey's arguments, philosophy rather carries its own past along with it too often and too much as a dead and deadening weight. However this may be, (it is an issue to be discussed later), certain it is that the historical development of philosophy contributes to the determination of philosophy's selectivity and sensitivity of response in interactive relations. And its selectivity and sensitivity, as of any one time, contribute to the determination of the influences it undergoes, how it accepts them or rejects them, to what extent in each case and to what frustrating or liberating end.

Hence the philosophic reason for and import of Dewey's constant excursions into historical analysis and evaluation. In his recent writings, analysis and criticism of classic Greek theories of nature, knowledge and mind figure ever more prominently. Forty and thirty years ago the Hegelian and Kantian philosophies and their derivatives were the main objects of his critical attention; thirty and twenty years ago, it was the then contemporary realisms of all varieties, American and English. But with *Experience and Nature* (1925), a great, though not unheralded, change took place: the foregoing receded into the background while into the focus of critical examination were placed the philosophies of Plato and Aristotle; and this interchange of position between modern and ancient philosophies has become more and not less marked with each succeeding volume.

The reasons for these two major changes in critical orientation are different. The earlier one occurred because of a change in the philosophic scene. For a number of causes, not the least of which was Dewey's own work, the Idealisms, in the first decade of this century, were fast disappearing from effective life and continued examination of them would have been socially as well as philosophically useless. The philosophies then in need of critical attention were the flock of Realisms, for with all the lustiness of the newly born they were disputing with experimentalism its claim to win the rising generation of philosophers.

The later change (1925) occurred for a far different reason. It was the result of a deepened insight on Dewey's part into the nature and sources of the basic ideas controlling the major movements of modern, and the newest movements of contemporary, philosophy. Dewey then saw that as long as these causative ideas continued to work in the bowels of the western mind, there would be no end to the forthcoming of new editions of the old unexamined assumptions. To struggle with each fresh variation on the ancient theme was an endless and hopeless task. It was like trying to conquer Antaeus by bouncing him on the ground.*

The validity of any thesis about the past must be established, in the first instance, by demonstrating its explanatory force for that past. And there is only one way of doing this when that past is the history of ideas, namely, by logical analysis and theoretical appraisal. This necessity weighs with equal force upon all philosophical investigators, be they experimentalists or not. But what distinguishes experimentalists from all others is their recognition that this is only in the first instance. If the analysis and appraisal are sound they must, in the second instance, be capable of experimental verification in the present and to be accepted must successfully pass this test.

For the past is not blocked off from the present by an impassable abyss; if it were, the abyss would also be impassable for us since, as abysses go in the empirical world, they cut off both sides and we would never be able to get over to the past to investigate it by any means, theoretical or otherwise. In fact, we would never know there was a past, and talk about being influenced by it would be impossible even as sheer hallucination. In some transcendental philosophies and also in some philosophies which claim to be realistic, for reasons only known to their authors, abysses, to be sure, do not operate in this pedestrian, empirical way. They cut off only one side, wiping out all routes from the past to the present while leaving always intact at least one route from the present to the past—a route through the transcendental air. Needless to say, Dewey has no such inspired conception of things, real or possible. No one has ever defended the cause of possibilities more vigorously and consistently than he. He has championed possibilities in season and out, along with others and alone. But to win his support they must be possibilities that can actually be realized. And this goes for abysses too. The fact that the only kind that can be introduced into Nature and experience are abysses that, in the very act of introducing them, must also be rendered congenitally incapable of operating with equal effectiveness on both sides of their job, is conclusive evidence for Dewey—and should be for any one—that the enterprise envisions not a real possibility but a chimera. Real possibilities are limited by the continuities in experience and Nature.

The validity of Dewey's thesis that in the elements of Greek thought carried along in the modern mind are to be found the generating causes both of the problems that have clogged and stultified modern philosophic intelligence and of the solutions which have repeatedly been proposed, often in sheer intellectual desperation—this thesis is experimentally proved to the hilt by the current revival of Idealism.† And further experimental verification, if such be needed, is supplied by the latest exhumation of medievalism.

---

*Since this volume presents the philosophy of Dewey and not the history of his development, selections of detailed historical criticism have been limited to those dealing with the Greeks.

†This revival, initiated chiefly by scientists—Eddington, Jeans, and others—began a few years after Dewey's first complete statement of his thesis in *Experience and Nature* (1925). Dewey's complete *development* of his thesis is in *The Quest for Certainty* (1929), which at one point specifically takes up for reply the work of Eddington.

IV

The theory of knowledge is pivotal in modern philosophy. All other issues have revolved around it and all problems peculiarly modern have been generated by it. It began, in Locke, as the universal solvent; it became, in Hume, the universal corrosive; and in Kantian and post-Kantian philosophies it ended up as the source of universal confusion. Judged by empirical standards of performance, there is nothing in the record of the theory of knowledge, or epistemology, to justify keeping it in its position of hegemony. On the contrary. Everything in its record necessitates, let alone justifies, that it be removed from its position if not, indeed, thrown out entirely.

The whole modern epistemological industry is principally supported and kept going by one fundamental assumption concerning the nature of mind and what it does when it knows. To be sure there are almost as many different kinds of elaborations and refined involvements of theory as there are philosophers, but when these secondary and tertiary outgrowths are cut away, what is left, as Dewey has shown, is an unmistakable identity. And this identity in conception of mind and its mode of operation has, in all essentials, been carried over without critical examination and often without even knowing it, from Greek speculation.

The Greek conception of mind and its mode of knowing, Dewey has aptly and accurately called "the spectator theory." The physical eye, according to the Greeks, is a positive source of emission of light and hence there inescapably takes place some sort of interaction between the eye seeing and the thing seen. Both being parts of the physical world which is in constant process of change, both are also necessarily involved in producing change. Although by a familiar figure, as familiar to Plato and Aristotle as to us, we speak of the mind "seeing" and even of the "mind's eye," the mind is, for them, distinguished from the eye in this basic and all-important respect: it does not, in knowing, interact with the object known. Plato and Aristotle, it is true, did not hold that the mind is like a slate which passively accepts what is written on it and has no activity at all. The mind, to know, has to act, has to envisage its object, to grasp it. But—and this "but" is crucial—the mind's activity is a "pure activity," that is, one which does not produce any change whatever in the object it acts upon. It is an actionless action—like the action of a spectator on the benches following the scene being played on the stage above or below.

The Greek philosophers did not arrive at their spectator theory by looking into the mind and thus finding out how it works. The method of introspection has been tried in the modern world for hundreds of years and with what uncertain, universally unestablishable results every one knows. The ancients made very little pretense of examining the mind by itself, whether by introspection or any other method. What they wanted were results absolute and certain and there was only one way, they knew, of getting them, namely, by logical reasoning, ratiocination, theoretical argument. They were all the more inclined, in this case, to find the answers to their questions by a process of inferential reasoning because the mind and knowing were for them of secondary, not primary, philosophic interest and concern. They came upon them in their search for something else.

The main objective controlling all Greek inquiry, scientific as well as philosophic, was what Dewey has again illuminatingly and accurately called "the quest for certainty." All human beings are implicated in the hazards and uncertainties of existence. And all human beings have been at least sufficiently practical, sufficiently motivated by mundane desire to want to eliminate hazard and uncertainty from life and to enjoy a state that is safe and sure. Since this objective can never be attained with absolute perfection in this world, the royal road that has time and again been sought by all peoples is the road of imagining another world wherein none of the hardships and at least all of the delights of this world are to be found and perpetually to be enjoyed. In these all-human respects, the Greeks were, of course, the same as others. What distinguishes them from all their forerunners and contemporaries is the epochal, world-revolutionary discovery made by their mathematicians and philosophers. Rivaling the poets and prophets (whom we now insignificantly call mythologists) the Greek mathematicians and philosophers discovered a new royal road to the heart's fondest and deepest desire—to the realm of eternal and immutable Being, replete with all that is good, true and beautiful and providentially devoid of everything else.

Now it takes no great wit to see that when the mind knows the eternal and immutable, it does not change what it knows in the act of knowing it. And it requires no greater wit to see that the eternal and immutable must have existed before the mind gained knowledge of it and that the mind, in knowing it, has no hand whatsoever in creating its Being. Take, then, as starting point, that Being, eternal and immutable, is the object of knowledge and by a line of inference as easy and compelling as the line demonstrating that the angles of an equilateral triangle are equal, you reach the Greek conclusion concerning the nature of the mind and its mode of knowing. To some, perhaps, this line of inference may seem far too simple, at any rate for philosophers to have really taken it. But Greek philosophers, unlike too many moderns, made a virtue of simplicity. They were quite thoroughly *en rapport* and in sympathy with their culture, and simplicity very profoundly characterizes it throughout—in sculpture and morals, architecture and politics, literature and religion, music and mathematics.

In fact, Greek philosophers were simple enough in their reasoning to be consistent along the main line. The natural world in which we live and act is a world of change, and from this position they never backed down. Knowledge of what constantly changes—the distinguishing trait of Becoming—can, obviously, never be eternal, absolute and certain; and from this position too they never backed down. Therefore knowledge of Becoming, of the world which practical action deals with, is not really knowledge but a bastard species of it which they called opinion. Real knowledge is of Being and only of Being. The realm of Being is eternal and unchanging and hence presents an object which once known is known forever and which when known at all is known with absolute certainty. That there is such a realm and also that we have knowledge of it they had proof in mathematics—new and absolutely certain proof, superseding the old and wavering proof of oracles and seers. Mathematics did not of course exhaust the contents of the realm of Being. It was, for the philosophers, rather a sign and symbol, an evangel of the happy tidings that a new road had been opened up which would take them where they always wanted to go and which made certain they would find what they always wanted to find. The new royal road was infinitely superior, in safety and comfort, to the one formerly used. And for philosophers, it was hardly a drawback that they alone could travel over it. Indeed, possession of exclusive right-of-way caused them none too secret pleasure and exultation. Plato, certainly, lost no time in giving public and peremptory notice to the poets and all other rivals that their day was over.

Now the Greek theory of eternal and immutable Being and its antiphonal spectator theory of mind entered into the bloodstream of modern thought at its very inception. They entered not only by way of philosophy and religion, in which fields they had luxuriantly flourished under the fervid care of medieval logicians and theologians; more importantly, they entered by way of science—more importantly because totally unsuspected and unacknowledged there. And since that time the attractive, ever-entrancing forms of Greek philosophy have circulated in every area, place and part of modern mentality and they create, wherever they are, an iridescent intellectual mirage.

The founders of modern science made a great show of being pure and uncontaminated philosophically. With one accord they attacked philosophy which meant, for them, medievalized Aristotelian logic and its stifling progeny. To free thought from the theological stranglehold, enforced in the temporal realm by the Church, was, they all recognized, the precondition of intellectual and scientific advance. In their march against the powers of darkness they were guided by the lamp of Euclid which they held aloft. But, alas, Euclid's mathematics and Aristotle's logic both involve the same basic presuppositions; they both rest on the same fundamental conceptions of knowledge, nature and mind; they are both results of the same type of metaphysical thinking and scientific method.

It was of undoubted advantage to fight the medievalized Aristotle with the unmedievalized Euclid, but as far as essentials are concerned it was tantamount to using a genuine form and product of Greek thinking as a weapon against a form that had been perverted. Aristotle's logic had been used by the medievalists to discourage and choke off all independent and original observation of nature; how much of a perversion this is and how serious can easily be appreciated when it is remembered that his logic was a presentation of the formal principles underlying the conduct of his own manifold and intensive naturalistic observations and was intended (and was so used by the Greeks) as a guide to further such.

The fathers of modern science, in addition to reclaiming mathematics and gaining unrestricted rights to observe nature—both strictly within the scientific limits of the genuine Greek tradition—also introduced, it is true, a new and non-Greek method of experimentation. This methodological novelty was destined to become all-important in the progress of scientific knowledge and the development of scientific ideas but, at the time of its introduction, it had only a supplementary intellectual value. It would be an exaggeration to say that it was first chiefly prized, not for its scientific but for its polemical power in the liberation movement. But the exaggeration would be nearer the truth than would be any contrary statement. Even in Newton's work, the most self-consciously "experimental" of all, the supreme right-of-way was given to mathematicians and not to experimental findings whenever the two came in conflict and blocked each other's path. In other words, the supreme right-of-way in the foundations of modern science was given to Greek ideas of method and science.

The anti-philosophical front presented by the fathers of modern science has been faithfully kept up by the vast and the most influential majority of their descendants. The latter are never so zealous as when they are warding off the sporadic advances of modern philosophy and never so firm and consistent as when rejecting its more formal proposals. This attitude of theirs has naturally been a continual source of irritation to philosophers and the ground for extensive, bitter and recurrent complaint. Only occasionally has the situation been relieved by offstage laughter. But, forsooth, the whole business—if we may ignore for the moment its tragic consequences—has all along been really a huge joke and the proper subject for unrestrained hilarity. For the philosophy modern scientists took from the Greeks they handed back to modern philosophers as science; and the latter, instead of at once spotting the intellectual sleight-of-hand and calling the game, were taken in by it to a man. Certainly there is some excuse for philosophers being thus easily deceived during the intellectually tumultuous years opening our epoch. But there is hardly excuse for philosophers who insist on staying deceived even now. Not that vulgar and childish delight in mystification entices and holds contemporaries in ludicrous trance. It is something more mature and for that reason more profoundly disturbing. As long as the deception is kept up, the staple and routine occupations of philosophers are not threatened; their intellectual habits, now easy, familiar and dear through long usage, can go on unreflectively grinding out philosophical reflections forever. And the by now standardized relation of philosophy to science—an unstable and explosive compound of envy and condescension—can be perpetuated, though not unperturbed.

One basic feature initially characterized all modern liberation movements: they advanced against the social and intellectual tyranny of Church and State, their oppressive authoritarianism, by appealing to the superior integrity, nature, authority and power of the individual.* In religion, this appeal took the form of Protestantism; in law and revolution the form of inalienable natural rights; in economics the form of *laissez faire;* in social ethics the form of the greatest good for the greatest number; in progressive politics the form of universal suffrage and representative, parlimentary government. In science and philosophy, the two predominantly theoretical areas of culture, the direction of the appeal was also essentially the same: the rationalist scientists and philosophers appealed to the natural light of reason brightly burning in the individual mind, while the empiricists appealed no less surely to the inextinguishable light and all-conquering power of individual experience. Although the differences between rationalists and empiricists developed to serious proportions later on, at the outset, expecially as far as concerns their common opposition to the manifold medievalisms and oppressions, the differences were tactical rather than strategic.†

It is possible, I think, to prove with reasonable surety that against oppression hardened in institutions and enforced by socially guarded and perpetuated dogmas only the assault of individualism, under whatever form it may be, has revolutionary power and effect. But whether or not this can be established as a principle, certain it is as a fact that modern culture started, and for nearly three hundred years won, all its greatest battles under this standard. Because of this prevailing cultural fact, to call it

---

*This feature, of course, continued to dominate until well into the nineteenth century.
†Dewey did not see this important point at the time of writing *"The Significance of the Problem of Knowledge"*; it is the one serious shortcoming in that great essay. It may also be pointed out that many still fail to see it.

no more, it was inevitable that in taking over Greek ideas, the spectator theory of mind should assume first and dominating place, acknowledged and unacknowledged, in the thinking of the period. Wherever it went, the spectator theory of mind necessarily brought along with it the theory of eternal and immutable Being, for they are an inseparable pair and play only together. But in modern thought, the original relation obtaining between them in Greek speculation was consistently reversed; their positions relative to each other were permanantly exchanged. This didn't matter very much in the sixteenth and seventeenth centuries, for their music, though badly off key, was sweet and melodious to any ear that, perforce, had hitherto heard only the harsh and grating, dull and offensive noise of medievalism. But by the eighteenth and nineteenth centuries, the consequences of the reversal or exchange had worked themselves through, far and wide; the simple tune for two had been amplified on orchestral scale and that it now mattered very much indeed was plain to every ear that could distinguish loud discord from quiet harmony, and to every eye that was not permanently shut and could see the difference between playing every which way and playing in unison.

V

When we look back over the histories of modern science and philosophy, from the vantage-point of the present time, one fact stands out clearly and boldly like the sphinx in the desert, and like the sphinx it too presents an enigmatical face. Although science and philosophy started out in community of effort, and with a common set of fundamental ideas, the courses they have run are not the same. They are not even parallel. They are divergent, so that the nearer we approach our own time the further apart they are. Now you may lightly say that there is really nothing strange about this, and certainly nothing enigmatical. For two activities to start from a common center and from that point onward to diverge in ever-widening degree is the sort of thing that happens every day and moreover the only kind of thing that can or should happen when their destinations lie in different directions. But this answer, true enough in what it says, does not at all meet the case in hand. The paths of philosophy and science have been progressively divergent not because their avowed, respective destinations lie in different quarters and each is intent on the shortest route. Just the contrary is the case. They are divergent *precisely because* they both have vowed to reach the *same* ultimate destination and by the quickest way. And this, surely, is an enigma.

But like all enigmas, the real explanation of it is not to be found in something more enigmatic and mysterious still—be it in the puniness of the finite reason strangely wrestling with the Infinitude of the Universe and being constantly thrown back; or in the inwardly palpitating Secret of Philosophy which cannot be exposed and live in the light of day because it is fetched from depths beyond all reckoning—in contrast to the dead contents of science which can be written in open script for any eye plainly to read because they are scraped off the surface of things; or vice versa, and so on. The real explanation, as Dewey has shown, is to be found in something really quite simple and historically to our human hand.

What is the common, ultimate destination of modern science and philosophy? To say it is the Truth is too vague to lead us anywhere. But to say it is the eternal and immutable Reality leads us straight home, into the

theoretical heart of modern science and philosophy—and back to the bosom of the Greeks.

The first success of science in its quest for certainty was wonderfully great, so wonderful that nearly three hundred years elapsed before science matured sufficiently to have serious doubts of its own as to whether or not it had exclusively and permanently captured eternal and immutable Reality first crack out of the box. It must, however, be said immediately on behalf of most scientists today that they have not allowed themselves to become too discouraged by these doubts. Recent revelations have set them back somewhat and shaken their early confidence, but most of them still hope to succeed in finally and exclusively cornering Reality in the next try, or in the try after that, and they steadfastly aim that way.

Nothing attracts like success. Why, then, were philosophers repelled and driven off at an obtuse angle by the first great victory of science? The answer is found in the field on which victory was won. Science discovered the eternal and immutable Reality in material masses and motion and the laws governing masses in motion. It was not, of course, just the fact that they were material masses and physical motion that assured scientists they had found what they were looking for. It was the eternal, indestructible nature of the constituent particles of the masses, and the eternal, unchangeable nature of the laws of motion that proved to them, with the inerrant simplicity and unshakable certainty of mathematics, that their conviction of success was true and not the fanciful product of their dream.

Now philosophers, like common men, have eyes and the eyes see colors—but colors, said the new science, are not ultimately real; philosophers have ears and ears hear sounds—but sounds, said the new science, are not ultimately real; philosophers have noses and noses smell smells—but smells, said the new science, are not ultimately real; philosophers have hands and hands feel surfaces, temperatures, and textures, rough and smooth, hot and cold, wet and dry, soft and hard—but soft and hot, wet and dry, cold and smooth and rough, said the new science, are not ultimately real; philosophers have tongues and tongues taste things bitter and sweet—but bitter and sweet, said the new science, are not ultimately real. All these, the new science said, are only words. And all the rest of them, and all like them, are also only words, or as Galileo formulated it, they "are not anything else than names." The only genuinely, ultimately real things are the atoms and their qualities of shape, size, hardness, motion, number, mass, inertia.* Really, in the last analysis, it is the tongue which is wholly at fault, the real obstruction blocking our path to the ultimately real! If it did not keep on foolishly uttering these words or names, misleading us, all would immediately accept with joyous heart and unquestioning mind the words of golden truth spoken by scientists!

Without exception the fathers of modern science had nothing to say destructive of the medievalized soul-substance or mind and its complement of ideas and sentiments, desires, purposes and plans. To have denied the ultimate reality of the mind at that early stage would have been intellectually suicidal for scientists, since it was on the power of the mind that they publicly rested their claim to have discovered the ultimately real. It would also have been socially suicidal, what with the powerful

---

*This list of "primary" or ultimately real qualities varies from scientist to scientist. The list above is a Galilean-Newtonian mixture.

backing the soul or mind enjoyed in secular prison and religious stake. As to which of these arguments, the intellectual or the social, was really effective and determined the scientists upon the course they took the reader should experience no great difficulty in deciding for himself. For the scientists without exception, Galileo as well as Newton, had also publicly rested their claim to have discovered the genuinely and ultimately real on the irrefragable nature of sensory evidence. How else, for example, did Galileo defend against the holy doctors of the church who for their "scientific" knowledge of Nature relied more insistently on the words of Aristotle than on the words of God—how else did Galileo defend against the holy doctors the reality of the satellites of Jupiter, except on the ground that he saw them with his very own sensory organs, namely, his eyes? How else did Newton prove that hardness, for instance, was an eternal, inalienable characteristic of the indestructible ultimately real atom, except by empirical inference which explicitly asserted that the testimony of the senses was of unimpeachable validity?*

Socially and religiously the senses were, in Galilean-Newtonian days, intrinsically sinful.† For Galileo and Newton, therefore, to turn completely round on their "scientific" axes and put the senses scientifically as well as socially and religiously in bad odor, was gratifying, not reprehensible, to State and Church. The State was pleased because Science by exposing the temporal, this-worldly vanity of the senses was contributing its strength, however much or little it might be, toward the fulfillment of the State's objectives: the restraint and control of the this-worldly desires of the people. And the Church, naturally, in its dual capacity of Temporal and Spiritual Power was doubly pleased because science denounced the senses to be as palpably false and treacherous means for gaining knowledge of this world, as the Church denounced them to be fatal and perditious means for gaining the knowledge and bliss of the next.

But with respect to the soul-substance or mind, some social-religious compromise was imperative, irrespective of what scientists might or might not think in the privacy of their cabinets. And Descartes hit upon the very neatest compromise, one which may be taken as accurately symbolic of the whole early philosophico-scientific movement.** He sequestered the soul-substance or mind in the pineal body, in the smallest area of the brain he

---

*Notice that "hardness" is a primary or ultimately real quality; but "softness" is only a "word" or "name" or—secondary quality. This accounts for what may have seemed to the reader an error in the preceding paragraph where the phrase "soft and hard" was used first and then in the repetition of the list only "soft" was included.

†The doctrine of "original sin" with its correlative "the intrinsic moral depravity of the senses" are by no means completely dead yet. At the present time, in fact, new life is being assiduously pumped into these doctrines by the strangest crew that ever manned the "idealistic" pumps in history. And the "pumps" employed are as strange as the assortment of "pumpers"—ranging from bombs and bullets to prayers and papal bulls designed to finish off what the bombs and bullets leave undone.

**The one exception is Spinoza—and of course a great one. The treatment Spinoza and his works received not merely from the illiterate but from his peers, is among the best evidences proving that *social*, not intellectual, considerations either consciously or unconsciously dictated the "compromise." In Descartes' case, the dictation was thoroughly conscious; he knew what he was doing and why. In Newton's case it was as completely unconscious; he was a "true believer" and his interest in the English translation of the alleged speech of God actually exceeded his interest in the works of creation. For ampler discussion of Spinoza's case, see my Introduction to *The Philosophy of Spinoza* (The Modern Library, 1927).

knew of, and central enough to comport with the dignity of its new and exalted occupant. That part of the physical world—the pineal body—Descartes said, in effect, belongs to you, O! defenders and guardians of the soul-substance or mind, and we shall never trespass within its holy bounds! But the rest of the human body—let us be perfectly clear about it—and the rest of the world, belong to us, to the mathematical and physical scientists!

Descartes' compromise obviously could not endure forever, nor indeed for very long. Just as soon as science felt its oats and felt sufficiently secure socially so that it need no longer fear the secular instruments of persecution used by the Church, science naturally and inevitably advanced its claim of divine right to investigate and explore everything and everywhere. You may think that mathematical physicists would always have the need of upholding the ultimate reality of the mind, if only out of sheer self-interest and self-preservation, else they lose their necessary and primary guarantee that what they discover, mathematically, is actually the ultimately real. But to think so is naive. For what further need of the mind have they, once Science itself, in all its glittering panoply of power, stands boldly in the field, its feet firmly planted on the eternal foundation-stones of the universe!

Because of Descartes' compromise, and the general socio-scientific situation it symbolizes, philosophers first acutely experienced the need of defending their sensory organs against the deprivations of science. How did it come about that all of us, scientists as well as philosophers and common persons, sensed such qualities as colors and sounds and what status, really, can be given them? Merely to call them names, as Galileo does, may be finally satisfactory for the scientist who is professionally so enraptured by his own work that he has no mind for anything else. But such summary disposal of secondary qualities can hardly be permanently satisfying for the philosopher whose professional interests are of universal scope and for whom there is nothing too small or too insignificant to merit and receive his thoughtful and tender consideration. Not that in the common life and world of man, colors and sounds and all other secondary qualities are a small item, like the hairs on his head or the sparrows in his fields. But compared to the soul or mind and all its longings and belongings, they are; and the soul or mind had been left untouched by science. Indeed, had not science solicitously placed it in an inviolable sanctuary beyond the reach of all molesting and harm from itself or others?

Call it clairvoyance, premonition or what you will. Or say, if you like, it is the fair and just reward that comes to those who take it upon themselves to defend the orphaned and the lowly. When the period of compromise was nearing its end, and the mind was in imminent danger of being imperiously sucked into the whirling stream of atoms forever, it was the secondary qualities that came to the rescue and saved the philosophers from falling into the dark and bottomless pit.

Very rarely are philosophers poets, even though, frequently, they robe their writings with rich and colorful imagery and make them reverberant with sonorous poetical effects. Only very rarely are they even like Wordsworth, hardy men of two senses, let alone like Coleridge and Keats, hardy men of five senses and for whom, without colors and sounds, without tastes and smells and textures, all places would be a blank hell though they be fastly secured in heaven. With few worthy exceptions, modern philosophers, especially since the time of Berkeley, have found secondary qualities of primary value because they can be used as means of argument to gain their own, and not the qualities' ends. Berkeley's great discovery was not that secondary qualities are the glorious garment of Nature and of which she cannot be deprived by any scientific means. What he joyfully discovered was that secondary qualities are the Achilles heel of science and struck there the monster could be slain. Incidentally, it is true, the earth we commonly enjoy would then be regained for man, but what was far more important to the good bishop and the motive forcing his attack, was that man would thus be reclaimed for God.

If science does not adequately account for this world—and the fact that science leaves secondary qualities inexplicably hanging in mid-air establishes this—then, reasoned Berkeley, the conclusion follows that philosophy, as an enterprise independent of and unconstrained by science is necessary. Merely to be necessary is something, of course, but not enough. Philosophy has to prove that it is also competent. And to do this, obviously, philosophy must first demonstrate that it can succeed where science has dismally failed. Philosophy, that is, must give secure and intelligible status to secondary qualities, give to them the reality they were denied by science. And at this point Berkeley was overcome by his most brilliant idea and made straight for a dazzling *coup*. It is not imperative, he saw, that the reality philosophy gives to secondary qualities be identical with the reality claimed by science for its atoms and such of their primary qualities as it could keep from changing into secondary ones. This was the great and egregious mistake Locke had made; because of it he was forced into the ludicrous position of vigorously accepting Newton's science and then feebly complaining about some of its results. Philosophy can escape from the insufferable and insoluble Lockean dilemma, can effectively demonstrate its competence, can firmly establish its complete independence of science and can, moreover, do all these marvelous things at once and in one by the simple expedient of giving to secondary qualities a reality *different* from that of the atoms and primary qualities, but a reality nonetheless securely founded and ultimately real. How can this be done? Very easily. Link secondary qualities not to the senses but to the as-yet-undisputed reality of the mind; the mind (also by still common consent) is linked to God and by divine devolution of power, secondary qualities are established as substantially real. (The atoms and primary qualities were, according to Descartes and Newton, the creation of God and their eternal reality was dependent upon and guaranteed by the Eternal Will.) Thus science, reasoned Berkeley, is overturned by its own mistake, confounded by its own distinction. Let science keep its primary qualities—if it can.* Philosophy has shown who is the true possessor of knowledge of this world, who the proud and faithful protector of the reality of God's visible creation.

But surely, you will say, the mind, in throwing its own mantle of reality about secondary qualities, could have been motivated only by a noble altruism. By thus sharing itself with others, how could it have been seeking merely benefits for itself? Surely it is an act unselfishly pure,

concerned solely with using fairly and justly what science had meanly and despitefully used, with charitably raising up what science had cast down. On the face of it the act does, assuredly, carry these benign features. But, alas, it is only on the face of it. Actually it was a method of using the indisputable natural reality of the secondary qualities to give body to the reality of the mind, a means for resuscitating the mind's fast failing spirit. Once the mind was revived and strengthened by this natural and wholesome food, what did the Idealists do? Did they acknowledge the mind's natural indebtedness and return natural good for natural good received? Not at all. They used the boost in spirit they had thus surreptitiously obtained for the mind to catapult the mind into a transcendental reality and then, with base ingratitude, they turned on the secondary qualities and spitefully made them blind wanderers in a phenomenal world.

Which is in all essential particulars doing to secondary qualities exactly what science had originally done. Instead of the Galilean-Newtonian "scientific" distinction between primary and secondary qualities, you have, with Kant, the "metaphysical" distinction between noumena and phenomena, between the empirical world and transcendental Reality, which became in some post-Kantian developments the distinction between what only appears to be real, but really isn't, and what is known by the mind's inner and unaided power to be exclusively and ultimately Real and Is. The "metaphysical" distinction undoubtedly sounds more soul-filling and grandiose than the "scientific," but deflate the artificial grandiosity and the two distinctions, though expanded in different ways and not always in one-to-one correspondence, are, for all intelligent purposes, the same.

You will say that this is only what Kant and the Kantians did but that Hegel, sensing the keen wrong, redressed it forthwith. Yes, Hegel did redress the wrong—by giving it another cloak. The mind, Kant thought, could get along entirely on its own in all its categorically imperative business, business having to do with its own soul's salvation. And for all lesser business, though the mind could not get along on its own entirely, it was sufficient if, from its transcendental seat, it occasionally looked down on the empirical world; by occasionally peering into the blind, phenomenal world, its noumenal eyes would see enough for thought. Hegel realized this wasn't so. The mind needed the natural world for all its business, ideal and real, for its own soul's categorical salvation no less than for everything else and that it could not get along without the natural world for a moment; it constantly needed it. Occasional contacts with it, occasional peerings down into it were not enough for thought. But did this recognition, on Hegel's part, of the constant need and dependency of the mind show him that the mind must be returned to its natural environment where it could and would naturally feed, and that leaving it where Kant had put it was leaving it, not in a transcendental heaven, but up a tree? Not a bit of it. To satisfy the constant and natural craving of the mind for natural food, Hegel tried, by an act of unnatural violence, to force the mind to swallow the natural world whole! Instead of permanently satisfying the hunger of the mind, this grotesque act of intellectual outrage gave it convulsive indigestion. And the spasmic regurgitations of Absolute Idealism are splattered over all the pages of subsequent cultural history.

---

*Berkeley's really significant and permanently valuable contribution was in showing that science can *not* keep its primary qualities, that is, that all qualities are in the same boat. However, the only proper conclusion this demonstration leads to is precisely the one Berkeley and subsequent philosophy (up to pragmatism and instrumentalism) did not make. To discuss it at this point would therefore be to complicate matters to no good end. It will be taken up later.

Because of the malicious exploitation made of secondary qualities by Idealism, the whole discussion of the theory of knowledge was subsequently narrowed down to a discussion of the theory of perception as the critical point of attack. To dissociate secondary qualities from the mind was the all-important thing to do. For just as soon as this dissociation is effected, the Absolute Mind is deflated and Idealism collapses like a punctured balloon.* This proves negatively what was asserted positively before, namely, that the attempt to give secondary qualities the reality of the mind was not really for the end of saving them but was a means whereby their natural reality could be used to bolster up and secure the vanishing reality of the "substantive" mind, or soul. William James, in a brilliant sentence written in 1904 said: "Those who still cling to 'consciousness' are clinging to a mere echo, the faint rumor left behind by the disappearing 'soul' upon the air of philosophy." The echoes of the "soul" would have completely died out in philosophy long before our century had not Idealism resorted to necromancy and by percussion instruments spread the deafening clamor of ghosts.

Pragmatists, realists and instrumentalists have approached and accomplished the critical task of dissociation in different ways. If the result of evacuating the mind of its universal pretensions were an isolated or isolable matter, the differences in methods used by these various schools would be practically irrelevant. But the consequences of the critical operation are neither isolated nor isolable. As with the operations performed by a surgeon on the human body, so with the critical operations performed by philosophers on ideas. Upon the instruments and skill with which a tumor is removed, the recovery of the patient depends. And the more malignant the tumor, the greater the skill and the better and finer the instruments must be. Done unskilfully and with the wrong instruments, it matters not with what nicety and dispatch the outer parts are sewn up again. The superficial wound heals, but at the real seat of trouble new and serious complications set in. There is a limit to the cuttings and removals a human body will undergo. A tumor, no matter how malignant, cannot grow forever, in new areas or old, because the body dies. With ideas, however, there is no limit—at any rate, none that has ever been reached and of which we with certainty know. And all ideas, malignant or benign, are in this respect alike: they spread from one area of the cultural body to every other and there is no stopping them.

If the *fact* of severing secondary qualities from the mind were of self-sufficient importance and the *method* of no philosophic consequence, then Dr. Johnson's method would be the very best, for it is so short and every one can use it. He kicked a stone and found it behaved as stones had always behaved and not as an idea.† That, for him, was refutation enough of Berkeley. And considered *sheerly as refutation* enough it is, final and complete, and no philosopher has improved upon it or can because all refinements of philosophical refutation must end up, when they do not begin, in the mode of Dr. Johnson. For his refutation is the *experimental coup de grace.* We all accept without question that a complicated piece of scientific apparatus kicking an electron out of its orbit can thereby, with rightful authority, kick a scientific theory in or out. Well, the human foot is itself not uncomplicated in structure—if you believe only complicated apparatus can perform operations of experimental test, which is, of course, not so. And the human foot too has the ability and rightful authority to kick a philosophic (or scientific) theory in or out. "I would rather," writes Dewey, "take the behavior of the dog of Odysseus upon his master's return as an example of the sort of thing experience is for the philosopher than trust to" philosophers' theorizings about experience.* A dog is a whole animal and the foot is only a part, but the point of reference and the contact are the same. To this extent, Dr. Johnson has certainly not received his proper philosophic due.

However, the *fact* of severance is not everything; it is necessary, it is indispensable, but it is not sufficient. The surgeon must remove the tumor, that is certainly clear; but the *method* he uses, *how* he removes it, is equally important, for the removal and the method of removal cannot be separated; they continually interact and it is the consequences of their interaction that determine the life and health of the patient. When it is so intricately complicated and delicate a matter as the life of the mind that is at stake, the importance of the consequences of the method employed is of course immeasurably increased, not lessened.†

The severance of secondary qualities from the mind reduces the mind to its natural size. It loses its *a priori* bigness and no longer needs a supernatural, transcendental realm to house its unnatural, swollen grandeur. The mind becomes something that *can* be included in the order of Nature, as a part having its natural function and place. The emphasis on *can* cannot be made too strong. For the mind does not automatically find its proper place as a soldier, when called to attention, automatically falls in line. Especially is this so after a bout of fever as wild and fearful as German Idealism. To return the mind to its place in the order of nature, to accustom it in the performance of its natural functions, to teach it to find its inner and highest joy in the fulfillment of its cultural obligations is a long and difficult task for deliberate art. This reconstructive, rehabilitative work, to be completely successful, requires, of course, the cooperative effort of all cultural forces, of all society. But the starting of this tremendously important and complex task, getting it under way, is peculiarly the philosopher's job—at any rate, as Dewey conceives it. Indeed, it is the one job which, when accepted by philosophy, intelligently legitimates or justifies its claim to universal (*i.e.* cultural) leadership. This peculiarly philosophic part of the greater undertaking no philosophy can adequately perform except it does its critical work properly.

Hence Dewey's insistence on the present need of philosophy to devote itself to criticism and the methods of criticism. For the ideas a philosopher uses in his critical operations necessarily become part of the foundational ideas for his reconstructive and constructive

---

*This is most neatly seen in G. E. Moore's thirty-page *Refutation of Idealism* (1903), which has been the originating and orienting point of practically all British Realism. Bertrand Russell is possibly Moore's very first disciple; he certainly is his foremost one .

†It should perhaps be explicitly stated that Dr. Johnson, as a man of letters, as a person whose profession was handling and dealing with ideas, was as competent an *authority* on the macroscopic behavior of ideas (or their general, overt characters) as, say, a scientist of comparable distinction would be an *authority* on the macroscopic behavior of his subject-matter (electricity or what-not). That Dr. Johnson was also as *qualified an expert* on the macroscopic behavior of stones as Bishop Berkeley (or any other philosopher or scientist) can, perhaps, be left as too obvious for statement.

**Experience and Nature*, (first edition), page 6.

†Notice, in the question above, that Dewey says "rather." The qualification is important.

follow-up work. This is true universally, a consequence flowing from the very nature and process of thinking. Wherein philosophers can and do differ from one another is in degree of awareness of this necessity, and in the competency of the critical apparatus they respectively possess. Of course, no philosopher has yet appeared who in his work shows that he is fully conscious of, and at every point alert to, all the implications of his critical work; nor is it possible to develop and construct a critical apparatus that will make any (let alone every) philosopher thus perfect. To be perfect philosophically is to be absolutely infallible—a possibility directly open only to God and indirectly open only to those erstwhile human thinkers who by retroactive edict (promulgated by His proper intermediaries) have been canonically uplifted into the ranks of angels.

On the other hand, it is also impossible for a philosopher to be totally unconscious of and unalert to the implications of such critical apparatus as he may possess, for such a one would not be a philosopher at all, and would not have any critical apparatus in the first place. Like all human works, the works of philosophers are to be found ranging between the extremes of perfection and imperfection and none of them reaching either. In this respect at least, they all exemplify with equal clarity and naturalness the universal cosmic characteristic William James never tired of celebrating: "Ever not quite."

Dewey, more persistently perhaps than any other contemporary, uses his critical apparatus and the results they yield as part of the basis for his constructive philosophy. This has caused a great deal of confusion on the part of many of his readers, whereas it should have enabled them to follow his thought with clear, if not easy, understanding. In saying this I have neither intention nor desire to shift all cause of confusion from Dewey's writings to his readers' minds. In view of what has already been said on this point, such would be ridiculous. But too much stress cannot be put upon the general principle at issue because unless it is firmly grasped, it is practically certain that any reader of Dewey will keep on being confused no matter how often and how studiously he reads him. Failure to understand the fundamental reason for and meaning of Dewey's persistent use of his critical results for constructive purposes has, beyond any doubt, been one of the most prolific sources of unenlightened and unenlightening attacks upon his work.* To the same cause may also be chiefly traced the failure of Dewey's twenty years' controversy with American and English realists, for that controversy was circumstantially concerned with this one basic point. That his fears about the competency of the Realists' critical apparatus were not unwarranted and his prophecies as to their eventual outcome were not without ground are proven conclusively by the way in which neo-Idealism has grown out of and sucked the strength from the Realistic movements.†

---

*The reader, if he so desires, may also consider the above paragraphs as stating the basic reason for my writing *this* Introduction and not another.

†This does not contradict what was said earlier. Neo-Idealism draws *immediately* from the Realisms, which is the point just made. It can do so because the Realisms carry in them the fundamental errors of Greek philosophy, which is the comprehensive point made before. That neither the older neo-Realisms nor the new neo-Idealisms are critically conscious of their historical sources and the basic repetitiousness of their views is a fact theoretically responsible for both.

VII

"A clash of doctrines," writes Whitehead, "is not a disaster—it is an opportunity." There are some "ifs" involved. It is an opportunity, or perhaps better said, it becomes (or is made) an opportunity *if* the clash stimulates intelligent response and not blind reaction; *if* the clash is taken not merely as a sign that something is wrong but is utilized as the starting point and control of inquiry into causes; *if* the inquiry into causes is not handicapped and stultified by impounding certain issues in a sacrosanct reservation, thus compelling inquiry to stop short at the first arbitrary point where a makeshift solution can be gerrymandered; *if* the inquiry is so conducted that it fearlessly re-examines and reconstructs everything necessary for a stable and fruitful solution, exempting nothing from scrutiny and reconstruction—above all not exempting the doctrinal foundations.

The clash of doctrine between science and philosophy on the all-determining, all-controlling issue considered in the foregoing pages is an excellent example not of an opportunity utilized but of a disaster prolonged. Consciously and unconsciously, philosophers exempted from critical examination and reconstruction certain conceptions both of science and philosophy. If you like to be excessively generous, you can say they inquired into and re-valuated everything—until they reached foundational principles. There they stopped short, with the necessary consequence that their solutions were makeshift and unstable, arbitrary and unfruitful in all but harmful ways.

Berkeley is one of the clearest illustrations of this and because of his pivotal influence, also one of the most important. He argumentatively demonstrated that all qualities are in the same boat, that it is impossible to classify qualities into two (or more) orders, distinguished from each other by quality of reality. If any quality is real, then all qualities are real. If any quality is of a modified or suspect reality, then in the same way and to the same degree, the reality of all qualities becomes modified or suspect. This conclusion is indubitably sound. But Berkeley unfortunately arrived at it by way of an argument the objective of which was to prove that all qualities are directly dependent for their very existence upon perception and therefore are creatures of mind. This tie-up was particularly vicious in its consequences because, although in theoretical purity it perhaps need not be so, it was practically inevitable that refutations of Berkeley—an idealism, of which there are legion—should almost automatically result in resuscitating and re-stabilizing (if not illegitimately re-validating) the "scientific" distinction between qualities his argument destroyed. But apart from this particular tie-up in Berkeley's argument, and without wishing in any way to minimize the extent and deleterious character of its historical influence—apart from this tie-up, Berkeley's whole method of approach, which became standard for subsequent philosophy, is vicious: it blinds the philosophic as well as scientific mind to the real issues and problems involved.*

---

*Berkeley's *method* of approach is of course still standard for Idealistic philosophers, whatever branch of Idealism they profess; but it is also standard for all British Realists, like Moore, Russell and Whitehead; and for all American philosophers who follow the British lead. The attack on Berkeley's *method* and so on all modern philosophy that led into it and all that developed out of it was initiated by C. S. Peirce—the logical father of pragmatism. The new *method* of philosophy, both as critical and constructive instrument, which Peirce began, was developed to some extent by William James and was carried out practically to its full critical and constructive *methodological* limits by Dewey. This method is a new contribution to philosophy and, we may well be proud, a distinctively, even exclusively, American contribution.

For when you develop a philosophic theory as ground for the assertion that all qualities are of equal reality, by that intellectual act you are also forced to assert, explicitly or implicitly, that the distinction between reality of qualities as it is made in science is valid within the domain of science. And from this tacit or overt admission only one consequence can follow, namely, what exactly has followed in modern philosophy: Inquiry into the legitimacy of and reason for the distinction as it is made in science is completely sidetracked and as substitute for this intelligent inquiry an unintelligent contest is staged between rival systems of "metaphysics" and in the other the (theoretically) endless succession of "philosophic" contenders for the crown. Dewey's detailed exposure of the tragi-comedy of this whole procedure is one of the great pieces of philosophic analysis.*

All the philosophic criticisms and attacks in the world will not persuade scientists to abandon any principle or distinction as long as they believe, whether rightly or wrongly, that the principle or distinction performs some scientific work. In this respect they are as loyal to their science as the best of philosophers are to philosophy; and the sooner all philosophers realize this and conduct themselves in accordance with such realization the better it will be for all. On the other hand, if scientists can be shown, or in some other way come to see for themselves, that a principle or distinction does not perform any scientific work, either in the form in which it is made or because it is inherently incapable, no matter how formulated, they will, without further argument, modify it immediately or drop it entirely as the case may be. And in this respect, alas, it must be said, scientists are vastly superior to the majority of philosophers—if one may fairly judge by the howl of protest which arose when James and Dewey made known† and began to develop "the principle of work" as the basic criterion or test for all philosophic ideas.

Einstein, about five years ago, opened a famous lecture by saying: "If you wish to learn from the theoretical physicist anything about the methods he uses, I would give you the following piece of advice: Don't listen to his words, examine his achievements."** This piece of advice loses none of its excellent qualities when it is extended to include the whole field of science; in fact, it considerably gains in excellence when applied to the basic methodology of all modern science. And in the latter sense, Dewey began following it some forty-odd years before it was given.†† In his analysis of scientific method, of what

science is, Dewey has been strictly guided and controlled by his examination of what scientists do. He has not, however, been able entirely to ignore what scientists say since so many influential philosophers have repeated their sayings, and their sayings and the repetitions of them have fostered the all-absorbing philosophic problem of modern times. Furthermore, it may be noticed in passing that even if science had not been a great determining force in modern philosophy, because of science's enormous influence in modern cultural life and the social standing scientists enjoy as examples, if not indeed paragons, of intelligence, it would be for the philosopher of first-rate importance to inquire into and find the answer to the great question: Why do scientists do one thing and say another and contrary thing?

Dewey's historical exposition of Greek philosophy and its modes of entrance into modern thought and his analysis of the theory of knowledge give the root-answer to this question. Starting from different loci in the western intellectual world, they converge toward this double objective, their critical forces uniting in this double point: they show that the distinction between primary and secondary qualities *as it is made in science* has no validity *in science* and performs *no scientific work:* and they show *how* this non-scientific, in fact anti-scientific, distinction, came to be made in the first place and *how* it has been fundamentally supported ever since. Dewey thus cuts under the ground of the whole modern controversy between philosophy and science; and by removing the source whence modern philosophers have drawn self-justifying reasons for setting up rival systems of "metaphysical" reality, he has also thereby removed the source of all modern intellectual jugglery in philosophy—as his own critical and constructive philosophy experimentally shows.

The real proof—and the only proof philosophers should adduce or resort to—that the "scientific" distinction is false in the form in which it is made in science is provided by science itself. Every time science performs an experiment it necessarily abrogates or invalidates the distinction—and if this is not conclusive proof of falsity, then there is no proof of falsity anywhere or anyhow to be found. Science abrogates or invalidates the distinction every time it performs an experiment because every experiment is carried out in the world of common experience and the *final authority*, for validation or invalidation of scientific theory, always rests with events in the "macroscopic" field of sense-perception, to use one of Dewey's analytical terms. For instance, the final verdict as to whether or not there was any ether-drift was rendered by the registration of visual effects on the interferometer employed in the famous Michelson-Morley experiment (1887). What the visual effects—the black-bands—meant, how they were to be interpreted, became a scientific, theoretical problem of the first magnitude. But be it noted—and this alone is of crucial importance here—a scientific theory was not needed to determine whether or not the black-bands, as seen with the eyes, were actually real. That they were real was unequivocally and unhesitatingly accepted by all scientists. Indeed, only because the reality of the black-bands was never in question and beyond all dispute, only because their reality could not be *scientifically* challenged in any way, was a theory necessary to explain—not their existence or occurrence—but their causes, or, what amounts to the same thing, their interconnections with other existences or events.

Because the Relativity Theory, among other things, did

---

*The Quest for Certainty is the *locus classicus.*
†Peirce was first as originator of "the principle of work" (pragmatism) and first in time of publication (1878). Not until James (1898) and Dewey (1897) took up the idea and began developing it, did it begin to claim attention and exert influence. Peirce didn't approve of many of the ways in which James developed "the principle of work" and violently criticized him. Neither did Dewey approve of all James' developments and he also criticized him—but without violence and to much greater effect.
**On the Method of Theoretical Physics, the Herbert Spencer Lecture, delivered at Oxford University, June 10, 1933.
††Peirce's essential contribution to the pragmatist movement consists precisely in giving "Einstein's advice" *and the underlying reason for it.* Einstein, grievous to say, in his capacity not as scientific practitioner but of philosophic interpreter of science has always been critically unable to follow "Einstein's advice."

this job of explanation it was in so far confirmed. One case of confirmation is rarely enough, especially when the theory is very complex, comes after the facts and is devised to explain those facts. Furthermore, the Relativity Theory required for its explanation of the black-bands a revolutionary change in the fundamental concepts of Newtonian physics. Hence the extraordinary significance of Einstein's prediction that if adequate visual attention were directed thereto, it would be observed that light rays are bent in passing the sun. This was an experimental test which the Relativity Theory proposed for itself and this is all that "prediction" of this sort scientifically means. The "foretelling" is scientifically nothing: what is everything scientifically is that by foretelling a hitherto unobserved event, the Theory proposes and provides an experimental test for itself wherefrom all possibility of "collusion" or *ex post facto* explanation is clearly and rigorously excluded. When, in 1921, the bending of the light rays was first photographically observed, Einstein's theory received its most exciting and dramatic confirmation. But human excitement and drama aside, this confirmation was in every scientific respect on all fours with the confirmation the theory received from the black-bands of the interferometer or the irregularities of Mercury's orbit*—both of which were known to science long before. If a theory were to come along to-morrow morning that satisfactorily explained all that the Relativity Theory explains and did not "predict" anything new at all, it would nevertheless displace the Relativity Theory if, in one significant way or another, preferably in its mathematics, it was also simpler or more consistent. A number of alternative theories have in fact already been proposed. None has succeeded in gaining acceptance not because none has made "predictions" of hitherto unobserved events, but because none has proved capable of doing all the explaining, or all the scientific work, the Relativity Theory does now.

What is true in the foregoing example as to the relation between scientific theory on the one hand and observation of macroscopic events on the other, is true in all cases. The reality or existential quality of what is observed in the experiment proves (or disproves) the validity of the theory; the theory does not and cannot confer reality or existential quality on what is observed. And what scientific theory has no power to confer, it naturally has no power to take away. The methods of scientific experimental practice unequivocally and definitely prove that all qualities are *in science and for science of equal reality.*

If scientific theory has no power to confer reality or existential quality on what is observed, still less—if the concept "less than nothing" be granted feasible for the moment—still less has scientific theory the power to confer reality or existential quality upon what is (presumably) *inferred* from the theory. One example should be sufficient. There was a time, and not so long ago, when all scientists firmly (even unshakably) believed in the existence of the ether *because* they believed its existence was not only a proper but a necessary inference from scientific theory then extant. To follow Whitehead's statement of the case: the wave theory of light, Clerk Maxwell's formulae for stresses in the ether, his equations of the electromagnetic field, his identification of light with electromagnetic waves—these major four conspired to give "concurrent testimony to its [the ether's] existence." Nevertheless, as was subsequently shown, the ether and its existence were not the consequences of weighty and legitimate scientific inference but were, in Whitehead's admirable phrase, "merely the outcome of metaphysical craving."* If the same may not yet be said of the current belief in the existence of atoms, in the precise form in which atoms figure in contemporary scientific discourse, at least the same may be reasonably suspected as probably being the case. It is quite proper, therefore, to reserve one's complete agreement with Bridgman for some future, unspecifiable date, when he writes: "It is one of the most fascinating things in physics to trace the accumulation of independent new physical information all pointing to the atom, until now we are as convinced of its physical reality as of our hands and feet."† It may well so happen that the fascination, rather than warranted scientific evidence and reasons, will, in the future, be held accountable for the experience of conviction Bridgman and his peers now enjoy.

How, then, did the false and scientifically untenable distinction between qualities (with respect to their reality) gain entrance into science, and once inside, what has spuriously perpetuated its residence there? Or, if you like, how were scientists made inveterate victims of "metaphysical craving?" The only answer ever proposed that satisfies all the facts to be explained is Dewey's analysis of Greek doctrine and his exposition of the peculiar way in which it has operated in modern times.

The Greek doctrine that scientific knowledge is knowledge of eternal and immutable Reality consistently functioned to make it inconceivable for the Greeks that the world of change could be scientifically studied and known. They wrote out their own prescription for science and their scientific activities were conducted in accordance with the directions they themselves prescribed. Consequently, through their science was restricted in fundamental character, and by our standards was hardly science at all,** they did not get into the muddle of contradictions, confusions and absurdities which has mired modern thought.

Modern scientists, however, began by taking precisely the world of change as their subject for scientific study, and to help them on their way, they introduced the method of experimentation which is no less and no other than a method whereby the natural changes going on can be further increased and complicated in manifold ways by changes deliberately made. From the Greek point of view (and in this case, *not* excepting any Greek), this is confounding confusion, science gone insane. But as events have fully demonstrated, it is science really come to its senses, and intelligence come into its own.

---

*The other fairly well-known achievement to the credit of the Relativity Theory: the Law of Gravitation which it yields, unlike Newton's Law accounts for certain irregularities in Mercury's orbit; the irregularities are macroscopic events observed by our senses supplemented by telescopes and other apparatus (which latter are constructed, manipulated and observed by our senses and are also, of course, macroscopic events). It need hardly be stated explicitly that the indubitable and indisputable reality of the irregularities confirms the theory and not the other way about.

The Principles of Natural Knowledge, pp. 20 and 25.
†*The Logic of Modern Physics,* p. 59. On the same page Bridgman says the atom "is evidently a construct." Hands and feet *may* be "constructs" (whether they are or not Bridgman does not tell), but they are not "evidently" so or if so at least evidently *not* in the same sense.
**Einstein and Whitehead for instance, agree with Dewey (independently, of course) that Greek science was hardly science as we understand it now. Einstein (as far as I know) is more sweeping than Whitehead who at least excepts Aristotle (the biologist) and Archimedes and an unnamed number of astronomers from his statement that the work of the Greeks "was excellent; it was genius. . . . But it was not science as we understand it." *(Science and the Modern World,* p. 10).

Unfortunately, in one crucially important respect modern scientists did not display anywhere near the intelligence the Greeks did. Instead of writing out a new prescription for science, one in accordance with their own new scientific practice, the moderns carried along the old Greek formula as self-evidently sacrosanct. To this one failure, to this original sin against intelligence, can be traced the generation of all our severest, purely intellectual ills. Since the moderns did not follow the directions prescribed by the Greek formula either as to which subject-matter could and should be scientifically studied and which not, or as to the method to be employed in scientific investigation, the only way the moderns could possibly use the formula was by applying it *ex post facto*, by giving it a reverse english. The prescription says: only the eternal and immutable Reality (or Being) can be scientifically studied and known. The moderns perforce had to read this backwards so as to make it retroactively mean: whatever we scientifically know is *ipso facto* eternal and immutable Reality. Do we know, scientifically, the shapes and sizes and motions of things? We do. Hence these are constituents of ultimate Reality, eternal and immutable. Do colors and sounds, etc., form part of our body of scientific knowledge? They do not. Hence they are not constituents of ultimate, eternal and immutable Reality. Of course the prescription was not exhausted by this one application. It could be and was reapplied every time science changed, whether by expansion or revision, so that at different periods in the history of modern science different things have been the ultimate constitutents of eternal and immutable "scientific" Reality; sizes and shapes and motions and atoms and phlogiston and waves and ether and quanta and rays and—that last infirmity of all eternal and immutable "Scientific Reality"—Eddingtonian pointer-readings on measuring-machines planned by human ingenuity and made by human hands!

The voluminous discussion of Eddington's pointer-readings has obscured rather than revealed his more important representative or symbolic significance. It is too often and too widely taken for granted that contemporary scientists, because they have wave mechanics and relativity theories, therefore must, in their basic theoretical orientation, be a breed far removed from their more simple-minded and scientifically unsophisticated modern classical forbears like Galileo and Newton. Such is far from being the case, and Eddington is unhappily the best proof of the distance. "Our chief reason," writes Russell, "for not regarding a wave as a physical object seems to be that it is not indestructible." Since science simply must have something indestructible, something eternal and immutable, consequently, as Russell goes on to say, "We seem *driven* to the view advocated by Eddington, that there are certain invariants [i.e. "mathematical invarants resulting from our formula for interval"] and that (with some degree of inaccuracy) our senses and our common sense have singled them out *as deserving names.*"[*] Shades of Plato and Galileo? Yes! But also, alas! for Galileo (though not for Plato). What Eddington and Russell (at that time) consider as deserving *names* are not secondary qualities but such *primary* things of science as (among others) energy and mass! The primacy of "mathematical invariants" originally served Eddington as basis for his Idealism. Later on, pointer-readings partially, but not wholly, superseded mathematical invariants and in consequence his Religious Mysticism partially, but not wholly, superseded his Idealism. This progress—or more accurately said—this progression is in a fairly natural line and as straight as can be expected in

the circumstances.[*] Russell's own "line of progression" which is neither straight nor single will be considered further on.

Philosophers too were all the time in possession of the same Greek formula and were not outdone by the scientists in assiduity and virtuosity in using it. If modern philosophers have frequently lagged behind the scientists it is because they have had to wait until scientists from time to time decided what they were going to include as constituents of ultimate Scientific Reality and what they were going to eject or discard therefrom; for it is out of the changing discards of scientists that philosophers in one way or another have had principally to build up and fill up their systems of Metaphysical Reality. As all-inclusive ideological receptacles of Ultimate Reality, scientists have used in various combinations Space, Time and Matter, and the so-called "materialistic" philosophers have followed them with simple devotion;[†] while "idealistic" philosophers, for corresponding purposes have, in various combinations, used Sensation, Perception, Consciousness, Idea and Mind. The straddling or half-and-half philosophers—who are in the vast majority—have had one intellectual foot precariously poised on Matter and the other precariously poised on Mind, their systems precariously swaying over the yawning gulf between.

With the impartiality possible only to theory purely and completely disconnected from actuality, the Greek formula has worked, in all cases and in all respects, for scientists and philosophers alike. When read backwards and applied *ex post facto*, it obligingly confers all highest "scientific" or "metaphysical" blessings if, when and as desired. It works with endless perfection, like a syllogism—indeed like a charm.

VIII

Philosophy, whatever else it may be, is an enterprise of thought. Whether a philosopher's reasons are good or bad, really reasons or only rationalizations is as the case may be—and *what* it may be in any given case is itself

---

*For a meticulous and merciless expose of the philosophic pretensions of Eddington, Jeans *et. al.*, see L. S. Stebbing, *Philosophy and the Physicists* (Methuen, 1937).

†One of the more bizarre of the absurdities now being given wide currency is that Communism is the source of Materialism, if not really identical with it. Whereas of course the illustrious fathers and founders of Materialistic Philosophy are none other than Galileo, Descartes and Newton. As for the differences between the Newtonian and Marxian varieties of Materialism they unquestionably redound to the favor of the latter. For Newtonian materialism, apart from its appalling intellectual poverty, is such a childishly dreary mechanical affair—an unimaginative push-and-pull business. But the Marxian Materialism goes along in ever more novel ways, developing itself and the universe (at the same time) in accordance with the magical antics of the Hegelian Idealistic dialectics secreted in its vitals. Whatever one may think of the philosophical value of Idealistic Magic (even when seemingly covered up with materialistic sober-sense), every fair mind must admit that the Magic does confer on the philosophy appropriating it the semblance if not the substance of organismic character. And almost any organismic philosophy, no matter how bad, is better than any mechanical philosophy no matter how good.

A cognate absurdity, and no less bizarre, sedulously being cultivated by the same people, in the same quarters and for the same ends, is that only Marx (or only a "materialist" like Marx) could believe and teach that "religion is the opium of the masses." This report is the product of nothing but ignorance and superstition, or if not of these two, then of something much worse: malice aforethought. For so high-minded and genuinely "idealistic" a thinker as John Ruskin believed and taught precisely the same thing: "Our national religion," says Ruskin in *Sesame and Lilies*, "is the performance of church ceremonies and preaching of soporific truths (or untruths) to keep the mob quietly at work, while we amuse ourselves."

Analysis of Matter (1927), pp. 82-83; italics mine.

discoverable only by using a method or process involving reasoning. To *establish* that it is the one or the other we must establish the *reasons* for *that* determination. Philosophy, James somewhere said, is an obstinate attempt to think clearly. If in the works of philosophers the obstinacy is sometimes more pronounced than the clarity, well, that is just additional evidence (if such be needed) that philosophers are human beings and that philosophy is not a Transcendental unveilment of Pre-embodied Thought, Eternal and Immutable, but is a thoroughly human *enterprise*, an *historical* activity, and like all human historical activities, consequently displays the temporal powers and deficiencies of persons, period and place.

Dewey is indefatigable in criticism of ancient classical philosophy. If repeating in one's own works what the ancients wrote in theirs makes one a classicist, then Dewey certainly isn't one. But if *doing* for and with the cultural material of one's own epoch what the classicals *did* for and with theirs makes one a classicist, then of all modern and contemporary philosophers Dewey has the best claim to the title. For what did Plato and Aristotle—considering their work from the fundamental standpoint of philosophic method—what did they *do*? They analyzed and evaluated the science of their time, and in terms of their analyses and evaluations constructed their philosophies, their theories of knowledge, mind and nature. About this there can be no sensible doubt. It is serially written on almost every page of Plato's *Dialogues* and it is the undebatable purpose of Aristotle's logic: the five books of the *Organon* systematically bring together, codify and amplify the series of principles, methods and rules the *Academy* and the *Lyceum* found in and developed out of the Greek sciences. Of course the philosophies of Plato and Aristotle are not the same as the sciences of that period. Philosophy is not identical with science. But the *reasons* and *methods* of reasoning, in Plato and Aristotle, are in part a direct transcription and in part a development and adaptation for the usages of philosophy, of the reasons and the methods of reasoning exemplified in ancient mathematics primarily and ancient medicine very secondarily.

Now Dewey's whole philosophic effort is concerned with doing for our epoch what the classicals did for theirs. Just as they took *their* science as exemplar of what knowledge is, and the *method* of their science as standard of the *method of knowing*, so Dewey takes *modern* science as exemplar of what knowledge is, and the *method* of *modern* science as standard of the *method of knowing*. But of course there is one great difference between Dewey and his classical forbears. They could approach most if not all their philosophic problems and "attack them" to quote James "as if there were no official answer preoccupying the field." *That* is an advantage no subsequent philosopher has ever been able to enjoy. After more than two millennia, Plato and Aristotle still preoccupy the philosophic field, and never so securely and completely as when unofficially.

Experimentalism is one of the two basic terms Dewey has used to designate his philosophy. The other is *instrumentalism*. The latter designation was the one first used, and though it has never been discarded or disavowed, it has, in recent years, been allowed to recede into secondary place. And that is where it rightfully or logically belongs because the basis of Dewey's constructive philosophy is his analysis and evaluation of experiment. The primary designation of Dewey's whole philosophy *is* experimentalism because its foundation is his *philosophy of the experiment.*

That the method of experimentation is the very essence of the method of modern science is the flesh and blood (not the bone) over which Dewey's whole philosophy contends. Grant that Dewey's analysis of scientific experimentation is in its principal contentions sound and valid and you will have to grant that pretty much everything else fundamental in his philosophy is sound and valid. Deny the general validity of his claim concerning the place and function of experimentation in scientific method and then, no matter how much else of his philosophy you may like and accept, it will be a liking and accepting of thises and thats. Which of course is the thing to do—if you are interested in doing that sort of thing. But to pick an idea up here and another down there and to bundle the disconnected pickings together with heterogeneous pieces of memorial string is *not* understanding a philosophy—whatever else it may be.

I do not of course mean that a philosophy *is* a philosophy only when everything in it is flawlessly interrelated or hangs perfectly together (and therefore can only be understood when understood this way). Still less do I mean that a philosophy *is* a philosophy only when it has a stated principle or set of principles from which everything else in it "deductively" comes down with the flowing inevitability of logical precision. As for the first, no philosophy ever written has been without errors, slips, gaps, obscurities, confusions, vaguenesses, mistakes, contradictions and other insufficiencies of one logical sort and another. And as for the second, it is seven parts myth and three parts folly to believe there is any system (philosophic, scientific or mathematical) that has, to start with, just so many axioms, so many postulates, so many definitions, and has, to end up with, just so many theorems—the latter "deduced" from the former by logical squeeze. What I do mean is this: every philosophy, if it is a philosophy to be taken seriously at all, has some vital organs and they, on the contents in the philosopher's system, perform functions very much like those performed on the contents of his body by its vital organs. And, naturally there are all sorts of malcoordinations, imbalances and waste products in both cases.

The fact that every great philosophy has some vitalizing and organizing center has of course nothing to do with determining whether the philosophy is *of* the organism (an "organismic philosophy") or *of* the mechanism (a "mechanistic philosophy"). A philosophy of the mechanism can be very vitally organic in the sense we here mean, and a philosophy of the organism can be quite fatally mechanical—just as there can be a living skeletology and a dead sociology.

From one point of view, it must undoubtedly seem that the issue concerning the place and function of experimentation in modern science is one that could be settled very easily. To determine what Greek scientists did we are reduced to the extremity of laboriously excogitating a few fragmentary records, and about such there can always be endless dispute. But to find out what modern scientists do, why that should be simple, indeed a cinch. There are hundreds, even thousands of scientists contemporaneously about and they do the same sort of scientific thing their modern forbears did—only better. To settle the issue, then, all philosophers need do is make a field trip to scientific laboratories and, if they don't understand anything they see the scientists doing, the scientists are there and if you ask them they will tell. If it were as simple as this, simple indeed would the settlement be. But, alas, it is far from being so simple. For one thing to get philosophers to make the field trips would itself be a difficult and complicated business. If they were willing and ready to do *that*, then perhaps half the battle would be over. And for another thing, even if you got them to go, the chances are too great that they *would* ask the scientists what they didn't know and, forgetting all about Einstein's advice, *would* listen to what they were told.

It is impossible to settle the *philosophic* issue over experimentation simply by visiting the laboratories because philosophers, like other mortals (scientists included), understand what they see as they have been accustomed to see and understand. Philosophers are *acquainted* with the fact that scientists have laboratories and make experiments, as well acquainted with this fact as are the scientists themselves. But acquaintance, like familiarity, breeds contempt rather than understanding. To understand the *meaning* of experimentation it is first of all necessary to get rid of an inherited set of philosophic ideas, to overcome the *set* those ideas have given to philosophic thinking.

The operation of the Greek formula which has dominated the interpretation of modern scientific inquiry has also, and to the same extent, quite naturally dominated the interpretation of the *method* of modern scientific inquiry. It is obvious why and how: a theory which holds that the objects of scientific knowledge are eternal and immutable constituents of Ultimate Reality must also hold that no procedure that is itself not conversant with eternality and immutability can possibly be an *integral* part of the *method* of science. A fixed habit of thinking, like any fixed habit of doing, maintains its own procedure by excluding all others. And a system of ideas or method of interpretation that has become traditional *is a social habit* and not a "purely intellectual" system. Wherefore the strange "intellectual" behavior, with respect to experimentation, that marks the great tradition in modern philosophy and also the philosophies of contemporaries—experimentalism excepted.

Whitehead and Russell are the two greatest of Dewey's contemporaries and their philosophies are, in general range, as comprehensive as his. Furthermore, just as Dewey's philosophy is fundamentally derivative from his analysis of scientific method, so their philosophies are derivative from their analyses of the same. There is therefore a genuine basis for comparatively studying these three philosophies. The method of contrast and comparison is particularly valuable in an introductory enterprise. By taking these three philosophies together we shall be able to see more clearly than any other way what are the fundamental issues involved and what are the main methodological consequences that result for philosophy from making practical experimentation an integral part of the procedure of scientific method.

Of course one doesn't have to be Dewey or a Deweyan to recognize that, as a matter of indisputable fact, experimentation has had *something* to do with the advance of modern science. This is one of those "stubborn facts"—to use James' phrase—that cannot be denied, though like everything else it can be minimized or ignored. And contemporary philosophers do persistently minimize and then ignore experimentation in science,* even when, as not too frequently happens, they *seem* intent on doing otherwise. Thus, for instance, Whitehead:

> The reason why we are on a higher imaginative level is not because we have finer imaginations, but because we have better instruments. In science, the

most important thing that has happened during the last forty years is the advance in instrumental design. This advance is partly due to a few men of genius such as Michelson and the German opticians. It is also due to the progress of technological processes of manufacture, particularly in the region of metallurgy. The designer has now at his disposal a variety of material of differing physical proper- ties. . . . These instruments have put thought onto a new level. A fresh instrument serves the same purpose as foreign travel; it shows things in unusual combinations. *The gain is more than an addition; it is a transformation. (Science and the Modern World* [1925], pp. 166-167; italics mine.)

Coming from a philosopher who is, by reputation and in fact, free of the double taint of experi- mentalism and instrumentalism, the above tribute is handsome, and taken by itself as tribute, is as handsome as any experimentalist-instrumentalist could pay. But philosophy is not a rhetorical enterprise of paying tributes—as no one knows better then White- head. It is a critical and evaluative enterprise of thought. This being the case, and it also being the case that recent experimentation has contributed not merely to the gross technological progress of science but to the development and uplifting of that finest and purest aspect of science, as of human experience, namely, its "imaginative thought," should not the analysis and evaluation of experimentation be a proper and significant part of the philosophic task? Are not the *implications* of the statement quoted as deserving of Whitehead's careful and faithful exploration as any other of comparable importance—for surely the uplifting of thought to a higher imaginative level is *very* important? And these questions are particularly relevant in White- head's case for he and not another wrote four years after the above citation:

> Whatever is found in 'practice' must lie within the scope of the metaphysical description. When the description fails to include the 'practice', the metaphysics is inadequate and requires revision. There can be no appeal to practice to supplement metaphysics, so long as we remain contented with our metaphysical doctrines. Metaphysics is *nothing but* the description of the generalities which *apply to all the details of* practice. *(Process and Reality*—An Essay in Cosmology [1929] p. 19; italics mine.)

That experimentation *is* the 'practice' of science is certainly not something for which we need argue. Moreover, since the 'practice' of science at least sometimes exerts a decisive, transformative influence upon the 'theory' of science there are *two* reasons (and both mandatory, one should think) for including experimenta- tion within one's metaphysical scope. For "in one sense Science and Philosophy are merely different aspects of one great enterprise of the human mind."* But despite his own doctrines metaphysically general, and his own observations scientifically particular, Whitehead most definitely believes otherwise—when it is the 'practice' of experimentation that is at issue. His conclusive (and comprehensive) staccato word on the subject is this: "Discussions on the method of science *wander off* onto the topic of experiment. But experiment is *nothing else* than a mode of cooking the facts for the sake of

---

*Science, hereinafter, means "modern science" unless otherwise qualified.

Adventures of Ideas (1933), p. 179.

exemplifying the law."* To damn with faint praise is an old established custom, in philosophy and out; but to dismiss, as beneath consideration, *after* making acknowledgments for extraordinary services rendered is rather new—in philosophy at least!

Certainly Michelson's genius and experiments, the work of the German opticians, the perfected processes of metallurgical manufacture, are constituent and contributory bases of the theoretical scientific achievements of the past forty years and more. But to select these instances of contributions made by experiments and instruments to the development of scientific theory, *as if they were unique and isolated cases*, is not only to be historically inaccurate but—what is much worse—to be philosophically, logically, metaphysically, cosmologically and scientifically unsound. Without historical antecedents, it will be granted, the existential *occurrence* of the experiments and instruments would be sheerly miraculous. But it is surely also as clear that were the birth of the instruments and experiments never so normal and natural, their contribution to the development of theoretical science—their raising of the level of imaginative thought—would be a sheer case of miraculous *intervention* (levitation, if you can pardon a pun) *unless* they were elements *integrally functioning in the methodology* of science. Miracle for miracle—occurrence or intervention—the one is as bad as, if not worse than, the other.

When Whitehead says: "Michelson's experiment could not have been made earlier than it was. It required the general advance in technology and Michelson's experimental genius," he avoids making the occurrence of the experiment a miraculous event. But he does nothing to relieve the miraculous quality of its consequences or effects. Indeed, he is forced, will-he, nill-he, to leave it as miraculous—forced because he isolates the past forty years' experimentation, because his tribute is nothing more than a tribute, of the character of a ceremonial compliment which, having been paid, he can then pass on to really serious things. That Whitehead does isolate recent experimentation (with respect to its "intervention"), does take it out of historical continuity and technical continuity in the development of the *method* of science is most clearly and briefly revealed in his statement concerning Galileo in the paragraph immediately preceding the last quotation:

> Galileo dropped heavy bodies from the top of the leaning Tower of Pisa, and demonstrated that bodies of different weights, if released simultaneously,

would reach the earth together. So far as experimental skill, and delicacy of apparatus were concerned, this experiment could have been made at any time within the preceding five thousand years. The *ideas involved merely* concerned weight and speed of travel, ideas which are familiar in ordinary life. The whole set of ideas might have been familiar to the family of King Minos of Crete, as they dropped pebbles into the sea from high battlements rising from the shore.*

Now it is simply not the case that in Galileo's experiment (taking the Pisa experiment to be his), "the ideas involved merely concerned weight and speed of travel which are familiar in ordinary life"—(though what is *merely* about such ideas, even if the experiment was merely about them is surely strange to tell; as strange as it would be to tell what is *merely* about the measurement of the velocity of light, or the *measurement* of anything.) Any more than the apocryphal Newtonian experiment involved ideas *merely* concerned with the falling of an apple on a human head—though the occurrence of falling apples *is* very familiar in ordinary life, something that obviously cannot be said even for the mere existential *occurrence* of light and heavy bodies falling at different rates. Galileo's Pisa experiment involved the fundamental ideas of Aristotelian physics and cosmology, and though familiar at that time to some, they were not of ordinary life. That the earth was the fixed center of the physical universe was then familiarly believed; but that the earth is "the *end* of motion for those things which are heavy, and the celestial spheres . . . The *end* of motion for those things whose natures lead them upwards"† are *not* ideas which ordinary life casts up in the routine course of the day. This teleological cosmology—the "rational" basis for the Aristotelian dogma that heavy and light bodies fall at different rates—required for its working out and logical perfecting the whole great line of Greek thinkers which Aristotle closed. And what Galileo's Pisa experiment did was to destroy the Aristotelian cosmology and physics and lay the foundations for the new physics and the new cosmology.**

Russell, who sometimes has more, and sometimes less but at no time any serious philosophic (or logical) use for the topic of experiment, has nevertheless, with his usual clarity, summarized the case for the Pisa experiment:

> Before Galileo, people believed themselves possessed of immense knowledge on all the most interesting questions in physics. He established certain facts as to the way in which bodies fall, *not very interesting on their own account, but of quite immeasurable interest as examples* of real knowledge and of a *new method* whose future fruitfulness he himself divined. But his few facts sufficed to destroy the whole vast system of supposed knowledge handed down from Aristotle, as even the palest morning sun suffices to

---

*Adventures of Ideas p. 111; italics mine. "Conclusive and comprehensive" not merely because in the last volume of his philosophic 'trilogy.' The *Preface* states that "the three books—*Science and the Modern World, Process and Reality, Adventures of Ideas*—are an endeavor to express a way of understanding things . . . each book can be read separately; but they supplement each other's omissions and compressions." However it may be in other respects, the Indexes to all three books concur in *omitting* the term experiment or any of its linguistic derivatives. This is not due to systematic carelessness but is indicative of Whitehead's estimation of experimentation as of negligible philosophic import. Experiment finds no place in *any* of Whitehead's philosophical books. Neither is the term experiment or any of its derivatives philosophic enough or relevant enough for discussions of the nature of science to find a place in the Indexes of: Russell's *Our Knowledge of the External World* (1914), *The Analysis of Mind* (1921), *Philosophy* (1927), *The Analysis of Matter* (1927); Broad's *Scientific Thought* (1923), Eddington's *The Nature of the Physical World* (1928); and, to come down to the positively logical revelation of the nature of science, Carnap's *The Logical Syntax of Language* (1937). These Indexes no more than Whitehead's are systematically careless; they are carefully and selectively philosophical as the reader can find out by reading the books.

---

*Science and the Modern World, pp. 167-168; italics mine. Scholarship since Whitehead wrote the above has pretty conclusively determined that Galileo didn't perform the Pisa experiment. But whether he did or not is of purely antiquarian interest. For if he didn't drop weights from the Tower of Pisa, he did roll balls down an inclined plane and did many other experimental things.
†*Ib.*, p. 11; italics mine.
**The falling of the apple is an apocryphal experiment but it popularly (and not too inaccurately) symbolizes Newton's scientific achievement as summed up in his Law of Gravitation. Similarly the Pisa experiment symbolizes Galileo's scientific work. Good scholarship is good, I know, but a good symbol is oftentimes better—of course when it does *not deny but reveals* the *meanings* disclosed by good scholarship.

extinguish the stars. (*Our Knowledge of the External World*, p. 240; italics mine.)

Making an experiment or fashioning an instrument may well be likened to foreign travel. But it is a commonplace that foreign travel is not even additive, let alone transformative, but vulgarly agglutinative *unless* the traveler, before his travels, has *learnt how* to make his ordinary, every day, common doings, *experiences* of value. Education, like so much else, if it does not begin at home, does not continue abroad. And the Pisa experiment very accurately symbolizes the period and the process of homely *learning how* modern scientific man had to first go through to make himself ready and to equip himself for the continuation of his education in ever new foreign parts unknown.

The family of King Minos or, for that matter, the families of cliff-dwellers could have dropped pebbles and the like, and most probably did—the cliff-dwellers at any rate. They could also have noted the weights of the pebbles, watched their speed of fall, and observed whether or not they landed together. They had the pebbles and cliffs, the hands to heft and the eyes to see. They had the equipment or apparatus necessary. But whatever ideas these doings could have involved for them, they could *not* have involved the ideas of the Pisa performance. For the Pisa event was not just a dropping or letting things fall. Nor was it just an observing of what took place or happened when they did fall. It was a deliberate, not a casual act; and it was deliberate, furthermore, not in the general sense that Galileo didn't act impulsively, on the spur of the moment, but had "thought about it" first; the act was deliberate in the *scientific* sense of being performed with a view to fulfilling an intellectual end: it was a dropping or letting fall of different weights for the purpose of *testing a set of ideas*. Because the Pisa performance *carried within itself* this scientific end-in-view, the two different weights could, as they fell, carry the Aristotelian physics and cosmology down with them; and could, when they touched ground, cause a new physics and cosmology to arise.

The Pisa performance was even more than *an* experiment: it was the introduction and establishment of the new *method* of experimentation. Pisa, as a symbol in the history of thought, marks the death of Aristotelianized medievalism and celebrates the birth of intelligence in the modern world.

IX

"The work of Galileo was not a development, but a revolution."* Like all revolutions it *started* something which led to further developments. Hence it is true to say, as does Russell, that Galileo's few facts sufficed *to destroy* the whole vast system of Aristotelian knowledge *if* you take the statement, not literally, but proleptically, *and if* you take "the few facts" to mean the *new method* of inquiry Galileo established (of which new method the few facts were the then results). And by taking the statement proleptically, I mean taking it that way *today:* after some three hundred years of continuous use and development, the new method of scientific inquiry has succeeded in finally destroying Aristotelianism in the technical fields of the most important natural sciences; but Aristotelianism (including the Platonism it both supports and is supported by) is very much alive and

*_The Quest for Certainty_, p. 94.

kicking in our current culture generally and in our social "sciences," philosophies and logics in particular.

Why is it that in the technical fields of science, the revolution in *method* initiated by Galileo has already been substantially completed, has, in our time, carried through its last fundamental reform, whereas in other fields, including fields as intellectual as philosophy and logic, the revolution is just about now seriously getting under way? The easy answer is to invoke a distinction between "natural" sciences and "social" sciences—leaving it up to philosophy and logic to "crash" into the one class or the other or stay out on the limb. This answer is easy, if not very neat; but it is just too easy to make any *explanatory* sense. The "distinction" simply repeats, as an explanation, the fact to be explained. It is the "logic of explanation" of ancient and medieval vintage working over their time: opium puts to sleep *because* of its dormative power; there is a difference in the development of scientific investigation of the natural and the social *because* the former is "natural" and the latter "social."

The backwardness of philosophy, logic and all social inquiries does not explain the *forwardness* of the natural sciences. It simply exposes and emphasizes the need for an explanation. De-socialize the natural sciences as much as you like, place them, if you will, completely outside the boundaries of human society in a realm or sphere apart, above or below, you have not thereby answered the question the contrast raises: you have only stated it; you have not thereby solved the problem it involves: you have only posed it. Let it be granted, for the sake of argument, that the natural sciences are *now* beyond the reach of influence or connection with social institutions, forces and all that goes with the latter. It is an undeniable fact of modern history—let alone of all human history—that they were not *always* there.* Hence the more you conceive the social to be retarding or inherently inimical to the development of science, the more must the "natural" sciences have been able to overcome in reaching their present estate. In so far as the "natural" sciences are *now* distinguished and distinguishable from the "social" sciences it is a *distinction* they have achieved; it is a *result*, not a gift ("something given" or a "datum"); it is a *consequence*, not a cause. The invocation of the "distinction" between "natural" and "social" *subject-matters* to explain the differences between "natural" and "social" *sciences* doesn't even explain the differences away. It just leaves them precisely where and as it *finds* them.†

A philosophy or logic of science cannot, without being foolish, take refuge in a "distinction" in subject-matter to explain the advance of the natural sciences in modern times. And the more the "distinction" is asserted to be *in rerum natura* as a ground for the explanation the greater the folly of the philosophy or logic becomes. The evidence of Aristotelian physics is sufficient to prove this conclusively. The subject-matter (of or *in* Nature) did not change when Galileo appeared. What did change was the *method* of investigating the subject-matter. And the employment and development of that new method resulted in the consequences which we now call the "natural" sciences. If Aristotle were suddenly recalled to life, he would think our "natural" sciences most

*It is also undebatable that they are not *there* always now; witness the transfer of "science" from the laboratories to the chancelleries of Germany and Italy.

†There is a reverse form of the argument criticized, which starts with the social (or economic) as omnipotent and then the "natural" sciences become a reflection or duplication of the "social." An absurdity stated in reverse is still an absurdity.

unnatural. In fact, it is not necessary to recall Aristotle to life to get this "test." When Eddington "goes Aristotelian" (unbeknownst to himself) we have the same result. And of course Eddington is not the only one who "goes" this way. The sphere or domain of operation of the Greek formula is far more embracive.

A discussion of the obstacles in the way of employing the method initiated by Galileo in the fields of philosophy and the social sciences is not here in point. Obviously, *that* question cannot be intelligently examined and discussed until judgment has been reached as to what *is* the method that has made the natural sciences so forward. And herein precisely lies the importance of the analysis of method in the physical sciences for the philosopher—for the philosopher, at any rate, of Dewey's sort. For Dewey, the "natural" (or physical) is not a realm disconnected and set apart from the "social" (or mental). There is not, for him, an abyss between the two. *If* there were an abyss, then the business of the philosopher would be to stay on his side of it—on whichever side he decided a philosopher belonged. And if, defying the impossible, he chose *both* sides and persisted in flitting across the abyss, to and fro, one could of course admire him for his miraculous versatility, but one could not learn anything from him, any time he came across from the "other" side. For though he, being a miraculous philosopher, could flit across the abyss, the abyss being *in Nature*, would absolutely prevent the transportation of goods.

Because Dewey does not believe there is an abyss between the 'natural' and the 'social' the study of the method of investigation employed in the so-called natural sciences is of primary philosophic concern. For *there* the method has been most consistently employed, most carefully and successfully developed, has resulted in the most important body of *tested* knowledge we possess. To study the *method* there is, consequently, the best place to study it. And since there is no abyss between the 'natural' and the 'social' it is not only possible, but certain, that *some* of the goods can be transported. And they need not be transportable bodily—like furniture in a van—to fulfill the necessary conditions of valuable transportation.

When a "distinction" in subject-matter between the "natural" (physical) and "social" (mental) is used as *ground* for explaining the differences between the "natural" (physical) and "social" (mental) sciences, the "distinction," if it does not start out as a variant term for "separation," is forced to grow into an assertion of an abysmal separation in order to maintain itself. And when the so-called natural sciences are separated from the social, are taken out of their context in human history, and out of relation to human activity, then an adequate and satisfactory explanation of the natural sciences themselves becomes impossible.

X

There are certain "gross or macroscopic" features, characteristic of scientific history from Galileo and Newton to Michelson and Einstein about which there is no serious disagreement.

All, for example, agree that science is in constant process of change. When this comprehensive feature is examined more closely or finely it is also universally agreed that the changes are continuous but irregular: that sometimes they succeed each other more rapidly, sometimes more slowly; sometimes they are more pronounced in one scientific area, sometimes in another;

sometimes they are fundamental—having to do with the foundations of science—and sometimes they are, not trivial, but concerned with details.

It is agreed also that changes in science, as overtly displayed in its history, are roughly of two general kinds: changes of addition or expansion; and changes of subtraction or correction (revision). And it is also agreed that these two kinds are not separate and distinct, having nothing to do with each other, but are related and interactive: sometimes the addition of new scientific knowledge or expansion of scientific inquiry into new areas reacts back into the old, requiring the making of corrections or revisions in the latter; sometimes it is the reverse interaction that takes place—correction or revision of old knowledge initiating or determining expansion into new fields.

It is also agreed that the corrections and expansions in science *to be scientific* must be developments proceeding or issuing from the methods of inquiry and the knowledge gained through inquiry. Any correction or expansion that is made in response to pressures exerted by non- or extra-scientific forces, or that is not submitted to the *tests* that have been developed through employment of the method of inquiry *is not science*. Keep science *in* its context of other human enterprises, and what this means is perfectly clear. It means that certain methods, rules, principles, standards have been developed by human beings and that these *define* science. There is no area or subject-matter that is inherently or "by Nature" non- or extra-scientific. But for any item of knowledge involving any subject-matter or area to be an item of *scientific* knowledge, *that item* must go through the processes and pass the tests that have themselves been tested and established through prior inquiries and that *define* what science is.

The integrity of science is *not* preserved, any more than it is established, by giving it an "autonomy" that separates it and makes it "independent" of human activity. It is not preserved—because science is changing, both in items of knowledge and in details of its methods. The methods are being developed, the tests are made more rigorous, the analyses more precise. If the body of scientific knowledge, or the method of scientific inquiry, is separated from human activity, given a trans-human "autonomy," then it loses, not gains, its integrity. For then how explain the changes? But as a human activity, scientific knowledge and scientific method have the integrity and "autonomy" of any enterprise tested by methods developed in the course of human experience. Baking bread is a method of treating and preparing materials for human consumption developed out of *baking* bread. The first bakers of bread were not as expert as bakers today. Looking backwards, bakers of today may indeed not consider them to have been *really* bakers of bread at all. However that may be, bakers of bread today have their methods, standards and tests. And anything *to be* bread has to pass those tests. In this sense, baking too is "autonomous."

If baking bread is too lowly and mean an "example"— or too far removed from the realm of the scientific— consider the case of geometry. Einstein, in his lyrical moments, can write:

> She [Greece] for the first time created the intellectual miracle of a logical system, the assertions of which followed one from another with such rigor that not one of the demonstrated propositions admitted of the slightest doubt—Euclid's geometry. *(Herbert Spencer Lecture)*

Now in so far as Euclid's geometry was considered thus "miraculous" like all miracles it stopped things, didn't start them, closed the road to better understanding, didn't open it. It took geometers just about two thousand years to get over all the stultifying consequences of *that* miracle. When in the second quarter of the nineteenth century, non-Euclidean geometries began to appear, they were, by the miracle-believing mathematicians, construed as outrageous, perhaps even insane, attacks on the eternal historicity of Euclid's miracle; but as the subsequent events have shown, the non-Euclidean geometries, by exposing the miracle, liberated the geometer, and contributed greatly to further the liberation of the philosopher, logician and physicist. So that the mathematical logician can write concerning the Euclidean geometry, as does Russell:

> The rigid methods employed by modern geometers have deposed Euclid from his pinnacle of correctness. Countless errors are involved in his first eight propositions. That is to say, not only is it doubtful whether his axioms are true, which is a comparatively trivial matter, but it is certain that his propositions do not follow from the axioms which he enunciates ... it is nothing less than a scandal that he should still be taught to boys in England. *(Mysticism and Logic*, pp. 94-95.)

It undoubtedly is a scandal that "the tedious apparatus of fallacious proofs for obvious truisms which constitutes the beginning of Euclid" (Ib., p. 62) should be taught contemporary learners when a much superior apparatus of geometry has already been developed. But in the history of man's learning geometry and the methods of rigorous mathematical proof, Euclid's geometry played its good, as well as bad part. The methods, standards and tests now employed in geometry were developed by using, among other things, Euclid's geometry as a method to be studied, revised, changed and reconstructed in terms and by means of methods which were the consequences of, or which were developed in the course of, mathematical and allied inquiries. And by the same general process, the methods and tests are being further developed almost daily. Set Euclid's geometry in a realm apart from human activity, and if correct, then its correctness is a miracle. And if not correct, as long as it is a miracle, there is no way of correcting it. Miracles do not submit to correction. For just as soon as they do they cease to be miraculous.

It would complicate matters to no good end if we tried to discuss here the way in which extra- or non-scientific, social, political or economic forces may influence *scientific* research. But one illustration may be helpful to the general point. A politician orders scientists to produce, let us say, synthetic rubber. Here, say some current thinkers, is a case of science receiving its direction (or perhaps I should say its "directive") from "politics." But where did the politician get the idea that there was a possibility of making synthetic rubber? To suppose he got that idea out of his own head is to suppose the miraculous. Also, because the politician issues an order that synthetic rubber should be made—his order does not convert the scientific possibility into a scientific actuality. To believe it does, or to argue in a manner that presupposes it does, is to believe the politican is what he wants others to believe he is—an omnipotent miracle-man. And, certainly, there is no intelligent reason for believing *that*. The *development* of science can not be *directed* by the politician, big businessman, etc., but it can be

exploited by them—which is a totally different thing. And science can of course also be starved, strangled or killed by the politician and big businessman, by starving, strangling or killing the scientists.

That *corrections*, to be *scientific* (or what is the same thing, to be worthy of intelligent acceptance), must be made by methods developed by inquiry, and in response to needs of *test* growing out of inquiry, is also best seen when science is placed in the social context and when contrasted, for instance, with the "method of correcting" science initiated and enforced by *political* demands.* "Nazi science" isn't something new; it is the revival of something, alas, very old. It is as old as religion and in the Western World the Nazi "scientific" model is the Catholic Church. The Catholic Church burnt books in the public place, and since losing secular control over that place, has continued to burn them in the silent fires of the *Index*. The Church also still outrageously falsifies history; witness, for example, its marginal notations to the Old Testament. The Church also coerced scientists into keeping quiet, and sometimes even succeeded in getting them to recant, witness, for example, the case of Galileo. But no scientist (or any person of intelligence) accepts Galileo's *recantation* as science. *That's* the difference.

XI

To recapitulate the macroscopic features about which there is complete agreement, or at any rate, no serious

---

*Heisenberg, in a lecture delivered in 1934 in Berlin, said that Michelson's experiments and Einstein's theory of relativity "belonged to the absolutely certain bases of physics." A Dr. Rosskothen (a high school teacher) heard the lecture and wrote in complaint to Reich Director Dr. Alfred Rosenberg, Commissioner appointed by the Führer to supervise the Philosophical Instruction of the National Socialist (Nazi) Movement: "should such a man [Heisenberg] occupy a chair at a German university? In my opinion, he should be given the opportunity to make a thorough study of the theories of the Jews of the Einstein and Michelson type, and no doubt a concentration camp would be an appropriate spot. Also a charge of treason against people and race would not be out of place." To which on November 24, 1934, Dr. Alfred Rosenberg, through his Staff Director, replied: "The Reich Director of the N.S.D.A.P., Commissioner appointed by the Leader to Supervise the Philosophical Instruction of the National Socialist Movement, states in answer to your communication that he shares your opinion in principle. He has taken steps to draw the attention of Professor Heisenberg to the reprehensible passages in his speech, and made clear to him, in the form of a reprimand, that he must refrain from remarks of this nature, which have to be regarded as an insult to the Movement. Unfortunately, in view of foreign opinion, it is not possible to administer a sharper reproof to Professor Heisenberg or, which would certainly be desirable, to dismiss him." (Quoted in *The Yellow Spot*, a collection of facts and documents relating to three years' persecution of German Jews, with an Introduction by The Bishop of Durham. Knight Publications, New York, 1936.) According to Dr. Rosenberg, the unique contributions of the "culture-creating" Nordic blood—of which Nazi blood is the highest culmination—the unique contributions of this blood are "the ideals of honor and spiritual freedom." What these Nazi "ideals" mean with respect to science, Dr. Rust, Reich Minister of Education, made clear on June 30, 1936 when, on the occasion of the 550th anniversary of Heidelberg he proclaimed: "The old idea of science based on the sovereign right of abstract intellectual activity has gone forever" [*sc*. "in Germany"]. And what Dr. Rust meant is completely clarified by *Deutsche Justiz*: publication of the German Ministry of Justice: "A handful of force is better than a sackful of justice." Most appropriately, before the Heidelberg Nazi celebration, the inscription "To the Living Spirit" was replaced by "To the German Spirit"; and the statue of Athena, Goddess of Wisdom, replaced by the German Eagle. The Nazis did what *they* could so that there may not be any Nazi-discordant note—and then some Harvard professors had to go there and accept "honorary degrees"!

disagreement among philosophers: Science is in constant process of change; the changes are not hit-and-miss, helter-skelter, sporadic innovations, interruptive and disconnected shifts from one position to another, but are changes consequent upon employing methods of inquiry, and to some extent always issue out of knowledge antecedently achieved and to the rest of the extent are new acquisitions which in turn, and to some extent, lead back into prior knowledge, both of content and method; the changes are determined by needs and established by methods of test developed by and in the process of inquiry; the events or changes of science are not a mere chronological succession but constitute an interconnected series indisputably exemplifying the characteristics of growth or development. In sum: the series of changes in science, from Galileo to Einstein, exhibit the continuity of a self-expanding and self-correcting *history*.

Up to this point agreement. But when you take the next step and assert that a philosophy or logic of science must be competent to explain (or account for) this *history*, must be able to give the reasons for its continuity and direction of development—all except Dewey and Deweyan experimentalists balk. No! they say. The history of science is one thing, and the nature of science is another. Science has a history, but it is not a history. Science is a *system*. And they mean by system a mathematical-logical system: so many axioms, so many postulates, so many definitions, so many theorems, all tied with inevitable deductibility together by so many principles (which latter may or may not also be in the system). Thus, for instance, Russell:

> There are three kinds of questions which we may ask concerning physics or, indeed, concerning any science. The first is: What is its logical structure, considered as a deductive system? What ways exist of defining the entities of physics and deducing the propositions from an initial apparatus of entities and propositions? This is a problem in pure mathematics, for which, in its fundamental portions mathematical logic is the proper instrument. *(The Analysis of Matter, 1927, pp. 1-2.)*

Now when you take this to be the first or primary question, the second question is bound to be one you can never answer—unless a series of contradictions be considered an answer. The second question is:

> ... the application of physics to the empirical world. This is, of course, the vital problem: although physics *can be pursued as pure mathematics*, it is not as pure mathematics that physics is important. What is to be said about the logical analysis of physics is therefore only a *necessary preliminary* to our main theme. The laws of physics are believed to be at least approximately true, although they are *not logically necessary*; the *evidence* for them is empirical. All empirical evidence consists, in the last analysis, of *perceptions*; thus the *world of physics must be*, in some sense, *continuous* with the *world of our perceptions*, since it is the latter which *supplies* the *evidence* for the *laws* of physics. (L.c.p.6; italics mine.)

By "application of physics to the empirical world" Russell does not of course mean "application" in the vulgar sense of "applied science"—making machines, telegraphy, radio, airplanes and so on. He means, in what sense, or to what degree can the mathematical-logical system of "physics" be said to be *about* the empirical world. Having *first* pursued physics, as pure mathematics (and the purity of mathematics *is* its *logical, non-*

empirical nature) the question as to how physics is *connected* with the empirical world necessarily becomes a *problem*. But if physics *can* be pursued as pure mathematics, then in what sense can it be true that the laws of physics are *not* logically necessary—since pure mathematics *is* pure logic (or vice versa) and the mathematical necessity of the one is the logical necessity of the other (or vice versa)? To say "it is not as pure mathematics that physics is *important*" is to misstate and confuse the case he propounds. For, obviously, it is not as pure mathematics that physics is *physics*—since its laws are not *logically necessary*, but rest on empirical evidence.

The logical systematization of the body of knowledge (known as "physics") results rather inevitably (and understandably) in a "logical" structure or *system*; but if the systematization is of *physics* (and not say, of anthropology, Egyptology, or whatever else), it must be because *physics* is, at the very least, *about* the *physical* (empirical) world—and, at the most, may be of it or *in* it—before ever the logical systematization was undertaken. Russell certainly *knew* when he was writing his *The Analysis of Matter* that he was not writing, say, his analysis of *politics*. One has only to read his *Freedom and Organization*, for example, and compare the two to have conclusive evidential proof. If, as Russell himself goes on to say, if the world of physics *must be*, in some sense, *continuous* with the world of our perceptions (the empirical world) how *can* there *be* a "second question" concerning (in Russell's sense of the term) "the *application* of physics to the empirical world?" To *apply* a world (the world of physics) to a world it is already and *necessarily continuous* with (the empirical world) is a very strange thing to try to do. And when you try to do it, you naturally find it presents a difficult, "mysterious" problem which even the strength of the mathematical-logical instrument is insufficient to solve. Why *raise* such a problem? Or why try to solve *such* if any one else raises it? True enough, Russell provides himself with a *verbal* reason for raising the problem when he says: the world of physics must be, *in some sense*, continuous with the world of our perceptions. But, it is also true, he doesn't provide himself for very long, for he goes right on to say *in what sense it must be continuous*, namely, it must be continuous in a sense *sufficient* to *supply* the *evidence for the laws of physics*. But a law of physics is *scientifically established* when the *evidence for it is supplied*. Surely, then, sufficient unto the laws *must be* the *evidence* thereof—as sufficient unto the *continuity* is the evidence of the *laws*.

Russell's third question, or problem, carries the self-contradiction to its logical conclusion. The third problem is presumably arrived at by combining the first and second problems together, by trying to *fuse* them in some way. But the "third" problem is, in fact, nothing more than a *repetition* of the "second" problem in a different form; the difference being due to the desire to get out of the "second" problem not the outcome for *physics* (i.e. the solution of the "first" problem) but the outcome for *metaphysics* (i.e. the solution of the "second" problem)—or as Russell puts it:

> ... The outcome for ontology—i.e. [to get the answer to] the question: What are the ultimate existents in terms of which physics is true (assuming that there are such)? And what is their general structure? And what are the relations of space-time, causality, and qualitative series respectively? ... We shall find, if I am not mistaken, that the objects which are mathematically *primitive in physics*, such

as electrons, protons, and points in space-time, are all *logically complex* structures composed of entities which are *metaphysically more primitive*, which may be conveniently called "events". It is a matter for mathematical logic to show *how to construct*, out of these, the objects required by the mathematical physicist. It belongs also to this part of our subject to inquire whether there is anything in the *known world* that is not part of *this metaphysically primitive material of physics*. Here we derive great assistance from our earlier epistemological inquiries, since these enable us to see *how physics and psychology can be included in one science*, more concrete than the former and more comprehensive than the latter. Physics, *in itself*, is exceedingly abstract, *and reveals only certain mathematical characteristics* of the material with which it deals. It does *not* tell us *anything* as to the *intrinsic character of this material*. Psychology is preferable in this respect ... by bringing *physics and perception together*, we are able to *include psychical events in the material of physics*, and to give to physics the greater concreteness which *results from our more intimate acquaintance* with the *subject-matter of our own experience*. To show that the traditional separation between physics and psychology, mind and matter, is not metaphysically defensible, will be one of the purposes of this work; but the two will be brought together, *not by subordinating* either to the other, but by *displaying each* as a *logical structure* composed of what, following Dr. H. M. Sheffer, we shall call "neutral stuff." (Ib., pp.9-10; italics mine.)

In saying that Russell, in his "third" problem carried the self-contradiction to its *logical* conclusion, I was, of course, very much in error. By tradition, "metaphysics" (or "ontology") is supposed to deal with the absolutely "first" or "ultimate" things (the "metaphysically *primitive*") and hence, when you use the terms, you cannot help but get the *feeling* that at last you have hit bottom. But to stop increasing the "number" of problems just because you *feel* the thud of finality is to come to a psychological *stoppage*, not to reach a *conclusion* of logic. When you start with a contradiction, as Russell's own mathematical-logic teaches in its Theory of Types, you can go on forever, carrying the contradiction in another "form" (or "formulation") from one "level" (or type) to the next, and never reach a *logical* conclusion because the hierarchy is without logical end.

And this is demonstrated, or at any rate, exemplified, in Russell's attempts at solving his "third" problem. "To bring physics and psychology together, *not by subordinating* either to the other" is his comprehensive purpose. And it is reasonable to understand that by "*not subordinating* either to the other," is meant that *with respect to each other* they will be given *some coordinate* status, though with respect to the "neutral stuff" anything might happen to them; but whatever does happen to them, because of the activity of the neutral stuff, will happen to both alike. For the "stuff" *is neutral*—and will do to "physics" what it does to "psychology" with an equal mind or will (or whatever else). Or, to put it in another way: it is Russell's explicitly avowed purpose to develop a philosophy which will bring psychology and physics *together* but which will *not allow* "physics" to swallow "psychology," either the one definitely the other, or both definitely in turn.

To go into the details of Russell's arguments is

impossible and also unnecessary. The statement of his "three questions" or problems *defines* the course his argument *must* take. Far from its being true that only in mathematical-physics is "prediction" possible it is possible to "predict" in philosophy as well. Given Russell's three problems, the *general* line (not the details—they vary from philosopher to philosopher, and from time to time in the same philosopher) is laid down.*

When Russell is dealing with his "first" problem—the world of physics taken by itself—he is predominantly (if not always) dealing with what the title of one of his books called: "Our knowledge of the external world." If, in dealing with the first problem, Russell cannot always and consistently stay "outside" that is not through lack of trying. But since he is after the mathematical-logical structure or system of "physics" it is inevitable that the sheer operation of his logical symbolism should every now and then drag him "inside."

When (having finished with the problem of the logical structure of "physics" as a "deductive system"), Russell passes on to his second problem we find what one could predict, namely, that the further he gets on with his second problem, the further "inside" he gets. And that he should finally wind up so far "inside" that everything is "inside the head" may appear shocking to some, but is no logical surprise. In the *statement* of his second problem Russell, true enough, tried to protect himself: "the world of physics must be, *in some sense*, continuous with the world of perceptions." But three words offer no real protection. They are no match against the logical force of his whole philosophic method:

We do not know much about the contents of any part of the world except our own heads; our knowledge of other regions, as we have seen, is wholly abstract. But we know our percepts, thoughts and feelings in a more intimate [i.e. "concrete"] fashion. Whoever accepts the causal theory of perception is compelled to conclude that percepts are in our heads, for they come *at the end* of a causal chain of *physical events* leading, *spatially*, from the *object* to the *brain* of the percipient. ... And with the theory of space-time as a structure of events, which we developed in the last two chapters, there is no sort of reason for *not* regarding a percept as being in the head of the percipient. ... It follows from this that what the physiologist sees when he examines a brain is *in* the physiologist, not in the brain he is examing. What is in the brain by the time the physiologist examines it if it is dead, I do not profess to know; but while its owner was alive, part, at least, of the *contents* of his brain consisted of his percepts, thoughts and feelings. Since his brain *also* consisted of electrons, we are compelled to conclude that an electron is a grouping of events, and that, if the electron is in a human brain, some of the events composing it are likely to be some of the "mental states" of the man to whom the brain belongs. ... I do not wish to discuss what is meant by a "mental state"; the main point for us is that the term must include percepts. Thus a percept is an event or a group of events, each of which belongs to one or more of the groups

---

*This, of course, is what Dewey proved to the hilt in his *Experience and Nature* and *The Quest for Certainty*. Given the "Greek formula" and the rest of philosophic discussion (with endless variations in detail) follows as a matter of inescapable logical course. In Russell's philosophy we have the "Greek formula" working in its latest (or almost latest, for there are the Logical Positivists) mathematical-logic dress.

constituting the electrons in the brain. This, I think, is the *most concrete* statement that can be made about electrons; everything else that can be said is more or less abstract and mathematical. (Ib., pp. 319-320; italics mine.)

To ask Russell what he means by the "percept" being at the *end* of a causal chain; what he means by the causal chain leading, *spatially*, *from* the object *to* the brain; what he means by "physiologist" who "examines" a "brain" of someone else; *how* he came to *know* that when the owner of a brain is dead, what the physiologist sees in it is different from what he sees when the owner is alive (though in both cases *all* the physiologist sees is in *his* head, and *he* is presumably alive both times); what he means by saying the brain "*also* consisted of electrons"—whether he means "electrons" concretely, or only abstractly and mathematically; and if he means that electrons are "concrete" groupings of events *in the same sense* in which he means "percepts, thoughts and feelings" are concretely or "intimately" known then why does the brain *also* consist of electrons? . . . to ask Russell these questions and dozens like them may have value as a "logical exercise" but he can give no answer to them other than the kind of answers he has already given. For the logical operation of his philosophic method will allow for no other sort of answers.

So when Russell comes to his "third" problem, we find him repeating on a more generalized plane, or in terms of more generalized formulations, precisely what his "solution" of the "second" problem would logically lead one to expect (or predict):

> On the question of the *material* out of which the *physical world* is constructed, the views advocated in this volume have, perhaps, more affinity with idealism than with materialism. What are called "mental" events, if we have been right, are *part of the material of the physical world*, and what is in our heads is *the mind* (with additions) *rather than what the physiologist sees through* his microscope [!] It is true that we have not suggested that *all* reality is mental. The positive arguments in favor of such a view, whether Berkeleyan or German, appear to me fallacious. The sceptical argument of the phenomenalists, that, whatever else there may be, we cannot know it, is much more worthy of respect. (Ib., pp. 387-388; italics mine.)

> While, on the question of the stuff of the world, the theory of the foregoing pages has certain affinities with idealism . . . the position advocated as regards scientific laws has more affinity with materialism than with idealism. . . . There are psychological laws, physiological laws, and chemical laws, which cannot *yet* be *reduced to* physical laws. (p. 388; italics mine.)

So far as causal laws go, therefore, physics seems to be *supreme among the sciences*, not only as against other sciences of matter, but also as against the sciences that deal with life and mind. There is, however, *one important limitation* to this. We need to know in what *physical circumstances* such-and-such a percept will arise, and we *must not neglect* [!] the more intimate qualitative knowledge which we possess concerning mental events. *There will thus remain a certain sphere which will be outside physics.* . . . It is obvious that a man who can see knows things which a blind man cannot know; *but a blind man can know the whole of physics.* Thus the knowledge which other men have and he has not is

not a part of physics. (p. 389; italics mine.)

Since "there is thus a sphere *excluded from physics*" (p. 389) the hasty reader, one insufficiently disciplined in the subtleties of mathematical logic, and insufficiently hardened by the rigors of fundamentally pure methods of symbolic-logical proof—such a reader might come to the conclusion that Russell has, at the end, at any rate, left *some* part of "psychology" unsubordinated and unsubordinatable to "physics" (or is it vice versa?); that with respect to each other there is a real difference between the two; and hence has (to some extent) fulfilled his comprehensive pledge given at the start: to show "how physics and psychology can be included in one science, more concrete than the former and more comprehensive than the latter." Although the last quotation (p. 389) was taken from *near* the end of the book, the *end* of it is, I take it, the last sentence, or two. And the penultimate sentence reads as follows:

> As regards the world in general, both physical and mental, *everything* that we know of its *intrinsic character* is derived from the *mental* side, and *almost* everything that we know of its causal laws is derived from the physical side. (p. 402; italics mine.)

The "mental side" has the edge so far because Russell (in 1927) is not quite certain but what there may also be some "causal laws" which are derived from the mental side—whatever it is that Russell *here* means by "derived." But the edge which the "mental side" enjoys in the penultimate sentence is very short-lived—as short-lived as the sentence.

Disregarding "the world in general," the ultimate sentence reads:

> But from the standpoint *of philosophy* the distinction between physical and mental is *superficial and unreal.**

XII

"The question of whether we should *begin* with the simple or the complex appears to me the most important problem in philosophic method *at the present time*"; the *complex* Dewey defines as "the gross, macroscopic, crude subject-matters in primary experience" and the *simple* he defines as "the refined, derived objects of reflection."† Russell's philosophy, of which we have just had a representative sample, is an illustrious contemporary exemplification of the *consequences* that unavoidably ensue when the method of beginning with refined, derived objects of reflection is followed. Dewey's philosophy is a *consequence* of following the other method. The contrast between the two philosophies, whatever else it does, should materially help the reader to understand how fundamentally serious for *philosophy* the issue over *scientific* method is. For Dewey and Russell both agree

---

*Other books of Russell's give variations of the same conclusion, though sometimes they may *seem* different. Thus in *The Analysis of Mind* (1921): 1. "One of the main purposes of these lectures is to give ground for the belief that the distinction between mind and matter is not so fundamental as is commonly supposed." (p. 108) 2. "I think that what has *permanent* value in the outlook of the behaviorists is the feeling that *physics* is the most *fundamental science* at present in existence." (preface) 3. "All our data, both in physics and psychology, are subject to *psychological* causal laws; but physical causal laws, strictly speaking, can only be stated in terms of matter, which is both inferred and constructed, never a datum. In this respect *psychology is nearer* to what *actually is*." (P. 308—last sentence of book; italics mine throughout.) In all Russell's treatments of the subject, the same with limited variations will be found. But enough, if not too much, has already been quoted here.

†*Philosophy and Civilization* (1931), p. 78; *Experience and Nature* (2nd ed.), p. 34; italics mine.

that philosophy, to be significant and intelligent, must be *scientific*, that is, must follow in its inquiries the *method of science*. With respect to these philosophies, the double-issue over method, scientific and philosophic, is squarely and explicitly joined.*

But furthermore, and more generally, the contrast should also help the reader toward understanding that fundamental differences in *philosophies* are *not explained* by the "personality" differences in *philosophers*. Just as the fundamental differences in the physics of Galileo and Aristotle, or of Einstein and Newton, are *consequences* of differences in *methods* of inquiry employed, so with respect to fundamental differences in philosophies. That philosophers have "personality" differences is not hereby denied. Neither is it denied that personality differences are dominant, even predominant, in many (but not all) philosophies. But to make "personality differences" the *ground of explanation* for the predominance of "personality differences" in philosophies, is to repeat as *explanation* of the fact the very fact *to be* explained. It is to convert a *consequence* into a cause; it is to set up a distinction *in subject-matter* as explanation of the *result of methods of inquiry*. For it is the *methods of inquiry* employed by philosophers that make it possible for their "personality" differences to achieve and retain predominance in their philosophies.†

To determine whether inquiry should begin with refined objects of reflection or macroscopic subject-matters in primary experience is the problem *philosophers* are faced with today. In this sense it is a *philosophic* problem. But not in any other. It is the fundamental methodological problem of all inquiry, irrespective of the field in which inquiry goes on. Galileo, for example, was faced with this problem when he undertook to inquire into the physical world and the motions of physical things. Should he begin with the refined objects of reflection which constituted the Aristotelian-medieval "science of physics"? Or should he begin with the macroscopic subject-matter in primary experience? When he decided to follow the latter method and climbed the Tower of Pisa to put his decision into practical effect, *modern* science was launched upon its career and a revolution got under way.

Contrariwise, the medieval scholastics remained medieval because they began with refined objects of reflection and insisted on staying with them.

> Within the sphere of dialectic debate, the Scholastics were supremely critical. They trusted Aristotle because they could derive from him a coherent system of thought. It was a *criticized* trust. *Unfortunately* they *did not reflect* that some of his main ideas depended upon his *direct acquaintance with experienced fact*. They trusted to the *logical* coherence of *the system as a guarantee* of the unrestricted relevance of his primary notions. Thus they accepted his confusion—where there was confusion—of superficial aspects with fundamental principles of widest generality. Their *method for the*

furtherance of natural knowledge was endless debate unrelieved by recurrence *to direct observation*. Unfortunately also their instrument of debate, Aristotelian logic, was a *more superficial weapon* then they deemed it. Automatically it kept in the background some of the *more fundamental topics for thought*. Such topics are the quantitative relations examined in mathematics, and the complex possibilities of multiple relationship within a system. All these topics, and others, were kept in the background by Aristotelian Logic. (Whitehead: *Adventures of Ideas*, pp. 149-150; italics mine.)

As heirs of twentieth century science we can look back to the period before Galileo and confidently speak as Russell does of "the whole vast system of *supposed* knowledge handed down from Aristotle." But *how* did it come about that that knowledge was *rendered* supposititious? As heirs of quantum physics and Relativity Theory of Space-Time we can also look back to the vast system of Newtonian science, with its indestructible, eternal billiard-ball atoms, and its Absolutism of Space and Time, and with equal confidence declare that the latter is a vast system of supposed knowledge. But again the same basic question is relevant: *How* did it come about that the Newtonian system, *in its fundamentals*,* was rendered supposititious?

If we search for an answer to either or both of these vital questions (they are really two continuous parts of one question) by pursuing "physics" as "pure logic" or as "pure mathematics" we are doomed to failure. Aristotle's physics, as a logical system, was as coherent as they come; whereas the system of Galileo was very much otherwise. Similarly with the change in science that was realized during the past fifty years. As Russell unambiguously points out: "The physics of Newton, considered as a deductive system, had a perfection which is absent from the physics of the present day."† But Galileo's badly systematized "few facts," not Aristotle's well-systematized many, are the "examples of *real* knowledge." And the ("purely") *mathematical inferiority* of present-day physics does not stand in the way of its being, for scientists and philosophers (Russell himself included), *scientifically superior knowledge*.

Of course Russell and Whitehead, when they face critical turning-points in the *history* of physical inquiry of the gigantic sort exemplified by the change from Aristotelian physics to Galilean, and from Newtonian to Einsteinian, find it necessary to abandon "pure logic" and "pure mathematics" and they surrender to necessity. They introduce at *such points* references to "direct observation," "new method," "experimental and technological [instrumental] progress," but having done so, they immediately pass back to the consideration of physics as "pure logic" or as "pure mathematics" leaving the intervenient preceding and succeeding *history* of physical inquiry to take care of itself. This method of wandering off and on the topic of experiment not only makes nonsense out of the history of scientific thought; it makes unintelligible the nature of science in particular and the nature of all knowledge in general.

Consider some of the things Whitehead says about the change from the medieval-Aristotelian to the Galilean method of scientific knowledge-getting:

---

*Those modern and contemporary philosophers who turn their backs in part or in whole on science and scientific method and claim another and totally different method for philosophy (like Berkeley, Hegel, Bergson, and latterly Whitehead, do so *because of* and *in terms of* their conception or *interpretation* of science and scientific method. Also, those who spurn science always claim their philosophy is a "Higher Science" or "Knowledge"). So that actually the fundamental issue in *all* philosophy from the time of Galileo is one and the same—whether frankly and explicitly faced, or left implicit and evaded.
†See *supra*, pp. 40-41.

---

*There are of course a vast number of *items* of Newtonian knowledge that are as good today as they ever were; and the same can be said for as *comparatively* large a number of *items* in the Aristotelian corpus of knowledge.
†*The Analysis of Matter*, p. 13.

"Galileo keeps harping on how things happen, whereas his adversaries had a complete theory as to why things happen. . . . It is a great mistake to conceive this historical revolt as an appeal to reason. On the contrary, it was *through and through* an anti-intellectualist movement. It was the return to *the contemplation of brute fact*; and it was based on a recoil from the *inflexible rationality of medieval thought.*" *(Science and the Modern World*, p. 12; italics mine.)

You may well ask what is so inflexibly *rational* about "thought" which uses "endless debate" as "a method for the furtherance of natural knowledge." You may also ask why it is that the Historical Revolt is a return to the *contemplation of brute fact* when the *consequence* of that "anti-rationalism"* is that

...although in the year 1500 Europe knew less than Archimedes who died in the year 212 B. C., yet in the year 1700, Newton's *Principia* had been written and the world was well started on the modern epoch. *(Ib.,* p. 8)

Are we to understand that Whitehead *means* that modern *science* really is anti-rational, anti-intellectualist? And if so, why call it *knowledge?* Of course not! When he passes over from his *contemplation* of the "logical perfection" of the "supremely critical dialectic debate" of the scholastics to his *contemplation* of modern science, his contemplation changes during the passage:

Aristotle by his Logic throws the emphasis on classification. The popularity of Aristotelian Logic retarded the advance of physical science throughout the Middle Ages. If only the schoolmen had *measured instead of classifying*, how much they might have learnt! Classification is a halfway house between the immediate concreteness of the individual thing and the complete abstraction of mathematical notions . . . in the *procedure* of relating mathematical notions to the facts of nature, by counting, by measurement, and by geometrical relations, and by types of order, *the rational* contemplation *is lifted.* . . . Classification is necessary. But *unless* you can progress from classification to mathematics, your *reasoning* will not take you very far. *(Ib.,* p. 43; italics mine.)

Did not Galileo and his co-workers of the seventeenth century have something to do with introducing the "procedure of relating mathematical notions to the facts of nature," and so have something to do with "*lifting the rational* contemplation"? Judging by Whitehead's statement that this Historical Revolt was not an appeal to reason but was anti-intellectualist, anti-rationalist, a return to the contemplation of brute fact, one might be tempted to think not; but to yield to the temptation would be irrational, illogical:

In the seventeenth century the influence of Aristotle was at its lowest, and mathematics recovered the importance of its earlier period [up to Archimedes]. It was an age of great physicists and great philosophers; and the physicists and philosophers were alike mathematicians. . . . In the age of Galileo, Descartes, Spinoza, Newton, Leibniz, mathematics was an influence of the first magnitude in the formation of philosophic ideas. But the mathematics, which now emerged into prominence, was a very different science from the mathematics of the earlier epoch. It had gained in generality, and had started upon its almost incredible modern career of piling subtlety of generalization upon subtlety of generalization; and of *finding*, with each growth of complexity, *some new application, to*

physical science, or to philosophic thought. *(Ib.,* p. 44; italics mine.)

That these statements, taken together and in relation, don't make sense is too obvious to need any demonstration. But Whitehead is not an irresponsible thinker; his contradictions and oscillations are not expressions of his "personality." The critical imbalance of his thought (on this topic and others) is a *consequence of his method* of philosophic inquiry which in turn determines and is determined by his conception (better, *pre*-conception) of logic and science (knowledge). A mind less original, less powerful and great than Whitehead's would easily find "rest" at one extreme or the other, or at that most precarious and delusively "restful" place of all—at the half-and-half point between.*

When philosophies of science (knowledge) and scientific method dismiss or neglect to take into central account "the topic of experiment" then are they doomed to wander off and wander about like the arguments of Shades in Purgatory who can look in both directions but can continue in neither.

XIII

If the schoolmen had measured instead of classified they would have learned much more. But they would have learned immeasurably more even with their classifying if only they had relaxed their "inflexible rationality of thought" and renewed "acquaintance with experienced fact." Or, to put it in Deweyan phraseology, the schoolmen would have vastly increased their real knowledge if they had not, by endless dialectic debate, kept themselves revolving in the circle of their refined objects of reflection and had instead turned for guidance and control to the gross, macroscopic, crude subject-matters in experience.

It was not the Aristotelian Logic that made the medievalists go round in dialectic circles. It was the way they *used* that Logic that caused them to do that. Aristotle's Logic is explicitly based upon and explicitly refers to experience of qualitative fact. Given the method of beginning with refined objects of reflection and staying with them, it makes no difference at all fundamentally whether you use a Logic of Classification or a Mathematics of Measurement. The chances are not only good, they are absolutely perfect, that if the schoolmen had "measured" instead of classified they would have remained schoolmen for all that. This is not a conjecture. It is a demonstrated certainty. For this is precisely what has happened with the "schoolmen" of modern and contemporary times. As Whitehead says: "*the sort of person* who was a scholastic doctor in a medieval university, today is a scientific professor in a modern university."† This is not to praise the scholastic doctor but to damn the scientific professor.** And foremost

---

Ib., p. 14.

*All this applies to Russell too.
†*Adventures of Ideas*, p. 149; italics mine. There was of course *another* "sort of person" too—for example, William of Occam *in* the university, and the far, far greater Roger Bacon *out*.
**The schoolmen "trusted Aristotle" not because he could help them make a coherent system of thought, but because the Church enforced upon them the task of "cohering" *its* doctrines. Wherefore their "criticized trust" involved no reflection on their part concerning the basic dependence of Aristotelian Logic upon experienced matter of fact. The Church has never been overly insistent upon turning to experience for guidance and knowledge. It has a supply of "eternal knowledge" ready-made. There is no insufficiency, among "mathematical schoolmen" today, of mystery-mongering, nor are "mathematical theologians" wanting, either.

among such "scientific" professors must of course be placed the "scientific professors of philosophy,"—those particularly who seek to make philosophy "scientific" by making it "mathematical-symbolical," "symbolical-formal," "positively logical." It doesn't have to work like a syllogism to be able to work like a charm!

Aristotle's Logic was a "superficial weapon" because Greek science was exclusively concerned with the *superficial* qualities of natural things and their *superficial* relations (hot, cold, wet, dry, light, heavy, up, down, etc.), the qualities and relations, namely, that are displayed on the *superficies,* the qualitative faces and relations of things that can be experienced by direct observation, that we can become acquainted with by simply *looking* at, by beholding as a spectator. There are many qualities, many combinations of qualities (natural things) and a few large relations that can be directly experienced. And with such as these, taken *as is,* as directly or immediately experienced, classification is the only logical thing that can be done. The schoolmen put Aristotle's syllogism to the fore; but it is his classification that is the "weapon" of natural science.* Anyway, whether they classified little of the natural world open to observation and syllogized much, the fundamental point Dewey makes over and over again is alone of commanding importance: no conflict was introduced between the world the Greeks and the schoolmen experienced (for even the schoolmen were alive, had eyes and ears, etc.) and their "science" of that world.†

Aristotle's Logic (even the syllogism alone) doesn't preclude recurrence to observation or to macroscopic subject-matter in primary experience; it encourages and fosters such recurrence. But it does absolutely, definitely *preclude* "quantitative relations." Whitehead is temporarily generous** to a great historical fault when he says Aristotle's logic "kept in the background" the "quantitative relations examined in mathematics." For Aristotle, "quantitative relations" are "accidents," of no metaphysical (cosmological) import, mutable and changeable, not eternal and unchanging. Hence they are not objects of *scientific* knowledge, and a logic of science need make no provision for them—except to "put them in their place" (which is "out").

The change from the method of classifying and syllogizing without observing, to observing and classifying without syllogizing, accounts for the modern progress made by such descriptive sciences of nature as "natural history." But the great change in modern science occurred when the change was made from the method of classifying to the method of quantitatively measuring.

Observation is involved in measuring; to measure the rate of fall, for instance, you have to observe the bodies falling. But observation is not all, and it is not enough. Eye-measurement of rate of fall is at best a rough estimate, not a quantitative measurement of any mathematical exactitude. One method and one method only makes possible the *modern* "procedure of relating mathematical notions to the facts of nature," namely, the *method of experimentation.* To be able to measure quantitative relations of change, it is absolutely essential to be able to *control* the changes, to stop them and start them, to accelerate them and retard them. If it is an accident, then it is a very happily symbolic one that at the outset of modern science of motion, *acceleration* was defined as a change in *direction* or velocity. For in *experimental* control of change, a control exercised for the objective of making mathematical, quantitative measurement, the two amount to the same. Otherwise they do not.

"The procedure of *relating* mathematical notions to facts of nature" is Whitehead's phrase. And when you refuse to make *experimentation* an integral, functioning element in scientific procedure, "relating" is the only term you can use. Take experimentation *out of* scientific method and leave mathematics *in,* and the procedure of "relating" mathematics to the facts of nature blossoms into the great "mysterious" problem of modern and contemporary philosophy, the problem, in Russell's phrase "of the *application* of physics to the empirical world." And the piling of "mathematical logic" on top of "mathematical physics" only deepens and darkens the "mystery" and increases and intensifies the *in*solubility of *that* problem.

By leaving experimentation out of modern scientific method, there is also created a mysterious *historical* problem which deserves far more attention than it has received. It deserves in fact the utmost philosophical attention because even a full recognition of the historical mystery might serve to stimulate universal solution of the modern "scientific" mystery. For "the procedure of *relating* mathematical notions to the facts of nature, by counting, by measurement, and by geometrical relations and by types of order" is an exact description of what *Greek scientists tried to do.* This, precisely, is what Pythagoras stated and what the *Academy* under Plato carried forward to the Greek *end.* To say that Aristotle was a biologist "though he was not thereby ignorant of mathematics"* and that Aristotle turned Greek scientific thought away from mathematical measurement and into the classificatory procedure, is to ascribe to Aristotle an extraordinary influence and, moreover, of the kind he could not possibly have exercised over his fellow Greeks. "Following the Leader" is *not* a philosophic (or intelligent) game. And Greek scientists *were* philosophers.

The plain historical matter of fact is that the Pythagoreans, the mathematicians, were, with respect to the development of Greek science and philosophy, on the ground floor. They were the most closely-knit Brotherhood of Scientists-Philosophers of the Greek world; Plato's *Academy* was nothing more than their Athenian home, after being driven out of Croton in Sicily and elsewhere. The only comparable society of scientists were the physicians, organized by Hippocrates and they came

---

*In Aristotle's natural science (physics etc.), *relations* (geometrical and spatial, such as up and down) figure. The syllogism can *relate* but cannot handle *relations.* A "logic of relations" is one of the achievements of modern logic. Aristotle perhaps would be surprised by this novel development. But he would be certainly surprised to learn that his syllogism was taken as the instrument of *investigation* and not what it obviously is (and is only fitted to be)—an *auxiliary* to classification.

†The schoolmen when they were wrong were still wrong about qualities and qualitative behavior, and since what they didn't know couldn't hurt them they never tried to find out whether they were wrong. The Church got terribly hurt when someone told them they were wrong—as Copernicus, Kepler and Galileo began to do. (They still get hurt in the same way and for the same reasons.)

**Aristotle's Logic "*entirely* leaves out of account the interconnections between real things ... [It] *renders* an interconnected world of real things *unintelligible.* The universe is shivered into a multitude of disconnected substantial things. ..But substantial thing *cannot* call unto substantial thing." (*Adventures of Ideas,* pp. 169-170; italics mine.) But the schoolmen, by following this logic by the method of endless dialectic debate, exhibited "the inflexible rationality of thought"!

Science and the Modern World, p. 43. As a matter of fact, as some scholars are coming to realize, Aristotle was at least as good a mathematician as Plato, and the chances are he was much better.

later. Only superficial reading of history backwards (making Greek philosophers and scientists into sheepish "scholastic doctors" and Aristotle into the Church, the omnipotent shepherd of the sheep) can yield the conclusion that Aristotle deflected the course of Greek scientific thought out of "relating mathematical notions to the facts of nature" into the halfway house of "classifying" those facts. Rather must the case have been that the Greek mathematical development, as a procedure of investigating nature, quickly reached an impasse and Aristotle's Logic was the only way out. And for this there is conclusive proof.

The original Pythagoreans did try to "relate" *quantitative* measurement to natural facts. But they very soon had to change their whole mathematical business. For they discovered early that there was a "number" that wasn't a whole number—namely, the square root of two. *Before* the discovery of the square root of two, Pythagoras could have said as Einstein did in 1933: "Our experience up to date justifies us in feeling sure that in Nature is actualized the ideal of mathematical simplicity. It is my conviction that pure mathematical construction enables us to discover the concepts and the laws connecting them which give us the key to the understanding of the phenomena of Nature." *But after*, Pythagoras and the Pythagoreans (and all Greeks) were considerably shaken in their feeling of conviction. They *had* an ideal of mathematical simplicity and they stuck by that ideal. Whole numbers and the relations between whole numbers were alone *ideally* simple; they alone were Rational, the object of Pure Thought and the object of Pure Thought was alone Ultimate Nature. Hence the square root of two could not *be* an object of Pure Thought, could not *be* an actualization of Nature, could not *be* Rational. It was an inexpressible, an unthinkable, without any *Reason* in it, without any *Measure* in it—incommensurable, in fact. It was "without measure" and hence not Rational, but Irrational. Hence also, it could not possibly be used for "measuring Nature" or any "Ultimate thing" in Nature—for Ultimate Nature was a Logos, a Rationality, and all *real* actualizations in Nature were "wholes," "measurables," "rational numbers."

Though Pythagoreans differed among themselves in details, though Aristotle differed from Plato in details, though Greek biologists and physicians differed from all the mathematicians and logicians in details, *all* Greek scientists and philosophers, physicians, biologists and mathematicians agreed with each other in fundamental principle: the rational, the measurable, the logical, the reasonable (they all mean the some thing) is the "whole." And why did they so agree? *Because* the qualities, combinations of qualities (natural things) and relations between qualities and combinations of qualities which are directly observed in experience, are always "wholes." To be able to "relate" mathematical notions to the "facts of nature," when those facts of nature are *taken as is*, taken an "something given," as we are directly and immediately acquainted with them, as macroscopic subject-matter in primary experience—to be able to "relate" mathematical notions to these facts of nature, the mathematical notions must be *qualitatively like* the facts to which they are to be "related": they also must be *qualitative wholes*.

It is a fact that with the "facts of nature" as directly experienced there is very little that can be done in the way of "relating" mathematical notions to them.* And

there is very much less that can be done when you hold to your ideal of mathematical simplicity that Nature *must* actualize. Hence the Greek scientists, mathematical-philosophical, and logically-mathematical, were, in their procedure of relating mathematical notions to the facts of nature, rapidly reduced to the level of *observing* (spectatorially beholding) such shapes and proportional relations of shapes and sizes as they could, and of classifying and systematically analyzing and developing their static and "whole" relationships. (Euclidean geometry and Eudoxian theory of ratio and proportion). The syllogism is the novelty which Aristotle contributed. But the fundamental procedure of his logic—observation and analytical-synthetic classification—is in essentials precisely what Greek mathematics had come to. Aristotle's Logic, from start to finish, is a logic of "wholes." Everything else is not *in* logic (or science), because nothing else is an eternal and immutable part of the Logos of Ultimate Nature. Everything else is an "accident": not an actualization of *Rational* Nature, but a manifestation of *irrational* Matter. And among these "accidentals"—manifestations of matter, not realizations of Form—are quantative relations, naturally.

Now the great historical mystery is this: if Greek science is *not* science as we understand it (Aristotle excepted, according to Whitehead) *how account* for the extraordinary difference between the success of modern science and the failure of Greek science in the procedure of relating mathematical notions to the facts of nature?

> Why did the pace suddenly quicken in the sixteenth and seventeenth centuries? . . . Invention stimulated thought, thought quickened physical speculation, Greek manuscripts disclosed what the ancients had discovered. (*Science and the Modern World*, p. 8)

If it is true, as Whitehead avers, that Greek science is *not* science as we understand it, then, surely, the discovery of Greek manuscripts could not have exercised a *positive* determining influence in *creating* the beginnings of modern science. Rather must the case have been that by the time of Galileo, the *practice* of modern science had gotten so well on its way that no *theory* of science, not even of the Greeks, could throw it off its *practical course*. (Though it could throw the *theory* of modern scientific practice off its natural theoretical course. Which it did.) When Whitehead writes further on:

> The history of the seventeenth century science reads as though it were some vivid dream of Plato or Pythagoras. In this characteristic the seventeenth century was only the forerunner of its successors (*Ib.*, p. 48.)

It is historically and scientifically impossible to agree with him. Since Pythagoras and Plato, while living, devoted all their energy to finding ways and means of circumventing, of stopping, precisely the sort of mathematical development (with respect to the facts of Nature and with respect to mathematics itself) that took place in the seventeenth century, had they read the science of that century and its successors it would have seemed to them, if the truth must be told, like some vivid nightmare. They would be much more inclined to agree with Whitehead's other statement about the seventeenth century: that is was a return to the contemplation of brute fact. Except that they would want to add: it was *not contemplation* and *not of fact*. "Facts" of Nature for the Greeks are rational, and when "contemplated" are seen to be such. Seventeenth century science, for them, would be a brutish distortion and mutilation of facts of nature. It would be a travesty and outrage of "mathematical contemplation of

---

*Even now—statistics apart. And to speak of 7 1/8 persons per square mile, etc., would have *horrified* the Greeks. What madness! What insane irrationality! What sacrilegious defiance of the Logos of Nature! In sum, how dreadfully *un*scientific, *un*mathematical.

nature." And hence for precisely the *opposite* reason—because of its mathematical aspects—they would agree with Whitehead in saying the Historical Revolt was anti-intellectualist, anti-rationalism.*

XIV

The double mystery—of the impotence of Greek mathematical science of Nature and the omnipotence of modern mathematical science of Nature—is solved at one and the same time when "the topic of experiment" is introduced into the theory of scientific method. Greek mathematics had a very brief and not very glorious career as an instrument of investigation of Nature, because Greek mathematical scientists tried to "relate" mathematical notions to the facts of nature, taking those facts as directly experienced, *as is*. Modern mathematics, on the other hand, has had a glorious and ever more wonderfully fruitful career as an instrument of investigating nature, precisely *in so far as* modern scientists abandoned the objective of "relating mathematical notions to the facts of nature" (taken as is) and began experimentally changing, controlling as-given facts of nature for the sole objective of instituting mathematical relations *between* the facts that were the resultants or consequences of their experimentation. And in pursuit of this dominant objective of establishing quantitative relations between facts, modern science has *in practice* more and more abandoned all pretense of holding to an "ideal of mathematical simplicity" and has less and less observed scruples in experimentally tearing apart the "given" (directly experienced) facts of nature and in experimentally bringing them into experimentally new relations. To the extent that science from Galileo onward *integrated* mathematics in experimental procedure, brought it under the control of experimentation and used it for instituting and formulating *relations between experimental findings*, it was successful and fruitful, and, to the extent that it did not, it blocked, retarded, distorted, obstructed the advance of scientific knowledge. The progress of physics from Galileo and Newton to Michelson and Einstein is the progress of effecting a more complete integration of experimental findings and mathematical formulations, bringing the latter under control of the former.

It was a consistent practice with Newton whenever there was a conflict between the then known experimental findings and the theoretical demands of the then known mathematics, always to enforce the latter. Because Newton did not summarily expropriate the basic rights of experimentation all at once, by a single comprehensive decree, but invariably cautiously argued them away as each specific occasion arose by means of the theory of mathematical priority and superiority, Newton has, perhaps not inappropriately, been held up as the paragon of purest British intelligence. However, it was the omnipresence of the "Greek formula" in modern scientific mentality, rather than the force of Newton's arguments, that established his "method of compromise" as the canonical procedure for mathematical-physicists

everywhere.*

Newton's "method of compromise" and the "scientific world" of eternal billiard-ball atoms, Absolute and separated Space and Time, and immutable (invariant) mathematical laws of Nature he set up by means of his method endured for approximately two and a half centuries. That the method was not disavowed sooner is not a tribute to its probity; it is a tribute to the overwhelming force of the Greek philosophic formula. That the "scientific world" stood up for as long as it did is not a tribute to its strength: it was kept going, at enormous intellectual and social price by the almost infinite ingenuity of modern scientific minds. And when new mathematical formulations enforced by new experimental findings could no longer be brought within the Newtonian system with any consistency at all, ingenuity lapsed into ingenuousness:

> In time, most physicists came to disbelieve in absolute space and time, while retaining the Newtonian technique, which assumed their existence. In Clerk Maxwell's *Matter and Motion*, absolute motion is asserted in one passage and denied in another, with hardly any attempt to reconcile these two opinions. *(The Analysis of Matter, pp. 14-15.)*

That the Newtonian reign should ever come to an end was simply inconceivable to Newtonians. The basic structure of the Newtonian system was eternal and immutable. If the alleged empirical certainty of indestructible Newtonian atoms ("the imperishable foundation-stones of the universe" according to Clerk Maxwell) had, by radioactivity and Rutherford's experimental bombardment of the atoms, become slightly less than absolute, then all the more reason for gradually shifting the eternality and immutability of the system back onto the original ground of the transempirical absolute certainty of Mathematics (the "invariant laws"). As a "deductive system" after all, the Newtonian had a high degree of perfection. But instead of solving the problems presented by experimental findings, the Newtonian method kept piling them up ever higher. And it is the last straw that breaks the camel's back. In this case, the black-bands in Michelson's interferometer.

What happened to the "eternal basic structure" of Newtonianism, to its immutable cosmological framework reputedly riveted "scientifically" to the three absolute pillars of Space, Time and Matter by eternally true and eternally enduring, non-corrodible struts and bolts of pure mathematics, every one knows. By reversing the Newtonian policy of giving to mathematics absolute authority to determine the meaning and to control the theoretical development of experimental findings, that is, by establishing the forthright and uncompromising procedure of giving to experimental findings first the authority to determine the meanings of mathematical-physical concepts and then the final authority to control their development and formulation in all respects relevant to the science of nature, Einstein accomplished in *scientific practice* the full enstatement of *experimentalism*. The verified success of Einstein's reversal of the Newtonian policy has demonstrated beyond all doubt and with a precision science alone is capable of, that for three hundred years Newtonianism had been driven from one extremity to another, and had latterly been forced to live ever more precariously from experimental hand to theoretical mouth

---

*"It follows from the Deism [of seventeenth century scientists like Newton] which is part of the whole conception, that the Laws of Nature will be exactly obeyed. Certainly, what God meant he did. When he said, Let there be light, there was *light* and not a mere imitation or a statistical average." *(Adventures of Ideas*, p. 145.) Without Deism, the Greeks believed light is *light* and not an imitation or statistical average. But for *scientific* knowledge of light, light *is* a "number of vibrations," a quantitative formula whether statistical or not. Hence the nightmare.

*"Except the blind forces of Nature nothing moves in this world which is not Greek in its origin." Sir Henry Sumner Maine's *words*, but representative of practically universal nineteenth century scientific and philosophic belief.

because of one basic methodological fault: it had literally upset the true relation between experimental findings and theoretical (mathematical) formulations. It had been living methodologically upside down.

Physics can be pursued as "pure mathematics" but it is not as pure mathematics that physicists have pursued it. If Nature actualizes the ideal of *mathematical* simplicity, and the pursuit of this ideal is the historic pursuit of modern physics, then physicists have gone about their pursuit in ever wilder and stranger ways. To discover the ideal of *mathematical* simplicity we should study mathematics, symbolic logic, perhaps even Logical Positivism, so that we may be able to settle upon that ideal, for it reveals itself only within a system of (mathematical) symbols. But to discover the simplest mathematical formulation of the complexities experimentally produced, though we must still study mathematics, our problem is significantly different. The ideal of mathematical simplicity in modern physics is the ideal of the simpl*est* formulation—no matter how complex from the standpoint of the ideal of mathematical simplicity—the actualities of Nature as experimentally discovered will allow. The ideal of the simplest is not the ideal of simplicity.

From the standpoint of pure mathematics neither the continuity nor the *direction* of change of modern physics can be accounted for. In multifarious ways, the system of modern physics has expanded, and from Newton's time onward physicists have certainly tried to preserve its theoretical systematic face. But the expansions, even within the limits of the theoretical system, were not in response to demands made by theoretical principles of the system. The expansions and revisions were necessitated by the need for bringing into the system new experimental findings as they were mathematically formulated. When it is forced to, mathematical physics keeps its theoretical face by adding supplementary laws, and even exceptions. In common-sense practice, we keep adding new exceptions to the old rule and think nothing of it. But in science, exceptions are scandalous, and the practice observed is that of reformulating old rules so that the exceptions will be included, and cease to be exceptions. The ideal of having one system in which all laws and rules belong is very powerfully operative with theoreticians. Just as Euclid took Greek mathematics, as is, and systematized it in accordance with certain principles of codification, so the science of physics, as it at any one time is, or any body of knowledge, can be taken as is, and formally arranged, systematized and codified according to certain rules, principles, methods and standards of pure mathematics or mathematical logic. That such efforts are important and valuable cannot be doubted; and that they involve dangers also cannot be doubted—witness Euclid's miracle. The practical emphasis on theoretical system in present day physics is a consequence of the fact that it is in many fundamental respects absent. Which is as good a demonstration as any that the *direction* of scientific change in physics is not due to considerations of "pure system."

The progress of physics from Galileo and Newton to Michelson and Einstein is the progress of effecting the complete integration of experimental findings and mathematical formulations, by bringing the latter under control of the former. The laboratory physicist does the experimenting; the theoretical does the theorizing. This is a social division of labor, not a separation of the one activity from the other. Michelson's experiment required the work of prior theoretical physicists just as much as it required the general advance in technological design and manufacture. Without theoretical developments and mathematical formulations, the interferometer experiment could neither be nor be conceived. Similarly with Einstein's theoretical formulation: without the prior experimental developments of physics, it could neither be nor be conceived. The *problems* of the theoretician are determined by the results obtained in the laboratory; and the solutions of the theoreticians have to solve *those* problems. Einstein had to develop scientific ideas or meanings that would satisfactorily or successfully solve the problem which the results of Michelson's experimental apparatus raised.*

In all theoretical physics, there is a certain admixture of facts and calculations; so long as the combination is such as to give results which observation confirms, I cannot see that we can have any *a priori* objection [to the "heterogeneity of space-time in Einstein's system]. Dr. Whitehead's view [which objects to the "heterogeneity"] seems to rest upon the assumption that the principles of scientific inference ought to be in some sense 'reasonable.' Perhaps we all make this assumption in one form or another. But for my part I should prefer to infer 'reasonableness' from success, rather than set up in advance a standard of what can be regarded as credible." (Russell: *The Analysis of Matter*, pp. 78-79.)

To attempt to assess the contribution of laboratory experimentation, taken by itself and of mathematical formulation and systematization, taken by itself, is to attempt the impossible. For the fruitfulness of modern scientific method is dependent upon the interactive union of the two. Now one, now the other, may be temporarily dominant in a specific case. But what gives continuity to modern scientific activity is their continuous interactivity; and what gives the direction to the continuity is the exercise of control by experimentation as the final authority for testing theory and pronouncing upon the validity of the mathematical formulations.

Aristotle and Plato die hard. The work of the experimental physicist and theoretical physicist, though interlocked and interwoven, can be for certain purposes distinguished. But the Greek Formula is not satisfied with making them distinguishable; it must make them separated and separable. "It is obvious," writes Russell, "that a man who can see *knows* things which a blind man can not know; but a blind man can *know the whole of physics.*" The whole of physics! Experimentation and all that laboratory experimentation involves has nothing to do with the "science of physics" and is not necessary for the knowledge thereof! For a blind man can "know" the abstract, mathematical propositions of "physics," its formulas and numbers, and that is *all* that *scientific* knowledge of the physical world is! Could a race of blind men *create* modern and contemporary physics? Could they *come* to know? Could they even find out which abstractions and which mathematical formulations and which entities and propositions they should select from the mathematical heaven as makings for their "deductive system of physics"? But why ask the Greek Formula *these* questions. *How* we come to know is a matter of trivial *history* and has nothing to do with the *nature* of the case. Knowledge has nothing to do with knowledge-getting; knowing has nothing to do with the process of coming to know. Knowing is the contemplation of the

---

*As Whitehead excellently remarks: "On the whole, it is better to concentrate attention on Michelson's interferometer, and to leave Michelson's body and Michelson's mind out of the picture." (*Science and the Modern World*, p. 173.)

object of knowledge. And contemplation is all the knowledge thereof.

In the actual conduct of scientific inquiry, the full and unhampered interactivity of mathematical thinking and experimental doing is now an accomplished fact. Leave out the element of experimental doing, and no matter what other elements you bring in, and from where and how many, the creation and development of modern science become an inexplicable mystery, an old-fashioned miracle in fact. Especially mysterious and miraculous does modern science become when the element of mathematics is made the determining one in its history and nature. Not that mathematics can or should be left out of modern account. Any more than the writer of *Hamlet* can or should leave out Ophelia. To compare the role of mathematics in the history of thought (and also in the history of science) to the part of Ophelia "is singularly exact. For Ophelia is quite essential to the play, she is very charming—and a little mad.''* But without Hamlet there is no modern play at all. Though Hamlet may sometimes make Ophelia desperate, without him she goes completely insane. The madness of mathematics is not an inherent characteristic; it is a consequence that results from failure of union with experiment. And the "divinity" of that madness—is just Plato's story. Mathematics is no more mad and no more divine than any other instrument of investigation and communication, than any other system of ideas; and when brought under the control and direction of experimental doing mathematics is as sensible in experience as the rest. And without the direction and control of experimental doing, when disunited and separated from practice, all ideas become mad. As the great Greek physician said: "All things are alike human or alike divine—it makes no difference how you call them."

The method of modern scientific inquiry is the method of experimentation: the functional integration of theory and practice. Separate and divorce theory from practice and you make the history of thought unintelligible and the progress and nature of modern scientific knowledge one unending and ever-increasing irrationality. But unite the two in your theory of scientific method as they are now completely united in the conduct of scientific inquiry and the unintelligibility disappears, and the nature and course of modern science become clear. We then see that the work of Galileo was not a development but a revolution; and the work of Einstein is not a revolution but a development. For the abiding significance of the work of Einstein is that it scientifically clarifies and fully enstates the meaning of the work of Galileo: that experiment is a method for developing theories and establishing evidences for theories, for bringing the findings of practice and the formulations of theory into continuous interactive relation the consequences of the interaction being scientific knowledge. For in this way experimental evidence continuously controls the formulation of the law and prevents it from ever cooking the facts.

By freeing the experimental method from the arbitrary and distorting limitations of "pure theory"—by freeing it from the operations of the Greek Formula—Einstein has made possible the full realization and actualization of the method of intelligence in the technical scientific domain. But the spirit of the Greek Formula is still actively abroad in the philosophic and cultural land, shackling the freedom of intelligence in the modern world.

---

*Science and the Modern World, p. 31.

Galileo's method of breaking through the self-enclosed circle of refined objects of reflection was not a *specific*. It was not a remedy capable only of breaking the magical spell of the Aristotelian-medieval dialectic "system of natural science." What Galileo discovered was a *general* method, available and adaptable for use by all, and when used proves a competent remedy against the circular charm of any dialectic-logicalism, any self-involved system of refined objects of reflection, no matter what the area or field of the system, no matter what the enclosure may be and by what name it is called—physics, chemistry, biology, psychology, economics, sociology, ethics, esthetics, religion, theology, philosophy, logic. Galileo's method is thus universally competent, not because it is itself a piece of counter-magic, an omnipotent *word* or *saying*, but because it is a quite thoroughly natural *deed* or *doing*.

The method of *beginning* with gross, macroscopic, crude subject-matter in primary experience performs in the conduct of philosophic inquiry qualitatively the same function as is performed in the conduct of scientific inquiry by the method of beginning with the subject-matter revealed in the laboratory experiment.*

The emphasis falls on *beginning* and cannot fall too hard. Dewey, who begins with the gross and macroscopic, does not *stop* there. That's where he *starts*. *From* there he goes *into* the realm of the refined, derived objects of reflection, and while in that realm, he is as analytical, dialectical, argumentative, ratiocinative, technically logical, logically synthetical, abstruse and abstract—*i.e.* to sum it up—as "purely theoretical" as the veriest philosophers of the opposite methodological school. But there is this enormous difference between Dewey and philosophers of the opposite method† even in the respect in which they most closely approach or cross each other: for Dewey, his dialectical, ratiocinative or formal-logical work is an *interval* or phase of his complete philosophic undertaking. Just as he does not begin, so he does not stay with the refined objects of reflection. Whatever subject-matter in primary experience he started from, he started *from* that subject-matter *because* that subject-matter raised a *problem*. The *objective* of his technical-philosophic excursion, or his formal-logical work, is to solve that problem. Hence, to be *through*, he must get back where he started from; to be through with that *philosophic* job the refined objects of reflection he has on hand after doing all the formal-logical work, *must lead back into* the subject-matter of primary experience, the gross and macroscopic subject-matter which constituted the starting point, the point of origin, of the inquiry. When they do so lead back into the gross and macroscopic subject-matter, then and only then does *Dewey* know that *that* philosophic task is done, for that leading back into the macroscopic subject-matter is the final or ultimate

---

**The Tower of Pisa is as much a piece of laboratory apparatus as a micrometer—cruder but qualitatively or functionally the same. Galileo's telescopic lens was comparatively as crude when compared with the lenses (photographic-telescopic) used by astronomers at Sobral. Nevertheless Galileo's telescopic observation in 1610 was as humanly dramatic and scientifically as significant and conclusive as the observations in 1921. Galileo's astronomical observation took one part of the Copernican theory out of the realm of theoretical speculation, and the observational expedition to Sobral could do no more for the Einstein theory.
†The fact that the up-to-the-minute practitioners of the opposite method use the symbolisms of mathematical logic or logical positivism and Dewey does not in a *technological* difference and not a difference in *fundamental* methods of philosophy.*

experimental test of the validity of the philosophic solution which *he*, in his professional capacity as philosopher, can give.

The emphases, in the last sentence, on "Dewey" and "he", are made not to call attention to the emergence of differences between experimentalism in philosophy and experimentalism in science, but to sharpen perception of the *identities* that obtain.

It is too obvious, I take it, to need any argument that the philospher *qua* philosopher—or in his strictly professional capacity—in aiming to become scientific (experimentalist) can aim to become so in a manner comparable to the scientific (experimentalist) *theoretician*, not the scientific (experimentalist) laboratorian. It is absolutely essential, therefore, in examining and evaluating any method or any element in a method that claims it can make philosophy *scientific*—in the sense of science as we understand it now—it is absolutely essential to keep constantly and centrally in mind that the philosopher, under this conception of science and scientific inquiry, can do only *half* of the total work of philosophic inquiry, and the more strictly only half, the more fully or completely scientific philosophic inquiry is. Failure to keep this central in mind, is partly responsible for the argument over "scientific method in philosophy" going on forever in circles of wilder and wilder amaze. Failure to keep this in mind is to be guilty of using, under guise of examining the validity of a method for making philosophy scientific, a conception of science that violates the fundamental nature of science *as we understand it now*. And hence violates also the fundamental nature of any philosophy that could possibly be scientific. It is to be guilty of using the Greek Formula again, uncritically and without acknowledgment, but this time in the form of the *absolute standard* that can automatically measure any method and infallibly determine whether or not it is capable of making *philosophy* scientific.

It is also too obvious, I take it, to need any argument to prove that the scientific theoretician—for example, the theoretical physicist—checks and rechecks every argument in his theory before *he* makes it public. After it is published, it has to undergo (and is thoroughly subjected to) public examination. And to be acceptable and accepted by the scientific public, theoreticians and laboratorians both, it has first successfully to pass a series of laboratory tests which the theoretician, as theoretician, couldn't possibly make.

However, there is always one kind of *practical test* that the theoretician not only can perform, but must perform and constantly does perform. Einstein, for example, *knew*, before ever he started, that his line of reasoning, his mathematical calculations, his formulations of refined objects of scientific reflection had to lead back into the black-bands of the Michelson interferometer.* He knew that any theory had to pass *that* experimental test. He knew that his theoretical-physics job was not done unless and until his theory did that. Any system of refined mathematical-physical objects of reflection that did not lead back into the black-bands but led away from them—that led to the conclusion, say, that the bands were not black nor bands—Einstein knew beforehand would not be worth the paper it was written on, no matter how infinitesimally small the piece. When his theory led him back into the black-bands, *Einstein* knew, as far as *he*, as theoretician could know, that *his* theoretical job was done.†

Einstein—as any theoretical scientist—knew *beforehand* that his theory must terminate in the consequences of the experiment as experienced in the laboratory, because the control of theoretical solution by laboratory consequences is established in the conduct of scientific inquiry. (Completely established, as we have seen, by Einstein's own work). The course of theoretical elaboration and solution in scientific inquiry is controlled by the subject-matter as experienced in the experiment. If philosophic inquiry is to become scientific, it too must be controlled in a qualitatively identical way. In proposing any methods to make philosophy scientific, or in reaching a judgment with respect to any methods proposed with this end-in-view, it is necessary to establish whether the method under evaluation, or undergoing judgment, does or does not enable the philosopher to be controlled in his inquiry in a way *qualitatively or functionally comparable* to the way in which the theoretician in science is controlled in his inquiry. The method of beginning with the gross and macroscopic subject-matter in primary experience performs this function. It is the beginning of experimentalism in philosophy, the beginning not everything. But it is the necessary beginning, and because necessary, is sufficient to disqualify as unscientific or anti-scientific any method of philosophic inquiry that begins, or pretends to begin, otherwise.

XVI

Scientific inquiry is "controlled inquiry."* To control, it is necessary to be controlled; to exercise *control over*, it is necessary to be *controlled by*.† *Controlling* without *being controlled* is possible only to creatures who are impotently omnipotent; being controlled without controlling is possible only to creatures who are omnipotently impotent. Both such kinds of creatures or beings are febrile figments of diseased imaginations, the one rationally indistinguishable from the other, except in the respect that each imagines the absurdity or impossibility of the other in reverse.** In the activities of Nature, as in the activities of human nature, *controlling* and being *controlled by* are each indispensable for the other, are interdependent or continuously interactive.

When science is taken in the gross and macroscopic, the general consequences of the interactivity of *controlling* and being *controlled by*, as that interactivity goes on between the theoretical and laboratory functions within inquiry, can be readily enough seen and in their generality easily enough denoted. This Whitehead does when he points out that "Every scientific memoir in its record of the 'facts' is shot through and through with interpretation."†† And Russell does the same when he points out that "In all theoretical physics, there is a certain admixture of facts and calculations."*** Each of these statements in its own way unambiguously denotes (points to) the *consequences* of interactivity, within inquiry,

---

*This is not all it had to lead back into, but it is enough for the purposes here.
†See footnote above.

*Dewey, *Logic: The Theory of Inquiry* (1938), p. 101.
†Dewey, *Essays in Experimental Logic*, p. 94-95, 176-178.
**The best theologians of the Church—following Aristotle and Plato—have realized that God cannot be so omnipotent that He is never *controlled by* anything. "What God *meant* He did." His doing was therefore *controlled by* His meaning. Since His meaning is Eternal and Immutable, it is *His* complete undoing when He is made to undo anything He has done. Miracles destroy God's nature without saving the world. For further discussion of this problem in terms of theological "miracles," see my Introduction to *The Philosophy of Spinoza* previously referred to. The discussion of the same topic, in terms of scientific law and continuity in Nature, comes into the argument further on.
††*Process and Reality*, p. 22.
***The Analysis of Matter*, p. 79; see fuller quotation, *ante*.

between laboratory fact and theoretical interpretation. The two statements quoted, separately and together, point to the general fact that within scientific inquiry the laboratorian is controlled by the theoretician and the theoretician in controlled by the laboratorian. And of course, in the respect that the one is *controlled by* the other, the other is, in that same respect, and from *his* standpoint, *controlling* the one. The *controlling* and *controlled by* do not take turn and turn about. They operate criss-crosswise and interweave.

The theoretician and laboratorian, although they are each controlled by the other's results—and in this respect may be said to be similar—are *controlled* by them in radically different ways, or to radically different ends—and in this respect they are basically dissimilar or functionally unlike.

The theoretician in searching for the solution of a problem taken from the laboratory is *controlled by* the facts the laboratorian obtained; that is, he is controlled by the consequences of the inter-actions which were set going in and through the organization of the physical-experimental apparatus in the experiment. And the laboratorian, in searching for an experiment that will put to the test the solution of a problem taken from the theoretician is *controlled by* the solution in constructing his apparatus and organizing the experiment. Obviously, and in both cases: the theoretician must solve *that* problem; and the laboratorian must test *that* solution. Precisely upon this interchange or cross-weaving of control (controlling and being controlled by) depends the existence and maintenance of the interactivity between the theoretical and laboratory functions in scientific inquiry. And the more precisely this interchange or cross-weaving of control, the greater and finer the precision in the results of scientific knowledge.

So much for the general similarity. Now for the specific and radical difference. The theoretician must *solve* that problem *as taken;* his solution must explain the facts *as they were found* in the laboratory—whence they were "taken" by him for solution or whence he received them as a "gift" ("something given," a *datum*). There are no limitations placed beforehand upon the theoretician* as to *how* he explains or solves the problem. He can make his solution simple or complex, new or old; but no matter how simple or how complex, how old or how new, one thing is absolute and final; his solution *must terminate* in those facts *as given or taken;* the *outcome* of his solution must leave those facts as found. If there are any methodological ultimates in scientific inquiry, then this is one of them.

The laboratorian is under a radically different obligation with respect to the solution "taken" or "given." He is under obligation to put it to the test and *not* to prove it right (or wrong). The *outcome* of the laboratory experiment is not something the laboratorian is under scientific obligation to contrive. Very much the opposite: his contrivances (apparatus) must be such that the *outcome* for the solution (as to whether it is right or wrong, correct or incorrect, true or false, acceptable or inacceptable scientifically) will be settled or determined by the *consequences* of the experiment. Michelson's

*I am speaking of the *current* practice in science. It was not always so nor do all contemporary philosophers of science think it should be so even now. See *ante*, p. 51.

interferometer put the ether-theory to the test; it did not prove it right. Michelson, in constructing the inter-ferometer and in organizing the experiment had to be (and was) *controlled by* the ether-theory; otherwise his experiment would have been irrelevant, or beside the point. But it would have been infinitely worse than an irrelevant experiment, it would have been a fraud, if Michelson had devised an instrument and organized an experiment so that the *outcome*, the *consequences* of it would be predetermined one way or the other. If Michelson had cooked an experiment for the sake of producing facts that would invalidate the ether-law (theory, solution), the outcome of his experiment would not have constituted a theoretical problem for scientists to solve; the problem scientists would then have been faced with would have been the very practical one of publicly disqualifying Michelson. Then, Michelson's body and Michelson's mind would have been the whole pseudo-scientific picture, and Michelson's interferometer would have made no scientific picture at all.

The *outcome* of the laboratory test does not have to prove the theory (solution) right; and it does not have to prove it wrong; the outcome may be such as to prove that the theory cannot as yet be put to decisive laboratory test. In which latter case, the issue as to the validity of the theory (solution) continues unsettled and undetermined, making further tests or further organizations of experiments necessary; and to accomplish such further laboratory experiments further elaboration and reformulation of the theory (solution) may be needed.

Experimentalism in science does not mean that every theory (solution) has to be such that it can be put to the decisive laboratory test immediately, or in its first formulation. The ether-theory (the solution of many scientific problems) had been kicking around in modern science for two hundred years or more before it was brought to the laboratory test. The amount of theoretical formulation and re-formulation that went into its development was simply enormous. Only because of the general advance in science during this historical period, advance in theoretical formulation and practical methods of laboratory experimentation, was Michelson (his genius thrown in) able to devise and instrument and organize an experiment that put the ether-solution to critical test. The only requirement fundamental in experimentalism, whether in philosophy or science, is that any solution to be acceptable as a *solution*, as a piece of scientific *knowledge*, must first pass the laboratory test. Only when at last it does or does not pass a decisive test, is it to be accepted as a *known* solution or rejected because *known not* to solve.

If there are any methodological ultimates in scientific inquiry, then this is one of them: the outcome of the laboratory test is *not* determined by the apparatus as organized in the experiment *taken by itself*. When the outcome is thus determined, you have either the honest manufacturing of contrivances or machines—which is not a case of *inquiry;* or else you have the dishonest manufacturing, or faking of evidences. Which is also not inquiry—though frequently called so—and now ever more frequently in certain parts now unknown. The outcome of the laboratory test, when the test is part of inquiry, is determined by the *consequences* of the natural subject-matter working in and through the organization of the experiment. This methodological ultimate of laboratory procedure is of course the *original* or the basis of the methodological ultimate in theoretical scientific procedure. Hence the primacy and ultimacy of laboratory experience in determining the *total* course and controlling the direction of inquiry.

Note. The interactivity of *controlling* and *controlled by* within the process of scientific or "controlled inquiry" is fundamental in Dewey's analysis of the logic of inquiry. Inquiry originates in a problem or difficulty and terminates in a test. We are *controlled by* the problem at the beginning of any case of inquiry and by the test at the ending. The phases through which the process of inquiry passes are phases of passing from one interactivity of *controlling-controlled by*, to another, the achievement or consequences of one phase being carried along into the next, giving any case of inquiry its self-corrective and self-expansive character. The complete *controlling*, with respect to *that* case, is the final issue or consequence of the complete process when the inquiry is successful. When the problem is solved, then we do control *that* problem and are no longer *controlled by* it. That is what solving means. We have *done* and so *know* how and what to do.

However, because the factor of *controlling* is the critical turning point in the history of scientific inquiry, and the determining issue in the philosophic controversy or debate, in the course of Dewey's specific discussions the factor of *controlled by* is often pushed into the background. Because it is in the background, it does not follow that it is not working *in* Dewey's discussion. It is just working in the shade. The reader should always bear this in mind. For with a few very rare exceptions Dewey always does. And where he does *not*, all the more reason why the reader *should*.

There are rare cases, portions of discussions, where *controlled by* gets pushed so far into the background that the consequences of its working are hardly appreciable, they have practically vanished. There is one passage included in this book in which, as far as I can find out, it has to all intents and purposes vanished altogether. And as far as I know it is the only passage in all Dewey's writings.* It's a *rara avis* which it would be a shame not to let the reader catch for himself. However, even this rare bird is only in passage, not in stoppage. And therefore having caught the bird, the reader should not stop but go on.

The term "controlled inquiry," like all terms, carries its dangers within it. Concentrate on and magnify the dangers, and the dangers can easily be converted into seeds of its own destruction. Since all terms carry dangers within them, by this ferocious method of conversion, the process of intelligent inquiry and rational life can be made into a passage from one destruction to another. If the reader likes to live that sort of "heroic life," he is welcome to do so. And if he further wants to call *that* "intelligent" (of a superhuman variety of course), there is no way of stopping him from doing so, nor would it be worth the effort to stop him. But it is *not scientific* intelligence.

I do not, of course, in any sense wish to imply that Dewey's lapses in the course of discussion are *solely caused* by any one term or combination of terms. Terms are themselves consequences, not aboriginal or "metaphysically primitive" causes. Terms also have their further consequences when used in further discussion or inquiry. They are therefore not entirely without blame. But to put all responsibility for error on terms is nonsensical. The reasons for Dewey's lapses are complex and many. Dewey, like all human thinkers, is quite human. And like

*I don't vouch that it is the only one. I go by the fact that it took me so by surprise, was so novel an experience to me. After that novel experience, I didn't go back and make a statistical research through all Dewey's writings. If the reader finds more than one in the text of this book, the keener he or she.

all human beings, in the course of a specific argument he is sometimes carried along too far in that course.

XVII

That the theoretician and laboratorian may be one person is of course to be understood all along. It usually happens that the laboratory genius is not the same person as the theoretical genius, just as the great musical composer is seldom the great performer. But this is as it may be. To speak of the theoretician and laboratorian (in physics or any other field) is a handy way of speaking of the theoretical and laboratory *activities* functioning in scientific inquiry. And it is also in part necessary, and in part eminently advisable to speak in this way because scientific inquiry is a human activity, undertaken and carried through by human beings. Whether the laboratorian and theoretician are, in any given case of inquiry, one person or two or many is another consideration. Because of the historical continuity of scientific inquiry— the involvement of problem in problem and solution in solution—they are always many, very many, neither one nor two. However, the fundamentally important point concerning the logical analysis of inquiry is that whether they be one, two or many, the laboratorian must, to some extent, be a theoretician and the theoretician must, to some extent, be a laboratorian because in the function of each, the activity of the other is internally involved.

Within inquiry, the theoretical and laboratory activities are constantly undergoing integration. When, with respect to any one case or problem inquired into, the integration is finished or completed, then in *that* respect inquiry reaches its *logical conclusion* and an item of tested, grounded, verified knowledge is added to the scientific store. But the macroscopic fact that scientific inquiry is still going strong, and going stronger today than at any other period, is all the evidence needed to prove that inquiry is in historical process, that the method of controlled inquiry does not deliver a once-and-for-all system wherein theoretical and laboratory activities are with finality integrated, wherein they are with respect to each other "under control" in the sense in which the military speak. The method of controlled inquiry is a method of controlling, a method of integrating, and as the method is pursued it systematically effects further integration between the two, progressively moving as each progresses.

Controlled inquiry—the method of experimentation— extends from the laboratory to the theoretical study and *includes* them both. The one without the other is not scientific inquiry as we understand it now. Experimentalism in science is impossible without the laboratorian and it is also impossible without the theoretician. Both are experimentalists, each performing, within the total process of experimental or controlled inquiry, a distinctive and distinguishable, but not separated and separable share of the work.

To some readers, perhaps, it may still seem that the last statement begs the whole issue. Such readers may think that if you start by putting the scientific theoretical and laboratory activities within scientific inquiry, it is not too amazing that you should find them each performing a distinctive but not separated, a distinguishable but not separable share of the work of scientific inquiry. It would be really too amazing only if you found them doing otherwise. This criticism would be valid and conclusively destructive *if* it were the case that *the statement puts* the theoretical and laboratory activities within inquiry, and *if* it were also the case that the statement, having first put them there, then offers itself as an *explanation* or accounting of their presence there.

If the statement were these two things, then, perhaps, it would be a "logical positivistic" statement on the order of Carnap's Logical Positivism.* But whether or not the latter, it would with certainty be a statement exemplifying the old dialectic whereby it is *explained* that opium puts to sleep *because* of its sleep-inducing powers. But the statement is neither of these two things.

As for the first, the statement does not *put* the scientific theoretical and laboratory activities any place. It *points* to where they are found. The statement is not a "definition" of what the term "scientific inquiry" is going to be used as meaning. It is a *report* of what scientific inquiry is existentially found to be. The statement is fundamentally denotative in logical function. That Dewey may possibly be a biased reporter, and his report be a product of his bias, is not at all too sinful a thought to harbor against any philosopher. Hence the great value of the reports handed in by such competent philosophers as Whitehead and Russell, who, though not without bias either, are certainly not biased in favor of Dewey's logic. "Every scientific memoir in its record of the 'facts' is shot through and through with interpretation." "In all theoretical physics, there is a certain admixture of facts and calculations." These two reports of eminently competent individual philosophers, of a competence within the technical fields of science and mathematical physics far superior to Dewey's, corroborate Dewey's report to the hilt. The "interpretation" that shoots through the memoirs of facts is *existential evidence*, gross and macroscopic, that *within* the results of the scientific laboratorian the consequences of the work of the scientific theoretician are *found*. The "admixture" of facts in the calculations of all theoretical physics is *existential evidence*, gross and macroscopic, that *within* the results of the scientific theoretician the consequences of the work of the scientific laboratorian are *found*.

Of course, Whitehead and Russell are not the only other philosophers who report the same findings as Dewey. But the multiplication of reports is of no philosophic value. As far as the particular report under discussion is concerned, all philosophers report the same. If they didn't there wouldn't be any philosophic controversy such as the modern and contemporary world displays. That's what all the shooting is precisely about.† The report is general.

As for the second point—whether Dewey's report is an explanation or offers itself as an explanation of what it reports. The statement "within inquiry, theoretical and laboratory activities each perform a share of the total work of inquiry" no more *explains* that state of affairs than do the statements of Whitehead and Russell *explain* what they respectively state. Our Deweyan statement is logically distinguishable from the two statements of Whitehead and Russell in that it comprehends them both. What their statements say separately, our statement says together. There is a logical gain, in explicitness and comprehension of statement, but not an *explanatory* gain. All three statements are logically of the same order— descriptive or denotative.

*"Perhaps" because I take Carnap's *The Logical Syntax of Language* as defining his Logical Positivism. And this book is quite old, dating way back to 1937.
†There is a great difference, of course, between "all the shooting being precisely about this," and "all the shooting about this being precise."

Without going into elaborate details, a descriptive statement is a description of what is found. If you want to rest on your description you may do so. But the description is not an explanation. It is a denotation of what is *to be* explained. If it is a description of the explanation, a denotation of what is found at the end of an explanatory inquiry, then it is customary to call that the conclusion. The statement that opium puts to sleep is a description. To present that description as an explanation is to convert an effect directly into a cause which is equivalent to taking the same thing twice over, once as "effect" and once as "cause," which is no gain at all except in confusion.

A description denotes *how* things are found. If *within* the *results* of the laboratorian and theoretician, taking their results separately, the consequences of the work of the other are found, then, in searching for an *explanation* of the one case or the other or both we must be *controlled by* this fact. Unless we believe in miraculous or supernatural intervention, or in some strange, inexplicable principle of transmigration, the gross and macroscopic subject-matter in this primary experience, the irreducible and stubborn fact that the *results* of laboratorian and theoretician are what they are, must be the consequence of some natural process of interactivity and must be evidence of some natural relation of continuity going on between the laboratory and theoretical functions. The disclosure of the nature of that interactivity and continuity is the disclosure of the explanation.

There is of course a vast difference between an explanation of *how* things as a matter of fact are, and an explanation of how they *should be* in order to meet certain desirable or desired specifications. But again, unless we believe in miracles (in which case we need believe in nothing else and have no *reason* for any inquiry into anything), the knowledge of the specifications desirable and the knowledge *how* to change things so they will fulfill the specifications are both consequences of learning first of all *how things are*.

By learning how things are in Euclid, geometers gradually learnt how things in geometry should be, and through knowledge of how they are and guided by the specifications of how they should be, geometry was reconstructed and is still on the advance. So in every case. If this were not so in every case of scientific advance, the method of scientific inquiry would not be self-corrective and self-expansive. That it is so is also a *report* of the existential facts of scientific history, not an explanation of those facts.

The *how* and the *should*, within any intelligent undertaking, mark a difference, a distinction, not a separation. You can know that things should be different only as a consequence of knowing how they are. You can know how to transform them into what they should be, only by first knowing how they are made as they are. It is a commonplace that modern science began when inquiry into *how* things are was undertaken. In any sense in which "Why?" is intelligent, then it is "How?" read backwards. In any sense in which "should be" is intelligent then it is "how they are" read forwards. In any sense in which "must" is intelligent then it is when "how things are" and "how they should be" are one and the same. Any other sense of "must" consists in taking the *de facto* "how" and converting it blindly into a *de facto* "should be." Because the latter is then called *de jure* doesn't make its mode or method of acquisition of that character any the less *de facto*.

Now with respect to the specific case in hand, namely,

how things are now in science, both Whitehead and Russell agree with Dewey in believing that that is the way they should be—from point of view, that is, of general methodology or way of scientific procedure.*

Whitehead does not believe that the "interpretation" *should* be taken out of the record of the facts, nor does Russell believe that the "facts" *should* be taken out of the calculations of theoretical physics. If you do the first, the record of the facts becomes not even a "contemplation of brute facts," not even a brutish contemplation of brute fact. It becomes no *record* at all. If you do the second—if you evacuate the calculations of theoretical physics of all facts—you may have the calculations left, but they are miserable, misshapen and bereft, meaningfully belonging nowhere, nowhere finding meaningful place no matter *how* they are then pursued: they cannot be *physics* any more—because the facts have all been evacuated; they cannot be "pure mathematics" any more—because the calculations were made in connection with the facts, and in that connection inevitably and irretrievably lost their purity.

Whitehead and Russell, that is, do not say: "It is true, within the record of scientific fact there are interpretations to be found; and within the record of scientific theories there are facts to be found, *but* this is only the way things are now and they *should* be different and our philosophies of science and scientific method—our logics—are dedicated to the task of bringing about this difference." Whitehead and Russell say, "*How* things are now inquired into by scientific method, is the way they *should* be inquired into; *how* scientific inquiry is now conducted is the way it *should* be conducted." They say, in short, how we (now scientifically) think is the way in which scientific thinking should be carried on.†

When Dewey says that theory and practice function within scientific inquiry, or that the theoretical and laboratory functions are interactive within inquiry, he is making a report, in his own terms, of the macroscopic fact upon which all reports agree. When he goes on further and says that all logics of scientific method or scientific inquiry *should be controlled by* this fact, and that the *outcome* of all logical calculations concerning scientific method *should terminate* in that fact—he is going beyond the reportorial to the scientific-philosophic function. He is laying down a *rule of method* that logicians of scientific method should follow. He is laying down the rule that all our logical analyses and theoretical calculations concerning scientific method must be *controlled by* our findings as to *how* scientific inquiry is done. To say that Dewey is "laying down this rule of method" is of course only a manner of speaking. What he is doing is saying that we *must* carry over into our method of logical inquiry the method of inquiry discovered in science *if* our logical inquiry is to be scientific. Since *how* the method of science is now, is the way scientific method *should be*, the "must" is an intelligent conversion of the *de facto* how things are into a *de jure* state of affairs.

When the whole course of our logical calculations is controlled by the gross and macroscopic findings in primary experience of scientific inquiry, it is of course

not amazing that the outcome of that course of logical reflection should terminate in those findings—that our explanation should end by explaining the findings we started out to explain. This is not amazing, but then the objective of philosophy—of scientific philosophy—is not to be amazing. And when our course of logical reflection is *not* controlled by the findings we started out to explain, it is also not amazing that we should *lose control over* the course of our logical calculations and that the outcome of that course should be any which irrelevant way—that the explanation should end by not explaining what we started out to explain but should end up by being an explanation that needs to be itself explained by an endless series of explanations. Although *this* endless outcome is not "in itself" amazing, but what one would naturally expect to result from the uncontrollable method pursued, the outcome is *how* amazing when it is presented not "as it is in itself" but "as in the logic of scientific method."

XVIII

Controlled inquiry involves exercise of control over, *controlling* as well as being *controlled by*. The theoretical experimentalist and laboratory experimentalist—who divide between them the total process of experimental (or controlled) inquiry—exercise *control over*, each in his own way. Each has his own distinctive means and methods, relevant and competent to handle his respective share of the total work, the means and methods of each being developed in the course of fulfilling or realizing the partial function, within inquiry, that each performs.

The laboratory experimentalist—to consider him first "by himself" for a spell—exercises *control over* by means of his laboratory apparatus and the methods of handling the apparatus that he progressively and cumulatively develops in and through the process of laboratory experimentation. "A technical description of the uses of the microscope in biology is not part of the philosophy of the sciences."* Nor of the philosophy or logic of scientific method. And what applies to the technological technicalities of the usages of the microscope applies equally to the technological technicalities of all laboratory equipment and methods. So nothing more need be said on *this* topic, else we would be in imminent danger of wandering off, and not staying on, the topic of experiment.

The laboratorian primarily exercises *control over* his instruments. That he can control them is neither accidental nor providential: he's constructed them that way. Taking an instrument by itself, control over it is pretty nearly absolute, in the sense that, taking a clock by itself, you can turn the hands at will. But a clock—in the laboratory at any rate—is not constructed as a plaything for passing the time of day, but as an instrument for telling the passage of time. The laboratorian never constructs a piece of apparatus so he can control *that*. He is not concerned with constructing apparatus for the pure and beautiful sake of constructing apparatus. He is not a toy-manufacturer or a manufacturer of any sort, not even of automobiles. The laboratorian wants to control something else; and it is with the purpose or end-in-view of controlling that something else that he turns to the making of instruments that will give him that control. That something else is his "material"—whatever it may be, physical, chemical,

---

*As for detailed procedures, and detailed results of procedures, there is always room for improvement, and on this point, too, our three philosophers fully agree.
†The title of Dewey's early book on logic, *How We Think*, has been a great stumbling block to "logicians." That is has been a sufficient indication of how seriously they believe that the method and objective of science are to find out the *how*.

---

*Whitehead, *The Concept of Nature*, p. 1.

biological, psychological, social.

Primarily, the laboratorian has control over his instruments, and through them, secondarily, control over his material. But his primary *objective, with respect to control,* is control over the latter and not over the former. With respect to control, control over the material is his *end;* control over his instruments, his *means* to that end.

The control the laboratorian exercises over his material by means of his instruments is a *secondary* control in another very important respect: taking one or more instruments by themselves, his control over them is pretty nearly absolute—in the sense explained above. But when the instruments are used for exercising control over the material being inquired into, the degree of control drops, the extent of the drop varying from case to case. This drop in control is also neither accidental nor providential. It is a *consequence* of the way in which the laboratorian *organized* his instruments into an experiment. For his inclusive or final end-in-view—inclusive or final because including the whole experiment and determining the end for which the experiment is organized—is not the perpetuating of his control over what he already has under control or (which is the same thing) the reproducing of what he can already produce. If the latter were his inclusive or final end-in-view he would not be setting up an *experiment,* he would be in a totally different business, the business of quantity manufacturing of one sort or another. The inclusive or final end of the laboratorian with respect to any experiment he organizes is *discovery,* the discovery, namely, of what the *consequences* will be of the *interactivities* set going *in* that experiment. Every experiment is a new experiment, a *new* organization, instituting new interactivities within that organization, and what the consequences of new interactivities will be one must perforce wait upon the issue to tell.

Experiments that methodically *repeat* experiments already performed are also experiments and have the function in scientific inquiry of testing or corroborating results previously obtained. If the *result* of a "repeater" were a predetermined or foregone conclusion, it would have no scientific corroborating power, and would not be an experiment at all. Although an experiment performed for corroboratory test is from point of view of methodical set-up—the organization of material and instruments—a "repeater" it is from the logical stand-point, which embraces its function in inquiry, a *new* experiment. For the consequences, the results of the methodical repetition are still problematic, have still—to use James's phrase—to be "cashed in."

The *partial end,* the control over the material is the "end" *within* the process of inquiry. The final or inclusive end is the end-in-view of which the whole inquiry is undertaken and for which *that* experiment is set up. Taking any experiment, it is of the utmost importance to distinguish between these two "ends." They are distinct, not separate. But they are functionally distinct. The end *within* any case of inquiry functions as one of the means in the conduct of that inquiry; the *end* which is the final consequence—the ending of that inquiry—is the attainment of the knowledge gained through that inquiry. The final consequence of any experiment, with respect to *that* experiment, is never a means but only an end. If it were not for the rampant confusion over means and ends, this point would be too commonplace even to whisper about. For obviously, what comes at the ending of any historical process cannot possibly be a *means* in that same historical

process. Effects are effects and not their own causes. And conclusions, endings of inquiry are effects. That a conclusion of one inquiry can be used as a means in another inquiry should also be too obvious to mention. If it couldn't, there would be no inquiry at all but a spasmic jolting from thing to thing, like the actions of grasshoppers or fleas.

The material the laboratorian is inquiring into, and the instruments by means of which he conducts his experimental inquiry are distinguishable from one another, but are *not* separated and disconnected. The laboratorian hasn't got "material" in one hand and "instruments" in the other, each unrelated and un-beknown to the other; and he doesn't "apply" the instruments "to" the material (or vice versa) the way, for example, the one hand "applies" soap "to" the other when the left hand doesn't know what the right hand is doing. The instruments are themselves organizations of the material, and the material is *in* the organization of the instruments constituting the experiment. The material is *in,* not metaphorically, but actually, that is, interactively. The whole business of laboratory experimentation is not to organize an experiment to show that the same causes produce the same effects,* but to find out what are the *consequences* when new *interactivities* are caused.

For the experimental laboratorian, the sky is the limit. Anything goes. According to Aristotelian logic (still widely used) the "nature" or "essence" of glass may be glass, wood wood, metal metal and so on pretty nearly forever; but for the laboratorian all things whatsoever are alike in only one fundamental respect: they are things to be brought together in new ways so that new interactivities may be set going and new consequences may ensue. And the consequences of the interactivities *are* the *natures* of things in and for science. As the laboratorian advances with the general advance in scientific knowledge, technological and theoretical, his instruments become more numerous, more precise, more powerful, and his organizations of the instruments into experiments become more elaborate, more delicate and more productive of new consequences because of the increased variety, in kind and degree, of the interactivities instituted in the material through and by means of the experiment. And so the laboratorian discovers that the same old "stuff" has ever more and ever different natures, or "essences."

The sky is the limit for the laboratorian. But where that limit is the laboratorian does not know. It is seriously to confuse matters therefore to say that the laboratorian is *controlled by* his *material.* In setting up any experiment, the laboratorian is controlled by his *knowledge of* the material as gained through prior experiments. He knows what the limits of the material are, as those limits were revealed in the consequences of prior experiments. But *those* limits are not *the* limits. Hence the new experiment. In passing from one experiment to the next, the laboratorian is *controlled by* his knowledge gained antecedently; and the consequences of any experiment that is the "next" are *limited by* the organization of interactivities which constitutes *that* experiment. Every experiment has its limitations; but what the limits of the material and instruments are (as interactive within that organization) the laboratorian does not know until the *consequences* of *that* experiment tell him. Since you cannot be *controlled by* anything you don't know—in any sense of "control" involved in

---

*This *is* the business of corroborating experiments, "repeaters."

"controlled inquiry" or intelligent method*—it is much better to speak of the laboratorian being *limited by* his material.

It is well known that we can't count up to the last possible number. In the case of counting numbers, we know pretty well that the next one will be one more than the one antecedent, and so there is no intelligent point in trying to count them even as far as we can go. But in the case of laboratory experiments, we do not know that the consequences of the next one will be just something more on the same line and to be added to the consequences of the antecedent experiment. Every experiment is a new experiment; it sets up a question or poses a problem; and what the answer will be is problematic until the consequences have been "cashed in." The consequences of the *next* experiment may not be in the same line at all; they may not be additive but transformative of knowledge; they may not add to the sum antecedently amassed but upset the whole previous account. The result will be an *increase* in our knowledge true enough, but it will be an increase not by way of addition and accretion but by way of subtracting and overturning. In simpler language, the *next* consequence may not be reformatory but revolutionary.

XIX

There is no special virtue or power, philosophic or scientific, internally resident in the *word* "interactivity." As far as words go, "interaction" is just as good; by usage they could be made identical. That precisely is the great danger. In the intellectual as in other worlds, possession is practically the whole of the law, and an idea or meaning long inhabiting the field will prevent any new idea from getting in or will swallow it up if and when perchance it does. Especially will this happen if the new idea comes clothed in an old word or a close derivative of it.

When a new idea comes clothed in old raiment, it practically invites its own annihilation. But new clothes, though they may slow up the pace of destruction, do not automatically function to prevent unwarranted demise. There is no royal road to enduring life, and no known method of insuring the life of a new idea. Some have an extraordinary gift for coining strange new words, like Whitehead and Peirce. But the philosophic procedure by coinage has its many drawbacks as well as advantages. There is also the great method of terminological coinage which goes by the name of symbolic logic. That the newest symbols may be just a disguise of the oldest ideas—a disguise which may deceive the symbolists more often than the non-symbolic—the history of the self-same symbolic logic is right there to tell.

The usage to which "interaction" has been put, the meaning it has become encrusted with during centuries of working in a Newtonian intellectual world, is all that is the matter with it. Newton's Third Law of Motion states that "An action is always opposed by an equal reaction, or, the mutual actions of two bodies are always equal and act in opposite directions." The phrase "action and

reaction are equal and opposite" has come to be the dominant definition of the meaning of "interaction."

If "action" and "reaction" were equal and opposite, there could be neither the one nor the other. For just as soon as any "action" started, supposing it to start somehow, it would be immediately estopped by the "reaction." It would always be an irresistible force meeting an immovable object. When such is the case, they might just as well not meet, but each stay at home. Meeting can do nothing for or to either, since they cannot *interact* when they meet. Their meeting in other words can only be purely formal, never real.

Newton claimed that his laws of motion were "derived from phenomena." He tells us that "the main business of natural philosophy[science] is to argue from phenomena without feigning hypotheses and to deduce causes from effects." Letting the term "deduce" pass for the time being, this procedure if followed would be equivalent to what Dewey calls the method of beginning with the gross and macroscopic subject-matter in experience. For effects are what we primarily experience, not the causes we "deduce." Newton also recognized that the "causes" we deduce should not be the "effects" over again. "To tell us that every species of things is endowed with an occult specific quality by which it acts and produces manifest effects, is to tell us nothing." The cause should be *equal to producing* the effect—but it should not be the same as the effect, if it is to tell us anything. However, as Newton goes on to say, "to derive two or three general principles of motion *from phenomena* and afterwards to tell us how the properties and actions of *all corporeal things follow* from those manifest principles, would be a very great step in philosophy [science]."* Such derived principles Newton considered his laws of motion to be.

Newton gives us the "experimental" phenomena whence he "derived" his Third Law.

"Whatever presses or pulls something else, is pressed or pulled by it in the same degree. If a man presses a stone with his finger, his finger is also pressed by the stone. If a horse *draws* a stone tied to a rope, the horse *will be* (so to speak) *drawn back equally* towards the stone: for the rope being stretched at both ends will by the same attempt to relax itself urge the horse towards the stone and the stone towards the horse; and *will impede the progress of one as much as it promotes the progress of the other. (Principia,* [Evans and Hain editions] 1871, Axioms, or Laws of Motion. Italics mine.)

Obviously, if it were true that the stretched rope impeded the progress of the horse *as much as* it promoted the progress of the stone, the horse and stone would neither of them be moving and the stretched rope, like Buridan's Ass, would be transfixed by its own immobility.

Of course it can be said that Newton meant his "experiment" to be taken this way; that he was not considering the motion of the rope relative to the earth, but was taking horse, stone and stretched rope "in isolation." The three constitute the experimental situation and he is concerned only with what goes on *within* that system. If this is so, then it is still a fundamental error on his part to say that the rope, within the system, "will impede the progress of the one as much as it promotes the progress of the other." If the system is stationary, is self-enclosed, then the rope, to maintain the "dynamic equilibrium," will have to *impede* the progress of the one as much as it *impedes* the progress of the other. But let

---

*Being pushed and pulled around by the law of gravitation, etc., is not being "controlled by" the law; it is being *limited by* the law. Human beings are *limited* to going around with the earth in its orbit; even when we fly we don't really fly away. The same thing applies to "human nature and conduct." When we "act" in terms of our "instincts," "intuitions," "inspirations," coming we know not whence and going we know not where, we are not "controlled by" them; we are pushed and pulled around by them; we *are* then animated *machines,* limited corporations; not intelligent, controlling, human beings.

*From Newton's *Opticks,* quoted in Burtt, *The Metaphysical Foundations of Modern Science,* pp. 219, 258; italics mine.

us consider this fundamental error as a temporary slip and pass it by too, for the time being.

Suppose we take it as a system in "dynamic equilibrium," the horse pulling, but not moving the stone with the rope stretched between. The rope is stretched because the pulls are equal and opposite. Now if we call the pull of the horse "action" and the pull of the stone "reaction," do we not get Newton's result? Does it not follow that throughout the length of the rope, action and reaction are equal and opposite? If at any point along the rope they were not, would not the rope, at that point sag and the whole stretch be gone?

But if we call the pull of the horse "action" and the pull of the stone "reaction," what are we going to call the rope? What *is* the rope? Action and reaction embrace within their "conjunctive union" the total world of motion. But within that world of motion the rope is not included. The rope is neither "action" nor "reaction." And *how* are we going to *cut* this mysterious rope that binds two physical pulls or "forces" together and is itself nothing physical? How are we going to cut it to find the "equality and opposition" of action and reaction which is supposed to be resident throughout its length? Action and reaction can never meet face to face within this mysterious rope. If they ever met head-on within the rope, they would stop each other dead in their tracks, and what would happen to the rope then? *Within* this rope, action and reaction must pass each other by. There is no use, therefore, in taking any one "point" along the rope unless we can *split* that point. Since splitting "points" is no occupation for anyone, let us say the rope is a "union" of two unsplittable lines and that action and reaction stream through these lines, each one in its own line and in the opposite direction. Within the rope, these two equal and opposite streams will be exactly opposite to each other along their routes only in the sense that two parallel lines are opposite each other. And any part of one stream will also be exactly opposite any part of the other stream, if "action" and "reaction" travel through their respective "lines" like two beams of light traveling *in vacuo* in an Einsteinian universe.

Newton, of course, was not thinking of an Einsteinian universe in which the velocity of light traveling *in vacuo* is "absolute." He started with an "absolute equality" instead of ending with one. For even if Newton's stone is a stone, and his horse a horse, his rope is not a rope. It is neither "action" nor "reaction." It does not participate in the goings on. It is absolutely neutral between the two. To be able to keep "action" and "reaction" equal and opposite, the rope must have two transmission lines, and both lines must be vacuous media of transmission. There is only one "rope" that is equipped to do this job with the mathematical perfection required in the Newtonian "world of physics": the two lines of an equality sign in a mathematical equation. The left side of an equation is equal and opposite to the right side. And transmission from one side to the other in either direction is done with mathematical simultaneity.

XX

When in the natural course of walking through a field we come across a horse, stone and stretched rope between, *that* organization of objects is not the gross and macroscopic subject-matter in our primary experience. In primary experience that organization is experienced as included within the field within which we are walking and within which we also are included. The field is not a uniform mathematical abstraction. It is a field full of other contents, and if our gross and macroscopic primary experience did not include the field and these other contents—if they also did not appear in our primary experience—we neither would nor could macroscopically discriminate the horse, stone and stretched rope between as a particular organization of objects *within* that field. The "field" would be a uniform mist, and when there are no differences in distribution within the mist, not even the mystical shapes of things come out. The total "field" would, in fact, be like Newton's rope. And without the horse and stone how would Newton have found that rope between?

Within the wider natural field—the environment—the organization of horse, stone and stretched rope is a "state of affairs," as Dewey sometimes calls it. As macroscopically experienced, it is a consequence, an effect, a result. It is, in his technical, logical language, a *situation*. As a consequence, an effect, a state of affairs, it is a phase of a history, having continuities with the past and moving into the future. There is something doing. The horse is pulling, the rope is stretching, and the stone is not going in the direction of the horse's pull because it is being pulled in another direction and is going in that.

Horse, rope and stone are organized together and are interacting with one another. The subject-matter in our primary experience of that situation is a consequence of the interactions going on, an effect, not the cause. Horse, stone and rope have not swallowed each other up and if unhitched they would not, because of that, vanish into nothingness. They are interacting with each other and also with other things. Each thing in the interactive system constituting that situation is an individual thing within its own distinguishable boundaries because of multitudes of interactivities going on within it. But it is a thing at all, and not a figment, because the boundaries within which its own individual interactivities are going on are not boundaries that shut it off from interactivity with other things within the greater environment. On the contrary. It would not be a natural thing at all, individualized or nonindividualized, if it were not for the multiple interactivities it sustains and is sustained by with things beyond its own boundaries. If it were in interactive relations with more things within a greater environment, it would be more fully individualized, not less. The extent of individualization of a thing is not determined by its boundaries, for every stone has its boundaries, and though stones are individualized, their extent of individualization is not notably great. The extent of individualization of a thing is determined by the range and qualitative kinds of interactive relations it enters into with other things.

The horse is distinguishable from the stone and the rope from both because each is a different organization of interactivities within its own boundaries. But the latter in turn are not primeval, aboriginal causes but consequences of prior interactivities and modes of interactivity which are maintained in the enduring present.

Taking any individual object, we may call it, within its own boundaries, an interactive continuum. But it is always a continuum, so to speak, partial not complete. For no continuum of interactions of an individual thing is self-sustaining. It can sustain itself only by interacting with other interactive continua. Modes and patterns of organization of interactions differ from one another, distinguishing kinds of things, and individuals or individual variations within each kind. But each kind and each individual within each kind interacts with other kinds of interactive continua. When the continuum of interactions constituting a horse interacts with the interactive continua constituting blades of grass, the pattern or organization of the horse is further maintained. When a horse interacts with the interactive continuum constituting a bolt of lightning, his pattern or organization of interactions is destroyed. Every individual thing, to paraphrase Dewey, is a qualitative whole of qualities in interactive qualitative relations. The pattern or organization within the individual is the qualitative consequence of qualitative interactions.

The consequences, effects, results are not dematerialized effluvia which the "causes" cast up and then recede to do something likewise on another occasion. The "causes" are always where the effects are; the causes of any situation or thing—as an interactive continuum (in the sense defined)—are *within* that situation or thing. But not *all* the causes. Only some. Those causes namely constituting the interactions going on within that interactive continuum, within the boundaries of that situation or thing.*

Since the boundaries of one thing do not cut it off from other things but each thing is embedded in its own interactive locality, we can *use* one thing, which is an effect, as a means for finding out the causes of another thing, by bringing the two into interactive relations. We can organize them, that is, into an experiment.

The horse-stone-and-rope is as good an organization of an experiment as any other. But *to be* an experiment, we cannot start by taking it as Newton did. That was the end of the experiment, the consequence. That was the phase of the "sequential order of events" constituting the experiment which *closed* the sequence of events. An experiment is in the experimenting; there has to be a sequence of events; a longitudinal section of a history; it must have beginning and middle period as well as an ending. If we take the "ending" alone, we can only contemplate it as a brute fact, or enjoy or suffer it as an event in our emotional or esthetic experiencing.

An experiment is an organization of interactions. It is an organization of "doings-undergoings" to use Dewey's homespun phrase. The resultant of the doings-undergoings is what is *done*. To get that done is the purpose or idea of the experimenting. Or, to put it more generally, to observe what the "done" will be. Before the rope is stretched there is a stretching; before the horse pulls, he is doing-undergoing something else; before the rope is tied around the stone, it is a stone without the rope tied around it. At what phase of the inclusive history we begin our experimenting is as the case may be, depending upon what we are experimenting for, what our idea is, our plan. But there must be *some* beginning of every experiment which leads into but is not the same as the ending.

Within the organization of the experiment, horse, stone and rope are all constitutive interactive members. If the objective, the end-in-view of the experiment, of organizing that system of interactions, is to find out how much pulling the rope will stand, then the rope is the material being inquired into, and the stone and horse are the instruments. If the objective is to move the stone, then the horse and rope are instruments and the stone is the material. If the objective is to determine how much the horse can pull, then the stone and rope are instruments and the horse the material. Within every experiment there are instruments and material, something used as the

means of inquiry and something inquired into. The consequences in the material of the interactions set going within the experiment constitute the objects of the closing, final, experimental observation. They are the "effects"; and when written up in scientific memoirs are the "final facts of the case." The consequences are the "reactions" of the interactions.

Of course, there are consequences in all the constitutive members of the interactive experimental system. If there weren't, it would not be an interactive system. Apart from this logical argument, which may appear circular, there is the existential, macroscopic evidence that objects which function as material in one experimental situation function as instruments in another. There is no reason to suppose that they undergo miraculous transformations in passing from one situation to the next. If there were any non-interacting member in any experimental situation, no inquiry would be possible and that situation would not be experimental. Witness the case with Newton's rope.

It takes two actions to make one interaction. And the consequence of an interaction is a new action. This is in logical formulation, not in existential fact. In Nature there are no actions which are stripped to the bare state of homogeneous oneness. All actions are themselves consequences of systems of interactions of various complexities of organization. However, for formal logical analysis, we may consider instrument and material as unitaries (however complex they may be) and the organization of instrument and material in an interactive system as constituting an experiment.

Every experiment is a doing-undergoing. Not metaphorically, but formal-logically speaking, the instrument is the "doing" and the material the "undergoing." For the instrument is an instrument to the extent that it is that by means of which we exercise *control over*; the material is what we are *limited by*— it is the undergoing.

No instrument is all compact with "doing" and no material we can in *any* way handle is all compact with "undergoing." Within experimental inquiry, instrument and material enter into interactive relations, and so they are interdependently doing-undergoing. The degree to which we are *controlling* the experiment going on, is the definition of the instrument functioning in *that* inquiry. It is always a degree, never an "absolute."

We are *doing* the inquiry by means of the instrument. The material is *undergoing* inquiry. The *consequence* of the interactivity of the doing-undergoing within the experimental situation is the *ending* of that inquiry. The consequence, as the ending of a controlled history, of events, is what is *known.** It is the effect, and effects are gross and macroscopic subject-matter in primary experience. The effect brings us back to primary experience. But the *effect* now in our primary experience is not a brute fact we can only stare at in dumb "contemplation" or in animal amazement. The *effect* is a state of affairs, a situation that is achieved through controlling a history. It is the terminal phase of a sequential order of events within which sequence we have had a guiding hand. As such an effect it carries the meanings of its history within it. It is a *conclusion*, and hence we can always look at it with understanding—and sometimes even with delight.

---

*Cause is an old name for a system of interactivities going on, and effect an old name for the qualitative consequences of the interactions.

---

*In terms of the Aristotelian logic, the *doing* is the predicate, the *undergoing*, the subject and the copula is the interaction of the doing-undergoing that results in the conclusion. The conclusion is what is *known* and constitutes the *judgment*. See, *Logic: The Theory of Inquiry,* pp. 124f.

XXI

"By the nature of the case, causality, however it be defined, consists in the sequential order itself, and not in the last term which as such is irrelevant to causality, although it may, of course, be, in addition, an initial term in another sequential order."* The nature of the case Dewey here speaks about is the case of Nature. However we define causes and effects they are both within Nature and constitute the same historical series. To discriminate causes in Nature from effects in Nature is to introduce distinctions within a continuously moving history. It is not to cut Nature into two halves, the first half the causes, the second half the effects and nothing but the cut in between. Still less is it to isolate one event, and set it up as the aboriginal beginning or the ultimate ending, as the "metaphysically primitive" mechanical or teleological "cause" of all. "The view held—or implied—by some mechanists, which treats an initial term as if it had an inherent generative force which it somehow emits and bestows upon its successors, is all of a piece with the view held by teleologists which implies that an end brings about its own antecedents. Both isolate an event from the history in which it belongs and in which it has its character. Both make a factitiously isolated position in a temporal order a mark of true reality, one theory selecting initial place and the other final place. But in fact causality is another name for the sequential order itself; and since this is an order of a history having a beginning and end, there is nothing more absurd than setting causality over against either initiation or finality."*

Nature is an inclusive history of multitudinous ongoing histories, the comprehensive interactive continuum consequent upon the interactivities of an infinite number of interactive continua of an indefinite number of general kinds. When the second law of thermodynamics will have brought about the heat-death of the universe, this will not be the case. The sequential order of Nature will have then reached a term that is irrelevant to all subsequent terms because no subsequent terms will ensue. Interactivity will then have ceased and continuity in Nature might just as well be a mathematical line. Passage from one end of Nature to the other will be an uneventful event, no undergoing at all, because every locality will be exactly like every other locality and nothing doing in any of them. To distinguish and interconnect within such a universe would be a waste of time—supposing there were any time left to waste. But there will then be no making of distinctions, for all differences will have been annihilated in the absolute uniformity and for the same reason there will be no making of interconnections; interactivity will have ceased and the selfsame absolute uniformity will be the only connection left. When that Nature eventuates, it will be the end of all events and as irrelevant to all its antecedents as to its non-eventuating succeedents. For it will be carrying none of its antecedents within it. Whether such a time will come to the universe is not the point here. What is in point is that we are able prospectively to write the history of that dead, indistinguishable and non-existential event because it has not yet arrived. We are able to include that non-historical event in the history of the universe because there is a contemporaneous, ongoing history within which to include it.

The Nature within which we live is an ongoing history of ongoing histories. When an event is connected with another event as cause-effect, that connection is the exemplification of the continuity between them. But that connection of continuity is the funded history of interactions and the effect is the funded consequence, the terminal phase of the inclusive history of cause-effect. "The two principles of continuity and interaction are not separate from each other. They intercept and unite. They are, so to speak, the longitudinal and lateral aspects"* of every history, of every situation, of every sequential order, of every connection of cause-effect.

The union and interception of continuity and interaction can be very simply illustrated. Half a dozen ivory billiard balls placed in line will transmit to each other the impulse the first ball receives from the cue. The impulse is not transmitted in a one-way linear series of pushes from next to next, starting with the first ball and going down the line, this one-way linear series being followed by a similar one-way linear series of pushes traveling in the reverse direction. The billiard balls are made of ivory; they are not Newtonian atoms. The cue is made of wood and is not external to the world within which the billiard balls are. Each ivory ball is "alive," as billiard experts rightly say. And so is the cue. When the cue hits the first ball, the ball hits right back. The consequence of that unequal contest is the reaction of the first ball and it crashes into the next, and so on with every ball down the line. At no point do you have cause following effect and effect following cause; at every point you have a cross-weaving or interweaving of actions, that is, interactions, and each interaction is an interweaving of cause-effect.

Wherever there is interaction there also is continuity. The interaction is the continuity taken laterally or cross-sectionally; the continuity is the interaction taken longitudinally or historically. The maintenance of the interception and union of continuity and interaction constitute or create an interactive continuum.

The billiard balls illustrate also how it is that interactivity is the creative matrix of individuality. Assume to start with that the billiard balls are all alike in every character of their composition. Just as soon as they are placed in line and the interactivity is set going by the interaction between the cue and the first ball, the balls become differentiated from each other by the variations in their actions and interactions. Initially, these differentiations or variations are solely due (by hypothesis) to their respective positions in the lineup, their nearness to or remoteness from the starting interaction. But suppose that the interactivity continues for an appreciable length of time. What happens as a consequence? The differentiations result in internal changes in the balls themselves. At the very least, they wear out at different rates (for they are in different positions in the lineup) and hence lose or gain resiliency at different rates. Two billiard balls differing in resiliency are no longer exactly alike in every character of their composition. They have become, in so far forth, individualized.

No two actual billiard balls are, of course, actually identical to start with. For theoretical purposes of illustration, their differences may be ignored, though for the practical purposes of playing billiards their differences may not be ignored—as any expert billiardist will tell.

Every actual billiard ball is an amazingly complex organization of interactivities within its own boundaries. Its "pervasive qualitative unity," to use Dewey's phrase, is a *consequence* of interactions, and is as stable as the stability of organization of interactions of which it is the

---

Experience and Nature (1st ed.), p. 99-100.

Experience and Education (1938), p. 42.

consequence. We may, following Whitehead, call the billiard ball a "society" instead of an interactive continuum, and the stable organization of interactions within the billiard ball, the "personal order" of that society or billiard ball. This "personal order" may also be called the "causal law" of that society.*

When the interactivities of the six billiard balls are kept going for a sufficiently long period of time, the consequence is the creation of a new qualitative whole, a new interactive continuum, a new society. In the process of creation of the new society, the members are changed, both as to their own individual boundaries and the "personal order" of each. The boundaries of the new society, or the new interactive continuum, are vastly different from the boundaries of each of the separate members before the new society was created. They are also vastly different from the boundaries of the members as constituent and interactive within the new society. The new society taking it as a totality has its "personal order," its "causal law," its pervasive qualitative unity.

If the reader finds this consequence of the interactivity of six billiard balls too strange and difficult to accept, let such reader substitute six astronomical bodies that begin to interact with one another and after an appreciable length of astronomical time settle down in a planetary system. Or if astronomical stretches of time are repugnant, for whatever personal reason it may be, any other substitution will serve as well, as long as the items substituted are actual things within Nature, that is, things capable of *interacting*.

The personal order, causal law, pervasive qualitative unity of any new society is not the arithmetical sum of the orders, laws or qualitative unities of the constitutive interactive members as they are *within* that society nor, of course, as they were before the emergence of the new society. Every interactive continuum or individual thing is a qualitative whole of qualitative interactions. There are quantities of qualities in Nature but nothing in Nature has the quality of a sheer quantity. The consequence of interactivity of two qualitative unities is a qualitative transformation. If qualitative transformation is not the consequence, then no interaction has taken place.

The constitutive interactive members within a society are never without their own individual orders, causal laws, or qualitative unities. If within a "society" there are no individual differences within the constitutive, interactive members, then you have no members and no society at all. You have a Newtonian "atom." That the latter is a pure figment of mathematical imagination fortunately no argument is at this date needed to prove. Within nature there are no "individual things" that are not members of a society and discriminable as members within that society because of natural differences. When theoretical physicists push their analytical reductions of experimental findings to the limit, they still have left on their hands an electron and proton within an atom, electron and proton differing from each other, and the "atom" within which they are, differing from both.

As Whitehead states it:

> A society is, for each of its members, an environment with some element of order in it, persisting by reason of the *genetic* relations between its own members. Such an element of order is the order prevalent in the society. (*Process and Reality*, p. 138; italics mine.)

When two interactive continua, two societies, each with

its own stable organization of interactions, its personal order, causal law, or pervasive qualitative unity begin to interact, they in that new interaction begin to suffer "disorder." The process of reaching a new stable organization of interactions is the process of creating a new order, or a new society. However, no more than effect *follows* cause does a new stable organization of interactions, a new order *follow* upon disorder. Cause-effect are interwoven; interaction and continuity intercept and unite. Likewise with "disorder-order." The new order, which is an ordering or organization of interactivities, is not an effluvium, a residue cast up, or supernatural excrescence which "supervenes" (to use Santayana's esthetic term) *after* the disordering is all over. The new ordering is constantly and continuously in the making throughout the period of so-called "disorder." Such periods are "transitional" only in the sense that the disordering-ordering process is in transit, is taking place. Isolate the beginning of the history, when the "disorder" (or interaction) began, and then isolate the ending of that same history when the new stable organization of interactivities has been established, and compare your two "isolated events" and the period in between becomes a "pure transition," a period when all is chaos, and wherein no "law and order" can be found.

When a new stable organization of interactions is achieved, that new order or causal law is the funded consequence of the total history. The new pervasive qualitative unity is the definite and emphatic emergence, in primary experience, of the consequence of internal stabilization of interactivities reached. Taking the whole period, from beginning to ending, longitudinally or historically there is a continuity of disordering-ordering interactivity, the new ordering emerging continuously through the genetic process. The "genetic process" is of course not something in addition to the history of interactions of disordering-ordering but just another name for that process. Any cross-section or lateral segment of the period will exhibit the disordering-ordering inter-activity with the qualitative characters or consequences emergent in that segment. For at no lateral section will there be sheer interactivity, without a qualitative consequence or product of that interactivity. To suppose the former is to suppose the absurd or miraculous—that is, that there is a period of interaction that is purely non-qualitative and then there "supervenes" a qualitative character. It is to suppose there are "causes" without "effects." Whether at any section we take we can discern the qualitative consequence emergent in that section is another matter. It is also another matter, or the same matter, whether we can discern the interactivity going on irrespective of the size of the segment we take.

When two interactive continua or two orders are beginning to interact, the new situation created by that initiation of interaction is a situation of conflict, disturbance, unsettlement—to use Dewey's terms. The *issue*, the consequence, is the outcome of the interactivity and what it will be, in any genuine experimental situation, is problematic. Whether the outcome will be progressive or regressive, whether the funded consequence will be an increase or decrease with respect to the original "investment" that went into the interactive situation is as the case may be. The history of change is "progressive or evolutionary" even when it runs down hill. It is one of the great misfortunes of the term "evolution" or "evolutionary" that it became identified with an "upward and onward" unidirectional meaning. Evolutionary development is evolutionary development irrespective of the

---

*Process and Reality, p. 137-139ff.

direction in which it is heading. The idea of an automatic upward trend was one of the supreme "scientific" absurdities of the nineteenth century, still persisting in the twentieth century. It was natural history with a shot of ancient theology in the arm. No idea is more fatal to human progress in the upward direction than the belief that the necessary and hence automatic functioning of "evolutionary processes" will take human beings up to the next step. Dewey's whole critical philosophy may be considered as one extensive exposé of the fallacy and the danger of letting Providence—in the modern scientific era variously nicknamed the "laws of Nature," the "laws of evolution," the "laws of dialectical materialism"—shape and take care of the destiny of human ends.

Of course no organization of interactions, no order or causal law of a society is so stable that it suffers no change. The stability is always a case of degree, never an absolute and eternal kind. Every system of ordered interactions is to some extent a disordered system. The element of order, as Whitehead says, is the order prevalent in that society. It is prevalent, not omnipotent. "A thing may endure *secula seculorum* and yet not be everlasting; it will crumble before the gnawing tooth of time, as it exceeds a certain measure. Every existence is an event."* What Dewey here calls the gnawing tooth of time is the disordering factor involved in any interaction. It's not the tooth but the gnawing of the tooth that counts.

We have so far been considering societies or interactive continua as within themselves, leaving the "environment" in the shadowy background. Analytically, inhabitant and habitat can be distinguished from one another, but it has sufficiently well been demonstrated that they are symbiotically related. Moreover, as is clear in the quotation from Whitehead above, the society constitutes the immediate environment of its members.

But every society is included within a larger environment. To quote Whitehead further on this point:

> There is no society in isolation. Every society must be considered with its background of a wider environment. . .the given contributions of the environment must at least be permissive of the self-sustenance of the society. Also, in proportion to its importance, this background must contribute those general characters which the more special character of the society presupposes for its members. But this means that the environment, together with the society in question, must form a larger society in respect to some more general characters than those defining the society from which we started. Thus we arrive at the principle that every society requires a social background, of which it is itself a part. In reference to any given society the world of actual entities [interactive continua, situations, occasions, individual societies] is to be conceived as forming a background in layers of social order, the defining characteristics becoming wider and more general as we widen the background. Of course the remote actualities of the background have their own specific characteristics of various types of social order. But such specific characteristics have become irrelevant for the society in question by reason of the inhibitions and attenuations introduced by discordance, that is to say, disorder. *(Process and Reality, p. 138.)*

The outline of fundamental, general agreement between

Experience and Nature (1st ed.), p. 71.

Dewey and Whitehead is the reason for this quotation just now. So we need not stop to consider the points in the passage which are indicative of specific differences in the two philosophies. They will emerge in the course of the subsequent argument—in case they have not emerged as yet.

An ivory billiard ball placidly resting on a billiard table in the Union Club in New York City is within the same Nature as an electron dancing around near the center of the sun. Billiard ball and electron are both included within the larger environment which includes the sun and New York City. However, by virtue of the "inhibitions and attenuations" introduced between the localities of the electron and billiard ball, the historical careers of these two are significantly irrelevant to each other. Within the locality where the electron is dancing, the dancing of the electron is productive of some consequences. Although it may be hard to believe it, yet it is true that within the locality where the ivory ball is placidly resting, it also is productive of consequences. Irrelevance with respect to each other of two restricted histories, by virtue of attenuation of interactivity between the localities within which these histories transpire, is not evidence of "discontinuity" or of breaks in nature, in the sense of abysses between. It is evidence of attenuation, of diminution of interactivity to the point where no appreciable consequences are the result. However, even the billiard ball and the electron can be made significantly relevant to each other by including them in an environment that is sufficiently wide, and by making the interactive functions within that embracive society sufficiently *narrow*. By the conjoint process of widening the environment and narrowing the range of interactivity, it might even be possible to make the billiard ball relevant in the political history of the United States. However, whether or not this can be done, it is by this double process of widening the environment and narrowing the range of interactivity that, scientifically, our solar system and Betelgeuse, for example, become connected minor histories within a larger history. And so on, illustrations without end.

XXII

Inquiry proceeds by making distinctions and every distinction is also a connection. If there were no natural differences in Nature and no natural interactions going on, inquiry could not proceed. The nature of Nature makes inquiry possible and not the other way about. Within the history of Nature, inquiry is one of the emergent histories, it is a proceeding included within the larger procession. It does not follow the whole procession after it has gone by, nor does it come into existence before the whole procession has started. It is neither trailing the rear nor leading at the point just above the head.

The existence of inquiry is an exemplification of one of the existential differences matured within Nature. As an existential difference within differences, inquiry can and does make further existential differences. This is saying the same thing said just before in another way. If the procedure of inquiry did not make differences within Nature, it would not be one of the goings on in Nature. It would be completely outside Nature. And if by some miraculous sleight-of-word we put it "inside," then our miracle—like all miracles—does nothing to alter the situation, except to make it worse if, while "inside" inquiry it is a non-interacting member, neither changing nor being changed nor making changes. The nature of the case before the miracle is the same as after, only now we have a miracle on our hands to "explain." Whatever cannot participate in the goings on, is completely outside even when you call it in. If the miracle makes inquiry interactive, the consequence of changes and productive of other changes, it is again a useless encumbrance. The goings on within Nature can take better care of that. On this point, too, Whitehead is in general agreement with Dewey: "The very possibility of knowledge should not be an accident of God's goodness; *it should depend on the interwoven natures of things.* After all, God's knowledge has equally to be explained."*

For inquiry to make existential differences within Nature, it is not of course necessary for inquiry to be competent to change everything in Nature, to make the whole procession different. To be competent to do this it would have to be outside the procession. Although both Archimedes and common experience have proved this with sufficient conclusiveness to satisfy ordinary intelligences, some idealistic philosophers (and scientists too) are still holding out against the proof. It is also not necessary for Dewey's fundamental proposition, that inquiry be able to make existential changes in the major divisions of the natural procession. The procedures of human inquiry are incompetent to change the monotonous rounds the earth makes about the sun. Because the planets go round in circles is, however, no reason why human beings should. Inquiry makes changes within the localities where inquiry goes on, and the extent and range of these localities is not determinable philosophically and is not fixed but is determined by and in the advance of inquiry itself.

When the philosophic or logical analysis of inquiry is controlled by the experimental procedure of inquiry as that appears in gross and macroscopic experience of inquiries, there is no difficulty at all in establishing the naturalness and validity of the foregoing. It becomes almost too obviously natural to bear mentioning. Newton to the contrary notwithstanding, experimental inquiry cannot come *after* the goings on in the experiment have ended nor, of course, can the inquiry start or finish *before* the process has begun. Experimental inquiry is a procedure that works within and while the processes inquired into are going on.

Furthermore, when laboratory activity is made an integral functioning element within inquiry, the validity of the proposition that inquiry is an existential procedure the consequence of which is the production of existential changes, is also quite naturally established. In the laboratory existential changes are made. No one questions that. Finally, when theoretical activity is also made an integral functioning element within inquiry, and hence interactive with the laboratory activity—the consequence of their interactivity constituting scientific or controlled inquiry—there is similar natural ease experienced in establishing the validity of the proposition that theoretical activity within inquiry is instrumental in the production of existential changes.

The difficulties come into philosophy and logic only when theoretical activity is separated from practical activity. And when that separation is made, and the sublime exaltation of theory maintained as a primitive metaphysical gift—and like all gifts, to be accepted without asking any questions and to be purely enjoyed—difficulties do not attach themselves merely to the

Process and Reality, p. 289; italics mine.

proposition or propositions just enunciated: they simply swarm over every topic and every part of every topic modern and contemporary philosophers inquire into. With a vengeful justice that may or may not be actually divine, the difficulties especially congregate thick and fast about the topic of theoretical activity, and are unceasingly harassing. The "purity" of theoretical activity which is originally invoked "to explain" the nature and progress of scientific inquiry has in turn to be "explained" by a still purer and hierarchically higher pure theoretical activity. The first pure theoretical activity being called "scientific," the second naturally becomes "metaphysical." That the hierarchical series ends abruptly with the latter is again a matter of psychological stoppage and not of logical conclusion. The term "metaphysics" is as wonderfully effective psychologically when the pure theoretical pursuit is hierarchically pursued in the line of ascent as when it is pursued in the line of descent: when it is introduced it just as effectively gives the feeling of having hit the ultimate ceiling as of having hit the ultimate floor.

The latter "metaphysical case" is exemplified in the philosophy of Russell, the former in the philosophy of Whitehead.

XXIII

"The concept of an ideally isolated system," writes Whitehead, "is essential to scientific theory." As an instance of such a system he cites Newton's First Law of Motion. In explanation of what he means by the concept of an ideally isolated system, Whitehead goes on to say:

This conception embodies a fundamental character of things, without which science, or indeed any knowledge on the part of finite intellects, would be impossible. The "isolated" system is not a solipsist system, apart from which there would be non-entity. It is isolated as *within* the universe. This means that there are truths respecting *this* system which require *reference* only to the remainder of things by way of a uniform systematic scheme of relationships. Thus the conception of an isolated system is not the conception of substantial independence from the remainder of things, *but of freedom from casual contingent dependence* upon detailed items within the rest of the universe. Further, this freedom from casual dependence is required only in respect to certain abstract characteristics which attach to the isolated system, and not in respect to the system in its full concreteness. (*Science and the Modern World*, p. 68; italics mine.)

It is hardly necessary to point out, after what has already been said, that there is no difference between Dewey and Whitehead as to the *general* issue that when we "isolate" a system within Nature we do not substantially tear it out of Nature. The continuity within Nature effectively prevents that. We have to isolate because scientific inquiry can get on only by attacking Nature piecemeal. But Nature is not in pieces.

There is also hardly any need for pointing out that there is no difference between Dewey and Whitehead as to the *general* issue whether or not this "conception"—without which, truly enough, no knowledge whatsoever could be acquired—"embodies a fundamental character of things." Things are iso*lable* within Nature because things within Nature are different. And they are theoretically isolable, but not substantially separable, because they are connected within Nature. The conception of an ideally isolated system, isolated as within Nature, is, therefore, an exemplification of fundamental characters in Nature. It exemplifies at the same time both the existence of differences and the existence of continuity.

So far so good and general agreement so far. But the agreement is general. Just as soon as we probe deeper all the *specific* differences between Dewey's philosophy and logic and Whitehead's begin to crop out, one after another. And they are all consequences of the one comprehensive fundamental difference between Dewey and Whitehead, namely, that Dewey does not substantially tear out experimental practice from scientific inquiry, whereas Whitehead does precisely this thing. This is *the* difference between these two philosophers, and the consequences of this difference appear and reappear in every specific context of their respective philosophies. Because of this fundamental difference the more closely these philosophies approach each other, the more clearly, emphatically and irreconcilably do they stand apart.

The theory of scientific theory is the formal meeting place of all philosophic differences. Here some differences have their formal-logical point of origin whence they issue to work their way through the macroscopic domains of philosophic inquiry, growing to ever larger macroscopic size as they proceed; here differences originally developed in the macroscopic fields are reduced to the microscopic size of derived, refined objects of logical reflection. All differences are gathered together, systematically or otherwise, in this central sheepfold of philosophy. That this should be so is of course natural. Philosophy itself is a professional theoretical activity and if it didn't do this sort of thing it could never pull itself together.

The main issue of difference between Dewey and Whitehead is also the main issue in the whole of Western philosophy. The only specific differences relevant for examination here are such as will help to lead the reader up to, if not into, the significance of Dewey's philosophy.

XXIV

In what sense does Whitehead mean that the concept of an ideally isolated system is *essential* to scientific theory? In the passage quoted his meaning is ambiguously implicit; in another passage he makes it quite explicit. "All scientific progress *depends* upon *first* framing a formula giving a general description of observed fact." That he means "first" in the sense that would make scientific theory *dependent* on the *prior* existence or operation of a metaphysical theory, he also in the same place makes quite explicit; in fact, his whole metaphysical construction is an elaborate exemplification and justification of this sense. ". . .speculative extension beyond direct observation spells some trust in metaphysics . . ." "Apart from metaphysical *presuppositions* there can be no civilization"—and science is of course a major ingredient in civilization. "Metaphysical *understanding guides* imagination and justifies purpose"*—in science as out.

Since the framing of the formula, according to Whitehead, is a generalization of observed fact, the observation of fact is of course "antecedent" to the theoretical formulation. But it is "antecedent" not in any logical sense, but in a brute existential sense; it existentially comes before, it is merely precedent, purely ancillary. That observation of fact can be thus brutally existential, without any logical significance, and without requiring some "metaphysical" precondition, pre-

Adventures of Ideas, pp. 163-164; italics mine.

supposition or precursive cause is itself not due to any logical theory but to the simple fact that Whitehead accepts observation as a *de facto* existential occurrence. This accords accurately with Whitehead's account of Galileo's contribution to scientific method—I mean with that one of his two accounts which described the Historical Revolt, of which Galileo was the scientific leader—as anti-rationalist, anti-intellectual, as a return to the "contemplation of brute fact."

That Whitehead was not able to keep consistently to that macroscopic description of the macroscopic *history* of modern scientific method we have already seen. And the vacillations, oscillations and contradictions there macroscopically manifested are repeated microscopically when he concerns himself with the *nature* of scientific method, taking the latter analytically, in terms of derived, refined objects of philosophic reflection.

Observation of fact comes first, merely first in the ancillary sense indicated above:

> Without the shadow of a doubt, all science bases itself upon this procedure. It is the first rule of scientific method—Enunciate observed correlations of observed fact. This is the great Baconian doctrine, namely, Observe and observe, until finally you detect a regularity of sequence. (*Adventures of Ideas*, p. 149.)

Hence the only trouble with the scholastics was that they trusted to the inflexible rationality of their metaphysical dialectic without renewing periodically their acquaintance with observable fact. By establishing the habit of brute contemplation of brute fact, by starting the anti-rationalist, anti-intellectual Revolt, Galileo, if he did not put modern science on its feet, at any rate put it under the necessary "restraint."

Science begins for Whitehead when theory begins, but to get scientific theory beginning, to get it on its feet and keep it going, "metaphysics" is necessary. It is "metaphysical *understanding* that *guides* imagination and *justifies* purpose." To go *beyond* the direct observation of fact is of course an act of imagination, and to frame a formula which is a generalization of observed fact is of course the fulfillment of scientific purpose. This purpose needs justification therefore in "metaphysical understanding" because it can find no justification in the facts directly observed. The latter are just brute facts and "observation" of them is apparently as brutish as the facts observed. But if this is so, what possible virtue can there be in the great Baconian doctrine? What profit to observe and observe until you finally detect a regularity of sequence? What is this regularity of sequence? When the enunciation is made of "the observed correlations of observed facts" is *that* the formula framed, or is the formula something else and beyond?

To accept "observation of fact" as a *de facto*, brute existential occurrence is one thing. But to accept the framing of a formula as an existential occurrence is however quite another. For a formula is a theory and a theory is a going beyond existence and cannot be "derived" from existence. The scientific formula cannot find justification in the facts because the latter are merely precedent to it; and of course it cannot find justification in itself. The Great Metaphysical Tradition is dead set against it. Otherwise how can it be that a "metaphysical understanding" justifies a scientific formula, but that metaphysical understanding does not require a super-metaphysical understanding to justify *it*? If the metaphysical understanding must be accepted on its *de facto* face, what reason is there for not accepting the scientific understanding on *its de facto* face? If metaphysics were introduced to explain something *further*, a complication or development which scientific understanding does not explain—then there might be some reason for the metaphysical extension. But that is not what Whitehead introduces metaphysical understanding for. He introduces it to *explain* the formulation of the observed correlations of observed fact. But the regularity of sequence, the observed correlations of observed fact, is presumably explained by the great Baconian doctrine: Observe and observe. The formula *is* the enunciation of the regularity observed. Do we need a metaphysical understanding to be able to "enunciate"? The metaphysical understanding is introduced to explain the explanation. And when so introduced, like the miracle discussed before, it does nothing to alter the situation except to make it worse: we have now a metaphysical understanding to explain.

If Whitehead were able to forget about the observation of fact, and the scientific necessity of renewing periodically the observation of fact, his philosophy would be able to exhibit the inflexible rationality of scholastic logic. His dialectic circle of scientific formula and metaphysical presupposition of scientific formula would be self-enclosed. But no philosopher of the logic of modern science can do this. "In all theoretical physics, there is a certain admixture of facts and calculations." "Every scientific memoir in its record of the 'facts' is shot through and through with interpretation." Even if a metaphysical philosopher desired to keep the interpretation and let the "facts" go, he could not do this as long as he has any regard for science. That Whitehead has a fundamental regard for science goes without saying. By "speculation" he does not mean irresponsible reverie that with so many passes as "deep metaphysical" thought. Nor does he mean the acrobatic juggling with metaphysical "categories" characteristic of German philosophy and carried to supreme heights of absurdity by Hegel. As Whitehead brilliantly remarks, Hegel's "procedure is [such] that when in his discussion he arrives at a contradiction, he construes it as a crisis in the universe." For Whitehead "speculative boldness must be balanced by *complete humility before logic and before fact.*"* And hence Whitehead repudiates the "belief that logical inconsistencies can indicate anything else than some antecedent errors."†

Complete humility before logic and before fact is a wise provision of restraint for speculative boldness. But this provision of restraint unfortunately only functions to intensify the problem as to how metaphysical understanding can "guide and justify" scientific theory. Humility before *what* logic? Restraint by *which* fact? "Every memoir in its record of the 'facts' is shot through and through with interpretation." Are we to be completely humble before *such* facts? Obviously not. Whitehead puts such "facts" in suspicion-engendering quotation marks. If we must *trust* metaphysical understanding to get beyond any direct observation of fact, whence issues the "logic" by which "speculative boldness" is to be restrained? Whitehead does not get self-enclosed in a dialectic circle only because he consistently involves himself in logical inconsistencies. The latter cannot indicate anything else than some antecedent errors—but they can indicate *that*.

And Whitehead's antecedent error is his dismissal of

laboratory experimentation as irrelevant for the philosophic understanding of the topic of scientific method. Scientific method then falls apart into two disconnected halves: direct observation of fact on the one hand and framing of formulas on the other.

The direct observation of facts *precedes* scientific theoretical formulation, but does not follow through. The framing of the formula *succeeds* the direct observation of fact, but does not follow from. It is impossible to leave the two in final and irrevocable unrelatedness not because the historic philosophic mission is to find some unity and this great purpose cannot be gainsaid—although in Whitehead this purpose is extraordinarily active. It is impossible to leave the unrelatedness final and irrevocable because of the indubitable macroscopic subject-matter in primary experience of modern science: the interpenetration of fact and theory. The natural tie between fact and theory being discarded at the outset, there is only one other way known to philosophy of establishing a tie: the introduction of a "metaphysical understanding" which serves like the stretched rope between Newton's horse and stone.

XXV

In Dewey's philosophy interaction and continuity intercept and unite. To give any special precedence or dominance to the one or the other *within* his philosophy would be to distort his philosophy and rob it of its unique strength. But within the history of philosophy, ancient, modern and contemporary, the weight of novelty and importance of contribution falls upon *interaction*. For there are a variety of "continuities" and various systems of philosophy have been developed in celebration of the varieties. To go no further afield than our two principal contrasting philosophies—those of Russell and Whitehead—they also are devoted to the end of establishing continuity. Some of Russell's strangest and most contradictory conclusions, in fact, are due to his efforts in search of "continuity" and the kind of "continuity" he uses. He manages to get everything "inside the head" because "causal continuity" makes that necessary, the causal continuity alleged to be firmly established by "physics." To get any other result, says Russell, we would have to suppose a "preposterous kind of discontinuity." *(Philosophy*, [1927] p. 140.) And rather than do that, a preposterous conclusion is apparently preferable from the standpoint of "logic" and "philosophy."

In Whitehead's philosophy also "continuity" is the dominant theme. And, despite the fact that in some domains of philosophic inquiry Whitehead has developed "continuities" with a wealth of detail and fineness of analytic precision superior to Dewey, Whitehead lands in the strangest and most contradictory conclusions—also because of the kind of conception of "continuity" he employs. Whitehead, unlike Russell, is not restrained by the kind of "causal continuity" alleged to be established by "physics"; as a creator of the "philosophy of the organism" the biological sciences carry great weight with him. However, in the philosophies of neither Whitehead nor Russell can it be said that "interaction" plays any significant role. At the risk of possible exaggeration, I would say that in the philosophies of neither do you find any interaction at all.* And at the same risk, I would say

---

Process and Reality, p. 25; italics mine.
†*Process and Reality*, p. viii. This is number ix and the last in the "list of prevalent habits of thought, which are repudiated" by Whitehead. It is also in some respects the most important.

*If their philosophies exhibited fundamental consistency, there would be no risk attached to making this statement. For obviously, there can be no *interaction* where practice is extruded *ab initio* as irrelevant.

that you will not find any interaction in any contemporary philosophy of equal rank—any philosophy that has not itself been produced under the influence of Dewey.

And where there are no interactions there are no consequences—both in the Deweyan sense explained, in the sense macroscopically experienced by everyone, in the sense that is emphatically evident in the performance of laboratory experiments. The failure of philosophers "to wander off on the topic of experiment" when dealing with the topic of scientific method and all other topics of philosophy is the root source of the failures of those philosophers. A contradiction or logical inconsistency in a philosophy is not evidence of a crisis in the universe. It is evidence of antecedent error in that philosophy. But a contradiction is also evidence of something more: a contradiction is evidence of discontinuity, of a break, or an irreconcilable conflict in the system. A philosophy that cannot proceed in its establishment of "continuity" except by going from one contradiction to another is not fulfilling its avowed purpose, is not realizing its acknowledged objective.

That Newton's "world of physics" should ever have been taken as the ultimate revelation of the reality of the physical world or of Nature was due, as we have seen, to the operation of the Greek Formula. But philosophers, who are living and thinking beings, could never rest satisfied with that acceptance. Instead of getting rid of the superstition, inherited from Plato, that any "theoretical science" is *ipso facto* the ultimate reality of the subject-matter it is alleged to be about, philosophers began constructing rival systems, aided in their efforts in the nineteenth and twentieth centuries by the new sciences of biology, psychology and latterly Einsteinian physics.

The fundamental constituents of Newton's "world of physics" were the atoms—mathematical points shot through and through with physical interpretation. The atom was a thing without internal differences, eternal and unchangeable. About the nature of the Newtonian atom no inquiry could therefore be made and nothing said. It had to be accepted as an ultimate and very brutish "fact." One atom could be distinguished from another by the purely external means of the something "between" them. But no atom ever got inside any "between" and no "between" ever got inside an atom. Like the stretched rope between horse and stone, all "relations" in the Newtonian world "tied things apart."

The "mechanics" of the Newtonian world of physics were mathematically perfect. The ultimate atoms could be put through all the known mathematical paces without ever an *interaction* taking place to disturb the perfect mathematical balance. Every now and then the physical world would seem to intrude and upset the calculations, but the ultimate metaphysical, mathematical balance of the system for several hundred years went along quite unperturbed. Of the general features of the Newtonian "world of physics" Whitehead well writes:

> . . .space and time, with all their current mathematical properties are ready-made for the material masses; the material masses are ready-made for the "forces" which constitute their action and reaction; and space, and time, and material masses, and forces are alike ready-made for the initial motions which the Deity impresses throughout the universe. (*Process and Reality*, p. 143-144.)

Newton had everything perfectly "related" in his "world of physics" but everything "related" was external to the relation between. "Is there, in the end," Bradley asked, "such a thing as a relation which is merely *between* terms? Or, on the other hand, does not a relation imply an underlying unity and an inclusive whole?"* Whitehead agrees with Bradley and so does Dewey—but for fundamentally different reasons and in different ways. Both Bradley and Whitehead ask whether *in the end* there can be such a relation. Dewey asks whether there can be such *in the beginning*. Bradley and Whitehead ask whether *in the end* all things are not *interrelated*; Dewey asks whether *in the beginning* all things are not *interactive*.

If you start out as Bradley did, by taking "external relations" exemplified in the Newtonian "world of physics" as the nature of relations in the physical world or empirical Nature, you will naturally "in the end" reach a philosophy wherein the situation is reversed. Instead of saying that the Newtonian "world of physics" was only a mathematicized "appearance" of the reality of the empirical world (and a very poor and distorted appearance at that), Bradley, as all other Idealistic philosophers, identified the "world of physics" with the physical world and then made the latter an "appearance" of a Supernatural Absolute Reality.

Russell escapes from Bradley's ending only because he abandons his project of "continuity" in the end. Whitehead does not abandon his project and though he is "in sharp disagreement with Bradley, the final outcome is after all not so greatly different."† And the reason for the approximate identity of outcome is rather simple. *In the beginning* they start with "relations," whereas "relations" are arrived at *in the ending* of philosophic as of scientific inquiry. *In the beginning*, in the gross and macroscopic subject-matter in primary experience, there are *interactions*.

"Once with internal relations, always with internal relations" says Whitehead (*Science and the Modern World*, p. 230). On behalf of Newton, if not of all mathematical physicists, it might be said "once with external relations, always with external relations." And on behalf of Dewey it can be said "once with interactions, always with interactions," *but* the relations between systems of interactions or interactive continua (societies, situations or local histories) may be "external" or "internal" *depending* upon the extent and quality of the *interactivity* between the systems of interactions or interactive continua involved. When by virtue of the "inhibitions and attenuations" the interactivity has decreased so that there are no appreciable *consequences* of the interactivity in the members in that interactive system, then the members *are* in so far *external* to each other. The billiard ball on the billiard table in New York City and the electron dancing round the center of the sun are, when judged by the *interactivity between them*, externally related. When we unite the billiard ball and electron by some system of relations, then *within that system* they are internally related and *that relation* is not merely between them but implies an inclusive whole and an underlying unity. That inclusive whole or underlying unity is precisely what the *system* provides.

For a philosophy not to be able to maintain distinctions between natural differences is as vicious and disastrous as not to be able to maintain connections between things that are interconnected. Every interaction is the interception and union of continuity and difference. The actions within any interaction are different—otherwise there would be no

*Quoted in Whitehead, *Adventures of Ideas*, p. 296; italics in original.
†*Process and Reality*, p. vii.

interaction. That actions within an interaction are internally related is obvious. That is what an interaction is. And the consequences of any interaction are internally related to the interaction of which they are the consequences. Things that are interactive are internally related. Things that are not interactive with respect to each other are externally related. Without "inhibitions and attenuations" of interactivity in some respect there could not be systems or organizations of interactivities in other respects. Without external relations there could be no differences. Without internal relations there could be no continuity. Interaction and continuity intercept as well as unite. Without the interception and union, there would be no world and no inquiry in that world. When we are controlled by our gross and macroscopic experience of doings-undergoings, of interactions and consequences, the doctrine of internal relations ceases to be "one of the dubieties of metaphysics"* and the doctrine of external relations ceases to be one of the absolute and self-sufficient certainties of mathematical science. Control of philosophic thought by the gross and macroscopic subject-matter in primary experience saves us from Hobson's nightmarish choice: *either* the suffocation of internal relations and nothing but internal relations; *or*, the vacuous extinction of external relations and nothing but external relations. In Nature there are lungs to breathe and air to be breathed. There is no *reason* why there shouldn't be the same in philosophy.

XXVI

In the beginning, in the laboratory experiment, in the experience of living beings, in the doings-undergoings of all things within Nature, there are *interactions and consequences*. In the ending, in the theoretical systems of derived, refined objects of reflection, whether scientific or philosophic, there are *relations and implications*.

In philosophies and sciences where there are no *interactions*, where the latter are never referred to, where they are never the objects of denotative reference, of pointing out, there are no *consequences*.

The corruption of the meaning of "interaction" which resulted from the workings of Newton's Third Law was bound to pass over into and corrupt the meaning of the term "consequences." One illustration of the use of "consequences" in the Newtonianesque sense is of special illuminating value. Whitehead writes:

> Mathematics can tell you the *consequences* of your beliefs. For example, if your apple is composed of a finite number of atoms, mathematics will tell you that the number is even or odd. But you must not ask mathematics to provide you with the apple, the atoms, and the finiteness of their number. There is no valid inference from mere possibility to matter of fact, or, in other words, from mere mathematics to concrete nature. (*Adventures of Ideas*, p. 161; italics mine.)

In no sense, no matter how you take the *belief*, does mathematics tell you a *consequence* of your belief when it tells you, under the conditions given, that the finite number of atoms is "even or odd."

Let us first consider a case which Whitehead clearly

does not intend. Let us suppose you make an actual count of the number of atoms in your apple and reach a finite number. That number is the *consequence* of your counting (doing-undergoing) and that number is the content of your belief. The consequence of your counting is that you *know* the finite number you reached and you know that it is even or you know that it is odd. You know which one.

By the process of counting you reached say the finite number 100. The *consequence* of your belief, says mathematics, is that your number is "even or odd." Suppose you take this "consequence" (as you should if it really is a consequence) as a questioning of the result reached and so you go back and count again. As a consequence of your second counting you reach say the number 101. The consequence of your belief, says mathematics, is that your number is "even or odd." After a time you might catch on, and after a further time still you might even get tired of that game. It's just like the game "true or false."

Now let us consider the case Whitehead clearly intends—when you have not made an actual count. Does mathematics in such case tell you a *consequence* of your belief when you provide it with a finite number of atoms and it tells you the number is "even or odd"? If you do not provide mathematics with a definite finite number, what can you provide it with? What can the *content* of your *belief* be? Why, it can only be that selfsame formula: a finite number "even or odd." The fact that you are providing mathematics with a finite number of *atoms* is only a joker in the case. For it is presupposed or assumed that atoms are always wholes, that they do not split, and hence any number of atoms will always be a finite whole number. You will never get 100 1/3 atoms, which is a number that is not "even or odd." A finite number of whole numbers (atoms) *is* a number that is "even or odd" when the number is not definitely specified. That is what a finite number of whole numbers mathematically means. Bring to mathematics a *belief* the content of which is "even or odd" and mathematics will tell you the "consequence" of your belief is "even or odd." We have the same thing taken twice over, once as "belief" and once as "mathematics." (Just as Newton took the same thing twice over, once as the Third Law of Motion and once as the horse, stone and stretched rope between.)

There is another possibility of interpretation to be briefly explored. Suppose you don't know anything about mathematics. Suppose you have, like some pre-literates, counted up only to four. Mathematics, that is, mathematicians who have gone beyond your limited researches in the divine art, can tell you what *consequences they have reached as a result of their inquiries*. In this sense, mathematics told Descartes Euclidean geometry and algebra—and then Descartes turned round and told mathematics what it didn't know before—analytical geometry. In this sense mathematics told Newton and Leibnitz what previous mathematicians had discovered as a *consequence of their inquiries* and then they turned round and told mathematics what it didn't know before—Fluxions or Calculus. In this sense, mathematics can tell you what mathematicians have learnt up to the time of the telling. But mathematics cannot tell you anything that has not yet been found out, any *consequence* of mathematical inquiry that has not yet been reached.

But even when thus pedagogically interpreted, mathematics is not telling you a consequence of your *belief*—if

---

**Adventures of Ideas*, p. 147.

you take the belief psychologically. Mathematics is the last inquiry on earth that can tell you what the consequences of any belief are—even when the content of the belief is mathematical. Witness the illustrious case of Pythagoras. He held the belief that numbers are "even or odd." When mathematics told him that there is a number that is neither even nor odd but both (the square root of 2) the *consequence of his belief* in "even or odd" was that he wouldn't believe what his own mathematical researches told him: he wouldn't believe that the square root of 2 was really a number at all, an object of rational mathematical thought and true mathematical belief. Pythagoras wasn't the first mathematician nor the last philosopher whose mathematical beliefs had consequences of this order. Whitehead himself is by far the greatest contemporary mathematical-philosopher whose mathematical beliefs have in this respect quite thoroughly Pythagorean consequences. Whitehead objects to the "casual" heterogeneity of space-time in Einstein's system because it is "inherent in my [Whitehead's] theory to maintain the old division between physics and geometry."* The details in the case between Whitehead and Einstein are much more complicated than in the case Pythagoras confronted. But the two cases are in principle identical. It was inherent in Pythagoras' theory to maintain the old division between even and odd. When a number came along that exhibited a casual heterogeneity of even-and-odd, Pythagoras would have nothing to do with it. And all good Greek mathematicians and philosophers followed his lead.

Mathematics can tell no one any consequence of any belief. And mathematics, pure mathematics, can tell no *consequences* of anything—probability theories notwithstanding. There is no valid inference from mere mathematics to concrete nature. And consequences are in concrete nature. Mathematics is the *consequence* of inquiry, and the consequences of mathematical inquiry *when used* have further consequences.

The practically inseparable complementary of the intellectual habit of calling "consequences" what are plainly not consequences, is the intellectual habit of not calling consequences what are plainly consequences. One illustration from Whitehead will amply exemplify the latter.

Whitehead is critical of "the Positivist doctrine concerning Law, namely, that a Law of Nature is *merely an observed* persistence of pattern in the *observed succession* of natural things: Law is then merely description."† In analytical description of this doctrine, Whitehead has the following essential things to say:

It [Positivism] presupposes that we have direct acquaintance with a succession of things. This acquaintance is analysable into a succession of things observed. But our direct acquaintance consists not only in distinct observations of the distinct things in succession, but also it includes a comparative knowledge of the successive observations. Acquaintance is thus cumulative and comparative. The laws of nature are nothing else than the observed identities of pattern persisting throughout the series of comparative observations. Thus a law of nature says something about things observed and nothing more.

The preoccupation of science is the search for simple statements which in their joint effect will express everything of interest concerning the observed occurrences. This is the whole tale of science, *that* and nothing more. It is the great Positivist doctrine, largely developed in the first half of the nineteenth century, and ever since growing in influence. It tells us to keep to things observed, and to describe them as simply as we can. That is all we can know. Laws are statements of observed facts. This doctrine dates back to Epicurus, and embodies his appeal to the plain man, away from metaphysics and mathematics. The observed facts of clear experience are understandable, and nothing else. Also "understanding" means "simplicity of description." (*Adventures of Ideas*, pp. 147-148; italics in original.)

The above account is an accurate description of the fundamentals of "the great Positivist doctrine" of which the current school of Logical Positivism is one of the hybrid offshoots. Although Whitehead is critical of Positivism, he is roundly critical—he turns right around and adopts their position. Thus on the heels of the above, he goes on to say:

Without doubt this Positivist doctrine contains a *fundamental truth about scientific methodology.* For example, consider the greatest of all scientific generalizations, Newton's Law of Gravitation:—Two particles of matter attract each other with a force directly proportional to the square of their distance. The notion of "force" refers to the notion of the addition of a component to the vector acceleration of either particle. It also refers to the notion of the masses of the particles. Again the notion of mass is also explicitly referred to in the statement. Thus the mutual spatial relations of the particles, and their individual masses, are required for the Law. *To this extent* the Law is an expression of the *presumed characters* of the particles concerned. *But the form of the Law,* namely the product of the masses and the inverse square of the distance, *is purely based upon description of observed fact.* A large part of Newton's *Principia* is devoted to a *mathematical investigation proving that the description is adequate for his purposes*; it collects many details under one principle. *Newton himself insisted* upon this very point. *He was not speculating; he was not explaining.* Whatever your cosmological doctrines may be, the motions of the planets and the fall of the stones, so far as they have been directly measured, conform to his Law. *He is enunciating a formula which expresses observed correlations of observed facts. ∗*

We have previously quoted in part the paragraph that immediately succeeds the above. For the benefit of completing the context of discussion, the paragraph is cited here entire:

Without the shadow of a doubt, all science bases itself upon *this* procedure. It is the first rule of scientific method—Enunciate observed correlations of observed fact. This is the great Baconian doctrine, namely, Observe and observe, until finally you detect a regularity of sequence. The scholastics had trusted to metaphysical dialectic giving them secure knowledge about the nature of things, including the physical world, the spiritual world, and the existence of God. *Thence they deduced* the various laws, immanent and imposed, which reigned supreme throughout Nature. (*Adventures of Ideas*, p. 149; italics mine.)

---

*Quoted from Whitehead's *The Principle of Relativity* in *Analysis of Matter*, p. 77.
†*Adventures of Ideas*, p. 147; italics mine.

∗Adventures of Ideas, p. 148; italics mine.

Whitehead's grievance against "the great Positivist school of thought [which] at the present time reigns supreme in the domain of science" is that its

aim . . .is to confine itself to fact, with a discard of all speculation. Unfortunately, among all the variant schools of opinion, it is the one which can least bear confrontation with the facts. It has never been acted on. It can never be acted on, for it gives no foothold for any forecast of the future around which purpose can weave itself. (L.c., p. 159.)

The great Positivist doctrine—and the Logical Positivist doctrine—cannot bear confrontation with the facts. But the facts it cannot bear confrontation with are the facts of scientific methodology. However, if you accept the Positivist doctrine as a true account of scientific methodology, if you accept, with them, the *consequences* of a long and complicated procedure of inquiry as the enunciation of direct observation of observed correlations of observed fact, the belated introduction of a "metaphysical understanding" of "speculative boldness" to guide imagination and justify purpose *in the future* does not help matters any. The future grows out of the present and the great trouble with Positivism is that it doesn't take care of the present. Just as the term "metaphysics" has the psychological effect of giving the user thereof the feeling of having hit the bottom of the bottom floor or the top of the top ceiling, so "Positivism" has the psychological effect of giving the user thereof the feeling of standing on the level.

"As long as man was unable by means of the arts of practice to direct the course of events, it was natural for him to seek an emotional substitute; in the absence of actual certainty in the midst of a precarious and hazardous world, men cultivated all sorts of things that would give them the *feeling* of certainty. And it is possible that, when not carried to an illusory point, the cultivation of the feeling gave man courage and confidence and enabled him to carry the burdens of life more successfully. But one could hardly seriously contend that this fact, if it be such, is one upon which to found a reasoned philosophy."*

Philosophic reasonings, like all reasonings, generate feelings of certainty. And no individual philosopher can ever escape from having those feelings engendered in him. A philosopher is at least as human as a scientist and usually he is more so. When a philosopher introduces a second explanation to explain what the first was introduced to explain and then introduces a third explanation to explain the second, we have gross and macroscopic evidence that something is seriously wrong with the first explanation and that the philosopher in question is aware that this is the case. When the philosopher stops with his third explanation because of the feeling of certainty engendered in the process, that may be a satisfactory way of enabling that philosopher to carry the burdens of his philosophy. But no subsequent philosopher ever feels under obligation to carry the burdens of his predecessors in the same old way. He wouldn't be a philosopher if he did. However, if he follows the same method of pyramiding explanations he will soon discover that he can escape the burdens of his predecessors only by adding new burdens. And they usually turn out to be reconstructed complications of the old.

The histories of Whitehead and Russell exemplify this to unfortunate perfection. They began by making a

*The Quest for Certainty, p. 33; italics in original.

clean sweep and then started out from symbolic-logical scratch. But what they failed to sweep out was the traditional method of philosophy. And that brought all that was so industriously swept out back in again.

As long as the old method prevails in philosophy—no matter what the symbolic disguise—the same forlorn history will be repeated. No burden will ever be removed and philosophy, instead of becoming lighter and clearer in its historical passage, will become weightier and weightier, a denser and denser mass of fiercely entangled "eternal problems."

XXVII

A formula is a formulation. In itself it is a finished and completed thing. Any given "ideally isolated system" such as the First Law of Motion is a formula. As such it is the final term of a sequential history of inquiry. Instead of all scientific progress depending upon *first* framing a formula, just the opposite is true: the framing of a formula is the fulfillment, the realization, the *consequence* of scientific progress made. As the last term of that sequential order of inquiry, it is irrelevant to scientific progress, although it may, of course, be, in addition, an initial term in another sequential order of inquiry. And when it is made such an initial term in another inquiry, then of course it does become relevant to *further* scientific progress. The Third Law of Motion, for example, is another of Newton's "ideally isolated systems," the final term, the consequence of another order of inquiry. And in reaching that conclusion, there can be no doubt Newton was helped by using the notion of the First Law. So, too, on a vastly more complicated scale, with the formula which constitutes the Law of Gravitation. It is the net consequence, the terminal result, of an elaborate history of inquiry involving the use of all the "notions" Whitehead specified—and many more besides.

When scientific theoretical inquiry is as highly developed as it is in modern times, the use of ideally isolated theoretical systems—formulae—is a matter of course. And to continue theoretical inquiry on the same high level, and to further the development of that high level the continued use of ideally isolated theoretical systems becomes a matter of necessity. *In this sense*, the concept of "ideally isolated systems" is *essential* to scientific theory.

However, there is a fundamental difference between the concept of *an* ideally isolated system, and the concept of "ideally isolated systems." The former is the concept of a specific system which is the consequence of inquiry undertaken and completed. The "truths" it contains are the truths attained. The latter is the general concept of a *method* of procedure. It contains no "truths" at all but is part of the *method* of attaining truths. And like every method of procedure, it is itself the *consequence* of proceeding. It is developed in the course of using that method of procedure.

Every specific isolation is the *consequence* of an activity involving the use of the method of isolating. When, a while back, we were considering the laboratory experimentalist "by himself," we *had* him "by himself" as a consequence of having performed an act of isolation. It was an achievement, a consequence of inquiry, not a datum or gift to inquiry. We are now so expert, so habituated to using the method of isolating—in some cases—we take the consequences for granted, as if they were naturally coming to us and we did not have to go out and get them.

Because laboratory and theoretical activities are interactive *within* inquiry, the consequence of performing the initial act of isolating did not separate the laboratory activity from the theoretical, but isolated it as *within* inquiry. If we had stopped after achieving that initial consequence, *that* consequence would have been the total content of our "ideally isolated system." The "truth" contained in *that* "ideally isolated system" would have been the sole "truth" that laboratory experimentation can be isolated as within inquiry. That would be a "truth" of that system because that system was the consequence of inquiry undertaken and carried to that completion. The enunciation of the proposition "laboratory experimentation can be isolated within inquiry" would be the formulation of the consequences achieved, of the scientific progress made.

However, we didn't stop there but continued with our inquiry. As a consequence of that continuance we acquired further "truths." The content of our initially ideally isolated system was increased. Some of the further consequences, some of the further truths, were, for example, that laboratory experimentation involves the use of instruments and materials; that an experiment is an organization of instruments and materials in accordance with a plan; that an experiment is performed with an end-in-view.

Now "plan" and "end-in-view" are theoretical elements, consequences of theoretical activity. Without going over the whole ground again, suffice to say that the *general* consequence of our inquiry into the conduct of laboratory activity was that the *consequences* of theoretical activity are internally involved at every point; that the continuity and interaction of the two intercept and unite.

If laboratory experimentation can be isolated within scientific inquiry, naturally, the same can be done with scientific theoretical experimentation. In fact, the consequence of isolating either one is that the other is also thereby isolated, since the two comprise the totality of scientific inquiry.

When we were analyzing the conduct of laboratory activity, we were constantly compelled to take into account the consequences of theoretical activity. Had we based our analytical inquiry on the antecedent presupposition that laboratory experimentation is "practice" and separated from "theory," the consequence of that presupposition would have been that no analysis would have been possible. Such an antecedent condition would have been a cause sufficient to have the effect of extruding laboratory activity from inquiry into scientific method.

When, having achieved the initial isolation of the scientific theoretical activity, we proceed to inquire further into it, we are likewise compelled to take into account the laboratory activity. In the one case as in the other, the compulsion is essentially inevadable because of the functional interactivity of theoretical and practical activities.

The compulsion is inevadable. But like all compulsions working in theoretical enterprise—of which philosophy is one—it is postponable. In fact, if it couldn't be postponed, there would be no theoretical activity of any sort. For all thinking or deliberation consists precisely in postponing what has to be done. In technical psychological terms, thinking, deliberation is consequent upon practical responses being delayed. Only by metonymy, however, is thinking itself a "delayed response." When thinking goes on, it delays the overt response further. But thinking is a *consequence* of a mode of socio-biologic organization of interactivity, not a metaphysically primitive condition or cause. There has to be a response delayed before thinking can come into existence and delay a response.

In the "purest" theoretical activities—in symbolist poetry, symbolic logic, pure mathematics, and some metaphysical philosophies—the compulsion of taking into account the consequences of practical activity can be indefinitely postponed. Just as soon as it seems imminent that the next turn will lead back into practice, all that it is necessary to do is to write another symbolist poem, develop another symbolic-logical distinction, inquire further into pure mathematics, excogitate some more metaphysical philosophy.

Such theoretical activities, or at least some of them, have a special fascination for the philosopher. A philosopher is occupationally, if not constitutionally, prone to dismiss the technological uses of the microscope in biology as irrelevant for a philosophy of the sciences or the logic of scientific method. And having dismissed the microscope at the very outset of his inquiry into scientific method, he thinks no more of it. Being a theoretician himself, he knows *he* is not going to use a microscope in the conduct of *his* inquiry. However, when he comes to the technological uses of the calculus in physics, he does not dismiss *that* as irrelevant for a philosophy of the sciences or the logic of scientific method—especially if the calculus is not the old one of Newton but the new tensor calculus of Einstein. And having started out by following the lead of his own bent, the philosopher can go on and on, indefinitely postponing taking into the account of his theoretical analysis the consequences of practical activity. He may even reach the point Russell early reached, of making the "philosophy of mathematics" the *whole* of the "philosophy of scientific method."

However, the ideally isolated systems Whitehead asserted to be essential to scientific theory were not of the "pure" theoretical sort. As an instance of the kind of system he meant, he cited Newton's First Law of Motion: "Every body continues in its state of rest, or of uniform motion in a straight line, except so far as it may be compelled by force to change that state." He did not, as an instance of the kind of system he meant, cite a formula like A is A or a+b=b+a. For Whitehead, as we have already seen, there is no valid inference from mere mathematics to concrete nature. The systems essential for scientific theory are such as are ideally isolated *within* concrete Nature. Hence Whitehead's concern to make clear that such systems are not substantially torn *out* of Nature.

Newton's Third Law of Motion is as much an instance of an ideally isolated system as his First Law. Formally considered, it is in fact much better because it has explicitly formulated two terms and a relation between, which is the barest minimum for any ideal system. Since we have had some dealings with the Third Law, we may as well continue with it. It is legitimate to make this substitution, for what Whitehead says has application to all ideally isolated systems. He is enunciating a general character of the conception. Also we may legitimately substitute the Third Law for the Law of Gravitation—as far as any general argument is concerned. If the form of the Law of Gravitation is "purely based upon description of observed fact," why, so is the Third Law of Motion. Newton himself insisted on this very point, as we have already seen. According to Newton, the Third Law—like all his Laws—is "derived from phenomena." He was not explaining, he was not speculating, he was simply enunciating a formula which expressed the observed correlations of observed facts. And like the Law of Gravitation, the Third Law "collects many details under one principle."

XXVIII

When we take a horse out of one field we can do so only by taking that horse, in the same process, into another field. We can take one thing in nature out of one sequential order of events, only by bringing that thing into another sequential order of events. In homely language, we can go *from* one place in nature only by going *into* another place in nature. That we can, as a matter of practical reality, take a number of things in nature from different places and bring them together in one place, is not a consequence of any theory about nature but an exemplification of one of the ways in which nature goes on.

Now when as experimental laboratorians we take a horse out of one field, a rope out of another, and a stone out of a third, and bring them all together into the laboratory we have, to quote Whitehead, "freed" those three things "from casual contingent dependence upon detailed items within the *rest* of the universe." That is, we have brought them into casual contingent dependence upon detailed items within the laboratory. For so far, we have only brought them *into* the laboratory—which, like every other place, is just another place in Nature. It is only by a futuristic figure of speech that we can say we have already "scientifically isolated" the horse, stone and rope. We have only gathered them together—as people gather together in a theater before the show begins.

The items within the laboratory (including the laboratory itself) are like all items within the universe. Within the laboratory there are items that are "casual" as far as the inquiry to be conducted is concerned; but there are also other items which are not casual, but *causal*, with the respect to the experiment to be performed.* The scale, pulley, meter or other piece of apparatus *when organized with the three objects into an experiment* is a necessary causal factor, and not a casual contingent one because we have it under control and know how it will behave when made an interactive member of the interactive system which the experiment constitutes.

If we had no experimental apparatus within the laboratory, things which we can exercise control over, bringing the objects into the laboratory would be of no scientific consequence. Horses can be looked at, contemplated, in the fields where they roam as well as in the laboratory—if not better.

A scale is a scientifically isolated physical system. A perfect scale is a perfect, ideally isolated physical system. Since there are no perfect scales, we may say that *in so far* as it is an instrument whose behavior we have standardized and regulated *to that extent* is it an *ideally* isolated physical system. And having made this qualification once, we need not make it again. It is taken for granted throughout the sequel.

The scale as an instrument of inquiry is the consequence of a long series of practical-theoretical investigations carried to completion. As a thing, within its own boundaries, the physical scale is no different from

*"From the standpoint of control and utilization, the tendency to assign superior reality to causes is explicable. A 'cause' is not merely an antecedent; it is that antecedent which if manipulated regulates the occurrence of the consequent." *Experience and Nature* (1st ed.), p. 109.

the stone within its boundaries. Within Nature both are interactive continua, neither superior to the other in this respect. When the stone is on the floor of the laboratory, scale and stone are as casually related, as externally related, as contingently related as the stone and the tree nearby the stone in the field out of doors.

When, however, we put the stone on the scale, the situation is radically changed. With respect to the progress of inquiry, the stone and scale are, within the experimental situation thus created, fundamentally different. The scale becomes an instrument of investigation, and the stone the material to be investigated.

The scale, let us say, is in perfect condition. It has been perfectly standardized as a consequence of a series of interactions. Although standardized, it is not standardized at one fixed point. The pointer of the scale is not fixed at the marking 0 or 100. The scale is so constructed that it can *interact* with things put on it, and the consequences of the interactions are different as the interactive things put on it differ.

Before the stone is placed on the scale, the scale is a settled, completed, finished thing. It rests in the bosom of the laboratory the way a stone rests in the bosom of the pasture and both rest in the bosom of Nature. When we put the stone on the scale, when we organize the two into an interactive system, the settled system of the scale is unsettled, and the final consequence of the new unsettlement is that a new dynamic equilibrium is achieved. When the pointer comes to rest, the interactions, the doings-undergoings within the experimental situation inclusive of the scale and stone have come to their conclusion. Stone and scale now constitute one interactive continuum, contained within the new boundaries which their interactivity has created. They have united their forces and face the world with a common or united front.

If the scale were so constructed that it entered into the interactivity completely, so that it put its whole soul and being into the doings-undergoings within the experimental situation, we would be no better off, with respect to inquiry *into the stone*, than we were before. Isolated hydrogen and isolated oxygen when made to interact with each other in the chemical laboratory do put everything they are and have into the doings-undergoings. Both are consumed in the interactivity and the consequence is something new. We know as result of that experiment that when hydrogen and oxygen interact completely, neither preserving a thought of saving itself, that *water* is the consequence. But what we know is equally divided between the two. Both went into the doings-undergoings and neither of them came out. We have learnt something about both, but nothing about either one of them alone through the *instrumentality* of the other.

The scale is constructed so that it won't do that sort of thing. It will enter into interactions but not so completely as to lose itself. It will let the stone upset its balance, but it keeps its head. The scale, in other words, is within two situations simultaneously, within the experimental situation and within its own situation within the universe. Hence the scale weighs the stone.

When the scale and stone have reached their conclusive adjustment and have settled down together, we look at the pointer of the scale and make the reading. We *know* how much the stone weighs, not because we "contemplate" the pointer on the face of the scale, but because the scale is a construction, an instrument, the net accumulated consequence of a history of inquiry, of doings-undergoings, practical and theoretical. Furthermore, the place where the pointer is at rest is a *new* consequence, the stable *effect* of the stable organization of interactivities the scale and stone have reached. If the pointer always stayed at the same place of rest it would give no reading. Whenever it gives a reading, it is because there has been a passage of the pointer from one place into another, the mode of passage controlled or regulated by the organization of the vital organs of the scale. The reading is the concluding phase of that passage, and that concluding phase includes within itself the historical sequence of events of which it is the net effect. The reading is a *fact* shot through and through with meaning. It is a refined object all ready for subsequent scientific reflection.

When we "ideally isolate" the pointer where it comes to rest from the historical passage through which it went before it reached that point of rest; and when we go further and "ideally isolate" the total scale from the history of inquiry of which it is the end-result, when we do these things* we get the Eddingtonian consequence. Eddington gratuitously condemns himself to the vacuous dizziness of going round from one meter to the next and finding nothing but numbers at every place he stops. And Eddington's fate is not unlike that of the gas-meter man who, having taken down meaningless numbers the livelong day, escapes, at the fall of darkness, into another world.

<h2 style="text-align:center">XXIX</h2>

Let us now consider the Third Law. It is an "ideally isolated system" isolated as within Nature. "All scientific progress depends upon first framing a formula." The Third Law is our formula. It is already framed and now we want to progress. As an ideal formula, it should be an ideal instrument of inquiry into things. That is its whole virtue, the reason for its exaltation. It tells us the way of the land. So let us consider the Third Law as an instrument, in the sense that a scale is, something with which we can take the measure of things and find something out about them with respect to their doings-undergoings, their actions and reactions.

Not unduly to prejudice the case, let us not take the Law to the horse, stone and stretched rope between, whence Newton "derived it" but to a contemporary horse, stone and stretched rope between. And what do we find? We find that when we take the Law purely theoretically, we always get "action and reaction are equal and opposite." No matter how we *theoretically* "apply" that Law "to" our contemporary state of affairs—applying it to the whole of it or any part of it, longitudinally or laterally—it cuts the same way. The horses are different, the ropes are different, the stones are different but the Third Law remains the same. The horse may be pulling and relaxing, the rope may be stretching and sagging, the stone may be moving and resting but the Third Law will tell us none of these things. As long as we listen to the Third Law we will hear it repeating "action and reaction are equal and opposite." As far as advancing our scientific knowledge is concerned, our application no more advanced our knowledge than Newton advanced his scientific knowledge when he brought the Third Law to his horse, stone and stretched rope between. He started with the Third Law and ended with the same.

With respect to Newton's experimental situation as with respect to ours, the Third Law, when theoretically applied, was as casual, as contingent an item within the

*They are really one thing done twice over: once microscopically, and once macroscopically.

universe as any blade of grass waving in the breeze at the antipodes.

The Third Law, like any ideally constructed refined object of reflection, is standardized to give one reading only; and when taken by itself gives only one reading—like a yardstick, which always says "one yard." But there is this great difference between a physical yardstick and a Law. If you apply a yardstick end-on, and proceed in an ongoing line, you get one yardstick, two yardsticks, three yardsticks. You get ahead. You count up. But when you count three applications of the Third Law, no matter how you proceed in your theoretical process of application, you get "action and reaction are equal and opposite" once, twice, thrice. You can count, but cannot count up. Since after three applications you get thrice, not three, once is enough.

The "application" of the yardstick is only by an unfortunate habit of language an "application *to*;" actually, it is an experimental doing-undergoing within Nature. The yardstick, as it moves through place into place, is interacting—as the scale interacts. Hence, unlike the Third Law when theoretically applied, it gives the measure of the land and not the measure of itself. Sometimes part of the interactive organization of the yardstick happens to be a human being. But mileage-meters, and meters of whatever sort, are experimental demonstrations that the human being is not needed for that kind of instrumental operation.

Any formula, any refined object of reflection, is the consequence of inquiry. Take it by itself, in "isolation," and it is a finished, completed thing. When you theoretically apply it to a situation you are still taking it by itself. And hence it will always give the reading of itself and not of the situation to which it is applied. The Third Law, or any other formula, is in this respect precisely like the formula "even or odd." Provide the mathematical formula "even or odd" with any number and it will always say "even or odd." Whether that number is even or is odd is something that you will never discover by means of theoretically applying the formula. To discover what that number is, you have to undertake the requisite inquiry.

A formula is the consequence of scientific progress made. Stop there, and that is where your science stops. The formula can become an esthetic object, a headache, a bore, a means of earning a livelihood by pedagogically putting it into the heads of others, and a possible variety of other things. But it does not become an instrument of scientific progress until something further is done with the formula in the course of another inquiry. In its solitary confinement, in its ideal isolation, the formula, if we may trust Aristotle, becomes either a god or a brute. For living experience, there is no genuine difference between the alternatives. The inevitable consequence of deification is the brutalization of human life.

<h2 style="text-align:center">XXX</h2>

There is only one way of finding out whether in an actual existential situation "action and reaction are equal and opposite," and that is by making an experimental laboratory test. For such a test, laboratory instruments are necessary, ideally isolated physical systems in the sense explained. Newton was no laboratorian and his test of the Third Law was no experimental test. The immortal achievement of Einstein fundamentally consists in making scientists realize the difference between theoretical scientific construction and experimental laboratory testing of the constructions theoretically reached. The experimental laboratory testing of conclusions reached by scientific theoreticians is now made by laboratorians competent and instrumentally equipped to do that part of the total scientific work. Einstein completed the revolution Galileo started.

The instruments of the theoretical physicist today are all of a mathematical quality because the *material* the theoretician receives from the laboratorian is already mathematicized. The material of the theoretical physicist is constituted by the refined objects of scientific-physical reflection which are consequences of laboratory experimentation and which he finds in the "memoirs." The records of the facts come from the laboratory and they come shot through and through with mathematical meaning. But obviously there must be something in the facts besides the mathematical interpretation, otherwise there would be no distinguishing between the two. This is what James called the "irreducible and stubborn facts" and what we have called, slightly modifying Dewey's phrase, the subject-matter in the primary laboratory experience.

Like the material and instruments of the laboratory physicist, so with the material and instruments of the theoretical physicist: they are distinguishable from one another but not separated and disconnected. Pure mathematics is separated and disconnected from the refined objects of scientific-physical reflection received from the laboratory. But pure mathematics is not the *instrument* the theoretical physicist uses. The mathematics already *in* the scientific-physical refined objects he receives from the laboratorian, is *de-purified* and the theoretical physicist in pursuing his inquiry into those objects and searching for a solution to the problem they raise, must constantly de-purify the mathematics he uses to be able to continue in his pursuit. "The mathematics in which the physicist is interested was developed for the *explicit* purpose of describing the behavior of the external world, so that it is certainly no accident that there is a correspondence between mathematics and nature."* This is not altogether so. The contemporary physicist is interested in the non-Euclidean geometries and they were not developed for the explicit purpose of describing the behavior of the external world. There is also the celebrated case of conic sections which for some eighteen hundred years was of no interest to the physicist. The physicist today is, generally speaking, interested in all pure mathematical systems and in the construction of more and more of them. Also, since physics became mathematical, the production of pure mathematics has vastly increased. In the world of intelligent activity, where there is a demand or need, there is concerted effort made to supply.

But the pure mathematical systems are, for the physicist, his instrumental *sources*. They are not his instrumental *resources* until by *using* them in his inquiry into refined scientific-physical objects—the facts or *data* he has received from the laboratorian—he has converted them into such. Pure mathematics may be as pure as the angelic hordes, but for the physicist, they are only half-raw material that must be further fashioned, by use, before they become finished instruments.† "In all

*Bridgman, *The Logic of Modern Physics*, pp. 60-61; italics mine.
†The theoretical physicist is not peculiar in having others supply *sources* of material for his instrumental uses. The laboratorian also has as *sources* the productions of industrial and fine arts—and for him too, the products, however finished and final whence they are taken, are only half-raw materials which become *resources* for laboratory experimentation only as they are used and changed in laboratory practice.

theoretical physics there is a certain admixture of facts and calculations" *because* the whole process of physical theorizing consists in continuously "admixing" the two in a certain way.

XXXI

Every instrument in the laboratory is the physical embodiment of the consequences of histories of interactivity of theoretical and laboratory functions in inquiry. The current microscope, for example, unites within itself the end-results cumulatively attained in the historical course of progressively integrating the consequences, corrective and expansive, of theory and practice. Every laboratory instrument is a (relatively) ideal physical system of interactions (or an interactive continuum) isolated within Nature. The more ideally organized, the more completely "isolated" within itself, the more carefully standardized, then the more adequate the instrument for further and furthering inquiry. The measure of our control in laboratory experimentation is measured by the range and quality and number of our ideally isolated physical systems—by our instrumental equipment.

The microscope today is constructed in accordance with the specifications of a formula. The formula of the microscope, *qua* formula, in its strictly professional capacity, is theoretical. But the formula is itself a product, the cumulative end-result of the same inclusive histories of inquiry of which the microscope is the physical product or end-result. The practical history of making glass, polishing, silvering, and so on, is as much *internally involved* in the finished product which constitutes the formula, as the theoretical history of *formulating the consequences* of polishing, silvering and so on, is *internally involved* in the finished product which constitutes the microscope.

The formula of the microscope is an ideally isolated theoretical-physical system, isolated as within Nature. The extent of our control in theoretical scientific experimentation is measured by the range and quality and number of our formulae (as just defined)—by our theoretical instrumental equipment.

The distinction between "material" and "instruments" is functional. There are no "materials in themselves" on the one hand and "instruments in themselves" on the other. The material is that which, within the history of inquiry going on, is under investigation, is being inquired into; the instruments are, within that same history, the means used in making the investigation, in making the inquiry. Both material and instruments are therefore within inquiry, the distinction between them existing only while the process of inquiry is going on. The distinction itself is a consequence of inquiry, not an antecedent, or a "cause" making inquiry possible. Outside of the process of inquiry, all things relapse into the "state of nature." A microscope, outside of use in inquiry, is no more an instrument than a boulder on the side of the Himalayas. A formula outside of use in inquiry is no more an instrument of scientific progress than the other side of the moon. We can "contemplate" both—and derive esthetic enjoyment from doing so, each enjoyment differing with the object enjoyed.

If there were things that were aboriginally and ineluctably just "material" and other things that were likewise just "instruments," that state of affairs would not *cause* inquiry, but stop it. The only thing then possible would be the footless process of externally "applying" the instruments "to" the material, and even that would be impossible if you had done a real job of separation, and had not left an ambiguous umbilical attachment somehow dangling between the two.

Within inclusive inquiry—including laboratory and theoretical functions—there is the distinction between material and means. Within each half of inquiry, there is the same distinction prevailing. Because within each half of inquiry there are both material and means (instruments), each half of inquiry can proceed in *partial* independence of the other. The theoretical experimentalist and laboratory experimentalist, taking each one by himself, has his own distinctive ways of exercising control over his material; each is controlled by the consequences of the activity of the other; and each is limited by the limitations of his instruments. There are morphological similarities and identities in the two activities because of the inevitable interactivity of the two.* They interweave and cross-weave, intercept and unite, each working *in* the territory of the other.

The laboratory and theoretical activities, taking them each in their own partial histories, are never exactly abreast. They are always shooting ahead or falling behind each other. The laboratorian, in performing a test or making a new experiment, often creates a new problem, and that is something for the theoretician. The theoretician in solving one problem often broaches another and that is something for the laboratorian.

When we take the macroscopic history of any modern science, the most obvious characteristic is the interweaving and cross-weaving, the interception and union, of new problem and old solution and new solution with old problem in an indefinite variety of ways and extents. It is this interdevelopmental process of inquiry that makes modern science progressive and cumulative, ever richer and more fruitful in consequences. When our logic of the nature of inquiry is controlled by the gross and macroscopic subject-matter presented in primary experience by the history of inquiry, the need for introducing, as Whitehead does, "metaphysical understanding" and "speculative boldness" to explain the *further* progress of science disappears.

In illustrating concretely the need for "metaphysics," Whitehead details Percy Lowell's calculations which led to the discovery of the new planet Uranus. Whitehead describes the complex calculations involved in the approved style of the Positivists. Then he goes on to say that the Positivists would claim that "we have only to look in the sky, towards Percy Lowell's moving point, and we shall see a new planet." And in reply to this Whitehead says:

Certainly we shall not. All that any person has seen is a few faint dots on photographic plates, involving the intervention of photography, excellent telescopes, elaborate apparatus, long exposures and favourable nights. The new explanation is now involved in the *speculative* extension of a welter of

physical laws, concerning telescopes, light, and photography, laws which *merely claim to register observed facts.*

However, continues Whitehead:

This narrative, framed according to the strictest requirements of the Positivist theory, is a travesty of the plain facts. The civilized world has been interested at the thought of the newly discovered planet, *solitary and remote,* for endless ages circling the sun and *adding its faint influence* to the tide of affairs. *At last it is discovered by human reason, penetrating into the nature of things* and laying bare the necessities of their interconnection. The *speculative* extension of laws, *baseless on the Positivist theory,* are the obvious issue of *speculative, metaphysical trust* in the *material permanences such as telescopes, observatories, mountains, planets,* which are behaving towards each other according to the necessities of the universe, including theories of their own natures. The point is, that speculative extension *beyond direct observation* spells some trust in metaphysics, however vaguely these metaphysical notions may be entertained in explicit thought. . . . Metaphysical understanding guides imagination and justifies purpose. Apart from metaphysical presupposition there can be no civilization.

There is a moral to be drawn as to the method of science. All scientific progress depends on first framing a formula giving a general description of observed fact. . . . At one stage, the method of all discovery conforms to the Positivist doctrine. There can be no doubt that, with this restriction of meaning, the Positivist doctrine is correct. (*Adventures of Ideas,* pp. 163-164; italics mine.)

Whitehead's account of the "real" nature of the method of discovery of Uranus is as much a "travesty of the plain facts" he himself recites as is the Positivist account. The Positivist theory is, true enough, baseless; but it is baseless throughout. At *no* stage does the method of *discovery* conform to the Positivist doctrine. It is only when the baseless Positivist theory is taken as point of departure that it becomes necessary to invoke "metaphysical extensions."

Whitehead is too great a mind to be satisfied with any easy solution, too great to accept any standardized scheme handed down. Hence his contradictions and oscillations. When he has the formula dominantly in mind, as the object of Rational Thought, the object discovered by "human reason penetrating into the nature of things" then it is the "facts" that are shot through and through with interpretation. The "facts" are then merely antecedent to the framing of the formula. When he follows the Positivist doctrine and has the facts dominantly in mind, then the formulas become replicas of the facts, enunciations of the observed correlations of observed facts, and as direct, as immediate, and as locally bound and restricted as the facts and the observation of the facts are assumed to be. In such case, obviously, the formula becomes as merely antecedent, as purely ancillary as the "facts" were in the first case. Antecedent to what? Antecedent, of course, to the Rationalists' future. Hence, just as the facts when they were purely antecedent had to be given a shot of interpretation, so now, the formula has to be given a shot of "metaphysical understanding." But the "metaphysical understanding"—which must be humble before both logic and fact—turns out to be, on examination, none other than the formula in a faint futuristic disguise. Speculative boldness empowers the

---

*The morphological similarities and identities are not here detailed, because it would involve a repetition of Sections XVIII and XX, substituting "theoretical" instruments and means for laboratory or practical instruments or means.

Because the consequences or products of theoretical activity are always *means* for guiding, regulating, practical activity—particularly so in scientific inquiry—Dewey calls all consequences of theoretical activity, *when taken by themselves, means,* and hence makes the distinction, within the theoretical activity of "material means and procedural means." (*Logic,* p. 136.) This is one way of emphasizing his fundamental doctrine. However, the same distinction between material means and procedural means can be made within the laboratory activity.

formula, when thus transformed, to reach back and collar itself so that it may enact its own purpose. All purposes being proleptic in nature, the self-captured formula is thus enabled to lead itself into its own future.

But when we keep our footing in the natural world, and are controlled in our philosophic reflections by the gross and macroscopic doings-undergoings in primary experience, the whole scheme of metaphysical apparatus becomes a useless, when not vicious, encumbrance. We can pass from situation to situation, with the passage of Nature, carrying along the consequences of our intelligent labor as we move from one task to the next, using the consequences already attained as means for further progress.

Our theories do not make knowledge possible. "The very possibility of knowledge...should depend on the interwoven nature of things." Not only should our knowledge so depend—it does. The historic development of scientific knowledge is not the consequence of Scientific, Philosophic or Logical Theory *furnishing* Nature with continuity—furnishing Nature therewith by assertive metaphysical fiat because continuities are indispensable for knowledge and "The Theory of Knowledge" wants knowledge to go on. Without there being natural existential continuities in Nature, there would be no knowledge at all—not even the knowledge that to have knowledge continuities are necessary.

But the continuities are not all. The interwoven nature of things is not interwoven in a system of eternal bonds, immutable and transcendent. The interweaving is the consequence of interactivity, and the interweaving changes as the interactivities change. Knowledge is an exemplification of both continuity and interaction in Nature and without either knowledge would perish, for Nature would stop.

The future grows out of the present activity and the present grows out of the past. When we are controlled by our gross and macroscopic primary experience we are able to bring under control our derived, refined objects of reflection—no matter how bold they are in their criticism. The bolder the better. When our philosophic and scientific understanding is controlled by experience, imaginative purpose has its natural roots and a natural mentor—no matter how far it leaps into the future. The further the better. Our purpose, being the consequence of controlled inquiry, does not weave around us in a beckoning haze, but leads through our history, carrying within itself the justification that that history can give. And as we act further upon our purpose, it gains or loses justification in the process of acting upon it because our acting is under our intelligent control. In science as out, guidance comes through undergoing, and justification is a consequence of doing.

XXXII

When Whitehead and Russell use, as they constantly do, such phrases as "mathematics tells" or "physics tells," they are not engaged in "personifying" mathematics or physics. The phrases, however, are not just "semantic" modes of speech, verbal or linguistic "conventions" of the English or philosophic language: they are indicative or revelatory of the fundamental logic or rationale of the traditional philosophic method they follow. That method consists in treating the *consequences* of inquiry—mathematical, physical, psychological or whatever—as if they were directly *given*, as if they were primitive gifts or data. This method of substituting derived, refined objects of reflection for the gross and macroscopic subject-matter

in primary experience does not result in any "personification" in the vulgar sense of the term—because Plato succeeded in taking all the vulgarity out of it. However, it is the refined philosophical or logical equivalent of personification, namely, the depersonalized personification that is technically known by the not too unambiguous term "hypostatization."

Plato put his refined objects of reflection *in rerum Supernatura.* Aristotle, except for his Moveless Mover, thought that was going a bit too far, and so he put his refined objects of reflection *in rerum Natura.* Dewey, in his criticism of Greek philosophy, has always been unduly partial toward Plato and unnecessarily harsh toward Aristotle on the ground that Plato, by putting his Ideas in a Transcendental Realm, at least left Nature alone, whereas Aristotle by putting his remodelled Platonic Ideas (species and genera) within Nature, immobilized the natural process of change within a fixed routine. This argument of Dewey's is far from well taken. It is making Aristotle shoulder the blame for the benighted centuries that succeeded the downfall of Greece. By putting his refined objects of reflection *in rerum Natura,* Aristotle put them where they could be empirically got at and tested. That they were not empirically tested before Darwin is no fault of Aristotle's. But it is to his credit—as against Plato—that he did put them where Darwin could empirically find them to be or not to be. Darwin was thus enabled empirically to explode Aristotelianism in natural history and do it once for all.

The natural inclination of every modern scientist is to be an Aristotelian—in the general sense that he puts his Laws, Formulae or whatever *in* Nature. It is the natural inclination, because every modern scientist is, when behaving normally, a naturalist. He wants, as Newton put it, to "deduce" causes from effects, to "derive from phenomena" all his knowledge. Newton put his Laws and Atoms *in* Nature and because he put them there, they were eventually dislodged from there, not by an "experiment" of the sort Newton performed, but by the consequences of actual laboratory experimentation.

Now in this general sense, Dewey is also an Aristotelian. His doctrine that knowledge is an exemplification of one of the ways of Nature; his doctrine that all knowledge must have passed experimental test before it can be considered knowledge—are sufficient proofs of this general statement. "Experience is *of* as well as *in* Nature."* And knowledge is one of the consequences of modes of experience. In this general sense, every naturalist, philosophic or scientific, is an Aristotelian. Not because he follows Aristotle, but because, in this general sense, Aristotle followed Nature. But Aristotle also followed Plato—and therein lies the difference.

Dewey's logic and philosophy are comprehensively directed against the fallacy of substituting refined, derived objects of reflection for the gross and macroscopic subject-matter in primary experience. And as far as this fundamental argument goes, it is a matter of secondary importance in what realm or part of what realm the transplantation is consummated: whether they are put deep down in the interior of natural things where only the penetrating eye of "human reason" can find them; or whether they are sprinkled on the surface of things where the great Positivist or the Logical Positivist can pick them up as he runs; or whether they are placed in a Supernatural Superstratosphere whereto only the "vision of contemplation" can, by gazing and gazing, ascend, and

Experience and Nature (2nd ed.), p. 4a.

there in its loftiest moment of transfixion momentarily behold, as through a glass very darkly, the faint Forms esthetically transfixed.

As far as concerns Dewey's theoretical doctrine, it is a matter of secondary importance in which of these three localities the substitution is allegedly effected. The substitution is always invalid. From the practical standpoint, however, the invocation of the Transcendental or Supernatural Realm has the most serious consequences of the three. And the Transcendental Platonic Realm—variously modernized and anaesthetized—is still the last, when not the first, refuge most frequently sought by philosophers in unnatural distress.

XXXIII

THE method of beginning with the gross and macroscopic subject-matter in primary experience is a *method* of beginning. Hence, like all methods, it works throughout the whole undertaking. Inquiry is not like a race and the beginning of inquiry is not the line that is left behind at the pop of the gun. With every step taken in the course of inquiry there is a new beginning issuing from a new ending; but beginning and ending do not follow upon each other—they intercept and unite. In walking along, the right foot does not follow upon the left—both are working through the whole stride. What is an ending or what a beginning depends upon the functional position as determined within that moment of inquiry. But every beginning is an ending and every ending is a beginning because both are always *in medias res.*

When Dewey says that the most important problem of philosophic method today is that of determining whether or not philosophers should begin with the gross and macroscopic or the derived and refined, he is not entirely correct. His statement is made within the context of philosophical discussion and is consequently already somewhat "refined" itself. As a matter of fact, all philosophers must start with the gross and macroscopic, and do. The gross and macroscopic *problem* is therefore that of getting philosophers to realize how they do as a matter of fact start and getting them to be controlled by their realization. Only when they are controlled by such realization can they exercise control over their philosophic reflections and proceed in their inquiry with understanding and intelligence.

It is obvious from the whole preceding discussion that the gross and macroscopic subject-matter changes as we pass from one area of inquiry to another within the same field and changes still more when we pass from one field of inquiry into another. It is also obvious that the gross and macroscopic subject-matter within any case of inquiry is not merely a penumbral field but is working within that activity of inquiry. When a laboratorian is weighing a stone, the gross and macroscopic subject-matter in that primary laboratory experience includes the scale as well as the stone and much more besides. Laboratorians take scales for granted, but in that grant are included as a minimum the whole laboratory and the history of inquiry of which the scale and the methods of using a scale are the consequences. How much of accumulated consequences of prior activities of inquiry is directly working in any specific case of inquiry under way, how much is in the background, how much is irrelevant is as the case may be. And what the case may be is never finally known until that inquiry is completed. Recall, for instance, Whitehead's impressive sketch of the consequences of

prior activities of inquiry involved in the laboratory testing of Percy Lowell's new planet. And Whitehead was giving just a general sketch—he did not go into the enumeration of details. Recall, also, his sketch of the consequences of prior activities of inquiry involved in the formulation of Newton's Law of Gravitation. And in this case too he was just giving a general sketch, he was by no means giving an exhaustive account.

The practice of substituting refined objects of reflection for the gross and macroscopic subject-matter in primary experience is also a *method*, that is, it is a procedure which involves making the substitutions at every point where a refined object of reflection previously obtained comes into the inquiry and at every point where a refined object of reflection is the consequence of the inquiry under way. A wholesale substitution can be made after the whole inquiry is over, but during inquiry substitutions must be continuously made throughout the process.

In the quotations from Russell the continuity of the process of substitution is well displayed.

At one moment Russell means by "physics" the empirical world, and at another moment, in the same argument, he means by "physics" the "logical constructs of the science of physics." When Russell descends from the generalized statements of his "problem" to specific cases, the situation doesn't improve but if anything becomes worse. Thus, for example, when he tries to bring together "physics" and "perception"—which unification is the objective of his whole undertaking—Russell suddenly begins talking, in the most "unrefined" fashion imaginable, about such crude, macroscopic, gross subject-matters as "brain" and "physiologist" and "microscope" and so on. However, Russell also keeps in mind that Science has discovered that it takes light from the Sun some eight minutes to travel the distance of 82,000,000 miles between the Sun and the Earth. Because of this fact, and others of similar nature, the "casual continuity" in "physics" makes it absolutely impossible to escape from the conclusion that "What the physiologist sees when he is examining a brain [by means of a microscope] is in the physiologist, not in the brain he is examining." Where the "physiologist" and where the "microscope" and where the "brain" are is a matter of some doubt. For by virtue of his same doctrine of "continuity" Russell also reaches the conclusion, "We do not know much about the contents of *any* part of the world except our *own* heads; our knowledge of other regions is *wholly abstract.*" From the last statement it follows that the "brain" and "physiologist" and "microscope"—in so far as Russell knows anything about them—are wholly abstract.* And so they are. At one moment of his argument they are abstractions perched on the mathematical point of Transcendental Peak. And so also they are not. For at the next moment of his argument, they are hurtling down the side of the Transcendental Mountain into the very depths of the "metaphysically primitive events" at the bottom of all. And so finally they are neither. For all during his argument, the brain, physiologist and microscope are also the gross and macroscopic objects that ordinary experi-

---

*It also follows that the distance between Sun and Earth and the time it takes light to travel are also wholly abstract. When this consequence of Russell's *conclusion* is given its full legitimate value, the scientific *ground* for his conclusion is completely destroyed. The *ground* of a conclusion is the *reason* of and for that conclusion. There can therefore be nothing more illogical or irrational than a "conclusion" which can be maintained only by destroying the "ground" upon which it is based or from which it is allegedly derived.

ence is familiar with. It is only by virtue of their being always the latter that Russell can keep up his "logical" argument at all.

In the course of a discussion of Berkeley's doctrine, Russell makes clearer than usual what his "logic" is:

> In spite of the logical merits of this [Berkeley's] view, I cannot bring myself to accept it, though I am not sure that my reasons for disliking it are any better than Dr. Johnson's. I find myself constitutionally incapable of believing that the sun would not exist on a day when he was everywhere hidden by clouds, or that the meat in a pie springs into existence at the moment when the pie is opened. I know the logical answer to such objections, and *qua logician* I think the answer a good one. The logical argument, however, does not even tend to show that there are *not* non-mental events; it only tends to show that we have no right to feel sure of their existence. For my part, I find myself in fact believing in them in spite of all that can be said to persuade me that I ought to feel doubtful.
>
> There is an argument, of a sort, against the view we are considering. I have been assuming that we admit the existence of other people and their perceptions, but question only the inference from perceptions to events of a different kind. Now there is no good reason why we should not carry our logical caution a step further. I cannot verify a theory by means of another man's perceptions, but only by means of my own. Therefore the laws of physics can only be verified by me in so far as they lead to predictions of *my* percepts. If then, I refuse to admit non-mental events because they are not verifiable, I ought to refuse to admit mental events in every one except myself, on the same ground. Thus I am reduced to what is called "solipsism", i.e. the theory that I alone exist. This is a view which it is hard to refute, but still harder to believe. I once received a letter from a philosopher who professed to be a solipsist, but was surprised that there were no others! Yet this philosopher was by way of believing that no one else existed. This shows that solipsism is not really believed even by those who think they are convinced of its truth. (*Philosophy* [1927], pp. 290-291; italics in original).

It is obvious that the solipsist made an enormous blunder writing to Russell. *Qua logician*, the solipsist had a very good case. "It is hard to refute." But then the solipsist went ahead and wrote a letter to another philosopher—and lo! he showed that he really did not believe, deep down in his solipsist heart, the strength of his "solipsist logic." Now the only important point about this episode is that Russell does not take the letter as constituting in any way an experimental, scientific invalidation of the "logic." For Russell himself, *qua logician*, the solipsist argument is still a hard one to refute—even after he received the letter. And, pray, what sort of letter was it? Where was it? Was it in Russell's head or in the head of the solipsist who sent it? Was it wholly abstract? No more than the solipsist "believed" his "hard-to-refute-logic" does Russell "believe" in his substitutions of refined objects of reflection—logical constructs—for the gross and macroscopic subject-matter in primary experience. *Qua logician*, Russell can, with an easy mind, go through the intellectual, mathematical-symbolic jugglery; but *qua* a human being he cannot believe it. Moreover, it is only by bringing into his "logical" exercise the allegedly non-logical, what he

believes but seems to have no rational argument for; it is only by constantly bringing this "extra-logical" within the operations of his dialectics that his dialectics can exhibit the semblance of moving along. Otherwise, Russell would be going around in a very narrow and self-enclosed circle. In sum, logic, for Russell, is precisely what logic was for the scholastics. His "inflexible rationality of thought" is of exactly the same order. The fundamental fact about Russell's "logic" is that experimental *test* has no place in it at all, has no *logical* standing whatsoever.

Whitehead, in his procedure of substituting refined objects of reflection for the gross and macroscopic subject-matter in primary experience, follows a different route and ends up at the opposite pole. Russell, as the reader remembers, finally reached the point where all distinctions between physics and perception, between mind and matter, were superficial and unreal. Since Russell's whole philosophic undertaking was devoted to the end of bringing the two together, without subordinating either to the other, his final conclusion (in *The Analysis of Matter*, of course) throws at least a glare of superficiality and unreality over his whole undertaking. In general terms, Russell's logical progress consists in making distinctions and then throwing them away so that at the end he is left with nothing at all. Whitehead proceeds in the reverse direction. He proceeds by making distinctions, and then internally involving them in each other so that at the end he has everything in everything else—which consequence also obliterates distinctions.

Thus, for instance, he starts:

> "Actual entities—also termed 'actual occasions'—are the final real things of which the world is made up. There is no going behind actual entities to find anything more real. They differ among themselves: God is an actual entity, and so is the most trivial puff of existence in far-off empty space. But, though there are gradations of importance, and diversities of function, yet in the principles which actuality exemplifies all are on the same level. The final facts are, all alike, actual entities; and these actual entities are drops of experience, complex and interdependent." ". . .actual entities are the only *reasons*; so that to search for a *reason* is to search for one or more actual entities." (*Process and Reality* pp. 27-28 and 37; italics in original).

Now this statement—leaving out the actual entity "God"—is on all fours with Dewey's fundamental position. It is another way of stating that the gross and macroscopic subject-matters in primary experience—the puffs of smoke and the stellar systems—are, with respect to existential quality, all on exactly the same level. This is also the fundamental doctrine as actually operative in the conduct of scientific inquiry. The black-bands in the interferometer are just as real as the super-galactic system. Since there is no going behind actual entities to find anything more real, the *ultimate* or metaphysically real is, precisely those actual entities themselves. And this too is thoroughly in accordance both with Dewey's doctrine of logic and the practice of science. Since the ultimate test of the validity of any theory is made by the laboratory experiment, the subject-matter as experienced in the laboratory experiment is the ultimately real, scientifically. Finally, since the actual entities or occasions are ultimately real, their differences, their gradations in importance, their qualitative characters, their existential extents must also be ultimate and real.

However, although Whitehead avers that speculative boldness must be humble before "logic" and "fact," the operation of his dialecticism carries him progressively away from both. Whitehead ends up by saying: "Each actual entity is a throb of experience *including the actual world within its scope.*" (Ib. p. 290; italics mine.) "No two actualities can be torn apart; each is all in all. Thus each temporal occasion embodies God, and is embodied in God." (Ib. p. 529)

A contradiction in a system of philosophy is evidence of some antecedent error. A fundamental contradiction is evidence of a fundamental error. It is obvious that Whitehead's final conclusion contradicts his basic doctrine as fundamentally as Russell's final conclusion contradicts his initial basic statement of doctrine (in *The Analysis of Matter*). And both conclusions are rather fantastic. It is fantastic to say that the most trivial puff of existence in far-off empty space *includes the actual world within its scope.* And the statement itself, in addition to its fantasticality, is self-contradictory. If each actual entity includes the actual world within its scope, there is no puff of existence, and there is no far-off empty space. Everything is in everything else: each is all in all. Whitehead does not get everything "inside the head" as Russell sometimes does, but as far as this aspect of his final doctrine is concerned he might just as well.

There is a world of difference between the final conclusions of Whitehead and Russell, taking the content of the conclusions by themselves. But they reach their diametrically opposed conclusions by using a "logic" or "method of philosophy" that is fundamentally the same. By using the same "logic" the doctrine of the one can be converted into the doctrine of the other by a simple dialectical twist. Russell, *qua logician*, is well aware of this, as we have seen. We have also seen that Russell is never always of the same mind as to which way the dialectical twist should be turned. Although Russell and Whitehead both humbly bow down before "logic" and are in the vanguard of those who uphold the "inflexible rationality of thought" when it comes to any critical juncture, when, in the course of their allegedly "rational" philosophic inquiry a showdown can no longer be postponed, it is always their "logic" that gives way and bows down to their "feeling" or their "metaphysical trust" or whatever.

A "logic" which makes it necessary constantly to resort to heroic, last-minute, extra-logical measures in order to keep the "logical" argument going and the philosophy afloat, is a "logic" that is not without its strong emotional appeal. It makes "philosophy" very exciting, quite a romantic adventure. But such a "logic," whatever its extra-logical merits may be, is *not* a scientific logic, for it does not display any of the fundamental characteristics of the logic of controlled inquiry.

XXXIV

THE refined objects of reflection—of whatever sort they may be—are consequences of inquiry. They are products not originals. They are the end-results that have come through the mill. Dewey's favorite metaphor for the "mill of inquiry" is that of a "refinery." Whence his technical term "refined." Our discussion of the "method of isolation" is an amplification of Dewey's doctrine of the "method of refining."

When you refine gold ore, for instance, the pure gold that is the end-result of the refining process was in the crude, raw materials. The refining process removed the dross or all extraneous matter and got the pure gold together. But the final product is qualitatively the same as the original. Now this metaphor adequately covers those cases of refined objects of reflection which go under the various names of primary and secondary qualities, percepts, sensations, and all natural qualitative objects. That is, it covers those refined objects of reflection which were the "elements" of Greek "scientific thought" and which appear in every descriptive and classificatory natural science. It is fundamental to Realist doctrine, and also Positivist, that these "elements" are "directly observed" that they appear in thought precisely as they are *in rerum Natura*, that they are not consequences of inquiry, but are the "given" or the data which a providential Nature hands out to inquiry. At one time, some Realists were fond of saying that the mind was like a search-light. As it flashed around, it immediately saw what was there and as immediately knew what it saw. When it happened to light upon an "ultimate simple" it had a case of "infallible knowledge" as Whitehead used to say. This flashlight theory of the mind is of course the Greek Formula disguised as a modern implement.

The metaphor of the "refinery" is adequate for illustrating the process whereby are obtained the refined, derived objects of reflection of the sort just mentioned, but it is no good at all when extended to the case of refined objects of modern *scientific* reflection. The mill of modern scientific inquiry—if a metaphor must be used—is like a chemical mill where alloys are made. In making an alloy, there is a double process involved: first, there is the process of "refining" or "isolating" the natural elements, and second, there is the process of bringing them into interactive relations, the consequence of the interactivity being a new object, qualitatively unlike either of the originals.

Dewey's use of the same metaphor for both cases is unfortunate. But he does not confuse the two kinds of results obtained; in fact, his whole argument is devoted to showing how fundamentally different these two kinds of refined objects of reflection are.

Newton's Third Law of Motion is a good example of a modern scientific object of reflection. The Third Law may be considered as a miniature "world of physics" and the horse, stone and s⁺retched rope between as a miniature empirical world, as the gross and macroscopic subject-matter in primary experience. One form of Russell's problem of the "application of physics to the empirical world" is the problem of the application of the Third Law to the horse, stone and rope. The High Rationalist Tradition in modern and contemporary philosophy, working with the Greek Formula up its sleeve, substitutes the Third Law for the horse, stone and rope and claims that the substituted article is the Ultimate Reality. Whitehead and Russell both follow this tradition but reluctantly; they cannot persuade themselves to follow it all the way. For Whitehead the horse, stone and rope are merely a cooking of the facts for the sake of exemplifying the Law. That Newton did cook the facts, there can be no doubt. But when Newton's cooking is all over, and he presents Whitehead with a complete "ready-made" world, Whitehead doesn't like it.

Of Newton's "ready-made world" Whitehead says that it cannot "survive a comparison with the facts." Neither can the Third Law survive a comparison with the facts. However, Whitehead goes on to say that "Biology is reduced to a mystery; and physics itself has now reached a stage of *experimental* knowledge inexplicable in terms of the categories of the *Scholium*."* Newton's world of physics, when substituted for the empirical world does, true enough, make a mystery of biology, but it makes a

————————
Process and Reality, p. 144; italics mine.

mystery of pretty much everything else. The trouble with Newton's "physics" is not that it reduces everything to the explanatory level of mechanical action. The great trouble with it is that it cannot even explain without involving itself in fundamental contradiction such an elementary mechanical action as the horse actually pulling the rope. It took close to two and a half centuries to prove to theoretical physicists that the standard and defining case of the nature of Nature is not the case where a stretched rope is transfixed between an immovable stone and immobile horse. Of course there are times when a horse cannot pull a stone along the surface of the earth, but even at such times the horse, if he is a horse at all, can move over the face of the earth himself. The Third Law of Motion apart from being symbolically a miniature "world of physics" is actually the standard, defining and ultimately controlling Law in Newton's complete "world of physics." The change from the Newtonian to the Einsteinian physics is the change that results from taking as the standard and defining case, the case where the horse is pulling the stone and moving along the face of the earth. If you take your position on that moving rope and begin to plot its mathematical formulation you fall head first into Einsteinian mathematics.

It is of course something to be thankful for that physicists now realize that the world is in motion, really and not just fictitiously. Unfortunately, they have come to that realization by way of such an incredibly circuitous route, they are still dazed by their journey and are afraid to believe it is true. And in their loftiest moments of "inspiration" they of course still desire to substitute Einstein's final equation for the real world, just as they formerly substituted Newton's initial equation for the real world.*

Now Dewey's philosophy is about a world that is actually in motion, that is really moving and not just playing at moving. And Dewey's logic is controlled by the fundamental fact that the horse is pulling the rope and horse, stone and stretched rope between are moving along the face of the earth. This fundamental fact is the fact Dewey's whole philosophy is controlled by. Dewey sometimes calls this fact "the practical character of reality."†

The application of the Third Law of Motion to the horse, stone and rope is one form of Russell's problem of the "application of physics to the empirical world." The other form of his problem is hidden away in his statement that "the world of physics must be, in some sense, continuous with the world of our perceptions." I say "hidden away" because what the "world of physics" is, in this case, depends entirely upon the course of the argument. Sometimes it is the empirical world, sometimes it is the "laws of physics," sometimes it is the electrons, protons, and whatnot which are identifiable with neither. However, if we take the summing up of his position on this form of his problem, it is fairly evident what it involves. "It is obvious that a man who can see knows things which a blind man cannot know; but a blind man can know the whole of physics. Thus the knowledge which other men have and he has not is not a part of

physics." A blind man who can know the whole of physics is, obviously, a person of great intelligence. There are millions of persons who are not blind who would experience the greatest difficulty in understanding any of physics. On the score of intelligence, there is no difference then between Russell's blind man and a seeing man who also can know the whole of physics. I suppose it is also fair to assume that the blind man in question has his other senses intact, that he can hear, touch, taste and smell. The *only* difference between the two men therefore is that the seeing man can see "secondary qualities," namely, those secondary qualities that require unimpaired vision. And it is on *this* difference that Russell rests his penultimate conclusion that "there is thus a sphere excluded from physics." *(The Analysis of Matter,* p. 389.)

We began by considering the problem of "secondary qualities" and this brings us back to the beginning. The genuineness of this problem is not the point here. Nor is it relevant to the point that Russell in the course of his discussion of the problem introduces many other differences and in the last three pages of his book temporarily reintroduces them again. What is in point here is that the second form of Russell's problem—the problem of secondary qualities (in an excessively simplified form, this time) continues to be the imperishable foundation-stone of the philosophic discussion.

The inextricable mixing up of the two forms of the problem—the shifting from one kind of refined object of reflection to a totally different kind—keeps the philosophic discussion alive, gives it an ever-changing and ever more complicated face. Russell's discussion of the same problem in *Our Knowledge of the External World* (1914) follows exactly the same general lines as his discussion in 1927. But it was vastly simpler, internally. At that time (1914) the Einsteinian "world of physics," with its manifold mathematical complications, and Space-Time had not yet come into its own. Hence, in that earlier volume, Russell could start off by dismissing Time as irrelevant for physics and therefore irrelevant for philosophy. The "temporal" then was *merely* temporal. The case with Time now is rather different. Just as in 1914 Russell took the Newtonian "world of physics" as something "given," so in 1927 Russell takes the Einsteinian "world of physics" as something "given." This method of procedure is not peculiar to Russell. It is part of the inherent methodology of the "rationalist" tradition in philosophy; it is the "logical" method that exhibits "inflexible rationality of thought."

It is fundamental in Dewey's analysis of the problem to maintain the distinction between the two general kinds of refined objects of reflection noted above. His extension of the metaphor of "refining" to cover both kinds of refined objects is unfortunate. But in view of the nature of philosophic discussion and controversy this extension is understandable. Furthermore, the standardized consequences of inquiry that persist as the stable foundations of the controversy are the "ultimate simples" that are obtained from the gross and macroscopic subject-matter in primary experience by the process of "refining." Hence Dewey's emphasis on this process is not only understandable but also justifiable within the context of the great debate.

*Purely formally speaking, the Einsteinian development consists in throwing the Third Law out at the "basis" of physics and bringing it back in at the "top." All the moving platforms, trains, etc., used in accounts of the Einstein Theory are really no better than the horse, stone and stretched rope between moving along the face of the earth.

†This is the title of an essay in *Philosophy and Civilization* (1931), originally written in 1908.

XXXV

"PHILOSOPHY" writes Whitehead "destroys its usefulness when it indulges in brilliant feats of explaining away.... Its ultimate appeal is to the *general conscious-*

ness of what in practice we experience. Whatever thread of presupposition characterizes social expression throughout the various epochs of *rational* society must find its place in philosophic theory. Speculative boldness must be balanced by complete humility before logic, and before fact. It is a disease of philosophy when it is neither bold nor humble, but merely a reflection of the temperamental presuppositions of exceptional personalities.

"Analogously, we do not trust any recasting of scientific theory depending upon a *single performance of an aberrant experiment*, unrepeated. The ultimate test is always widespread, recurrent experience; and the more general the rationalistic scheme, the more important this final appeal."*

Dewey's statement on the same general topic is as follows: "A first-rate test of the value of any philosophy which is offered us is this: Does it end in conclusions which, when they are referred back to *ordinary* life-experiences and their predicaments, render them more significant, more luminous to us, and make our *dealings* with them more fruitful? Or does it terminate in rendering the things of *ordinary* experience more opaque than they were before, and in depriving them of having in 'reality' even the significance they had previously seemed to have? Does it yield the enrichment and increase of power of ordinary things which the results of physical science afford when applied in every-day affairs? Or does it become a mystery that these ordinary things should be what they are, or indeed that they should be at all, while philosophic concepts are left to dwell in separation in some technical realm of their own? It is the fact that so many philosophies terminate in conclusions that make it necessary to disparage and condemn primary experience, leading those who hold them to measure the sublimity of their 'realities' as philosophically defined by remotensss from the concerns of daily life, which leads cultivated common sense to look askance at philosophy."†

These two statements very closely approach each other, and yet, as in cases already considered, the more closely they come together the further apart they are. And in this instance, as in all others, for one and the same reason: Whitehead never actually reaches the point where he is ready to consider "practice" as a functioning, integral factor in inquiry. Whitehead, as Russell, will on occasion recognize that an appeal must be made to "experiment" or "practice" and that such appeal is "ultimate" but he will never "wander off on the topic of experiment" to the extent of effecting an integrative, interactive union of theory and practice.

Dewey says that the test of a philosophy is whether or not the conclusions when referred back to *ordinary* life-experiences make the latter more significant and our *dealings* with them more fruitful. Whitehead says that the ultimate appeal or test is to "the general consciousness of what in practice we experience." This test is altogether different from Dewey's. For "the *general consciousness* of what in practice we experience" is more likely than not to turn out to be, not actual, practical, or experimental behavior, but simply a "philosophy of practice" over again. So that Whitehead's "test" will really be of the kind Newton performed when he "tested experimentally" his Third Law, by "relating" it to the horse, stone and stretched rope between. And that this is so is evident in Whitehead's next sentence, that philosophic theory must include or find a place for all the threads of presupposition that are found in the various epochs of

"rational society." "Rational society" consists of the various systems of ideas, philosophic, cultural, and scientific that are found to be rational. Although Whitehead does not believe that Newton's "ready-made world of physics" can survive a comparison with the facts, he also believes that *that* "world of physics" must nevertheless be included in any cosmology, or philosophy of Nature. Now there can be no doubt that Newton's "world of physics" merits some sort of inclusion in a comprehensive philosophy. But no theory of philosophy can be *tested* by reference to that "world of physics" any more than that "world of physics" can be *tested* by a theory of philosophy. The *test* of any theory scientific or philosophic is experimental in the practical sense, in the sense of doing-undergoing.

Whitehead gives another statement of his conception of the method of philosophy which more sharply points up the fundamental difference we have been considering:

"...the true method of philosophical construction is to frame a scheme of ideas, the best that one can, and unflinchingly to explore the interpretation of experience in terms of that scheme." (Ib., p. x)

By following this method, it is obvious that "experience" will always turn out to be a replica of the "scheme of ideas" in terms of which "experience is unflinchingly explored." Newton constructed his scheme of ideas contained in his Third Law and then unflinchingly explored the interpretation of "experience" in terms of that scheme. And he found that "experience" and the Third Law agreed with one another, that they were in one-to-one correspondency, that the harmony between them was perfect. When you take the same thing twice over, once as "experience" and once as "scheme of ideas," you will always get Newton's perfect results. And this taking of the same thing twice over is what Realists staunchly hold to be the fundamental method of discovering Truth!

Any scheme of ideas is already the interpretation of "experience"—of the experience of which that scheme of ideas is the formulated consequence. When that scheme of ideas is the unflinching formulation of the consequences of *that* experience, then that genuine occasion for being unflinching is over. The *next* occasion for being unflinching is when we *test* that scheme of ideas by practical, experimental doings-undergoings whether the practical experimentation be in the laboratory or in ordinary life-experiences, in our daily *dealings* with things. It is necessary to be unflinching on *this* next occasion because the scheme of ideas which is thus undergoing genuine test, may not survive the trial.

It cannot be denied of course that it also requires a high degree of "unflinchingness" to follow the "method" that Whitehead prescribes for philosophers. The "inflexible rationality of thought" he advocates is not easily acquired. When we use any given scheme of ideas for the interpretation of experience in terms of that scheme, there are bound to arise many occasions—when we adventure abroad and our "explorations" are wide enough—that may well cause the stoutest philosophic heart to quail. I doubt whether there is, in the world today, a philosopher of stouter heart than Russell. And yet Russell "flinched" when it came to accepting some of the "answers" which he *qua* logician (or *qua* schematizer of ideas) believed were "good logical answers." However, it must be said on behalf of Russell, that his "flinchings" were not final, but only temporary twinges. When, in *The Analysis of Matter*, he reached the very last sentence, Russell had to make his final interpretation of experience

---

Process and Reality, p. 25: italics mine.
†*Experience and Nature* (2nd ed.), p. 7-8; italics mine.

in terms of his scheme of ideas. And then, on the very pin-point standpoint of philosophy, Russell unflinchingly made his last stand (in that book, of course). Likewise with Whitehead. During the course of his philosophic "interpretations" of experience in terms of his scheme of ideas, there are many occasions when he "flinches." But when the last stand has to be made, he unflinchingly makes the last stand.

One of Whitehead's great contributions to philosophy is his discovery of an oft-repeated and widespread fallacy in modern thought which he calls "the fallacy of misplaced concreteness." But a far greater philosophic fallacy, and in its consequences infinitely more destructive of what Dewey calls intelligence, is "the fallacy of misplaced unflinchingness."

XXXVI

THE mathematics in which the physicist is interested was developed for the explicit purpose of describing the behavior of the external world, so that it is certainly no accident that there is a correspondence between mathematics and nature." This statement of Bridgman's is correct only when it is interpreted to mean that there is a "correspondence" between the mathematics used in describing nature and *the nature that is the consequence of using that mathematics*. Thus it is no accident certainly that there is a "correspondence" between Newton's Third Law and Newton's horse, stone and stretched rope between. In any other sense than this, there is no "correspondence" at all.

It is also no accident that there is a "correspondence" between a microscope constructed in accordance with the specifications of a formula and the formula in accordance with which the microscope is constructed.

If you take the microscope in one hand and the formula of the microscope in the other and examine them alternately you will find, as Spinoza would say, that the order and connection of ideas in the formula are the same as the order and connection of things in the microscope. Now the formula of the microscope is what Whitehead calls an "ideally isolated system." And, says Whitehead, "This means that there are truths respecting *this* system which require *reference* only to the remainder of things by way of a uniform systematic scheme of relationships." If you develop a systematic scheme of relationships with the consistency and perfection exhibited by Spinoza, you will get Spinoza's result. The "correspondency" of microscope and formula of the microscope, when extended or referred to the remainder of things within the universe, becomes the doctrine that there are two orders, one the order of ideas (Mind, Formulae) and the other the order of things (Matter, Bodies); the two orders running in parallel lines or in one-to-one correspondency.

Of course, Spinoza did not leave the two orders each alone by itself. Just as soon as you bring on the one hand and on the other in juxtaposition, you are philosophically bound to "unite" them. And so Spinoza included them in one comprehensive order of Nature. But if there are two such orders in Nature, and they parallel each other, they parallel each other. That's that, and that is all there is to it. It is an "irreducible and stubborn fact." Comprehending them in one inclusive embrace doesn't make their parallelism more parallel, and leaving them without the embrace doesn't make their parallelism any less parallel. Precisely the same holds true, for example, of Newton's Absolute Space and Absolute Time. They also were two "orders" and they "paralleled" or "corresponded" in

one-to-three-and-three-to-one mathematical formal perfection and including them in One System of Nature didn't change their Newtonian relations one bit. It also didn't help matters very much as far as the progress of scientific theory is concerned.

When you take two end-results, like the microscope and the formula of the microscope, two consequences which are the products of the self-same historical process of inquiry, they are each bound to contain characteristics which "exemplify" or "parallel" or "correspond" to the characteristics of the other. The interactivity of which they are the joint product has taken care of that. It could not be otherwise. When you take two such products, and "compare" them with one another you will always find, says Dewey, that they will be in one-to-one harmony. Then the "existence" of the one will reflect the "essence" of the other; the "mind" of the one will portray the "matter" of the other; the "form" of the one will reveal the "body" of the other; the "law" of the one will express the "conduct" of the other; the "fact" of the one will exemplify the "proposition" of the other; the "refined object of reflection" of the one will mirror the "subject-matter in experience" of the other; and so on in every field and in every case. And of course also in every case vice versa if not also versa vice. For as Leibniz put it, the "harmony is pre-established."

Although the harmony between a microscope and the formula of the microscope is entrancingly perfect when "pre-established," the differences between the two are enormous. And it is only by neglecting the differences in the first place, that the one-to-one correspondency can be obtained. When, after having made the correspondency, an appeal to "experience" is inadvertently made, all the "eternal problems" of philosophy begin to crop out again. And as long as the same "method of philosophy" is pursued, these problems will never be solved.

XXXVII

"WE do not trust," writes Whitehead, "any recasting of scientific theory depending upon a single, aberrant experiment, unrepeated. The ultimate test is always widespread, recurrent experience; and the more general the rationalistic scheme, the more important is this final test." But if a single experiment is aberrant, we would not trust any recasting of scientific theory depending upon it, no matter how widespread and recurrent that experiment had become through sheer repetition. Newton's "experiment" with the Third Law was, for example, an aberrant experiment if ever there was one. Repetition of that experiment would perpetuate, not test, the aberration.

And this particular Newtonian aberration has been "tested" by making it widespread and recurrent. Newton has been the model "experimental" scientist and his system the model of all scientific systems. Theoreticians in all fields, possessed of a "modern classical" cast of "scientific" mind have, with studied envy and anxiety, followed the lead of Newton. By carrying his "method" into their fields of inquiry, they were certain that their results would be truly scientific. And following Newton's method has meant starting with a formulation patterned after the Third Law.

And so we have, for example, in "classical, scientific economics" the fundamental Law that "supply equals demand"—equal and opposite. Supply follows demand and demand follows supply and this "iron law" of economic nature—like the rope between horse and stone—holds the economic world perfectly togeth-er—providing you only let it alone, let it go and let it pass. And in "classical psychology" we have a similar exemplification of the same "iron law." Idea equals sensation or idea follows sensation and sensation follows idea and this keeps Mind and Body together—providing again that you only let it alone, let it go and let it pass.

For such a psychology, it was of course an inestimable boon when the neurological system was discovered. For the neurological system was, obviously, the very conduit needed. Like Newton's rope, it could be the vacuous go-between. When, under the inspiration of greater scientific exactitude, the shift was made to the terms "stimulus" and "response" the same fundamental Law prevailed; stimulus equals response, equal and opposite.

Of course, no scientific theory is ever dependent upon a single experiment. It may conceivably happen that a scientific theory has to be changed because of the consequences of one experiment. But the scientific theory, both before and after the change, is not dependent upon that one experiment alone—any more than the laboratory experiment is self-dependent. As Whitehead himself so clearly described, every scientific theory (or formula) is an organization of accumulated consequences of prior activities of inquiry. And likewise with every laboratory experiment. When philosophic theory of scientific method is controlled by the indubitable, gross and macroscopic fact that theories and laboratory experiments are the funded consequences of histories of inquiry, the significance of the appeal to widespread and recurrent experience is radically clarified. For then it is seen that an appeal to a single laboratory experiment is, by the very nature of the case, a concentrated appeal to widespread and recurrent experi-ence. A single experiment no matter how extensive and internally complex it may be is, to be sure, a limited experiment. It does not encompass the totality of the universe within its scope. There is, therefore, need for recurring to further experimental laboratory tests as new formulations are reached or as old formulations are carried into new fields. The need is a constant and progressive one. In scientific inquiry it is not the case that the more general the rationalistic scheme (or the more comprehensive the theory) the more important is the final or experimental test. The process of experimental testing is continuous throughout the development of scientific theory; it occurs at every stage. One can make a distinction of "importance" such as Whitehead makes, only at the expense of violating the basic continuity and interactivity of the developmental process of scientific or controlled inquiry.

Every practical or theoretical instrument—from the crudest practical tool to the most highly refined mathematical symbol—is inherently a *social* product. Every case of experimental testing is an appeal to "widespread and recurrent experience." According to some philosophic theories of experience, human experi-ence is a private, convulsive, peristaltic movement occurring inside an aboriginally individualized psyche or soul; according to others, it is the automatic registration of private effects on a private brain inside a private head. Whether such extremely diseased modes of human experience are possible or not, we need not stop to inquire. But such modes of experience—supposing, for the argument, that they may occur—do not define the *rational* mode, the standard mode of experience which constitutes the ultimate test of theory. Rational experi-ence is experience as organized and realized in the performance of an experiment. Dewey's philosophy of the experiment is his philosophy of experience. The method of experimentation defines the nature of the method of socialized intelligence.

Dewey's recasting of philosophic theory depends upon his theory of the experiment. Originally, Dewey's philosophy acquired the designation "instrumentalism." Although by usage the term "instrumentalism" could be made equivalent in meaning to the term "experi-mentalism," in the current intellectual epoch it is practically impossible to do so. By commonsense standards of thinking and judging, an "instrument" necessarily implies something for which it is an "instrument"; an "instrumental theory of knowledge" would therefore by the same standards imply that knowledge was instrumental, not to "instrumental knowl-edge" (which is an absurdity) but to consummatory modes of experience, which are non-instrumental.

But the absurd interpretation of "instrumentalism" as the "philosophy or logic of the instrument" was inevitable. For the "logics of the instrument" are the dominant unending varieties of "rationalistic logics." The inflexible rationality of scholastic thought was, precisely, an inflexible idolatry of the "logical" instrument then available to their hand. The most popular idolatry of the instrument now current is that exhibited in the Logical Positivist movement. Carnap's Logical Positivism very closely nears the ultimate philosophic apotheosis of Esperantism.

When the term instrumentalism is made secondary to experimentalism there remains no terminological ground for confusing Dewey's philosophy with any "philosophy of the instrument."

XXXVIII

IF the problems of philosophy were inherently, and not just formally, technical, their "eternal" perpetuation would not matter so very much. But the "eternal" problems of philosophy are the social problems *par excellence*. In the process of technical formulation they have lost all the obvious features and characteristics of the social. It is not true that nothing can rise higher than its source; witness every case of development. But it is true that nothing can rise so high above its source that it becomes entirely disconnected therefrom and after its disconnection first begins to live a real and flourishing life off its own transcendental vitals. Technical terms, linguistic forms, symbolic devices, can make a problem look like nothing else on earth. They can do wonders in facial transformation. But even the most potent of these devices and instrumentalities cannot perform miracles.

Contemporary philosophers are of course distin-guishable in many ways from medieval scholastics. But in so far as contemporaries accept standardized problems and seek for their solution by dialectically arranging standardized parts, they are every whit as medieval as the veriest scholastics of ten centuries ago. And from point of view of fundamental method of philosophy it matters very little by what names such philosophers designate their philosophies, nor whence nor how they obtain their standardized equipment.

There can be no intelligent objection to standardizing instrumental equipment, theoretical and practical. Stand-ardization is necessary for efficiency and precision in control. But there is fundamental cause for intelligent objection when control over the standardized equipment is substituted for control in the solution of an actual problem which the use of the standardized equipment can give. When such substitution is made, the use of the equipment, instead of enriching experience and helping its growth, stunts and distorts it.

The multiplication of theoretical instrumentalities widens the mental horizon and increases the possibilities that can be entertained in thought. The multiplication of practical instrumentalities increases power for trying out possibilities, for changing and reconstructing existential events. When practical and theoretical instruments are developed in interactive relation with each other, we have the cumulative and progressive advance exhibited in the history of modern science. When the practical and theoretical activities are separated from each other, we have the kind of "advance" exhibited in the tragic history of modern society.

Some form and degree of separation of theory from practice is to be found in every field of modern thought and in every area of social life. Theoretical solutions of the problems generated by the separation of theory from practice in the fields of thought do not, of course, automatically function to solve the problems that are everywhere to be found in contemporary society. The actual solution of actual social problems can be accomplished only by employment of actual social instrumentalities.* The philosopher, in his professional capacity, is a theoretician, not a laboratorian. This does not relieve him of social responsibility, but defines the kind of responsibility he can be legitimately expected professionally to assume.

The fundamental problem in philosophy is the problem

---

*In the Editor's Note, pp. 525-566 of *Intelligence in the Modern World*, devoted to a discussion of the Outlawry of War, some of the difficulties and problems involved in using social instrumentalities for the solution of an actual social problem are concretely considered. Nothing more is therefore said on this topic here.

of scientific method. With respect to some specific problems, alternative solutions are possible, but with respect to the basic problem of scientific method there is no valid alternative to Dewey's solution. If this Introduction has any one comprehensive purpose, then it is to indicate the reasons why this is so.

Of course I do not mean that the whole world—not even the whole world of philosophy—is to be found in Dewey's works. Nor do I mean that whatever is in his works is perfect, that every solution he offers is the right solution and every analysis he makes is the final and correct analysis. Such is far from being the case. Some criticisms of Dewey have been explicitly made in the foregoing pages and others are implicit. And many needful criticisms of Dewey the reader can undoubtedly make for himself; and the foregoing may possibly help the reader in this direction.

But the all-important problem, social as well as philosophic, is the problem of method. There is nothing inherent in the nature of things that makes it possible for the method of experimentation—or of controlled inquiry—to be employed in certain fields and nowhere else. From the fact that Dewey's analysis of controlled inquiry is fundamentally the correct analysis and no valid alternative is possible—from this fact it does not follow that the body of knowledge in Dewey's philosophy (or in any one else's) is crystallized and fixated as "eternal and immutable." Just the opposite follows. Galileo started a revolution in method which has proved its singular validity, not by immobilizing a body of knowledge, but by making it possible for that body to change and grow. The employment of controlled inquiry—or the method of intelligence—in the fields of philosophy and the social sciences, and all human affairs, can prove its validity only in the same way.

Section C

KNOWING AND THE KNOWN

JOHN DEWEY

and

ARTHUR F. BENTLEY

Editorial Note:

We reprint here the entire text of *Knowing and the Known,* including the Index. The only changes are as follows:

1) In the original printing, the footnotes were grouped at the end of each chapter and were numbered consecutively throughout the chapter. To facilitate the reader's reference to the footnotes, many of which are highly important to the text, we have placed the notes at the bottom of each page and have re-numbered them accordingly.

2) On a few occasions, Dewey and Bentley made cross references in one footnote to other footnotes. We have revised those references to conform to the numbering system of the present printing.

R. H.
E. C. H.

KNOWING AND THE KNOWN

by

John Dewey and Arthur F. Bentley

PREFACE

The difficulties attending dependability of communication and mutual intelligibility in connection with problems of knowledge are notoriously great. They are so numerous and acute that disagreement, controversy, and misunderstanding are almost taken to be matters of course. The studies upon which report is made in this volume are the outgrowth of a conviction that a greater degree of dependability, and hence of mutual understanding, and of ability to turn differences to mutual advantage, is as practicable as it is essential. This conviction has gained steadily in force as we have proceeded. We hold that it is practicable to employ in the study of problems of knowing and knowledge the postulational method now generally used in subjectmatters scientifically developed. The scientific method neither presupposes nor implies any set, rigid, theoretical position. We are too well aware of the futility of efforts to achieve greater dependability of communication and consequent mutual understanding by methods of imposition. In advancing fields of research, inquirers proceed by doing all they can to make clear to themselves and to others the points of view and the hypotheses by means of which their work is carried on. When those who disagree with one another in their conclusions join in a common demand for such clarification, their difficulties usually turn out to increase command of the subject.

Accordingly we stress that our experiment is one of cooperative research. Our confidence is placed in this method; it is placed in the particular conclusions presented as far as they are found to be results of this method.

Our belief that future advance in knowledge about knowings requires dependability of communication is integrally connected with the transactional point of view and frame of reference we employ. Emphasis upon the transactional grew steadily as our studies proceeded. We believe the tenor of our development will be grasped most readily when the distinction of the transactional from the interactional and self-actional points of view is systematically borne in mind. The transactional is in fact that point of view which systematically proceeds upon the ground that knowing is co-operative and as such is integral with communication. By its own processes it is allied with the postulational. It demands that statements be made as descriptions of events in terms of durations in time and areas in space. It excludes assertions of fixity and attempts to impose them. It installs openness and flexibility in the very process of knowing. It treats knowledge as itself inquiry—as a goal *within* inquiry, not as a terminus outside or beyond inquiry. We wish the tests of openness and flexibility to be applied to our work; any attempts to impose fixity would be a denial—a rupture—of the very method we employ. Our requirement of openness in our own work, nevertheless, does not mean we disregard or reject criticisms from absolute points of view. It does, however, require of such criticisms that the particular absolute point of view be itself frankly, explicitly, stated in its bearing upon the views that are presented.

We trust that if these studies initiate a co-operative movement of this sort, the outcome will be progress in firmness and dependability in communication which is an indispensable condition of progress in knowledge of fact.

The inquiry has covered a period of four years and the material has had preliminary publication in one or other of the philosophical journals. We have not undertaken to remove from our pages the overlappings arising out of the protracted inquiry and of the varied manners of presentation. Since new points of approach are involved, along with progress in grasp of the problems, even the repetitions, we may hope, will at times be beneficial. We have taken advantage of this opportunity to make a number of small changes, mostly in phrasings, and in the style and scope of inter-chapter references. Some additional citations from recent discussions have been made. In only one case, we believe, has a substantive change in formulation been made, and that is exhibited in a footnote.

As continuance of our present work we hope the future will see the completion of papers on the transactional construction of psychology; on the presentation of language as human behavior; on the application of mathematical symbolism to linguistic namings and to perceivings; and on the significance of the wide range of employment, both philosophically and in practical life, of the word "sign" in recent generations.

The reader's attention is called to the Appendix containing a letter from John Dewey to a philosopher friend. He who fails to grasp the viewpoint therein expressed may find himself in the shadow as respects all else we have to say.

We owe our thanks to Joseph Ratner and Jules Altman for their many suggestions in the course of this study, and to the latter particularly for his careful work in preparing the Index.

June, 1948

INTRODUCTION

A SEARCH FOR FIRM NAMES

A YEAR or so ago we decided that the time had come to undertake a postponed task: the attempt to fix a set of leading words capable of firm use in the discussion of "knowings" and "existings" in that specialized region of research called the theory of knowledge. The undertaking proved to be of the kind that grows. Firm words for our own use had to be based on well-founded observation. Such observation had to be sound enough, and well enough labeled, to be used with definiteness, not only between ourselves, but also in intercourse with other workers, including even those who might be at far extremes from us in their manner of interpretation and construction. It is clear, we think, that without some such agreement on the simpler fact-names, no progress of the kind the modern world knows as scientific will be probable; and, further, that so long as man, the organism, is viewed naturalistically within the cosmos, research of the scientific type into his "knowings" is a worth-while objective. The results of our inquiry are to be reported in a series of papers, some individually signed, some over our joint names,[1] depending on the extent to which problems set up and investigations undertaken become specialized or consolidated as we proceed. We shall examine such words as fact, existence, event; designation, experience, agency; situation, object, subjectmatter; interaction, transaction; definition, description, specification, characterization; signal, sign, symbol; centering, of course, on those regions of application in which phrasings in the vaguely allusive form of "subject" and "object" conventionally appear.

The opening chapter arose from the accumulation of many illustrations, which we first segregated and then advanced to introductory position because we found they yielded a startling diagnosis of linguistic disease not only in the general epistemological field, where everyone would anticipate it, but also in the specialized logical field, which ought to be reasonably immune. This diagnosis furnishes the strongest evidence that there is a need for the type of terminological inquiry we are engaged in, whether it is done at our hands and from our manner of approach, or at the hands and under the differing approach of others. We are in full agreement as to the general development of the chapter and as to the demonstration of the extent of the evil in the logics, its roots and the steps that should be taken to cure it.

One point needs stress at once. In seeking firm names, we do not assume that any name may be wholly right, nor any wholly wrong. We introduce into language no melodrama of villains all black, nor of heroes all white. We take names always as namings: as living behaviors in an evolving world of men and things. Thus taken, the poorest and feeblest name has its place in living and its work to do, whether we can today trace backward or forecast ahead its capabilities; and the best and strongest name gains nowhere over us completed dominance.[2]

It should be plain enough that the discussions in the first chapter, as well as in those that are to follow, are not designed primarily for criticizing individual logicians. In view of the competence of the writers who are discussed, the great variety of the confusions that are found can be attributed only to something defective in the underlying assumptions that influence the writers' approach. The nature of these underlying defects will, we trust, become evident as we proceed; and we hope the specific criticisms we are compelled to make in order to exhibit the difficulty will be taken as concerned solely with the situation of inquiry, and not with personalities.[3]

---

[1] Of the papers chosen for incorporation in this book, those forming Chapters I, VIII, and IX are written by Bentley. That forming Chapter X is written by Dewey. The rest were signed jointly. The original titles of some of the papers have been altered for the present use. Places of original publication are noted in an appended comment.

[2] In later development we shall grade the poorer namings as Cues and Characterizations; and the better and best as Specifications.

[3] As a preliminary to further appraisal, one may profitably examine Max Wertheimer's discussion of the vague uses of leading terms in the traditional deductive and inductive logics, due to piecemeal dealings with "words" and "things" in blind disregard of structures. *Productive Thinking*, (New York, 1945), pp. 204-205.

I.

VAGUENESS IN LOGIC[1]

I

LOGICIANS largely eschew epistemology. Thereby they save themselves much illogicality. They do not, however, eschew the assumed cosmic pattern within which the standardized epistemologies operate. They accept that pattern practically and work within it. They accept it, indeed, in such simple faith that they neglect to turn their professional skills upon it. They tolerate thereby a basic vagueness in their work. Sometimes they sense such defects in their fellow logicians, but rarely do they look closely at home, or try to locate the source of the defects found in others. Perhaps a tour of inspection by inquirers who use a different approach may indicate the source from which the trouble proceeds and suggest a different and more coherent construction.

The logical texts to which we shall give especial attention are the work of Carnap, Cohen and Nagel, Ducasse, Lewis, Morris, and Tarski. To economize space citations in our text will be made by use of initials of the authors, respectively, C, CN, D, L, M, and T.[2]

The cosmic pattern to which we have referred is one used by Peirce as an aid to many of his explorations, and commonly accepted as characteristic of him, although it does not at all represent his basic envisionment. It introduces for logical purposes three kinds of materials: (1) men; (2) things; (3) an intervening interpretative activity, product, or medium—linguistic, symbolic, mental, rational, logical, or other—such as language, sign, sentence, proposition, meaning, truth, or thought. Its very appearance in so many variations seems of itself to suggest a vagueness in grasp of fundamentals. A crude form of it is well known in Ogden and Richard's triangle (The Meaning of Meaning, p. 14) presenting "thought or reference," "symbol," and "referent." Similarly we find Cohen and Nagel remarking (CN, p. 16) that "it seems impossible that there should be any confusion between a physical object, our 'idea' or image of it, and the word that denotes it...." Lewis, claiming the authority of Peirce, holds that "the essentials of the meaning-situation are found wherever there is anything which, for some mind, stands as sign of something else" (L, p. 236). Carnap sets up "the speaker, the expression uttered, and the designatum of the expression," altering this at once into

"the speaker, the expression, and what is referred to" (C, pp. 8-9), a change of phrasing which is not in the interest of clarity, more particularly as the "what is referred to" is also spoken of as that to which the speaker "intends" to refer. Morris introduces officially a "triadic relation of semiosis" correlating sign vehicle, designatum and interpreter (M, p. 6), sometimes substituting interpretant for interpreter (M, p. 3), sometimes using both interpreter and interpretant to yield what is apparently a "quadratic" instead of a "triadic" form, and always tolerating scattered meanings for his leading words.

We view all the above arrangements as varieties of a single cosmic pattern—an ancient patchwork cobbling, at times a crazy quilt. The components shift unconscionably. Anyone who has ever tried to make them lie still long enough for matter-of-fact classification has quickly found this out.

We may not take time to show in detail here how radically different all this is from Peirce's basic procedure—our attention will be given to that at another time[3]—but since Peirce is continually quoted, and misquoted, by all parties involved, we shall pause just long enough to illuminate the issue slightly. Such words as Lewis takes from Peirce do *not* mean that minds, signs and things should be established in credal separations sharper than those of levers, fulcrums, and weights; Peirce was probing a linguistic disorder and learning fifty years ago how to avoid the type of chaos Lewis's development shows. Similarly Cohen and Nagel (CN, p. 117) quote a sentence from Peirce as if in their own support, when actually they depart not merely from Peirce's intent but from the very wording they quote. In his *Syllabus of Certain Topics of Logic* (1903) Peirce wrote:

"The woof and warp of all thought and all research is symbols, and the life of thought and science is the life inherent in symbols; so that it is wrong to say that a good language is *important* to good thought, merely; for it is of the essence of it."[4]

Peirce here makes flat denial of that separation of

---

[1] This chapter is written by Bentley.

[2] The titles in full of the books or papers specially examined are:
C: Rudolf Carnap, *Introduction to Semantics,* Cambridge, 1942.
CN: Morris R. Cohen and Ernest Nagel, *An Introduction to Logic and Scientific Method,* New York, 1934. (References are to the fourth printing, 1937.)
D: C. J. Ducasse, "Is a Fact a True Proposition?—a Reply." *Journal of Philosophy,* XXXIX (1942), 132-136.
L: C. I. Lewis, "The Modes of Meaning," *Philosophy and Phenomenological Research,* IV (1943), 236-249.
M: Charles W. Morris, *Foundations of the Theory of Signs,* Chicago, 1938. (*International Encyclopedia of Unified Science* I, No. 2.)
T: Alfred Tarski, "The Semantic Conception of Truth and the Foundations of Semantics," *Philosophy and Phenomenological Research,* IV (1944), 341-376.

Other writings of these logicians will be cited in footnotes. To show the scope of these materials as a basis for judgment, it may be added that the seven logicians examined represent, respectively, The University of Chicago, The College of the City of New York, Columbia University, Brown University, Harvard University, The University of Chicago and The University of California.

[3] Peirce experimented with many forms of expression. Anyone can, at will, select one of these forms. We believe the proper understanding is that which is consonant with his life-growth, from the essays of 1868-1869 through his logic of relatives, his pragmatic exposition of 1878, his theory of signs, and his endeavors to secure a functional logic. Recent papers to examine are: John Dewey, "Ethical Subjectmatter and Language," *The Journal of Philosophy,* XLII (1945) and "Peirce's Theory of Linguistic Signs, Thought, and Meaning" *Ibid.,* XLIII (1946), 85; Justus Buchler, review of James Feibleman's *An Introduction to Peirce's Philosophy Interpreted as a System, Ibid.,* XLIV (1947), 306; Thomas A. Goudge, "The Conflict of Naturalism and Transcendentalism in Peirce" *Ibid.,* XLIV (1947), 365. See also p. 105, footnote 3, and p. 167, footnotes 8 and 9 of this volume.

It is of much interest with respect to this issue to note that in a late publication (October, 1944) Otto Neurath, the editor-in-chief of the *International Encyclopedia of Unified Science,* of which Carnap and Morris are associate editors, expressly disavows the threefold position the others have taken and thus makes an opening step towards a different development. "There is always," he writes, "a certain danger of looking at 'speaker,' 'speech,' and 'objects' as three actors...who may be separated....I treat them as items of one aggregation....The difference may be essential." ("Foundations of the Social Sciences," *International Encyclopedia of Unified Science,* II, No. 1, 11.)

[4] *Collected Papers of Charles Sanders Peirce,* ed. by Charles Hartshorne and Paul Weiss (Cambridge, 1931) 2.220. See also p. 99, footnote 5.

word, idea and object which Cohen and Nagel employ, and which they believe "impossible" to confuse. The two world-views are in radical contrast.

Consider again what Peirce, cutting still more deeply, wrote about the *sign* "lithium" in its scientific use:

"The peculiarity of this definition—or rather this precept that is more serviceable than a definition—is that it tells you what the word 'lithium' denotes by prescribing what you are to *do* in order to gain a perceptual acquaintance with the object of the word."[1]

Notice the "perceptual"; notice the "object" of the "word." There is nothing here that implies a pattern of two orders or realms brought into connection by a third intervening thing or sign. This is the real Peirce: Peirce on the advance—not bedded down in the ancient swamp.

The cosmic pattern we shall employ, and by the aid of which we shall make our tests, differs sharply from the current conventional one and is in line with what Peirce persistently sought. It will treat the talking and talk-products or effects of man (the namings, thinkings, arguings, reasonings, etc.) as the men themselves in action, not as some third type of entity to be inserted between the men and the things they deal with. *To this extent* it will be not three-realm, but two-realm: men and things. The difference in the treatment of language is radical. Nevertheless it is not of the type called "theoretical," nor does it transmute the men from organisms into putative "psyches." It rests in the simplest, most direct, matter-of-fact, everyday, common sense observation. Talking-organisms and things—there they are; if there, let us study them as they come: the men talking. To make this observation and retain it in memory while we proceed are the only requirements we place upon readers of this first chapter. When, however, we undertake hereafter a changed form of construction, we must strengthen the formulation under this observation, and secure a still broader observation. The revelatory value of our present report nevertheless remains, whether such further construction is attempted or not.

In the current logics, probably the commonest third-realm insertion between men and things is "proposition," though among other insertions "meaning" and "thought" are at times most active rivals for that position. In the first two logics we examine, those of Cohen and Nagel, and of Carnap, we shall give attention primarily to "proposition." Our aim will be to find out what in logic—in these logics, particularly—a proposition *is*, where by "is" we intend just some plain, matter-of-fact characterization such as any man may reasonably well be expected to offer to establish that he *knows what he is talking about* when he names the subjectmatter of his discussion. We shall ask, in other words, what sort of fact a proposition is taken to be.

In the logics, in place of an endeavor to find out whether the propositions in question are facts, we shall find a marked tendency to reverse the procedure and to declare that facts are propositions. Sometimes this is asserted openly and above board; at other times it is covert, or implied. Cohen and Nagel flatly tell us that facts are propositions—"true" propositions, this is to say. Their book *(CN)* is divided between formal logic and scientific method. Under the circumstances we shall feel at liberty to bring together passages from the two portions of the work, and we shall not apologize—formal logic or no formal logic—for a treatment of the issues of fact and proposition in common. Following this we shall examine the manner in which Carnap *(C)*, though always

seeming to be pushing fact behind him with the flat of his hand, makes his most critical, and possibly his most incoherent, decision—that concerning sentence and proposition—with an eye upon the very "fact" he disguises behind a tangle of meanings and designations.

The issue between proposition and fact is not minor, even though it enters as a detail in logical systematization. It is apparently an incidental manifestation of the determined effort of logicians during the past generation to supply mathematics with "foundations" through which they could dominate it and make further pretense to authority over science and fact as well. (The whole tendency might be shown to be a survival from antiquity, but we shall not go that far afield at this time.) We shall simply stress here that if fact is important to the modern world, and if logic has reached the point where it declares facts to be propositions, then it is high time to reverse the operation, and find out whether *propositions* themselves, as the logicians present them, are facts—and if so, what kind.

II

Cohen and Nagel's *Logic (CN)* is outstanding, not only for its pedagogical clarity but for the wide-ranging competence of its authors going far beyond the immediate requirements of a collegiate textbook. The index of their book does not list "fact," *as* "fact," but does list "facts," directing us among other things to a six-page discussion of facts and hypotheses. We are frequently told that a "fact" *is* a "proposition" that is "true." Thus *(CN*, p. 392): "The 'facts' for which every inquiry reaches out are propositions for whose truth there is considerable evidence." Notice that it is their own direct choice of expression, not some inference from it or interpretation of it, that sets our problem. If they had said, as some logicians do, that "fact" is truth, or propositional truth, that might have led us on a different course, but they make "true" the adjective and "proposition" the noun, and thus guide us to our present form of inquiry.

As the case stands, it is very much easier in their work to find out what a "proposition" is *not*, than to find out what it *is*. Propositions are:

> *not* sentences *(CN*, p. 27, No. 1)
> *not* mental acts *(CN*, p. 28, No. 4)
> *not* concrete objects, things, or events *(CN*, p. 28, No. 5).[2]

What, now, are propositions, if they are neither physical, mental, nor linguistic? It takes more ingenuity than we have to make sure; it is a strain even to make the attempt. A form of definition is, indeed, offered thus: "a *proposition* may be defined as anything which can be said to be true or false" *(CN*, p. 27). This is fairly loose language, to start with, and how it operates without involving either the mental or the linguistic is difficult to see. A variant, but not equivalent, phrasing is that a proposition is "something concerning which questions of truth and falsity are significant" *(CN*, p. 28, No. 3). Unfortunately the words "something," "anything," "said" and "significant" in these citations—just dictionary words here, and nothing more—are hard to apply in the face of all the negations. We are no better off from incidental phrasings such as that a proposition is "information conveyed by sentences" *(CN*, p. 17), or that it is

---

[1] *Ibid.*, 2.330.

[2] The Cohen-Nagel indexing differs here from the text. It distinguishes propositions from sentences, judgments, resolutions, commands and things. Compare the old "laws of thought" which *(CN*, p. 182) take modernistic dress as laws of propositions.

"objective meaning" *(CN*, p. 28, No. 4), or that it is what a sentence "signifies" *(CN*, p. 27). If sentences are actually, as they tell us, just marks or sounds having a "physical existence" on surfaces or in air waves *(CN*, p. 27), just how such marks "convey" or "signify" anything needs elucidation; as for "objective meaning," the words rumble in the deepest bowels of epistemology. We also note other difficulties when we take their language literally, not impressionistically. While the proposition "must not be confused with the symbols which state it," it cannot be *expressed* or *conveyed* without symbols" *(CN*, p. 27); while it is not "object, thing, or event," it may be "relation," though relations are "objects of our thought," and, as such, "elements or aspects of actual, concrete situations" *(CN*, pp. 28-29); while a proposition is what is "true or false," there is no requirement that anyone, living or dead, *"know* which of these alternatives is the case" *(CN*, p. 29, No. 6).[1]

Literally and with straight-faced attention we are asked by Cohen and Nagel to concern ourselves with propositions that are not physical, not mental, not linguistic, and not even something in process of being expressed or conveyed, but that nevertheless have a tremendous actuality wherein they possess truth and falsity on their own account, regardless of all human participation and of any trace of human knowing. All of which is very difficult to accomplish in the Year of Our Lord, 1944. It is even more troublesome factually, since everything we are logically authorized to know about *facts* (apart from certain "sensations" and other dubieties residing on the far side of the logical tracks) must be acquired from such "propositions." Our "knowledge," even, the authors tell us, "is *of* propositions" *(CN*, p. 29); and what a proposition *that* is, unless the "of" by some strange choice is a synonym of "through" or "by means of."[2]

Supplementing their position that facts are propositions—while propositions are, at the same time, stripped of all the characteristics research workers since Galileo would accept as factual—Cohen and Nagel offer a free account of "facts" *(CN*, pp. 217-218). This, however, clears up nothing. They note "different senses" of "fact" which they proceed at once to render as "distinct things" "denoted" by the word. Apparently they do not intend either four different dictionary meanings of the word, as "senses" would imply, or four distinct "classes of objects," as "denotes" would require *(CN*, p. 31), but something uncertainly between the two. The passage in question reads:

"We must, obviously, distinguish between the different senses of 'fact.' It denotes at least four distinct things.

1. . . .certain discriminated elements in sense perception. . . .

2. . . .the propositions which *interpret* what is given to us in sense experience.

3. . . .propositions which truly assert an invariable sequence or conjunction of characters. . . .

4. . . .those things existing in space or time, together with the relations between them, in virtue of which a proposition is true."

Two of these four do not enter as propositions at all. The other two use the word "propositions" but involve interpretations and technical assertion of types which evidently run far into the "mental" region from which "proposition" is excluded. Whether we have here "senses" or "classes of objects," some kind of organization of the "things" should be offered if the passage is to have any logical relevance whatever. Such organization is conspicuously lacking,[3] and the total effect of the passage is to take advantage of the very confusion that so greatly needs to be cleared away.

We get no help by going back to the word "meaning," for meaning is as badly off as "proposition" is. Some logicians employ the word heavily—we shall note one of them later—but in the present work, so far as the index indicates, the word merely yields a change of phrasing. The "meaning of a proposition" is something we must know before deciding whether it is "true" *(CN*, p. 9); no matter how formal our implication, it must not ignore "the entire meaning" *(CN*, p. 12); universal propositions have meanings that require "at least *possible* matters of fact" *(CN*, p. 43).

Nor do we get any help when we try the words "true" and "false." No direct discussion of "true" has been observed by us in the book. It enters as the essential "is-ness" of propositions: "if a proposition is true it must always be true" *(CN*, p. 29). Apparently neither truth nor proposition can survive without an eye on the other, but when emphasis is desired we hear of "true in fact" (as *CN*, p. 7, p. 76), so that even the axioms must have their truth empirically established *(CN*, p. 132). This is the only variety of "true" we have noticed, even though we are told that "truths" may be proved out of other "truths." We have the curious situation (1) that facts are propositions; (2) that propositions are truth (or falsity) assertions; (3) that under pressure "true" turns out to be "true in fact"—just like that, no more, no less—and "false," no doubt, the same.

We are about half through with our exhibit, but we shall omit the rest of it. It all comes to the same thing. A word is officially introduced and assigned a task. Turn around once, and when you look back it is doing something else. You do not even need to turn around; just let your direct gaze slip, and the word is off on the bias. Cohen and Nagel believe their logic to be in tune with the infinite, this being a standard convention among logicians. "Its principles," they say, "are inherently applicable because they are concerned with ontological[4]

---

[1] Note that a proposition is first "not an object," then that it is an "object of thought," finally that it is an "aspect of the concrete," and that the first assertion and its dyadic belying all occur in a single paragraph. What the writers "really mean" is much less important logically than what they say (what they are able to say under their manner of approach) when they are manifestly *doing their best to say what they mean.*

[2] The word "knowledge," incidentally, is unindexed, but we learn that it "involves abstraction" *(CN*, p. 371); that it does not cover merely the collecting of facts *(CN*, p. 215); that true knowledge cannot be restricted to objects actually existing *(CN*, p. 21); and that many open questions remain as to immediate knowings *(CN*, p. 5)—nothing of which is significantly treated.

[3] Casual comments do not organize. As to the first item, we learn: "All observation appeals ultimately to certain *isolable* elements in sense experience. We search for such elements because concerning them *universal agreement among all people* is obtainable" (italics for "isolable" are theirs, the others ours). Again, a fact in the second or third sense "states" a fact in the fourth. And a fact in the fourth sense is not "true"; it just "is" *(CN*, p. 218). Separately such comments are plausible. Together they scatter like birdshot.

[4] More recently, however, Professor Nagel has written a paper, "Logic without Ontology," which will be found in the volume *Naturalism and the Human Spirit* (1944), edited by Y. H. Krikorian. Here he advances to an operational position approximating that of the instrumental logic of the nineteen twenties, which he at that time assailed in a paper entitled "Can Logic Be Divorced from Ontology?" *(Journal of Philosophy*, XXVI [1929], 705-712), written in confidence that "nature must contain the prototype of the logical" and that "relations are discovered as an integral factor in nature." Also of great interest for comparison is his paper "Truth and Knowledge of the Truth" *(Philosophy and Phenomenological Research*, V [1944], 50-68), especially the distinction as it is sharply drawn (p. 68).

traits of utmost generality" *(CN*, p. v). We, on the contrary, believe their "principles" are inherently defective because they are concerned with verbal traits of the utmost triviality. The practical work of discussing evidence and proof is admirably done in their work. Theoretical construction defaults altogether. But the very deficiencies are valuable—if one will but look at them—as clues to the kind of research that, under our present manner of examination, is most important for the immediate future.

III

When Professor Nagel reviewed Carnap's *Introduction to Semantics (C)* and came to its "propositions," he felt impelled to shake his head sadly at such "hypostatic Platonic entities."[1] Now Carnap's "propositions" may be more *spirituelles* than Cohen-Nagel's—which are hopefully of the earth earthy, even though nothing of the physical, mental, linguistic or communicative is allowed them—but what little difference there is between the two types is one of philosophical convention rather than of character. Nevertheless, such is logic that we are not greatly surprised, while Nagel is grieving over Carnap, to find Carnap placing Cohen-Nagel in the lead among his fellow-travelers, with evidence attached *(C*, p. 236).

Fact, in Carnap's work, is farther away around the corner than it is in Cohen-Nagel's. It is something logic is supposed never quite to reach, but only to skim past at the edges, with perhaps a little thought-transference on the way. It has a sort of surrogate in "absolute concepts" which are to be recognized as being present when all words agree, and which therefore, somewhat surprisingly, are said to be totally unaffected by language *(C*, pp. 41-42; p. 89, Convention 17-1). Nevertheless, when Carnap distinguishes proposition from sentence he does it with a hazy eye upon a certain unity of organization which must some way or other, some time or other, be secured between the formal and the factual.

In his thirteen-page terminological appendix which cries "Peace, peace" where there is no peace, Carnap notes two main uses—two "different concepts," he says—for the word "proposition" *(C*, p. 235). He distills these out of a welter of logical confusions he finds well illustrated in Bertrand Russell. These outstanding uses are first "for certain expressions" and then "for their designata." His elaboration—we cite meticulously, and in full, since this is the only way to make the exhibit plain—runs:

"'Proposition'. The term is used for two different concepts, namely for certain expressions (I) and for their designata (II).

I: As 'declarative sentence'. Other terms: 'sentence'*, 'statement' (Quine), 'formula' (Bernays).

II*: As 'that which is expressed (signified, formulated, represented, designated) by a (declarative) sentence' (§§ 6 and 18). Other terms: 'Satz an sich' (Bolzano), 'Objectiv' (A. Meinong), 'state of affairs' (Wittgenstein), 'condition'."

The asterisks are used by Carnap to mark the terminology he himself adopts. In I, he states he will use the word "sentence" for what others might call declarative sentence, statement, or formula. In II*, he adopts the word "proposition" for whatever it is he there sets forth. 'Sentence' (I) and 'Proposition' (II) together make up what the man in the street would call a sentence: roughly, this is to say, an expression of meaning in words. A reader who merely wants a whiff of

[1] *The Journal of Philosophy,* XXXIX (1942), 471.

characterization while the semantic march proceeds may be satisfied with the passage as we have cited it. It offers, however, serious difficulty to the man who wants to grasp what is involved before he goes farther. We propose to take this passage apart and find out what is in it; for nothing of the semantic construction is safe if this is defective. Since Carnap offers us "pure" semantics—free from all outer influence, practical or other—*we shall give it "pure" linguistic analysis, staying right among its sentences*, and dragging nothing in from the outside. He is meticulous about his definitions, his theorems and his conventions; we shall be meticulous about the verbal materials out of which he builds them. This will take much space, but no other course is possible. One great hindrance is the way he slips one word into the place of another, presumably in synonymic substitution, but usually with so much wavering of allusion that delivery becomes uncertain. Such shifting verbal sands make progress slow. For our immediate purposes, we shall employ *italics* to display precisely the wordings we quote as we dissect them.

The word "proposition," if used without quotation marks, would be an "expression (sign, word)." Supplied with single quotation marks—thus *'Proposition'*—it becomes "a name for that expression. . .in the metalanguage for that language" *(C*, p. 237). Having written down *'Proposition'*, he then proceeds: *The term is used for. . . .*Here "term" is an evasive word, unindexed, unspecified and undiscussed in his text. (It, together with certain other evasive words, will be given separate attention later.) In the present passage it represents either "proposition" or *'proposition'* or possibly a mixture of both. Look at it, and it should represent the latter. Read it, and you will think it represents the former. We shall risk no opinion, more particularly because of the vagueness of what follows.[2] Taking the *is used for*, however, we may venture to guess we have here a substitute for "names" (as the word "names" is used in *C*, p. 237), with an implication of variety in namings, and this evasively with respect to "current" uses on the one side, and names as they "ought to be" used on the other. Our criticism here may look finical, but it is not. When the word "term" is used in a vital passage in a logic, we have a right to know exactly how it is being used.

If we add the next three words, the declaration thus far seems to be to the effect that the name of the expression, or perhaps the expression itself, names variously, for various people, *two different concepts.*

The word "concept" dominates this sentence and produces its flight from simplicity and its distortion. What follows is worse. We face something undecipherable and

[2] A competent critic, well acquainted with Carnap, and wholly unsympathetic to our procedure, attacks the above interpretation as follows: Since Carnap *(C*, p. 230, line 16) writes "Concept. The word is. . . .," it is evident that to Carnap *'concept'* is here a word, not a name for a word; it is evident further that under even a half-way co-operative approach the reader should be able to carry this treatment forward five pages to the case of *'proposition,'* accepting this latter frankly as "word" not "term," and ceasing to bother. Unfortunately for our critic this course would make Carnap's treatment in both instances violate his prescription and thus strengthen our case. All we have done is to exhibit an instance of vagueness, drawing no inference here, and leaving further discussion to follow. To consider and adjust are (1) proposition-as-fact; (2) "proposition" as a current logical word; (3) 'proposition' in the metalanguage; (I) Carnap's prescription for 'sentence'; (II) Carnap's prescription for 'proposition'; *(a)* factual adequacy for 'sentence'; *(b)* factual adequacy for 'proposition'; *(c)* general coherence of the textual development within the full syntactic-semantic-pragmatic construction. It is this last with which we are now concerned. Partial or impressionistically opinionative analyses are not likely to be pertinent.

without clue. Balanced against "concept" in some unknown form of organization we find *certain expressions (I) and. . .their designata (II).* Here concept introduces (presents? represents? applies to? names? designates? includes? covers?) certain expressions *and* their (certain) designata. If he had said in simple words that "proposition" is currently used in two ways, one of which he proposes to call 'sentence,' and the other, 'proposition,' the reader's attention might have been directed to certain features of his account, in which something factually defective would have been noted.[1] What concerns us, however, is not this defect but his elaborate apparatus of terminological obscurity, and to this we shall restrict ourselves. Holding for the moment to the three words "concept," "expression" and "designatum," and noting that the "certain" designata here in question are "propositions," we turn to his introductory table *(C*, p. 18) in which he offers his "terminology of designata." Applying our attention to this we are led to report that for Carnap:

1. concepts are one variety of designata, the other varieties being individuals and propositions;

2. designata enter as entities, with which, so far as we are told, they coincide in extension;

3. expressions (signs, terms), in the functions they perform in the *Semantics,* are not entities, but are balanced theoretically over against entities; they live their lives in a separate column of the table, the whole distinction between syntactics and semantics resting in this separation of the columns;

4. propositions, though entities, are most emphatically not a variety of concept; they are collateral to the whole group of concepts;

5. despite (3) and (4) the important terminological passage before us (from *C,* p. 235) reads: *for. . .concepts, namely, for certain expressions. . .and[2] for their designata. . .;*

6. there is a curious shift of phrasing between the paragraphs of our citation *(C,* p. 235), where "the term" is the expressed or implied subject for each sentence: in the introductory statement it is used "for" concepts, in I "as" an expression, and in II "as" a designatum; in loose colloquial phrasing such shifts are familiar, but where the whole technique of a logic is at stake they make one wonder what is being done.[3]

There is a marked difference in allusion and in verbal "feel" between "entity" and "designatum" in the above procedure, so that a report on the extension and intension of these two words would be helpful. Such a report, however, would require adjustments to the word "object," which is one of the vaguest in Carnap's text—an adjustment that we may well believe would be wholly impracticable for him under his present methods. It would be helpful also, as we shall see, if we could distinguish the cases in which a concept enters as an "entity" from those in which it is used as a sign or expression. In the present instance we have already found much room for suspicion that it is used, in part, "as" a sign and not "for" a designatum.[4] It seems to have never occurred to him that the "concept" that runs trippingly throughout the text requires terminological stability with respect to the "concept" that enters among the materials, objects or objectives, of his inquiry.

The case being as it is, our report on the nineteen-word sentence comprising the first paragraph of the citation must be that it tells us that a certain expression, or its name, is used to name concepts which in their turn either are or name certain expressions *and* their designata, although neither the expressions nor their designata are officially concepts.

Having thus made his approach to "proposition" in a characteristic mixture of allusions, he now turns to the distinctions he himself intends to display. Earlier *(C,* p. 14), and as a legitimate labor-saving device, he had said that the word "sentence" was to stand for "declarative sentence" throughout his treatise. His desire and aim is to study the coherence of certain types of connective signs (calculus) in such declarative sentences in separation from the substance of the declaration (semantics). To do this he splits the common or vulgar "sentence" of the man in the street into two separate "things." This sort of "thing-production" is, of course, the outstanding feature of his entire logical attitude. The coherence-aspect now presents itself as the first "thing" (I), even though under his preliminary tabulation (as we have already seen) it is not listed among the "entities." The "meaning" portion, or substance of the declaration (II), is no longer to be called "sentence" under any circumstance whatever,[5] but is to be named 'proposition.' These names, it is to be understood, themselves belong in the metalanguage as it applies to the object language. As before, we shall not argue about the merits of the position he takes but

---

[1] Carnap reports his distinctions I and II as appearing in the literature along with mixed cases *(C,* p. 235). His illustrations of his II, and of the mixed cases, fit fairly well. However, the wordings of Baldwin, Lalande, Eisler, Bosanquet, etc., cited for I, though they have some superficial verbal similarity, would not come out as at all "the same," if expanded in their full expressive settings, *viz:* American, French, German and British. Certainly none would come out "the same as" Carnap's completely meaningless "expression" which, nevertheless, expresses all that men take it to express.

[2] Carnap, if memory is correct, once displayed five varieties of "and," to which Bühler added two more. One wonders whether this "and" is one of them. Another illustration, an unforgettable one, of his libertine way with little connectives is his impressive advance from "not" to "especially not" in setting up the status of "formal" definition *(International Encyclopedia of Unified Science,* I, No. 3, 16).

[3] Again, the welcome comment of a critic unsympathetic to our procedure is of interest. As to (3) he asserts that since expressions consist of sign-events and sign-designs, the former being individuals and the latter properties, and since both individuals and properties are entities, therefore expressions are themselves entities. We have no breath of objection to such a treatment; only if this *is* the view of the *Semantics,* why does the classification (p. 18) conflict? Or, alternatively, if the great technical advance rests on separating expressions from entities, what does it mean when we

are told in answer to a first simple question that, *of course,* expressions *are* entities too? As to (4) and (5) our critic in a similar vein asserts that for Carnap propositions are properties of expressions, that properties are concepts, and hence that propositions *are* concepts. Here again, one asks: If so, why does Carnap classify them differently in his table? Dissecting our critic's development of his thesis we find it to contain the following assertions:

1. *Being a proposition* is a property of entities.
2. *Being a proposition* is therefore a concept.
3. The *property* (being a proposition) is named 'proposition.'
4. The *property* (being a proposition) is not a proposition.

From which we can hardly avoid concluding:

5. That which is named 'proposition' is not a proposition.

We leave these to the reader's private consideration, our own attention being occupied with the one central question of whether double-talk, rather than straight-talk, is sanitary in logic.

[4] Our phraseology in the text above is appalling to us, but since we here are reflecting Carnap it seems irremedial. The indicated reform would be to abandon the radical split between sign-user and sign with respect to object, as we shall do in our further development.

[5] However, before he concludes his terminological treatment he introduces *(C,* p. 236) certain sentences that he says are "in our terminology sentences in semantics, not in syntax." This is not so much a contradictory usage as it is an illustration of the come-easy, go-easy dealing with words.

confine ourselves to the question: how well, how coherently, does he develop it?

Since the sentence in question is a declarative sentence, one might reasonably expect that any "proposition" carved out of it would be described as "that which is declared." It is not so described. Carnap shifts from the word "declare" to the word "express," and characterizes *'proposition'* as *that which is expressed.* "Expressions" (inclusive of "sentences") had previously, however, been separated from meaningfulness, when "meaning" was closely identified as "proposition." (We shall later display this in connection with "language" and with "meaning.") Despite this, the verb "expressed" is now used to establish that very meaningfulness of which the noun "expression" has been denied the benefit. Thus the word "express" openly indulges in double-talk between its noun and verb forms.[1] For any logic such a procedure would rate as incoherent. Yet before we recover from it, whether to make outcry or to forgive, we find ourselves in worse. We at once face four synonymic (or are they?) substitutes for *expressed,* namely: *signified, formulated, represented, designated.* Each of these words breathes a different atmosphere. "Signified" has an internally mentalistic feel, sucking up the "signs," so to speak, into the "significance"; "formulated" wavers between linguistic embodiment and rationalistic authority; "designated" has its origins, at least, among physical things, no matter how it wanders; "represented" holds up its face for any passing bee to kiss that is not satisfied with the other pretty word-flowers in the bouquet.[2]

At this point we should probably pause for a discussion of "designation." Designation is not a chance visitor, but a prominent inmate of the system. As such it certainly ought not to be tossed around as one among several casual words. Neither it nor any of its derivatives, however, has gained place in the terminological appendix. Full discussion would take much time and space. We shall here confine ourselves to a few hints. At its original entrance (*C,* p. 9) the status of designatum is so low that it is merely "what is referred to," possibly something outside the logic altogether. We have seen it gain the status of "what is expressed" in substitution for "what is declared" in a fast company of "meanings" that run far beyond the range of the usual official identification of meaning with designatum. Designation is sometimes a "relation" of a type that can "apply" (*C,* p. 49) to expressions; again "having a certain designatum" may be "a semantical property of an expression";[3] still again it tells what the speaker *intends* to refer to (*C,* p. 8); and there are times when Carnap inspects an open question as to whether the designata of sentences may not be "possible facts. . .or rather thoughts" (*C,* p. 53). *Officially* he decides that the designatum of a sentence (I) is a proposition (II*), much as the designatum of an object-name is an object (*C,* p. 45; p. 50 Des-Prop; p. 54; p. 99). Suppose the proposition is the designatum of the sentence; suppose the proposition (as we shall note later) may be called "true" as well as the sentence (which latter is officially what is "true" or "false") (*C,* p. 26, p. 90, p. 240); and suppose that "true" is built up around designation. It would then appear that the proposition which "is" the designatum of its own sentence must have somewhere beyond it certain sub-designata which it sub-designates directly instead of by way of its master (or is it servant?) sentence. This is far too intricately imaginative for any probing here. It looks plausible, but whether it makes sense or not we would not know.

The three-realm pattern of organization Carnap uses includes speakers (I), expressions (II) and designata (III). It is now in desperate state. We are not here arguing its falsity—we shall take care of that in another place—but only showing the *incoherence it itself achieves.* Expressions (II) are meaningful or not, but on any show-down they presumptively take speakers (I) to operate them. The meaning of an expression (II) is a designatum (III), but soon it becomes in a special case an expression-meaning that has not moved out of realm II. This designatum (as object) in II is presumptively given justification by comparison with an object in III, although the object in III is so void of status of its own in the logic (other than "intuitively" nominal) that it itself might do better by seeking its own justification through comparison with the proposition-object in II.

The soil in which such vegetation grows is "language" as Carnap sees it. Here he seems to have become progressively vaguer in recent years.[4] We found Cohen-Nagel asserting flatly that language consists of physical things called "signs." Carnap proceeds to similar intent part of the time, but differently the rest of the time, and always avoids plain statement. Consider the first sentence of his first chapter (*C,* p. 3):

"A *language,* as it is usually understood, is a system of sounds, or rather of the habits of producing them by the speaking organs, for the purpose of communicating

---

[1] The source of tolerance for such contradictions is well enough known to us. It lies in the reference of the "meanings" to a mental actor behind the scenes. This is apart, however, from the immediate purpose of discussion at the present stage. Consider "adequacy" as intention (*C,* p. 53); also "sign" as involving intent (*International Encyclopedia of Unified Science,* I, No. 3, p. 4; and similarly *C,* p. 8).

[2] Alonzo Church, referring to this passage in its original magazine appearance, holds that the charge of inconsistency against Carnap's switch from "designation" to "expression" fails because the various alternatives Carnap suggests for "expression" refer, partially at least, to the views of others. This, at any rate, is the way we understand him. Church's words are: "The charge of inconsistency to Carnap because he says "officially" that a sentence *designates* a proposition but on page 235 writes of sentences as *expressing* propositions (along with a list of alternatives to the verb 'express') fails, because it is obvious that in the latter passage Carnap is describing the varied views of others as well as his own" (*The Journal of Symbolic Logic,* X (1945), p. 132). The situation here seems to be about as follows: (1) Carnap's "designate" and "express" do not separate into an earlier official and a later casual or descriptive use, but both appear in a single passage of eight lines (*C,* p. 235) which is as "official" as anything in his text, and which we have already cited in full. (2) The pseudosynonyms for "express" are not attributed to other writers, but are run in without comment apparently as current usages. (3) In the succeeding page and a half of discussion he gives to other writers only one of these words, namely "represent," enters as employed by a specific other writer—in this case by Bosanquet. (Compare footnote 1, p. 95). (4) The alternatives for "express" do not appear in the portion of the passage dealing with 'sentence,' but strangely enough in that portion dealing with 'proposition,' that is to say with "that which is expressed by a sentence." (5) Even if Church were correct in identifying here "the varied views of others," the point would be irrelevant for use as keystone in a charge of default in proof; our passage in question might be called irrelevant or flippant, but certainly never a determining factor. (6) The charge in our text is one of abundant chaos in Carnap's linguistic foundations, and never of a particular inconsistency. We strongly recommend the careful examination of the texts of Carnap and Church alongside our own in this particular disagreement, and equally of the other positions Church attributes to us in comparison with the positions we actually take in our examination. Only through hard, close work in this field can the full extent of the linguistic chaos involved become evident.

[3] Rudolf Carnap, *The Formalization of Logic* (Cambridge, 1943), pp. 3-4.

[4] However, to his credit, he seems to have largely dropped or smoothed over the older jargon of physical language, physical thing-language, and observable thing-predicates (as in *International Encyclopedia of Unified Science,* I, No. 1, 52).

with other persons, i.e., of influencing their actions, decisions, thoughts, etc."

Does "usually" give *his* understanding? If the sounds are physical, in what sense are they in system? Can physics set up and discuss such a "system"? How do "habits"[1] of producing differ from "producing," especially when "speaking organs" are specified as the producers? Does the "i.e." mean that "communicating" is always an "influencing"? What range have the words "purpose," "actions," "decisions," "thoughts"? Sounds are perhaps physical, habits physiological, communications and influencings broadly behavioral, and the other items narrowly "psychical." May not, perhaps, any one of these words—or, indeed, still more dangerously, the word "person" under some specialized stress its user gives it—destroy the presumable import of many of the others?

Even if we accept the cited sentence as a permissible opening, surely better development should at once follow. Instead we find nothing but wavering words. We are told (*C,* pp. 4-5) that utterances may be analyzed into "smaller and smaller parts," that "ultimate units" of expressions are called "signs," that expressions are finite sequences of signs and that expressions may be "meaningful or not." We are not told whether signs are strictly physical sounds or marks, or whether they are products, habits or purposes. Later on (*C,* p. 18) we find sign, term and expression used as equivalents. We suspect as the work proceeds that the word "sign" is used mostly where physical implications are desired, and the word "term" mostly for the logical, while the word "expression" is waveringly intermediate—the precision-status being more that of campaign oratory than of careful inquiry. When the accent mark on a French *é* is viewed as a separate sign from the *e* without the accent (*C,* p. 5), "sign" seems clearly physical. When expression is "any finite sequence of signs," "sign" is certainly physical if the word "physical" means anything at all. Still, an expression may be a name, a compound or a sentence (*C,* p. 25, p. 50). And when an expression expresses a proposition, what are we to say? Again the issue is evaded. We get no answer, and surely we are not unreasonable in wanting to find out before we get too far along. Not knowing, not being able to find out—this is why we have here to search into the text so painfully.

All in all, the best that we are able to report of Carnap's procedure is that '*proposition*' or *proposition* appears as or names an entity, this entity being the certain meaning or designatum that is meant or designated by a non-designating and meaningless, though nevertheless declarative, sentence, representing, whether internally or externally, certain other designata besides itself, and manipulated through a terminology of "concepts" under which it at times is, and at times is not, itself a concept.

It is difficult to tell just where the most vicious center of terminological evil lies in Carnap's procedure. Probably, however, the dubious honor should go to "concept," a word that is all things to all sentences. We shall exhibit a few samples of his dealings with this word, and then quote what he once said in a moment when he stopped to think about it—which is not the case in the book in hand. The word, as he uses it, derives, of course, from *Begriff,* which among its addicts on its native soil can

without fatigue insert itself a dozen times on a page for any number of pages. In the present book (*C*) "concept" is employed in thirteen of the thirty headings of the constructive sections lying between the introductory chapter and the appendix, without in any case having determinable significance. The appendix (*C,* p. 230) lists three types of current uses for the word "concept": (1) psychological; (2) logical; (3) "as term or expression." The first and last of these uses he rejects. Among variations in the logical use he accepts the "widest," using asterisks (see *C,* p. 229 n.) to make the word "concept" cover properties, relations, functions, all three.

One could show without difficulty that Carnap's own practical use of "concept" is heavily infected with the psychological quality, despite his disavowal of this use; one can likewise show that he frequently uses the word for "term" or "expression," and this perhaps as often as he uses it for some form of "entity." We find him (*C,* p. 41) treating concepts as being "applicable" to certain attributes in almost precisely the same way that in another passage (*C,* p. 88) he makes terms "apply."[2] On pages 88 and 89 all semantical concepts are based on relations; some concepts are relations, and some are attributed to expressions only, not to designata. We get glimpses of such things as "intuitive concepts" (*C,* p. 119) and heavy use of "absolute concepts" of which a word later. Endless illustrations of incoherent use could be given, but no instance in which he has made any attempt to orient this word-of-all-work either to language, to thing, or to mind.

The passage in which he once stopped for an instant to think about the word may be found in his paper "Logical Foundations of the Unity of Science,"[3] published a few years before the present book. He wrote:

"Instead of the word 'term' the word 'concept' could be taken, which is more frequently used by logicians. But the word 'term' is more clear, since it shows that we mean signs, e.g., words, expressions consisting of words, artificial symbols, etc., of course with the meaning they have in the language in question."

The vagueness of his position could hardly be more vividly revealed. It is as if a microscopist could not tell his slide from the section he mounted on it, and went through a lot of abracadabra about metaslides to hide his confusion. Not until the words "concept" and "term" are clarified will a metalanguage be able to yield clear results.

"Term" runs "concept" a close second. One finds an interesting illustration (*C,* p. 89) where Carnap finds it convenient to use "the same term" for a certain "semantical concept" and for its corresponding "absolute concept." He goes on to remark, though without correcting his text, that what he really meant was "the same word," not "the same term," but in Convention 17-1 he goes back to "term" again. Thus a single "term" is authorized by convention to designate (if "designate" is the proper word) two meanings (if "meanings" is the proper word) at a critical stage of inquiry. Carnap

---

[1] In an earlier paper (*Ibid.,* I, No. 3, 3) such "habits" were called "dispositions," and we were told both that language is a system of dispositions and that its elements are sounds or written marks. Whether Carnap regards dispositions as sounds, or sounds as dispositions, he does not make clear.

[2] An interesting case of comparable confusion (superficial, however, rather than malignant) appears in the word "function," which is listed (*C,* p. 18) among the "entities," although "expressional function" and "sentential function" (both non-entitative) appear in the accompanying text. Terminological discussion (*C,* pp. 232-233) strongly favors the entitative use but still fails to star it as Carnap's own. The starring gives endorsement to the expressive uses cited above. In place of expression and entity consider, for comparison's sake, inorganic and organic. Then in place of a function among entities we might take a rooster among organisms. Carnap's "expressional function" can now be compared to something like "inorganic rooster."

[3] *International Encyclopedia of Unified Science,* I, No. 1 (1938), 49.

considers the ambiguity harmless. Indeed he says "there is no ambiguity." The use of an admittedly wrong word in his convention was apparently the lesser of two evils he was facing, since if one takes the trouble to insert what he says is the right word (*viz.,* word) for what he says is the wrong word (*viz.,* term) in the convention and then skeletonizes the assertion, one will somewhat surprisingly find oneself told that "a word...will be applied ...without reference to a language system."[1] Similarly a term may apply both to attributes and to predicates that designate attributes, i.e., both to designata and to expressions (*C*, p. 42).

For a mixture of terms and concepts his defense of his "multiple use" of term (*C*, p. 238) is worth study. A "radical term" may "designate" relations between propositions or relations between attributes (both cases being of "absolute concepts"), or between sentences or between predicates (these cases being "semantical"). In other words every possible opening is left for evasive manipulation.

"Definition" gets into trouble along with "term" and "concept." It enters, not by positive assertion, but by suggestion, as a matter of abbreviations, equalities and equivalences (*C*, p. 17). However, we find concepts that are entities being defined as liberally as terms that are expressions (*C*, p. 33). The absolute concepts are heavily favored in this way (*C*, p. 41, p. 90). One may even seek definitions to be in agreement with intuitive concepts for which only vague explanations have been given (*C*, p. 119). So many experiences has definition had *en route* that, when the calculus is reached, the assurance (*C*, p. 157) that definition may be employed there also seems almost apologetic.[2]

An excellent illustration of the status of many of the confusions we have been noting—involving also the mystery of "object" in the logic—is found in the case of Function (*C*, pp. 232-233) a brief notice of which is given in footnote 2 on page 97. Here a certain designatum is referred to as "strictly speaking, the entity determined by the expression." The word "determined" interests us, but is difficult to trace back to its den. The "entity" is what gets determined. Surely the "expression," taken *physically* as a sign, cannot be the determiner, nor can it, as a word of record, label, or tag, have initiative assigned it. Designation appears frequently as a "relation" between entity and expression, but we are told nothing to indicate that the expression is the active, and the entity the passive, member of the "relation." Back in its hide-out a "determiner" doubtless lurks, as soul, or intellect, or mind, or will—it can make little difference which, so long as something can be summoned for the task. Our

objection at the moment is not to such a soul—that issue lying beyond our immediate range—but to the bad job it does; for if the *expression*, with or without such a proxy, determines the *entity*, it gives the lie to the whole third-realm scheme of relational construction for expression, sentence, proposition and designatum.

We have written at length about expression and concept, and briefly about term, designation, definition and object. The word "relation" (presumptively entitative) is found in suspicious circumstances, similar to those of concept and the others. Thus (*C*, p. 49) you can "apply" a relation to a system. The word "meaning" deserves further mention as it is involved with all the rest. Most frequently "meaning" stands for designatum (*C*, p. 245); wherever a "sentence," as in the calculus, appears as meaningless, it is because designation (as "meaning") is there excluded from consideration. However, if one examines the passages in which meaning is casually spoken of, and those in which sense or meaning is brought into contact with truth-conditions (*C*, p. 10, p. 22, p. 232), the case is not so simple. In *The Formalization of Logic* (p. 6) it occurs to Carnap that he might let pure semantics abstract from "the meaning of descriptive signs" and then let syntax abstract from "the meaning of all signs, including the logical ones." This manner of observation could be carried much farther, and with profit, since one of the first practical observations one makes on his work is that six or eight layers of "meaning" could be peeled apart in his materials, and that he is highly arbitrary in establishing the two or three sharp lines he does.

We have said nothing about "true" in Carnap's procedure, for there is almost nothing that can be said dependably. He introduces it for "sentences" (and for classes of sentences), but takes the privilege at times of talking of the truth of "propositions," despite the sharp distinctions he has drawn between the two on the lines we have so elaborately examined (*C*, p. 26, p. 90; and compare p. 240 on "deliberate ambiguity"). He has C-true, L-true, F-true, and 'true,' distinguished (and legitimately so, if consistently organized and presented); he might have many more.

The situation may be fairly appraised in connection with "interpretation," an important word in the treatise. Leaving pragmatics for others, Carnap considers syntax and semantics as separate, with an additional "indispensable" distinction between factual and logical truth *inside* the latter (*C*, p. vii). A semantical system is a system of rules; it is an interpreted system ("interpreted by rules," p. 22); and it may be an interpretation of a calculus (p. 202). It also turns out, though, that interpretation is not a semantical system but a "relation" between semantical systems and calculi, belonging "neither to semantics nor to syntax" (*C*, p. 202, p. 240).

Fact does not enter by name until the work is more than half finished (*C*, p. 140), except for slight references to "factual knowledge" (*C*, p. 33, p. 81) and possibly for a few rare cases of presumptively positive use of "object" such as we have already mentioned (*C*, p. 54). However, it has a vociferous surrogate in "absolute concepts," the ones that are "not dependent upon language" and merely require "certain conditions with respect to truth-values" (*C*, p. 35)—"conveniences" (*C*, p. 90)—which are able to be *much less important* than the L- and C-concepts and, at the same time, to serve chiefly as *a basis* for them (*C*, p. 35).

We repeat once more that the significance we stress in our inquiry lies entirely in the interior incoherence of current logical statement it exhibits. While (as we have

---

1 Carnap has, as is well known, a standing alibi in all such cases as this. It is that he is not talking about an actual language, but about an abstract system of signs with meanings. In the present case there would seem to be all the less excuse for vacillating between word and term. If the distinctions are valid, and are intended to be adhered to, exact statement should not be difficult. It is, of course, understood that the general problem of the use of "word" and "term" is not being raised by us here; no more is the general problem of the entry of "fact," whether by "convention" or not, into a logic. For further comparison, and to avoid misinterpretation, the text of Convention 17-1 follows: "A term used for a radical semantical property of expressions will be applied in an absolute way (i.e. without reference to a language system) to an entity *u* if and only if every expression $\mathfrak{A}$, which designates *u* in any semantical system *S* has that semantical property in *S*. Analogously with a semantical relation between two or more expressions."

2 Again, we are not assailing Carnap's actual research into linguistic connectivities. The point is the importance of talking coherently about them.

intimated) we believe the source of such incoherence is visible behind its smoke-screens, the weight of our argument does not rest upon our opinion in this respect.

We find it further only fair to say of Carnap that in many respects he is becoming less assertive and more open to the influence of observation than he has been in the past. He recognizes now, for example (C, p. 18), something "not quite satisfactory" in his namings for his designata. He is aware that his basic distinction between logical and descriptive signs (C, p. vii, p. 56, p. 59, p. 87) needs further inquiry. He sees an *open* problem as to extensional and intensional language systems (C, p. 101, p. 118). He notes the "obviously rather vague" entry of his L-terms (C, p. 62). At one point he remarks that his whole structure (and with it all his terminology) may have to change (C, p. 229). More significant still, he has a moment when he notes that "even the nature of propositions" is still controversial (C, p. 101).

If he should come to question similarly his entitative concentrations he might have a better outlook, but in his latest publication he still feels assured that certain critical semantical terms can be "exactly defined on the basis of the concept of entities satisfying a sentential function," and that "having a certain designatum is a semantical property of an expression,"[1] though just how he would build those two remarks together into a coherent whole we do not know. His confidence that his own semantics is "the fulfillment of the old search for a logic of meaning which had not been fulfilled before in any precise and satisfactory way" (C, p. 249) needs modification, it would thus appear, under the various qualifications we have considered.

IV

Let us next glance at three specialized treatments of proposition, meaning and designation: those of Morris, Ducasse and Lewis.[2]

Morris attaches himself to Carnap. His contribution (apart from the verbal chaos of his semiotic) lies in the "pragmatics" he has added to the earlier "semantics" and "syntactics" (M, p. 6, p. 8) to yield the three "irreducibles," the "equally legitimates" (M, p. 53) that form his rotund trinity. Carnap gratefully accepts this offering with qualifications (C, p. 9). It enables him to toss all such uncomfortable issues as "gaining and communicating knowledge" to the garbage bucket of pragmatics, while himself pursuing unhampered his "logical analysis" (C, p. 250) in the ivory tower of syntactics and in the straggling mud huts of semantics scattered around its base. Neither Carnap nor Morris seems to be aware—or, if aware, neither of them is bothered by the fact—that pragmatism, in every forward step that has been taken in the central line from Peirce,[3]

has concentrated on "meanings"—in other words, on the very field of semantics from which Carnap and Morris now exclude it. To tear semantics and pragmatics thus apart is to leap from Peirce back towards the medieval.[4]

As for the "semiotic" which he offers as a "science among the sciences" (M, p. 2), as underlying syntactics, semantics and pragmatics, and as being designed to "supply a language...to improve the language of science" (M, p. 3), we need give only a few illustrations of the extent to which its own language falls below the most ordinary standards of everyday coherence. He employs a "triadic relation" possessing "three correlates": sign vehicle, designatum and interpreter (M, p. 6). These, however, had entered three pages earlier as "three (or four) factors" where "interpretant" was listed with the parenthetic comment that "interpreter" may be a fourth. Concerning each of these three (or four) factors in his "triadic relations," he writes so many varying sentences it is safe to say that in simple addition all would cancel out and nothing be left.

Consider the dramatic case of the birth of an interpretant.[5] You take a certain "that which" that *acts* as a sign and make it produce an *effect* (called interpretant) on an interpreter, *in virtue of which* the "that which" becomes, or "is," a sign (M, p. 3, lines 23-25). Four pages later the sign may *express* its interpreter. The words are incoherent when checked one against another. As for the signs themselves, they are "simply the objects" (M, p. 2); they are "things or properties...in their function" (M, p. 2); they are something "denoting the objects" (M, p. 2); they are something to be determined for certain cases by "semantical rule" (M, pp. 23-24); they are something of which (for other cases) one can say that "the sign vehicle is only that aspect of the apparent sign vehicle in virtue of which semiosis takes place" (M, p. 49) etc., etc. Some signs designate without denoting (M, p. 5);[6] others indicate without designating (M, p. 29). Some objects exist without semiosis (M, p. 5), and sometimes the designatum of a sign need not be an "actual existent object" (M, p. 5). Comparably a man may "point without pointing to anything" (M, p. 5), which is as neat a survival of medieval mentality in the modern age as one would wish to see.[7]

In Morris' procedure language is one thing, and "using it" is another. He may talk behaviorally about it for a

---

1 *The Formalization of Logic*, p. xi, p. 3.

2 Procedure should be like that of entomologists, who gather specimen bugs by the thousands to make sure of their results. It should also be like that of engineers getting the "bugs" (another kind, it is true) out of machinery. Space considerations permit the exhibit of only a few specimens. But we believe these specimens are significant. We trust they may stimulate other "naturalists" to do field work of their own. Compare the comment of Karl Menger when in a somewhat similar difficulty over what the "intuitionists" stood for in mathematics. "Naturally," he wrote, "a sober critic can do nothing but stick to their external communications." "The New Logic," *Philosophy of Science*, IV (1937), 320. Compare also our further comment on this phase of inquiry in Chapter III, p. 117, footnote 8.

3 In "How to Make our Ideas Clear" (1878) where "practical" bearings and effects are introduced, and where it is asserted that "our conception of these effects is the whole of our conception of the object" (*Collected Papers*, 5.402).

4 That even Morris himself has now become troubled appears from a later discussion in which—under the stimulus of a marvelously succulent, syllabic synthesis applied to "linguistic signs," namely, that they are "transsituationally intersubjective" —he votes in favor of a "wider use of 'semantics'" and a "narrower use of 'pragmatics'" hereafter (*Philosophy of Science*, X [1943], 248-249). Indeed, Morris' whole tone in this new paper is apologetic, though falling far short of hinting at a much-needed thorough-going house-cleaning. No effect of this suggested change in viewpoint is, however, manifested in his subsequent book, *Signs, Language, and Behavior* (New York, 1946), nor is his paper of 1943 as much as listed in the bibliography therein provided.

5 Where Morris allots a possible four components to his "triadic relation" he employs the evasive phrase-device "commonly regarded as," itself as common in logic as outside. (*Cf.* Carnap's "language as it is usually understood," which we have discussed previously.) The word "interpretant" is of course lifted verbally, though not meaningfully, from Peirce, who used it for the operational outcome of sets of ordered signs (*Collected Papers*, 2.92 to 2.94, and *cf.* also 2.646). The effect (outcome or consequences) of which Peirce speaks is definitely *not* an effect upon an interpreter. There is no ground in Peirce's writings for identifying "interpretant" with "interpreter."

6 A demonstration of the meaninglessness of Morris' treatment of denotation and designation—of objects, classes and entities—has been published by George V. Gentry since this paper was prepared. (*The Journal of Philosophy*, XLI [1944], 376-384).

7 For Morris' later development of "sign" (1946) see Chapter IX.

paragraph or two, but his boldest advance in that direction would be to develop its "relation" to the "interpreter" ("dog" or "person") who uses it. Sometimes, for him, science *is* a language; at other times science *has* a language, although semiotic has a better one. A "dual control of linguistic structure" is set up *(M,* pp. 12-13) requiring both events and behaviors, but independently physical signs and objects that are not actual find their way in. Similarly, in the more expansive generalizations, at one time we find (as *M,* p. 29) that syntactic or semantic rules are only verbal formulations within semiotic, while at other times (as *M,* p. 33) we learn that syntactics must be established before we can relate signs to interpreters or to things. The net result is such a complete blank that we find it almost exciting when such a venturesome conclusion is reached as marked an earlier paper by Morris: that "signs which constitute scientific treatises have, to some extent at least, a correlation with objects."[1]

V

Ducasse has labored industriously to discover what a proposition actually "is," if it is the sort of thing he and Cohen-Nagel believe it to be. We do not need to follow him through his long studies since, fortunately, he has recently provided a compact statement. Rearranging somewhat his recipe for the hunting of his snark *(D,* p. 134), though taking pains to preserve its purity, we get:

Catch an assertion (such as "the dog is red"). Note it is "the verbal symbol of an opinion." Pin it securely on the operating table.

Peel off all that is "verbal" and throw away. Peel off all "epistemic attitude" (here "belief") and throw away also.

The remainder will be a *proposition.*

Dissect carefully. The proposition will be found to have two components, both "physical entities": the first, a "physical object"; the second, a "physical property."

Distill away from these components all traces of conscious process—in especial, as to "object," all that is perceptual; as to "property," all that is conceptual.

When this has been skilfully done you will have remaining the pure components of the pure proposition, with all that is verbal or mental removed.

Further contemplation of the pure proposition will reveal that it has the following peculiarities: *(a)* if its two components cleave together in intimate union, the first "possessing" the second, then the proposition is "true," and the "true proposition" is "fact"; *(b)* if the second component vanishes, then what remains (despite the lack of one of its two essential components) is still a proposition, but this time a "false proposition," and a false proposition is "not a fact," or perhaps more accurately, since it is still an important something, it might be called a "not-fact."

This is no comfortable outcome. The only way it can "make" sense, so far as one can see is by continuous implied orientation towards a concealed mental operator, for whom one would have more respect if he came out in front and did business in his own name.[2]

---

[1] *International Encyclopedia of Unified Science,* I, No. 1 (1938), p. 69.

[2] A later attempt by Ducasse is found in a paper, "Propositions, Truth, and the Ultimate Criterion of Truth" *(Philosophy and Phenomenological Research,* IV [1944], 317-340), which became available after the above was written. In it the confusion heightens. For Ducasse, now, no proposition has either a subject

VI

Lewis illustrates what happens when words as physical facts are sharply severed from meanings as psychical facts, with the former employed by a superior agency—a "mind"—to "convey" the latter *(L,* p. 236). He makes so sharp a split between ink-marks and meanings that he at once faces a "which comes first?" puzzler of the "chicken or egg" type, his sympathies giving priority to the meanings over the wordings.

He tells us *(L,* p. 237) that "a linguistic expression is constituted by the association of a verbal symbol and a fixed meaning." Here the original ink-spot-verbal is alloted symbolic quality (surely it must be "psychic") while the meaning is allegedly "fixed" (which sounds very "physical"). Our bigamist is thus unfaithful in both houses. He is doubly and triply unfaithful, at that, for the last part of the cited sentence reads: "but the linguistic expression cannot be identified with the symbol alone nor with the meaning alone." First we had physical words and mental meanings; then we had verbal symbols and fixed meanings; now we have symbol alone and meaning alone, neither of them being expressive. He uses, it is true, a purportedly vitalizing word—or, rather, a word that might vitalize if it had any vitality left in it. This word is "association," outcast of both philosophy and psychology, a thorough ne'er-do-well, that at best points a dirty finger at a region in which research is required.

So slippery are the above phrasings that no matter how sternly one pursues them they can not be held fast. The signs are physical, but they become verbal symbols. A verbal symbol is a pattern of marks; it is a "recognizable pattern"; it becomes a pattern even when apart from its "instances"; it winds up as an "abstract entity" (all in *L,* pp. 236-237). Expression goes the same route from ink-spots on up (or down), so that finally, when the symbol becomes an abstract entity, the expression (originally a physical "thing") becomes a "correlative abstraction" *(L,* p. 237).

A term is an expression that "names or applies to" (one would like to clear up the difference or the identity here) "a thing or things, of some kind, actual or thought of" (again plenty of room for clarification); it changes into something that is "capable" of naming, where naming is at times used as a synonym for "speaking of" *(L,* p. 237); in the case of the "abstract term," however, the term "names what it signifies" *(L,* p. 239). One would like to understand the status of proposition as "assertable content" *(L,* p. 242); of a "sense-meaning" that is "intension in the mode of a criterion in mind" *(L,* p. 247); of signification as "comprehensive essential character" *(L,* p. 239). One could even endure a little information about the way in which "denote" is to be maintained as different from "denotation," and how one can avoid "the awkward consequences" of this difference by adopting the word "designation" (apparently from

---

or a predicate (p. 321). Many varieties of "things" or "somethings" are introduced, and there is complete absence of information as to what we are to understand by "thing" or "something." Thus: "the sort of thing, and the only sort of thing, which either is true or is false is a proposition" (p. 318); it is to be sharply discriminated from "other sorts of things called respectively statements, opinions, and judgments..." (p. 318); "the ultimate...constituents of a proposition are some *ubi* and some *quid*—some *locus* and some *quale*" (p. 323); "a fact is not something to which true propositions 'correspond' in some sense...a fact *is* a true proposition" (p. 320). Incidentally a proposition is also the *content* of an opinion (p. 320) from which we may infer that a fact, being a true proposition, is likewise the content of an opinion. It is very discouraging.

Carnap and Morris, and apparently in a sense different from either of theirs)—an effort which Lewis himself does not find it worth his while to make *(L*, p. 237). Finally, if "meaning" and "physical sign" cannot be better held apart than Lewis succeeds in doing, one would like to know why he tries so elaborately.[1]

VII

We shall discuss Bertrand Russell's logical setting in Chapter VIII. His terminology, as previously noted, appears confused, even to Carnap, who finds Russell's explanations of his various uses of the word "proposition" "very difficult to understand" *(C*, pp. 235-236). The voluminous interchanges Russell has had with others result in ever renewed complaints by him that he is not properly understood. Despite his great initiative in symbolic formulation in the border regions between logic and mathematics, and despite the many specializations of inquiry he has carried through, no progress in basic organization has resulted from his work. This seems to be the main lesson from logical inquiry in general as it has thus far been carried on. We may stress this highly unsatisfactory status by quoting a few other remarks by logicians on the work of their fellows.

Carnap, in his latest volume,[2] regrets that most logicians still leave "the understanding and use of [semantical] terms. . .to common sense and instinct," and feels that the work of Hilbert and Bernays would be clearer "if the distinction between expressions and their designata were observed more strictly"—and this despite his own chaos in that respect.

Cohen and Nagel in their preface pay their compliments to their fellows thus:

"Florence Nightingale transformed modern hospital practice by the motto: Whatever hospitals do, they should not spread disease. Similarly, logic should not infect students with fallacies and confusions as to the fundamental nature of valid or scientific reasoning."

Tarski, whose procedure is the next and last we shall examine, writes *(T*, p. 345):

"It is perhaps worth-while saying that semantics as it is conceived in this paper (and in former papers of the author) is a sober and modest discipline which has no pretensions of being a universal patent medicine for all the ills and diseases of mankind whether imaginary or real. You will not find in semantics any remedy for decayed teeth or illusions of grandeur or class conflicts. Nor is semantics a device for establishing that every one except the speaker and his friends is speaking nonsense."

VIII

Tarski's work is indeed like a breath of fresh air after the murky atmosphere we have been in. It is not that he has undertaken positive construction or given concentrated attention to the old abuses of terminology, but he is on the way—shaking himself, one might say, to get free. His procedure is simple, unpretentious, and cleared of many of the ancient verbal unintelligibilities. He does not formally abandon the three-realm background and he

occasionally, though not often, lapses into using it—speaking of "terms," for example, as "indispensable means for conveying human thoughts"[3]—but he seems free from that persistent, malignant orientation towards the kind of fictive mental operator which the preceding logicians examined in this chapter have implicitly or explicitly relied upon. He sets "sentences" (as expressions) over against "objects referred to" *(T*, p. 345) in a matter-of-fact way, and goes to work. He employs a metalanguage to control object-languages, not as an esoteric, facultative mystery, but as a simple technical device, such as any good research man might seek in a form appropriate to his field, to fixate the materials under his examination.[4]

In his latest appraisal of "true" under the title "The Semantic Conception of Truth," Tarski concludes that for a given object-language and for such other formalized languages as are now known *(T*, p. 371, n. 14)—and he believes he can generalize for a comprehensive class of object-languages *(T*, p. 355)—"a sentence is true if it is satisfied by all objects, and false otherwise" *(T*, p. 353). The development, as we appraise it, informs us that if we assume *(a)* isolable things (here we make explicit his implicit assumption of the "thing") and *(b)* human assertions about them, then this use can be consistently maintained. In his demonstration Tarski discards "propositions," beloved of Cohen-Nagel, Carnap and Ducasse, saying they are too often "ideal entities" of which the "meaning. . .seems never to have been made quite clear and unambiguous" *(T*, p. 342). He establishes "sentences" with the characteristics of "assertions," and then considers such a sentence on the one hand as in active assertion, and on the other hand as designated or named, and thus identified, so that it can be more accurately handled and dealt with by the inquirer. After establishing certain "equivalences of the form *(T)*" which assure us that the sentence is well-named (*x* is true if, and only if, *p*) *(T*, p. 344), he sharpens an earlier formulation for "adequacy," the requirement now becoming that "all equivalences of the form *(T)* can be asserted" *(T*, p. 344). (For all of this we are, of course, employing our own free phrasing, which we are able to do because his work, unlike the others, is substantial enough to tolerate it.) "A definition of truth is 'adequate' if all these equivalences follow from it." Given such adequacy we have a "semantic" conception of truth, although the expression *(T)* itself is not yet a definition.

To demonstrate his conclusion Tarski identifies as primarily semantic: (1) designation (denoting), (2) satisfaction (for conditions), (3) definition (unique determining); he calls them "relations" between "sentences" and "objects." "True," however, he says, is not such a "relation"; instead it expresses a property (or denotes a class) of sentences *(T*, p. 345). Nevertheless it is to be called "semantic" because the best way of defining

---

[1] Professor Baylis finds some of the same difficulties we have found in Lewis' procedure, and several more, and regards portions of it as "cagey" *(Philosophy and Phenomenological Research*, V [1944], 80-88). He does not, however, draw the conclusion we draw as to the radical deficiency in the whole scheme of terminology. Professor Lewis, replying to Professor Baylis *(ibid.*, 94-96), finds as much uncertainty in the latter as the latter finds in him.

[2] *Formalization of Logic* pp. xii, xiii.

[3] Alfred Tarski, *Introduction to Logic and to the Methodology of Deductive Sciences* (New York, 1941), p. 18. Compare also his remark about "innate or acquired capacity," *ibid.*, p. 134.

[4] In the preface to the original (Polish) edition of his *Logic* he had held that "the concepts of logic permeate the whole of mathematics," considering the specifically mathematical concepts "special cases," and had gone so far as to assert that "logical laws are constantly applied—be it consciously or unconsciously—in mathematical reasonings" *(ibid.*, p. xvii). In his new preface *(ibid.*, p. xi, p. xiii) he reduces this to the assurance that logic "seeks to create. . .apparatus" and that it "analyzes the meaning" and "establishes the general laws." Even more significantly he remarks *(ibid.*, p. 140) that "meta-logic and meta-mathematics" means about the same as "the science of logic and mathematics." (Compare also *ibid.*, p. 134.)

it is by aid of the semantic relations *(T*, p. 345). His outcome, he thinks, is "formally correct" and "materially adequate," the conditions for material adequacy being such as to determine uniquely the extension of the term "true" *(T*, p. 353). *What he has done is to make plain to himself at the start what he believes truth to be in everyday use, after which by prolonged study he advances from a poorer and less reliable to a richer and more reliable formulation of it.* We do not say this in deprecation, but rather as high praise of the extent of progress in his standpoint. We may quote his saying that his aim is "to catch hold of the actual meaning of an old notion" *(T*, p. 341; compare also p. 361, bottom paragraph), where, if one strikes out any remaining sentimentality from the word "actual" and treats it rigorously, the sense becomes close to what we have expressed.

We must nevertheless, to make his status clear, list some of the flaws. He does not tell us clearly what he intends by the words "concept," "word," "term," "meaning" and "object." His applications of them are frequently mixed.[1] "Word" shades into "term," and "term" into "concept," and "concept" retains much of its traditional vagueness. Designation and satisfaction, as "relations," enter as running between expression and thing (the "semantic" requirement), but definition, also a relation, runs largely between expressions (a very different matter).[2] "True," while not offered as a "relation," is at one stage said to "denote," although denoting has been presented as a relating. The word "meaning" remains two-faced throughout, sometimes running from word (expression) to word, and sometimes from word to thing.[3] Lacking still is all endeavor to organize men's talkings to men's perceivings and manipulatings in the cultural world of their evolution. The ancient non-cultural verbal implications block the path.

IX

Along with proposition, truth, meaning and language, "fact" has been in difficulties in all the logics we have examined. We displayed this in Section II through the development of a curious contrast as to whether a fact is a proposition or a proposition a fact. The answer seemed to be "Neither." In various other ways the puzzle has appeared on the sidelines of the logics throughout.

Now, "fact" is not in trouble with the logics alone; the philosophies and epistemologies are equally chary of looking at it straight. Since direct construction in this field will occupy us later on, we shall here exhibit the character of this philosophical confusion by a few simple illustrations from the philosophical dictionaries and from

current periodical essays.[4] Consider first what the dictionaries report.

The recently published *Dictionary of Philosophy*[5] limits itself to three lines as follows:

"Fact (Lat. factus, p.p. of facio, do): Actual individual occurrence. An indubitable truth of actuality. A brute event. Synonymous with actual event."

Any high-school condensation of a dictionary should do better than that. This is supplemented, however, by another entry, allotted three times the space, and entitled "Fact: in Husserl" (whatever that may literally mean). Here unblinking use is made of such locutions as "categorical-syntactical structure," "simply is" and "regardless of value."

Baldwin's definition of a generation ago is well known. Fact is "objective datum of experience," by which is to be understood "datum of experience considered as abstracted from the experience of which it is a datum." This, of course, was well enough among specialists of its day, but the words it uses are hardly information-giving in our time.

Eisler's *Wörterbuch* (1930 edition) makes *Tatsache* out to be whatever we are convinced has objective or real *Bestand*—whatever is firmly established through thought as content of experience, as *Bestandteil* of the ordering under law of things and events. These again are words but are not helps.

Lalande's *Vocabulaire* (1928 edition) does better. It discusses fact to the extent of two pages, settling upon the wording of Seignobos and Langlois that "La notion de *fait*, quand on la précise, se ramène à un jugement d'affirmation sur la réalité extérieure." This at least sounds clear, and will satisfy anyone who accepts its neat psychology and overlooks the difficulties that lie in *jugement*, as we have just been surveying them.

Turning to current discussions in the journals for further illustration we select three specimens, all appearing during the past year (1944). Where mere illustration is involved and all are alike in the dark, there is no need to be invidious, and we therefore omit names and references, all the better to attend to the astonishing things we are told.

1. "Fact: a situation having reality in its own right independent of cognition." Here the word "situation" evidently enters because of its indefiniteness; "reality in its own right" follows with assertion of the most tremendous possible definiteness; and "independent of cognition," if it means anything, means "about which we know nothing at all." The whole statement is that fact is something very vague, yet most tremendously certain, about which we know nothing.

2. "There is something ultimately unprovable in a fact." Here a rapturous intellectualism entertains itself, forgetting that there has been something eventually uncertain about every "truth" man has thus far uncovered, and discrediting fact before trying to identify it.

3. "A fact can be an item of knowledge only because the *factual* is a character of reality. . . . Factual knowledge means the awareness of the occurrence of events felt, believed, or known to be independent of the volitional self. . . . The sense of fact is the sense of the self

---

1 Thus *Logic*, p. 18, p. 139. For "object" see *T*, p. 374, n. 35. He recognizes the vagueness in the word "concept" *(T*, p. 370) but continues to use it. His employment of it on page 108 of the *Logic* and his phrasing about "laws. . .concerning concepts" are of interest. His abuses of this word, however, are so slight compared with the naive specimens we have previously examined that complaint is not severe.

2 For "definition," consider the stipulating convention *(Logic*, p. 33) and the equivalence (p. 150) and compare these with the use of "relation" *(T*, p. 345) and with the comments *(T*, p. 374, n. 35). It is not the use of the single word "definition" for different processes that is objectionable, but the confusion in the uses.

3 Thus *Logic*, p. 133; one can discard first of all "independent meanings," and then the customary meanings of "logical concepts," and finally, apparently, "the meanings of all expressions encountered in the given discipline. . .without exception." The word "meaning" is, of course, one of the most unreliable in the dictionary, but that is no reason for playing fast and loose with it in logic.

4 The only considerable discussion of fact we have noted is the volume *Studies in the Nature of Facts (University of California Publications in Philosophy*, XIV [1932]), a series of eight lectures by men of different specializations. An examination of the points of view represented will reward anyone interested in further development of this field.

5 D. D. Runes, editor (New York, 1942).

confronting the not-self." The outcome of this set of warring assertions is a four-fold universe, containing: *(a)* reality; *(b)* truth; *(c)* a sort of factuality that is quasi-real; *(d)* another sort of factuality that is quasi-true. Poor "fact" is slaughtered from all four quarters of the heavens at once.

The citations above have been given not because they are exceptional, but because they are standard. You find this sort of thing wherever you go. No stronger challenge could be given for research than the continuance of such a state of affairs in this scientific era.

X

Enough evidence of linguistic chaos has been presented in this paper to justify an overhauling of the entire background of recent logical construction. This chaos is due to logicians' accepting ancient popular phrasings about life and conduct as if such phrasings were valid, apart from inquiry into their factual status within modern knowledge. As a result, not only is logic disreputable from the point of view of fact, but the status of "fact" is wretched within the logics. The involvement both of logic and fact with language is manifest. Some logics, as anyone can quickly discover, look upon language only to deny it. Some allot it incidental attention. Even where it is more formally introduced, it is in the main merely tacked on to the older logical materials, without entering into them in full function.

Our understanding thus far has been gained by refusing to accept the words man utters as independent beings—logicians' playthings akin to magicians' vipers or children's fairies—and by insisting that language is

veritably man himself in action, and thus observable. The "propositions" of Cohen and Nagel, of Ducasse and of Carnap, the "meanings" of Lewis, the "sign vehicles" and "interpretants" of Morris and the "truth" of Tarski all tell the same tale, though in varying degree. What is "man in action" gets distorted when manipulated as if detached; what is "other than man" gets plenty of crude assumption, but no fair factual treatment.

We said at the start that in closing we would indicate a still wider observation that must be made if better construction is to be achieved. The locus of such widened observation is where "object," "entity," "thing" or "designatum" is introduced. "Things" appear and are named, or they appear as named, or they appear through namings. Logics of the types we have been examining flutter and evade, but never attack directly the problem of sorting out and organizing words to things, and things to words, for their needs of research. They proceed as though some sort of oracle could be issued to settle all puzzles at once, with logicians as the priests presiding over the mysteries.

This problem, we believe, should be faced naturalistically. Passage should be made from the older half-light to such fuller light as modern science offers. In this fuller light the man who talks and thinks and knows belongs to the world in which he has been evolved in all his talkings, thinkings and knowings; while at the same time this world in which he has been evolved is the world of his knowing. Not even in his latest and most complex activities is it well to survey this natural man as magically "emergent" into something new and strange. Logic, we believe, must learn to accept him simply and naturally, if it is to begin the progress the future demands.

II.
THE TERMINOLOGICAL PROBLEM

SCIENCE uses its technical names efficiently. Such names serve to mark off certain portions of the scientific subjectmatter as provisionally acceptable, thereby freeing the worker's attention for closer consideration of other portions that remain problematic. The efficiency lies in the ability given the worker to hold such names steady—to know what he properly names with them —first at different stages of his own procedure and then in interchange with his associates.

Theories of knowledge provide their investigators with no such dependable aids. The traditional namings they employ have primitive cultural origins and the supplemental "terms" they evolve have frequently no ascertainable application as names at all.

We have asserted that the time has come when a few leading names for knowings and knowns can be established and put to use. We hold further that this undertaking should be placed upon a scientific basis; where by "scientific" we understand very simply a form of "factual" inquiry, in which the knowing man is accepted as a factual component of the factual cosmos, as he is elsewhere in modern research. We know of no other basis on which to anticipate dependable results—more particularly since the past history of "epistemology" is filled with danger-signs.

What we advocate is in very simple statement a passage from loose to firm namings. Some purported names do little more than indicate fields of inquiry—some, even, do hardly that. Others specify with a high degree of firmness. The word "knowledge," as a name, is a loose name. We do not employ it in the titles of our chapters and shall not use it in any significant way as we proceed. It is often a convenience, and it is probably not objectionable—at least it may be kept from being dangerous—where there is no stress upon its accurate application and no great probability that a reader will assume there is; at any rate we shall thus occasionally risk it. We shall rate it as No. 1 on a list of "vague words"[1] to which we shall call attention and add from time to time in footnotes. Only through prolonged factual inquiry, of which little has been undertaken as yet, can the word "knowledge" be given determinable status with respect to such questions as: (1) the range of its application to human or animal behaviors; (2) the types of its distribution between knowers, knowns, and presumptive intermediaries; (3) the possible localizations implied for knowledges as present in space and time. In place of examining such a vague generality as the word "knowledge" offers, we shall speak of and concern ourselves directly with knowings and knowns—and, moreover, in each instance, with those particular forms of knowings and knowns in respect to which we may hope for reasonably definite identifications.

I

The conditions that the sort of namings we seek must satisfy, positively and negatively, include the following:

1. The names are to be based on such observations as are accessible to and attainable by everybody. This condition excludes, as being negligible to knowledge, any report of purported observation which the reporter avows to be radically and exclusively private.

2. The status of observation and the use of reports upon it are to be tentative, postulational, hypothetical.[2] This condition excludes all purported materials and all alleged fixed principles that are offered as providing original and necessary "foundations" for either the knowings or the knowns.

3. The aim of the observation and naming adopted is to promote further observation and naming which in turn will advance and improve. This condition excludes all namings that are asserted to give, or that claim to be, finished reports on "reality."

The above conditions amount to saying that the names we need have to do with knowings and knowns in and by means of continuous operation and test in work, where any knowing or known establishes itself or fails to establish itself through continued search and research solely, never on the ground of any alleged outside "foundation," "premise," "axiom" or *ipse dixit*. In line with this attitude we do not assert that the conditions stated above are "true"; we are not even arguing in their behalf. We advance them as the conditions which, we hold, should be satisfied by the kind of names that are needed by us here and now if we are to advance knowledge of knowledge. Our procedure, then, does not stand in the way of inquiry into knowledge by other workers on the basis either of established creeds or tenets, or of alternative hypotheses; we but state the ground upon which we ourselves wish to work, in the belief that others are prepared to co-operate. The postulates and methods we wish to use are, we believe, akin to those of the sciences which have so greatly advanced knowledge in other fields.

The difficulties in our way are serious, but we believe these difficulties have their chief source in the control exercised over men by traditional phrasings originating when observation was relatively primitive and lacked the many important materials that are now easily available. Cultural conditions (such as ethnological research reveals) favored in earlier days the introduction of factors that have now been shown to be irrelevant to the operations of inquiry and to stand in the way of the formation of a straightforward theory of knowledge—straightforward in the sense of setting forth conclusions reached through inquiry into knowings as themselves facts.

The basic postulate of our procedure is that knowings are observable facts in exactly the same sense as are the subjectmatters that are known. A glance at any collection of books and periodicals discloses the immense number of subjectmatters that have been studied and the various grades of their establishment in the outcome. No great argument is required to warrant the statement that this wide field of knowledge (possessed of varying depths in its different portions) can be studied not only in terms of

---

[1] Even the words "vague," "firm" and "loose," as we at this stage are able to use them, are loosely used. We undertake development definitely and deliberately within an atmosphere (one might perhaps better call it a swamp) of vague language. We reject the alternative—the initial dependence on some schematism of verbal impactions—and propose to destroy the authoritarian claims of such impactions by means of distinctions to be introduced later, including particularly that between specification and definition.

[2] The postulations we are using, their origin and status, will be discussed in a following chapter. See also Dewey, *Logic, the Theory of Inquiry* (New York, 1938), Chap. I, and Bentley, "Postulation for Behavioral Inquiry" (*The Journal of Philosophy*, XXXVI [1939], 405-413).

things[1] known, but also in terms of the knowings.

In the previous chapter we pointed out instances, in the works of prominent contemporary logicians, of an extraordinary confusion arising from an uncritical use in logic, as theory of knowledge, of forms of primitive observation; sometimes to the utter neglect of the fuller and keener observation now available, and in other cases producing such a mixture of two incompatible types of observation as inevitably wrecks achievement. It was affirmed in that chapter that further advance will require complete abandonment of the customary isolation of the word from the man speaking, and likewise of the word from the thing spoken of or named. In effect, and often overtly, words are dealt with in the logics as if they were a new and third kind of fact lying between man as speaker and things as spoken of. The net result is to erect a new barrier in human behavior between the things that are involved and the operating organisms. While the logical writers in question have professedly departed from the earlier epistemological theories framed in terms of a mind basic as subject and an external world as object, competent analysis shown that the surviving separation their writings exhibit is the ghost of the seventeenth-century epistemological separation of knowing subject and object known, as that in turn was the ghost of the medieval separation of the "spiritual" essence from the "material" nature and body, often with an intervening "soul" of mixed or alternating activities.

Sometimes the intervening realm of names as a new and third kind of fact lying between man as speaker and things as spoken of takes the strange appearance of a denial not only of language as essential in logic, but even of names as essential in language. Thus Quine in a recent discussion of the issue of "universals" as "entities" tells us that "names generally. . .are inessential to language" and that his "suppression of names is a superficial revision of language." The world in which he operates would thus seem comparable with that of Whitehead in which "language" (including apparently that which he himself is using) is "always ambiguous," and in which "spoken language is merely a series of squeaks."[4] One may admire the skill with which Quine uses his method of abstraction to secure a unified field for symbolic logic in which "all names are abstract," and in which the bound variables of quantification become "the sole vehicle of direct objective reference," and still feel that the more he detaches his symbolic construction from the language he is referring to through the agency of the language he is using, the more he assimilates his construction to the other instances of "intervening" language, however less subtly these latter are deployed.

The importance we allot to the introduction of firm names is very quickly felt when one begins to make observation of knowledge as a going fact of behavioral activity. Observation not only separates but also brings together in combination in a single sweep matters which at other times have been treated as isolated and hence as requiring to be forced into organization ("synthesized" is the traditional word) by some outside agency. To see language, with all its speakings and writings, as man-himself-in-action-dealing-with-things is observation of the combining type. Meaningful conveyance is, of course, included, as itself of the very texture of language. The full event is before us thus in durational spread. The observation is no longer made in terms of "isolates" requiring to be "synthesized." Such procedure is common enough in all science. The extension as observation in our case is that we make it cover the speaker or knower along with the spoken of or known as being one common durational event. Here primary speaking is as observable as is a bird in flight. The inclusion of books and periodicals as a case of observable man-in-action is no different in kind from the observation of the steel girders of a bridge connecting the mining and smelting of ores with the operations of a steel mill, and with the building of bridges, in turn, out of the products. For that matter, it is no different from observation extended far enough to take in not just a bird while in flight but bird nest-building, egg-laying and hatching. Observation of this general type sees man-in-action, not as something radically set over against an environing world, not yet as something merely acting "in" a world, but as action *of* and *in* the world in which the man belongs as an integral constituent.

To see an event filling a certain duration of time as a description across a full duration, rather than as composed of an addition or other kind of combination of separate, instantaneous, or short-span events is another aspect of such observation. Procedure of this type was continuously used by Peirce, though he had no favorable opportunity for developing it, and it was basic to him from the time when in one of his earliest papers he stressed that all thought is in signs and requires a time.[3] The "immediate" or "neutral"

1 "Thing" is another vague word. It is in good standing, however, where general reference is intended, and it is safer in such cases than words like "entity" which carry too great a variety of philosophical and epistemological implications. We shall use it freely in this way, but for more determinate uses shall substitute "object" when we later have given this latter word sufficient definiteness.

2 Alfred North Whitehead: *Process and Reality* (New York, 1929), p. 403; W. V. Quine, "On Universals," *The Journal of Symbolic Logic*, XII (1947), p. 74. Compare also W. V. Quine, *Mathematical Logic*, Second Printing, (Cambridge, 1947), pp. 149-152 *et al.*

3 "The only cases of thought which we can find are of thought in signs" (*Collected Papers*, 5.251); "To say that thought cannot happen in an instant but requires a time is but another way of saying that every thought must be interpreted in another, or that all thought is in signs" (*ibid.*, 5.253). See also comment in our preceding chapter, Sec. I. For a survey of Peirce's development (the citations being to his *Collected Papers*) see "Questions Concerning Certain Faculties Claimed for Man" (1868), 5.213 to 5.263, "How to Make Our Ideas Clear" (1878), 5.388 to 5.410, "A Pragmatic Interpretation of the Logical Subject" (1902), 2.328 to 2.331, and "The Ethics of Terminology" (1903), 2.219 to 2.226. On his use of leading principles, see 3.154 to 3.171 and 5.365 to 5.369; on the open field of inquiry, 5.376n; on truth, 5.407, 5.565; on the social status of logic and knowledge, 2.220, 2.654, 5.311, 5.316, 5.331, 5.354, 5.421, 5.444, 6.610; on the duplex nature of "experience," 1.321, 5.51, 5.284, 5.613. For William James's development, see his essays in *Mind, a Quarterly Review of Psychology and Philosophy* in the early eighteen-eighties, Chapter X on "Self" in *The Principles of Psychology* (New York, 1890), the epilogue to the *Briefer Course* (New York, 1893) and *Essays in Radical Empiricism* (New York, 1912). For Dewey, see *Studies in Logical Theory* (Chicago, 1903), *How We Think* (Boston 1910, revised 1933), *Essays in Experimental Logic* (Chicago, 1916), *Experience and Nature* (Chicago, 1925), *Logic, the Theory of Inquiry*, and three psychological papers reprinted in *Philosophy and Civilization* (New York, 1931) as follows: "The Reflex Arc Concept in Psychology" (1896, reprinted as "The Unit of Behavior"), "The Naturalistic Theory of Perception by the Senses" (1925) and "Conduct and Experience" (1930). See also "Context and Thought" (*University of California Publications in Philosophy* XII [1931], 203-224), "How Is Mind to Be Known?" (*The Journal of Philosophy*, XXXIX [1942], 29-35) and "By Nature and by Art" (*ibid.*, XLI [1944], 281-292). For Bentley, see *The Process of Government* (Chicago, 1908), *Relativity in Man and Society* (New York, 1926), *Linguistic Analysis of Mathematics*, (Bloomington, Indiana, 1932), *Behavior, Knowledge, Fact* (Bloomington, Indiana, 1935), three papers on situational treatment of behavior (*The Journal of Philosophy*, XXXVI [1939], 169-181, 309-323, 405-413), "The Factual Space and Time of Behavior" (*ibid.*, XXXVIII [1941], 477-485), "The Human Skin: Philosophy's Last Line of Defense" (*Philosophy of Science*, VIII [1941], 1-19), "Observable Behaviors" (*Psychological Review*, XLVII [1940], 230-253), "The Behavioral Superfice" (*ibid.*, XLVIII [1941], 39-59) and "The Jamesian Datum" (*The Journal of Psychology*, XVI [1943], 35-79).

experience of William James was definitely an effort at such a form of direct observation in the field of knowings. Dewey's development in use of interaction and transaction, and in presentation of experience as neither subjective nor objective but as a method or system of organization, is strongly of this form; his psychological studies have made special contributions in this line, and in his *Logic, The Theory of Inquiry* (1938), following upon his logical essays of 1903 and 1916, he has developed the processes of inquiry in a situational setting. Bentley's *Process of Government* in 1908 developed political description in a manner approaching what we would here call "transactional," and his later analysis of mathematics as language, his situational treatment of behavior and his factual development of behavioral space-time belong in this line of research.

If there should be difficulty in understanding this use of the word "observation," the difficulty illustrates the point earlier made as to the influence of materials introduced from inadequate sources. The current philosophical notion of observation is derived from a psychology of "consciousness" (or some version of the "mental" as an isolate), and it endeavors to reduce what is observed either to some single sensory quality or to some other "content" of such short time-span as to have no connections—except what may be provided through inference as an operation outside of observation. As against such a method of obtaining a description of observation, the procedure we adopt reports and describes observation on the same basis the worker in knowledge— astronomer, physicist, psychologist, etc.—employs when he makes use of a test observation in arriving at conclusions to be accepted as known. We proceed upon the postulate that *knowings* are always and everywhere inseparable from *the knowns*—that the two are twin aspects of common fact.

II

"Fact" is a name of central position in the material we propose to use in forming a terminology. If there are such things as facts, and if they are of such importance that they have a vital status in questions of knowledge, then in any theory of knowings and knowns we should be able to characterize fact—we should be able to say, that is, that we know what we are talking *about* "in fact" when we apply the word "fact" to the fact of Fact.[1] The primary consideration in fulfilling the desired condition with respect to Fact is that the activity by which it is identified and the *what* that is identified are both required, and are required in such a way that each is taken along with the other, and in no sense as separable. Our terminology is involved in fact, and equally "fact" is involved in our terminology. This repeats in effect the statement that knowledge requires and includes both knowings and knowns. Anything named "fact" is such both with respect to the knowing operation and with respect to what is known.[2] We establish for our use, with respect to both fact and knowledge, that we have no "something known" and no "something identified' apart from its kno*wing* and identify*ing*, and that we have no kno*wing* and identify*ing* apart from the somewhats and

somethings that are being known and identified. Again we do not put forth this statement as a truth about "reality," but as the only position we find it possible to take on the ground of that reference to the observed which we regard as an essential condition of our inquiry. The statement is one about ourselves observed in action in the world. From the standpoint of what is observable, it is of the same straightforward kind as is the statement that when chopping occurs something is chopped and that when seeing takes place something is seen. We select the name "fact" because we believe that it carries and suggests this "double-barrelled" sense (to borrow a word from William James), while such words as "object" and "entity" have acquired from traditional philosophical use the signification of something set over against the doing or acting. That Fact is literally or etymologically *something done or made* has also the advantage of suggesting that the knowing and identifying, as ways of acting, are as much ways of doing, of making (just as much "behaviors," we may say), as are chopping wood, singing songs, seeing sights or making hay.

In what follows we shall continue the devices we have in a manner employed in the preceding paragraph, namely the use of quotation marks, italics, and capitalized initials as aids to presentation, the two former holding close to common usage, while the third has a more specialized application. We shall also freely employ hyphenization in a specialized way, and this perhaps even more frequently than the others. Thus the use of the word "fact" without quotation marks will be in a general or even casual manner. With quotation marks "fact" will indicate the verbal aspect, the word, sometimes impartially, and sometimes as held off at arm's length where the responsibility for its application is not the writer's. With initial capitalization Fact may be taken to stand for the full word-and-thing subjectmatter into which we are inquiring. Italicising in either form, whether as "*fact*" or as *Fact* will indicate stress of attention. Hyphenization will indicate attention directed to the importance which the components of the word hyphenized have for the present consideration. The words *inter-action* and *transaction* will enter shortly in this way, and will receive a considerable amount of hyphenizing for emphasis throughout. No use of single quotation marks will be made to distinguish the name of a thing from the thing, for the evident reason that expectantly rigid fixations of this type are just what we most need to avoid. All the devices mentioned are conveniences in their way, but only safe if used cautiously. Thus in the third preceding sentence (as in several others) its most stressed words, there inspected as words, should have quotation marks, but to use such marks would in this case destroy the intended assertion. Rather than being rigorous our own use will be casually variable. This last is best at our present stage of inquiry.

For the purpose of facilitating further inquiry what has been said will be restated in negative terms. We shall *not* proceed as if we were concerned with "existent things" or "objects" entirely apart from men, nor with men entirely apart from things. Accordingly, we do not have on our hands the problem of forcing them into some kind of organization or connection. We shall proceed by taking for granted human organisms developed, living, carrying on, of and in the cosmos. They are there in such system that their operations and transactions can be viewed directly—including those that constitute knowings. When they are so viewed, knowings and knowns come before us

---

[1] The wretched status of the word "fact" with respect to its "knowing" and its "known" (and in other respects as well) was illustrated in Chapter I, Section IX.

[2] It may be well to repeat here what has already been said. In making the above statement we are not attempting to legislate concerning the proper use of a word, but are stating the procedure we are adopting.

differentiated within the factual cosmos, not as if they were there provided in advance so that out of them cosmos—system—fact—knowledge—have to be produced. Fact, language, knowledge have on this procedure cosmic status; they are not taken as if they existed originally in irreconcilably hostile camps. This, again, is but to say that we shall inquire into knowings, both as to materials and workmanship, in the sense of ordinary science.[1]

The reader will note (that is, observe, give heed to) the superiority of our position with respect to observation over that of the older epistemological constructions. Who would assert he can properly and in a worth-while manner *observe* a "mind" *in addition to* the organism that is engaged in the transactions pertinent to it in an observable world? An attempt to answer this question in the affirmative results in regarding observation as private introspection—and this is sufficient evidence of departure from procedures having scientific standing.[2] Likewise, the assertion or belief that things considered as "objects" outside of and apart from human operations are observed, or are observable, is equally absurd when carefully guarded statement is demanded of it. Observation is operation; it is human operation. If attributed to a "mind" it itself becomes unobservable. If surveyed in an observable world—in what we call cosmos or nature—the object observed is as much a part of the operation as is the observing organism.

This statement about observation, in name and fact, is necessary to avoid misinterpretation. It is not "observation," however, to which we are here giving inquiry; we shall not even attempt to make the word "firm" at a later stage. In the range in which we shall work—the seeking of sound names for processes involving naming—observation is always involved and such observation in this range is in fusion with name-application, so that neither takes place except in and through the other, whatever further applications of the word "observation" (comparable to applications of "naming" and of "knowing") may in widened inquiries be required.

If we have succeeded in making clear our position with respect to the type of name for which we are in search, it will be clear also that this type of name comes in clusters. "Fact" will for us be a central name with other names clustering around it. If "observation" should be taken as central, it in its turn could be made firm only in orientation to its companionate cluster. In any case much serious co-operative inquiry is involved. In no case can we hope to succeed by first setting up separated names and then putting them in pigeonholes or bundling them together with wire provided from without. Names are, indeed, to be differentiated from one another, but the differentiation takes place with respect to other names in clusters; and the same thing holds for clusters that are differentiated from one another. This procedure has its well-established precedents in scientific procedure. The genera and species of botany and zoology are excellent examples—provided

they are taken as determinations in process and not as taxonomic rigidities.[3]

III

In certain important respects we have placed limitations on the range of our inquiry and on the methods we use. The purpose is to increase the efficiency of what we do. These decisions have been made only after much experimentation in manners of organization and presentation. The main points should be kept steadily in mind as we now stress them.

As already said, we do not propose to issue any flat decrees as to the names others should adopt. Moreover, at the start we shall in some cases not even declare our permanent choices, but instead will deliberately introduce provisional "second-string" names. For this we have two sound reasons. First, our task requires us to locate the regions (some now very largely ignored) that are most in need of firm observation. Second, we must draw upon a dictionary stock of words that have multiple, and often confusedly tangled, applications. We run the risk that the name first introduced may, on these accounts, become involved in misapprehensions on the reader's part, sufficient to ruin it for the future. Hence the value of attempting to establish the regions to be named by provisional namings, in the hope we shall secure stepping stones to better concentration of procedure at the end.

We do not propose in this inquiry to cover the entire range of "knowledge"; that is, the entire range of life and behavior to which the word "knowledge," at one time or another and in one way or another can be applied. We have already listed "knowledge" as a vague word and said we shall specify "knowings" and "knowns" for our attention. Throughout our entire treatment, "knowledge" will remain a word referring roughly to the general field within which we select subjectmatters for closer examination. Even for the words "knowings" and "knowns" the range of common application runs all the way from infusoria approaching food to mathematicians operating with their most recondite dimensions. We shall confine ourselves to a central region: that of identifications under namings, of knowing-by-naming—of "specified existence," if one will. Time will take care of the passage of inquiry across the border regions from naming-knowing to the simpler and to the more complex forms.

We shall regard these naming-knowings directly as a form of knowings. *Take this statement literally as it is written.* It means we do not regard namings as primarily instrumental or specifically ancillary to something else called knowings (or knowledge) except as any behavior may enter as ancillary to any other. We do not split a corporeal naming from a presumptively non-corporeal or "mental" knowing, nor do we permit a mentaloid "brain" to make pretense of being a substitute for a "mind" thus maintaining a split between knowings and namings. This is postulation on our part; but surely the exhibits we secured in the preceding chapter of what happens in the logics under the separation of spoken word from speaking man should be enough to justify any postulate that offers hope of relief. The accept-

---

1 It is practically impossible to guard against every form of misapprehension arising from prevalent dominance of language-attitudes holding over from a relatively pre-scientific period. There are probably readers who will translate what has been said about knowings-knowns into terms of epistemological idealism. Such a translation misses the main point—namely, that man and his doings and transactions have to be viewed as facts within the natural cosmos.

2 "Conceptions derived from. . .anything that is so occult as not to be open to public inspection and verification (such as the purely psychical, for example) are excluded" (Dewey, *Logic*, p. 19).

3 Other defects in the language we must use, in addition to the tendency towards prematurely stiffened namings, offer continuous interference with communication such as we must attempt. Our language is not at present grammatically adapted to the statements we have to make. Especially is this true with respect to the prepositions which *in toto* we must list among the "vague words" against which we have given warning. Mention of special dangers will be made as occasion arises. We do the best we can, and discussion, we hope, should never turn on some particular man's personal rendering of some particular preposition in some particular passage. The "Cimmerian" effect that appears when one attempts to use conventional linguistic equipment to secure direct statement in this region will be readily recalled.

ance of this postulate, even strictly during working hours, may be difficult. We do not expect assent at the start, and we do not here argue the case. We expect to display the value in further action.

IV

Thus far we have been discussing the conditions under which a search for firm names for knowings and knowns must be carried on. In summary our procedure is to be as follows: Working under hypothesis we concentrate upon a special region of knowings and knowns; we seek to spotlight aspects of that region that today are but dimly observed; we suggest tentative namings; through the development of these names in a cluster we hope advance can be made towards construction under dependable naming in the future.

1. *Fact, Event, Designation.* We start with the cosmos of knowledge—with nature as known and as in process of being better known—ourselves and our knowings included. We establish this cosmos as *fact*, and name it "fact" with all its knowings and its knowns included. We do *not* introduce, either by hypothesis or by dogma, knowers and knowns as prerequisites to fact. Instead we observe both knowers and knowns as factual, as cosmic; and never—either of them—as extra-cosmic accessories.

We specialize our studies in the region of naming-knowings, of knowings through namings, wherein we identify two great *factual aspects* to be examined. We name these *event* and *designation*. The application of the word "fact" may perhaps in the end need to be extended beyond the behavioral processes of event-designation. Fact, in other words, as it may be presumed to be present for animal life prior to (or below) linguistic and proto-linguistic behaviors, or as it may be presumed to be attainable by mathematical behaviors developed later than (or above) the ranges of the language behavior that names, is no affair of ours at this immediate time and place. We note the locus of such contingent extensions, leave the way open for the future, and proceed to cultivate the garden of our choice, namely, the characteristic Fact we have before us.

Upon these namings the following comments will, for the present, suffice:

(a) In Fact-Event-Designation we do not have a threefold organization, or a two-fold; we have instead one system.

(b) Given the language and knowledge we now possess, the use of the word "fact" imposes upon its users the necessity of selection and acceptance. This manifest status is recognized terminologically by our adoption of the name "designation."

(c) The word "aspect" as used here is not stressed as information-giving. It must be taken to register—register, and nothing more—the duplex, aspectual observation and report that are required if we are to characterize Fact at all. The word "phase" may be expected to become available for comparable application when, under the development of the word "aspect," we are sufficiently advanced to consider time-alternations and rhythms of event and of designation in knowledge process.[1]

(d) "Event" involves in normal use the extensional and the durational. "Designation" for our purposes must likewise be so taken. The Designation we postulate and discuss is not of the nature of *a* sound or *a* mark applied *as* a name *to* an event. Instead of this it is the entire activity—the behavioral action and activity—of naming through which Event appears in our knowing as Fact.

(e) We expect the word "fact" to be able to maintain

itself for terminological purposes, and we shall give reasons for this in a succeeding chapter, though still retaining freedom to alter it. As for the words "event" and "designation," their use here is provisional and replacement more probable. Should we, for example, adopt such words as "existence" and "name," both words (as the case stands at this stage) would carry with them to most readers many implications false to our intentions—the latter even more than the former; understanding of our procedure would then become distorted and ineffective.

(f) "Fact," in our use, is to be taken with a range of reference as extensive as is allotted to any other name for cosmos, universe or nature, where the context shows that knowledge, not poesy, is concerned. It is to be taken with its pasts and its futures, its growings-out-of and its growings-into; its transitions of report from poorer to richer, and from less to more. It is to be taken with as much solidity and substantiality as nature, universe or world, by any name whatsoever. It is to be taken, however, with the understanding that instead of inserting gratuitously an unknown something as foundation for the factually known, we are taking the knowledge in full—the knowings-knowns as they come: namely, both in one—without appeal to cosmic tortoise to hold up cosmic elephant to hold up cosmic pillar to hold up the factual cosmos we are considering.

(g) In a myopic and short-time view Event and Designation appear to be separates. The appearance does no harm if it is held where it belongs within narrow ranges of inquiry. For a general account of knowings and knowns the wider envisionment in system is proposed.

(h) Overlapping Fact, as we are postulating it within the range of *namings*, are, on one side, perceptions, manipulations, habituations and other adaptations; on the other side, symbolic-knowledge procedures such as those of mathematics. We shall be taking these into account as events-designated, even though for the present we are not inquiring into them with respect to possible designatory, quasi-designatory or otherwise fact-presenting functions of their own along the evolutionary line. Our terminology will in no way be such as to restrict consideration of them, but rather to further it, when such consideration becomes practicable.

(i) If Designations, as we postulate them for our inquiry, are factually durational-extensional, then these Designations, as designat*ings, are* themselves Events. Similarly, the Events as events are designational. The two phases, designating and designated, lie within a full process of designation. It is not the subjectmatter before us, but the available language forms, that make this latter statement difficult.[2]

---

[1] "Aspect" and "phase" may stand, therefore, as somewhat superior to the "vague words" against which we give warning, though not as yet presenting positive information in our field.

[2] This paragraph replaces one noted in the Preface as deleted. As first written it read, after the opening sentence, as follows: "Similarly, the Events as designational, *are* Designations. It is not the subjectmatter before us, but the available language forms that make this latter statement difficult. The two uses of 'are' in the sentence 'Events are Designations' and 'Designations are Events' differ greatly, each 'are' representing one of the aspects within the broader presentation of Fact. To recognize events as designated while refusing to call them designations in the activity sense, would be a limitation that would maintain a radical split between naming and named at the very time that their connective framework was being acknowledged. Our position is emphatic upon this point. It is clear enough that in the older sense events are not designations; it should be equally clear and definite that in our procedure and terminology they are designational—designation—or (with due caution in pluralizing) Designations. To control the two uses of the word 'are' in the two forms of statement, and to maintain the observation and report that 'Designations are Events,' while also 'Events are Designations'—this is the main strain our procedure will place upon the reader. Proceeding under hypothesis (and without habituation to hypothesis there will be no advance at all) this should not be too severe a requirement for one who recognizes the complexity of the situation and has an active interest in clearing it up."

(j) Most generally, Fact, in our terminology, is not limited to what any one man knows, nor to what is known to any one human grouping, nor to any one span of time such as our own day and age. On the designatory side in our project of research it has the full range and spread that, as we said above, it has on the event side, with all the futures and the pasts, the betters and the poorers, comprised as they come. In our belief the Newtonian era has settled the status of fact definitely in this way, for our generation of research at least. First, Newtonian mechanics rose to credal strength in the shelter of its glorified absolutes. Then at the hands of Faraday, Clerk Maxwell and Einstein, it lost its absolutes, lost its credal claims, and emerged chastened and improved. It thus gained the high rating of a magnificent approximation as compared with its earlier trivial self-rating of eternal certainty. The coming years—fifty, or a thousand, whatever it takes—remain quite free for change. Any intelligent voice will say this; the trouble is to get ears to hear. Our new assurance is better than the old assurance. Knowing and the known, event and designation—the full knowledge—go forward together. Eventuation is observed. Accept this in principle, not merely as a casual comment on an accidental happening:—you then have before you what our terminology recognizes when it places Fact-in-growth as a sound enough base for research with no need to bother over minuscular mentals or crepuscular reals alleged to be responsible for it.

2. *Circularity.* When we said above that designations are events and events designations, we adopted *circularity* —procedure in a circle—openly, explicitly, emphatically. Several ways of pretending to avoid such circularity are well known. Perhaps at one end everything is made tweedledum, and perhaps at the other everything is made tweedledee, or perhaps in between little tweedledums and little tweedledees, companionable but infertile, essential to each other but untouchable by each other, are reported all along the line. We have nothing to apologize for in the circularity we choose in preference to the old talk-ways. We observe world-being-known-to-man-in-it; we report the observation; we proceed to inquire into it, circularity or no circularity. This is all there is to it. And the circularity is not merely round the circle in one direction: the course is both ways round at once in full mutual function.

3. *The Differentiations That Follow.* Given fact, observed aspectually as Event and as Designation, our next indicated task is to develop further terminological organization for the two aspects separately. We shall undertake this shortly and leave the matter there so far as the present preliminary outline is concerned. To aid us, though, we shall require firm statement about certain tools to be used in the process. We must, that is, be able to name certain procedures so definitely that they will not be confounded with current procedures on a different basis. Events will be differentiated with respect to a certain range of plasticity that is comparable in a general way to the physical differentiations of gaseous, liquid and solid. For these we shall use the names Situation, Occurrence and Object. As for Designation, we shall organize it in an evolutionary scheme of behavioral sign processes of which it is one form, the names we apply being Sign, Signal, Name and Symbol. The preliminary steps we find it necessary to take before presenting these differentiations are: first, steady maintenance of a distinction among the various branches of scientific inquiry in terms of selected subjectmatters of research, rather than in terms of materials assumed to be waiting for research in advance; second, a firm use of the word "specification" to designate the type of naming to be employed as contrasted with the myriad verbal processes that go by the name of "definition"; third, the establishment of our right to selective observational control of specific situations within subjectmatters by a competent distinction of *trans*-actions from *inter*-actions.

4. *Sciences as Subjectmatters.* The broad division of regions of scientific research commonly recognized today is that into the physical, the biological and the psychological. However mathematics, where inquiry attains maximum precision, lacks any generally accepted form of organization with these sciences; and sociology, where maximum imprecision is found, also fails of a distinctive manner of incorporation.[1] Fortunately this scheme of division is gradually losing its rigidities. A generation or two ago physics stood aloof from chemistry; today it has constructively incorporated it. In the biological range today, the most vivid and distinctive member is physiology, yet the name "biology" covers many gross adaptational studies not employing the physiological techniques; in addition, the name "biology" assuredly covers everything that is psychological, unless perchance some "psyche" is involved that is "non-" or "ultra-" human. The word "psychological" itself is a hold-over from an earlier era, in which such a material series as *"the* physical," *"the* vital" and *"the* psychic" was still believed in and taken to offer three different realms of substance presented as raw material by Nature or by God for our perpetual puzzlement. If we are to establish knowings and knowns in a single system of Fact, we certainly must be free from addiction to a presumptive universe compounded out of three basically different kinds of materials. Better said, however, it is our present freedom from such material enthrallment, attained for us by the general advance of scientific research, that at long last has made us able to see all knowings and knowns, by hypothesis, as in one system.

Within Fact we shall recognize the distinctions of the scientific field as being those of subjectmatters, not those of materials[2] unless one speaks of materials only in the sense that their differences themselves arise in and are vouched for strictly by the technological procedures that are available in the given stages of inquiry. Terminologically, we shall distinguish *physical*, *physiological* and *behavioral*[3] regions of science. We shall accept the word

---

1 We shall deal with the very important subject of mathematics elsewhere. Sociological inquiries, with the exception of anthropology, are hardly far enough advanced to justify any use of them as subjectmatters in our present inquiry.

2 An extended consideration of many phases of this issue and approaches to its treatment is given by Coleman R. Griffith in his *Principles of Systematic Psychology* (Urbana, Illinois, 1943). Compare the section on "The Scientific Use of Participles and Nouns" (pp. 489-497) and various passages indexed under "Science."

3 Our use of the word "behavioral" has no "behavioristic" implications. We are no more behavioristic than mentalistic, disavowing as we do, under hypothesis, "isms" and "istics" of all types. The word "behavior" is in frequent use by astronomers, physicists, physiologists and ecologists, as well as by psychologists and sociologists. Applied in the earlier days of its history to human conduct, it has drifted along to other uses, pausing for a time among animal-students, and having had much hopeful abuse by mechanistic enthusiasts. We believe it rightfully belongs, however, where we are placing it. Such a word as "conduct" has many more specialized implications than has "behavior" and would not serve at all well for the name for a great division of research. We shall be open to the adoption of any substitutes as our work proceeds, but thus far have failed to find a more efficient or safer word to use. In such a matter as this, long-term considerations are much more important than the verbal fashions of a decade or two.

"biological" under our postulation as covering unquestionably both physiological and behavioral inquiries, but we find the range of its current applications much too broad to be safe for the purposes of the present distinctive terminology. The technical differentiation, in research, of physiological procedures from behavioral is of the greatest import in the state of inquiry today, and this would be pushed down out of sight by any heavy stress on the word "biological," which, as we have said, we emphatically believe *must* cover them both. We wish to stress most strongly that physical, physiological and behavioral inquiries in the present state of knowledge represent three great distinctive lines of technique; while any one of them may be brought to the aid of any other, *direct* positive extension of statement from the firm technical formulations of one into the information-stating requirements of another cannot be significantly made as knowledge today stands. Physical formulation does not directly yield heredity, nor does physiological formulation directly yield word-meanings, sentences and mathematical formulas. To complete the circle, behavioral process, while producing physical science, cannot directly in its own procedure yield report on the embodied physical event. This circularity, once again, is in the knowledge—in the knowings and the knowns—not in any easy-going choice we are free and competent to make in the hope we can cleave to it, evidence or no evidence.

5. *Specification.* The word "definition," as currently used, covers exact symbolic statements in mathematics; it covers procedures under Aristotelian logic; it covers all the collections of word-uses, old and new, that the dictionaries assemble, and many still more casual linguistic procedures. The word "definition" must manifestly be straightened out, if any sound presentation of knowings and knowns is to be secured.[1] We have fair reason to believe that most of the difficulty in what is called the "logic of mathematics" is due to an endeavor to force consolidation of two types of human behavior, both labeled "definition," (though one stresses heavily, while the other diverges from, the use of namings) without preliminary inquiry into the simpler facts of the life linguistic. In our terminology we shall assign the word "definition" to the region of mathematical and syntactical consistency, while for the lesser specimens of "dictionary definition" we shall employ the name "characterization." In our own work in this book we shall attempt no *definition* whatever in the formal sense we shall assign the word. We shall at times not succeed in getting beyond preliminary characterization. Our aim in the project, however, is to advance towards such an accuracy in naming as science ever increasingly achieves. Such accuracy in naming we shall call "specification." Consider what the word "heat" stood for in physics before Rumford and Joule, and what it tells us in physical specification today. Consider the changes the word "atom" has undergone in the past generation. Modern chemical terminology is a highly specialized form of specification of operations undertaken. However, the best illustration for our purposes is probably the terminology of genera and species. In the days when animals were theological specialities of creation, the naming level was

that of characterization. After demonstration had been given that species had natural origins, scientific specification, as we understand it, developed. We still find it, of course, straining at times towards taxonomic rigidities, but over against this we find it forever rejuvenating itself by free inquiry up even to the risk of its own obliteration. Abandonment of the older magic of name-to-reality correspondence is one of the marks of specification. Another will be observed when specification has been clearly differentiated from symbolic definition. In both its aspects of Event and Designation we find Fact spread in "spectrum-like" form. We use "specification" to mark this scientific characteristic of efficient naming. Peirce's stress on the "precept that is more serviceable than a definition"[2] involves the attitude we are here indicating. Specification operates everywhere in that field of inquiry covered by affirmation and assertion, proposition and judgment, in Dewey's logical program. The defects of the traditional logics exhibited in Chapter I were connected with their lack of attention to the accurate specification of their own subjectmatters; at no point in our examination did we make our criticisms rest on consistency in definition in the sense of the word "consistency" which we shall develop as we proceed through the differentiation of symbol from name and of symbolic behavior from naming behavior.

6. *Transaction.* We have established Fact as involving both Designation and designated Event. We have inspected inquiry into Fact in terms of subjectmatters that are determinable under the techniques of inquiry, not in terms of materials presented from without.[3] Both treatments make selection under hypothesis a dominant phase of procedure. Selection under hypothesis, however, affects all observation. We shall take this into account terminologically by contrasting events reported in interactions with events reported as transactions. Later chapters will follow dealing with this central issue in our procedure: the right, namely, to open our eyes to see. Here we can only touch broadly upon it. Pre-scientific procedure largely regarded "things" as possessing powers of their own, under or in which they acted. Galileo is the scientist whose name is most strongly identified with the change to modern procedure. We may take the word *"action"* as a most general characterization for events where their durational process is being stressed. Where the older approach had most commonly seen *self-action* in "the facts," the newer approach took form under Newton as a system of interaction, marked especially by the third "law of motion"—that action and reaction are equal and opposite. The classical mechanics is such a system of interaction involving particles, boundaries, and laws of effects. Before it was developed—before, apparently, it could develop—observation of a new type differing from the pre-Galilean was made in a manner essentially transactional. This enters in Galileo's report on inertia, appearing in the Newtonian formulation as the first "law of motion," namely, that any motion uninterfered with will continue in a straight line. This set up a motion, directly, factually, as event.[4] The field of knowings and

---

1 The task of straightening out proved to be more complex, even, than we had estimated. It led us to drop the word "definition" altogether from technical terminology, thus reducing it for the time being to the status of a colloquialism. We nevertheless permit our text in this passage to appear unrevised, since we are more interested in the continuity of inquiry than we are in positive determinations of word-usage at this stage. *See* the introductory remarks to Chapter VII, and the summary in Chapter XI.

2 See Chapter I, Section I.

3 Again, a very vaguely used word.

4 In the psychological range the comparable fundamental laboratory experiments of import for our purposes are those of Max Wertheimer upon the direct visual observability of motions. See "Experimentelle Studien über das Sehen von Bewegung" (*Zeitschrift für Psychologie,* LXI [1912], 161-265). In a much weakened form his results are used in the type of psychology known as "Gestalt," but in principle they still await constructive development.

knowns in which we are working requires transactional observation, and this is what we are giving it and what our terminology is designed to deal with. The epistemologies, logics, psychologies and sociologies today are still largely on a self-actional basis. In psychology a number of tentative efforts are being made towards an *interactional* presentation, with balanced components. Our position is that the traditional language currently used about knowings and knowns (and most other language about behaviors, as well) shatters the subjectmatter into fragments in advance of inquiry and thus destroys instead of furthering comprehensive observation for it. We hold that observation must be set free; and that, to advance this aim, a postulatory appraisal of the main historical patterns of observation should be made, and identifying namings should be provided. Our own procedure is the *transactional*, in which is asserted the right to see together, extensionally and durationally, much that is talked about conventionally as if it were composed of irreconcilable separates. We do not present this procedure as being more real or generally valid than any other, but as being the one now needed in the field where we work. In the same spirit in which physicists perforce use both particle and wave presentations we here employ both interactional and transactional observation.[1] Important specialized studies belong in this field in which the organism is made central to attention. This is always legitimate in all forms of inquiry within a transactional setting, so long as it is deliberately undertaken, not confusedly or with "self-actional" implications. As place-holders in this region of nomenclature we shall provisionally set down *behavior-agent* and *behavior-object*. They represent specialized interactional treatments within the wider transactional presentation, with organisms or persons or actors named uncertainly on the one hand and with environments named in variegated forms on the other.

7. *Situation, Occurrence, Object.* We may now proceed to distinguish Situation, Occurrence and Object as forms of Event. Event is durational-extensional; it is what "takes place," what is inspected as "*a taking place.*" These names do not provide a "classification," unless classification is understood as a focusing of attention within subjectmatters rather than as an arrangement of materials. The word "situation" is used with increasing frequency today, but so waveringly that the more it is used the worse its own status seems to become. We insist that in simple honesty it should stand *either* for the environment of an object (interactionally), *or* for the full situation including whatever object may be selectively specified within it (transactionally), and that there be no wavering. We shall establish our own use for the word *situation* in this latter form. When an event is of the type that is readily observable in transition within the ordinary spans of human discrimination of temporal and spatial changes, we shall call it *occurrence*. The ordinary use of "event" in daily life is close to this, and if we generalize the application of the word, as we have provisionally done, to cover situation and object as well as occurrence, then we

require a substitute in the more limited place. Occurrence fairly fills the vacancy. *Object*[2] is chosen as the clearly indicated name for stabilized, enduring situations, for occurrences that need so long a span of time, or perhaps so minute a space-change, that the space and time changes are not themselves within the scope of ordinary, everyday perceptual attention. Thus any one of the three words Situation, Occurrence and Object may, if focusing of attention shifts, spread over the range of the others, all being equally held as Event. We have here a fair illustration of what we have previously called a word-cluster. The Parthenon is an object to a visitor, and has so been for all the centuries since its construction. It is nevertheless an occurrence across some thousands of years. While for certain purposes of inquiry it may be marked off as object-in-environment, for thoroughgoing investigation it must be seized as situation, of which the object-specification is at best one phase or feature. There is here no issue of reality, no absolute yes or no to assert, but only free determination under inquiry.

8. *Sign, Signal, Name, Symbol.* When we turn to Designation, our immediate problem is not that of distinguishing the variety of *its* forms. Specification, the form most immediately concerning us, has already been noted. What we have to do instead is to place designation itself among behavioral events. Circularity is again here strikingly involved. Our treatment must be in terms of Event as much as in terms of Designation, with full convertibility of the two. The event is behavioral. Designation (a behavioral event) can be viewed as one stage in the range of behavioral evolution from the sensitive reactions of protozoa to the most complex symbolic procedures of mathematics. In this phase of the inquiry we shall alter the naming. Viewing the behavioral event, we shall name it directly Name instead of replacing "name" by "designation" as seemed necessary for provisional practical reasons on the obverse side of the inquiry. At a later stage we shall undertake to establish the characteristic behavioral process as *sign*, a process not found in either physical or physiological techniques of inquiry. We shall thus understand the name "sign" to be used so as to cover the entire range of behavioral activity. There are many stages or levels of behaviors, but for the greater part of our needs a three-level differentiation will furnish gross guidance. The lower level, including perceptions, manipulations, habituations, adaptations, etc., we shall name *signal* (adapting the word from Pavlov's frequent usage). Where organized language is employed as sign, we shall speak of *name*. In mathematical regions (for reasons to be discussed fully later) we shall speak of *symbol*. Signal, Name and Symbol will be the three differentiations of Sign, where "sign" indicates most broadly the "knowledge-like" processes of behavior in a long ascending series. Vital to this construction, even though no development for the moment may be offered, is the following statement: The name "Sign" and the names adjusted to it *shall all be understood transactionally*, which in this particular case is to say that they do not name items or characteristics of organisms alone, nor do they name items or characteristics of environments alone; in every case, they name the *activity* that occurs *of both together.*

---

[1] The word "field" is a strong candidate for use in the transactional region. However, it has not been fully clarified as yet for physics, and the way it has been employed in psychological and social studies has been impressionistic and often unscrupulous. "Field" must remain, therefore, on our list of vague words, candidates for improvement. When the physical status of the word is settled—and Einstein and his immediate associates have long concentrated on this problem—then if the terminology can be transferred to behavioral inquiry we shall know how to make the transfer with integrity. See p. 128, footnote 4.

[2] "The name *objects* will be reserved for subjectmatter so far as it has been produced and ordered in settled form by means of inquiry; proleptically, objects are the *objectives* of inquiry" (*Logic, the Theory of Inquiry*, p. 119). For "situation" see *ibid.*, pp. 66 ff. The word "occurrence" is, as has been indicated, provisionally placed.

V

By the use of Sign-Signal-Name-Symbol we indicate the locus for the knowing-naming process and for other behavioral processes within cosmos. By the use of Fact-Event-Designation we specify the process of event-determination through which cosmos is presented as itself a locus for such loci. The two types of terminology set forth different phases of a common process. They can be so held, if we insist upon freedom for transactional observation in cases in which ancient word-forms have fractured fact and if we lose fear of circularity. It is our task in later chapters to develop this terminology and to test it in situations that arise.

For the present our terminological guide-posts, provisionally laid out, are as follows:

SUGGESTED EXPERIMENTAL NAMING

Fact: Our cosmos as it is before us progressively in knowings through namings.

Event:[1] "Fact" named as taking place.

Designation: Naming as taking place in "fact."

Physical, Physiological, Behavioral: Differentiations of the techniques of inquiry, marking off subjectmatters as sciences under development, and not constricted to conformity with primitive pre-views of "materials" of "reality."

Characterization: Linguistic procedure preliminary to developed specification, including much "dictionary-definition."

Specification: Accuracy of designation along the free lines through which modern sciences have developed.

Definition:[2] Symbolic procedure linguistically evolved, not directly employing designatory tests.

Action (Activity): Event stressed with respect to durational transition.

Self-Action: Pre-scientific presentation in terms of presumptively independent "actors," "souls," "minds," "selves," "powers" or "forces," taken as activating events.

Interaction: Presentation of particles or other objects organized as operating upon one another.

Transaction:[3] Functional observation of full system, actively necessary to inquiry at some stages, held in reserve at other stages, frequently requiring the breaking down of older verbal impactions of naming.

Behavior-Agent: Behavioral organic action, interactionally inspected within transaction; agent in the sense of re-agent rather than of actor.

Behavior-Object: Environmental specialization of object with respect to agent within behavioral transaction.

Situation: Event as subjectmatter of inquiry, always transactionally viewed as the full subjectmatter; never to be taken as detachable "environment" over against object.

Occurrence:[4] Event designated as in process under transitions such as are most readily identifiable in everyday human-size contacts.

Object: Event in its more firmly stabilized forms—never, however, as in final fixations—always available as subjectmatter under transfer to situational inspection, should need arise as inquiry progresses.

Sign: Characteristic adaptational behavior of organism-environment; the "cognitive" in its broadest reaches when viewed transactionally as process (not in organic or environmental specialization).

Signal: Transactional sign in the perceptive-manipulative ranges.

Name: Specialized development of sign among hominidae; apparently not reaching the full designational stage (excepting, perhaps, on blocked evolutionary lines) until *homo sapiens.*

Symbol: A later linguistic development of sign, forfeiting specific designatory applications to gain heightened efficiency in other ways.

The above terminology is offered as provisional only. Especially is further discussion needed in the cases of Event, Occurrence, and Definition. Later decisions, after further examination, are reported in Chapter XI, with several footnotes along the route serving as markers for progress being made.

We regard the following as common sense observation upon the manner of discourse about knowledge that we find current around us.

The knowledge of knowledge itself that we possess today is weak knowledge—perhaps as weak as any we have; it stands greatly in need of de-sentimentalized research.

Fact is notoriously two-faced. It is cosmos as noted by a speck of cosmos. Competent appraisal takes this into account.

What is beyond Fact—beyond the knowing and the known—is not worth bothering about in any inquiry undertaken into knowings and knowns.

Science as *inquiry* thrives within limits such as these, and science offers sound guidance. Scientific specification thrives in, and requires, such limits; why, then, should not also inquiry and specification for knowings and the known?

Knowings are behaviors. Neither inquiry into knowings nor inquiry into behaviors can expect satisfactory results unless the other goes with it hand in hand.[5]

---

[1] The word "existence" was later substituted for "event" in this position. See Chapter XI.

[2] The word "definition" later dropped from technical terminological use, so far as our present development goes.

[3] For introductory uses of the word see John Dewey, "Conduct and Experience," in *Psychologies of 1930* (Worcester, Massachusetts). Compare also his *Logic, the Theory of Inquiry,* p. 458, where stress is placed on the *single continuous event.*

[4] The word "event" was later substituted for "occurrence" in this usage. See Chapter XI.

[5] Attention is called in summary to the "vague words" one is at times compelled to use. "Knowledge," "thing," "field," "within" and "without" have been so characterized in text or footnotes; also all prepositions and the use of "quotes" to distinguish names from the named; even the words "vague" and "firm" as we find them in use today. "Aspect" and "phase" have been indicated as vague for our purposes today, but as having definite possibilities of development as we proceed. It will be noticed that the word "experience" has not been used in the present text. No matter what efforts have heretofore been made to apply it definitely, it has been given conflicting renderings by readers who among them, one may almost say, have persisted in forcing vagueness upon it. We shall discuss it along with other abused words at a later place.

III.

POSTULATIONS

I

IN the search to secure firm names for knowings and knowns, we have held, first, that man, inclusive of all his knowings, should be investigated as "natural" within a natural world;[1] and, secondly, that investigation can, and must, employ sustained observation akin in its standards—though not, of course, in all its techniques—to the direct observation through which science advances.

Scientific observation does not report by fiat; it is checked and rechecked by many observers upon their own work and the work of others until its report is assured. This is its great characteristic. From its simplest to its most far-reaching activities it holds itself open to revision in a degree made strikingly clear by what happened to the Newtonian account of gravitation after its quarter millenium of established "certainty." The more scientific and accurate observation becomes, the less does it claim ultimacy for the specific assertions it achieves.

Where observation remains open to revision, there is always a certain "if" about it. Its report is thus conditional, and the surrounding conditions, under careful formulation, become the postulation under which it holds place. In the case of problems of limited range, where conditions are familiar to the workers (as, for example, in a physical laboratory, for a particular experiment under way), an unqualified report of the verified results as "fact" is customary and meets no objection. Where, however, assertions that run far afield are involved, the postulational background must be kept steadily in view, and must be stated as conditional to the report itself; otherwise serious distortions may result.

This is emphatically required for a search such as ours in the case of knowings and knowns. Our procedure must rest on observation and must report under postulation. Simply and directly we say that the sciences work in nature, and that any inquiry into knowings and knowns must work in the same nature the sciences work in and, as far as possible, along the same general lines. We say observation is the great scientific stronghold. We say that all[2] observations belong in system, and that where their connections are not now known it is, by postulation, permissible to approach them as if connection could be established. We totally reject that ancient hindrance put upon inquiry such as ours by those who proclaim that the "knower" must be in some way superior to the nature he knows; and equally by those who give superiority to that which they call "the known." We recognize that as observers we are human organisms, limited to the positions on the globe from which we make our observations, and we accept this not as being a hindrance, but instead as a situation from which great gain may be secured. We let our postulations rise out of the observations, and we then use the postulations to increase efficiency of observation, never to restrain it. It is in this sense of circularity that we employ those very postulations of nature, of observation and of postulation itself, that our opening paragraphs have set down.[3]

The dictionaries allot to the word "postulate" two types of application. One presents something "taken for granted as the true basis for reasoning or belief"; the other, "a condition required for further operations." Our approach is manifestly of this second type.[4] We shall mark this by speaking of postulations rather than of postulates, so far as our own procedures are concerned. This phrasing is more reliable, even though at times it will seem a bit clumsy.

What we have said is equivalent to holding that postulations arise out of the field of inquiry, and maintain themselves strictly subject to the needs of that field.[5] They are always open to re-examination. The one thing they most emphatically *never* are is unexaminable.

To this must be added a further comment that postulation is double-fronted.[6] It must give as thorough a consideration to attitudes of approach in inquiry as it does to the subjectmatter examined, and to each always in conjunction with the other.[7]

It is very frequently said that no matter what form of inquiry one undertakes into life and mind one involves himself always in metaphysics and can never escape it. In contrast with this hoary adage, our position is that if one seeks with enough earnestness to identify his attitude of workmanship and the directions of his orientation, he can by-pass the metaphysics by the simple act of keeping

---

[1] By "natural world" with man "natural" within it, the reader should understand that background of inquiry which since Darwin's time has become standard for perhaps all fields of serious scientific enterprise with the single exception of inquiry into knowings and knowns. We shall not employ the words "naturalism" or "naturalistic." We avoid them primarily because our concern is with free research, where the word "nature" specifies nothing beyond what men can learn about it; and, secondarily, because various current metaphysical or "substantial" implications of what is called "naturalism" are so alien to us that any entanglement with them would produce serious distortion of our intentions.

[2] The word "all" is, of course, one more vague word. Heretofore we have avoided it altogether—or hope we have. An adequate technical language for our purposes would have one word for the "all" of scientific specification, and another for the "all" of symbolic definition. As we have previously said, our discussions limit themselves strictly to the former use.

[3] Compare the three conditions of a search for names set down at the start of Chapter II, Section I, and accompanied by the three negations: that no purely private report, no "foundations" beyond the range of hypothesis, and no final declaration secure from the need of further inquiry can safely be accepted or employed.

[4] Max Wertheimer, *Productive Thinking* (New York, 1945), p. 179 reports a conversation with Einstein concerning the latter's early approaches to relativity. In answer to a direct question Einstein said: "There is no...difference...between reasonable and arbitrary axioms. The only virtue of axioms is to furnish fundamental propositions from which one can derive conclusions that fit the facts."

[5] Dewey: *Logic, the Theory of Inquiry* (New York, 1938), pp. 16-19.

[6] Bentley: *Behavior, Knowledge, Fact* (Bloomington, Indiana, 1935), Chap. XXX. It is in the behavioral field particularly that this characteristic must never for a moment be neglected.

[7] One further comment on the word "postulation" is needed. We are not here attempting to determine its final terminological status, but merely specifying the use we are now making of it. In the end it may well be that it should be assigned to the region of Symbol (Chapter II, Section IV, No. 8) and a different word employed in such territory as we are now exploring. We are choosing "postulation" instead of "hypothesis" for the immediate task because of its greater breadth of coverage in ordinary use. Freedom, as always, is reserved (Chapter II, Section III, and Chapter XI) to make improvements in our provisional terminology when the proper time comes.

observation and postulation hand-in-hand; the varied "ultimates" of metaphysics become chips that lie where they fall. Our postulations, accordingly, gain their rating, not by any peculiarity or priority they possess, but by the plainness and openness of their statement of the conditions under which work is, and will be, done. If this statement at times takes categorical verbal form, this is by way of endeavor at sharpness of expression, not through any desire to impose guidance on the work of others.

In the course of our preliminary studies for this series of reports we assembled a score or two of groups of postulations. These experiments taught us the complexity of the problem and the need for a steady eye upon all phases of inquiry. Instead of obtaining a single overall postulation, as we might have anticipated, we found that the more thorough the work became, the more it required specializations of postulations, and these in forms that are complementary. We shall display certain of these postulations, primarily as aids to our further discussion, but partly because of the interest such exhibits may have for workers in collateral fields. We further hope the display may stimulate co-operation leading to better formulation from other experimenters with similar manners of inquiry.

In approaching the examination let the reader recall, first, that we have previously selected namings as the species of knowings most directly open to observation, and thus as our best entry to inquiry;[1] and, secondly, that we have taken the named and the namings (being instances of the known and the knowings) as forming together one event for inquiry[2]—one transaction[3]—since, in any full observation, if one vanishes, the other vanishes also. These things we observe; we observe them under and through the attitudes expressed in our opening paragraphs; as such observations they form the core of the postulatory expansion to follow.[4]

II

In order to make plain the background against which our postulations can be appraised, we start by exhibiting certain frequently occurring programs for behavioral inquiry,[5] which are to be rated as postu*lates* rather than as postu*lations* under the differentiation we have drawn between the two words. Characteristic of them is that they evade, ignore, or strive to rid themselves of that "circularity"[6] in knowledge which we, in contrast, frankly accept as we find it. Characteristic, further, is that their proponents take them for granted so unhesitatingly in the form of "truths" that they rarely bring them out into clear expression. It is because of this latter characteristic that we cannot readily find well-organized

specimens to cite but are compelled to construct them as best we can out of the scattered materials we find in works on epistemology, logic, and psychology. Because their type is so different from the postulations we shall develop for our own use, we label them with the letters X, Y, and Z in a series kept separate at the far end of the alphabet.

X. Epistemological Irreconcilables

1. "Reals" exist and become known.
2. "Minds" exist and do the knowing.
3. "Reals" and "minds" inhabit irreconcilable "realms."[7]
4. Epistemological magic[8] is required to reveal how the one irreconcilable achieves its knowing and the other its being known.

Y. Logical Go-Betweens

1. "Reals" exist ("objects," "entities," "substances," etc.).
2. "Minds" exist ("thoughts," "meanings," "judgments," etc.).
3. "Thirds" exist to intervene ("words," "terms," "sentences," "propositions," etc.).
4. Logical exploration of "thirds"[9] will reconcile the irreconcilables.

Z. Physiologic-Psychologic Straitjackets

1. "Reals" exist as matter, tactually or otherwise sensibly vouched for.
2. "Minds" exist as mentaloid manifestations of organically specialized "reals."[10]
3. Study of organically "real" matter (muscular, neural, or cortical) yields knowledge of matter, including the organic, the mentaloid, and the knowledges themselves.
4. The "certainty" of matter in some way survives all the "uncertainties" of growing knowledge about it.

These three groups of postulates all include non-observables; that is, through the retention of primitive namings surviving from early cultures they adopt or purport to adopt certain materials of inquiry that can not be identified as "objects" under any of the forms of observation modern research employs.[11] X is in notorious disrepute except among limited groups of epistemological specialists. Y works hopefully with linguistic devices that

---

[1] Chapter II, Section IV and Chapter I, Section I.

[2] Chapter II, Section III.

[3] Chapter II, Section IV and Chapters IV and V.

[4] One of the authors of this volume (J.D.) wishes to make specific correction of certain statements in his *Logic, the Theory of Inquiry* about *observation*. As far as those statements limit the word to cases of what are called "sense-perception"—or, in less dubious language, to cases of observation under conditions approaching those of laboratory control—they should be altered. For the distinction made in that text between "observation" and "ideation" he would now substitute a distinction between two phases of observation, depending on comparative temporal-spatial range or scope of subjectmatter. What is called observation in that text is only such observations as are limited to the narrower ranges of subjectmatter; which, however, hold a distinctive and critical place in the testing of observations of the more extensive type.

[5] For the word "behavior," see p. 109, footnote 3.

[6] Chapter II, Section IV, No. 2.

[7] With variations of "more or less" (though still "irreconcilable"), and with special limiting cases on one side or the other in which winner takes all.

[8] "Magic" (dictionary definition): "Any supposed supernatural art."

[9] Though always with the risk of other thirds "to bite 'em; And so proceed *ad infinitum.*"

[10] Watson's early "behaviorism" (far remote, of course, from the factual behavior of our inquiry) included an identification of linguistic procedure as physiological process of vocal organs—an identification that lacked not merely the transactional view we employ, but even an interactional consideration of the environment. An excellent recent illustration of much more refined treatment is that of Roy Woods Sellars, *The Journal of Philosophy*, XLI (1944) who writes (p. 688): "I think we can locate the psychical as a *natural isolate* in the functioning organism and study its context and conditions." The issue could hardly be more neatly drawn between the "process" we are to investigate and the purported "things" the X, Y, and Z postulates offer for examination.

[11] Cf. postulations B5 and B6, below. For "objects," see Chapter II, Section IV.

our preceding examination has shown to be radically deficient.[1] *Z* is serviceable for simple problems at the level of what used to be called "the senses," and at times for preliminary orderings of more complex subjectmatters, but it quickly shows itself unable to provide the all-essential direct descriptions these latter require. All three default not only in observability, but also in the characteristics of that manner of approach which we have here called "natural" (though *Z* has aspirations in this latter direction).[2] Beyond this, as already indicated, all three are employed rather as articles of faith than as postulations proper.

III

In contrast with the approaches *X*, *Y*, and *Z*, we shall now write down in simple introductory statement what we regard as the main features of the postulations which, inspired by and in sympathy with the progress modern sciences have made, are most broadly needed as guides to inquiry into behaviors as natural events in the world.

A. Postulations for Behavioral Research

1. The cosmos: as a system or field of factual inquiry.[3]
2. Organisms: as cosmic components.
3. Men: as organisms.
4. Behavings of men: as organic-environmental events.
5. Knowings (including the knowings of the cosmos and its postulation): as such organic-environmental behavings.

The above postulations are to be taken literally and to be scrupulously so maintained in inquiry.

So important is the italicized sentence, and so common and vicious that manner of lip-service to which hands and eyes pay no attention, that we might well give this sentence place as a sixth postulation.

Entry No. 1 accepts positively the cosmos of science as the locus of behavioral inquiry. This acceptance is full and unqualified, though free, of course, from the expansive applications speculative scientists so often indulge in. No. 2 and No. 3 are perhaps everywhere accepted, *except for inquiry into knowings and knowns*. No. 4 differs sharply from the common view in which the organism is taken as the locus of "the behavior" and as proceeding under its own powers in detachment from a comparably detachable environment, rather than as a phase of the full organic-environmental event.[4] No. 5, so far as we know, is not yet in explicit use in detailed research of the sort we are undertaking, and its introduction is here held to be required if firm names for knowings and knowns are to be achieved.

Following postulations *A* for behavioral events, as subjectmatters, we now set forth postulations *B* for inquiry into such behavioral subjectmatters. The type of inquiry we have before us is that which proceeds through Designation. Long ago we chose naming-events as the particular variety of knowings upon which to concentrate study.[5] Now we are selecting Designation[6] as the specialized method of inquiry we are to employ. Before proceeding to more detail with postulations *A*, we complement them with postulations *B*, as if we set a right hand over against a left somewhat in the manner we have already spoken of as "double-fronted." *A* and *B* together offer us instances of that "circularity" we find wherever we go, which by us is not merely recognized, but put to work—not deplored but seized upon as a key to observation, description, and controlled inquiry.[7] The procedure looks complex but we cannot help it any more than the physicists of three generations ago could "help it" when electricity (to say nothing of electromagnetic waves) refused to stay in locations or submit to a mathematics that had sufficed, until that time, for the mechanics of particles.

Given complementary postulations *A* and *B*, one may expect to find the components of one postulation reappearing in the other, but differently stressed, and under different development. Thus postulation *A*1 views Fact in the aspect of Event, whereas *B*1 views it in the aspect of inquiry under or through Designation ("event" being here understood with the range given in Chapter XI to the word "existence"). Similar cases appear frequently; they are typical and necessary.

B. Postulations for Inquiry into Subjectmatters Under Designation[8]

1. A single system of subjectmatters is postulated, to be called cosmos or nature.
2. Distribution of subjectmatters of inquiry into departments varies[9] from era to era in accordance with variation in the technical stage of inquiry.
3. Postulations for each of the most commonly recognized present departments (physical, physiological, and behavioral) are separately practicable, free from the dictatorship of any one over another, yet holding all in system.[10]
4. The range of the knowings is coextensive with the range of the subjectmatters known.
5. Observation, such as modern technique of experiment has achieved, or fresh technique may achieve, is postulated for whatever is, or is to be, subjectmatter. Nothing enters inquiry as inherently non-observable nor as

---

1 Chapter I, Section X.

2 One of our earlier experimental formulations may be mentioned: *(a)* existing epistemologies are trivial or worse; *(b)* the source of the trouble lies in primitive speech conventions; *(c)* in particular, the presentation of a "mind" as an individual "isolate," whether in "psychical" or in "physiological" manifestation, is destructive.

3 The system is named Fact (Chapter II, Section IV, No. 1).

4 For legitimate procedures in provisional detachments, see postulations *D*8 and *G*3.

5 Chapter II, Sections II and III.

6 Chapter II, Section IV, No. 1. For the distinction provisionally employed between the word "naming" and "designation," see Chapter II, Section IV, No. 8.

7 No priority is assumed for *A* over *B* or *vice versa*. Postulations *A* enter first into our immediate treatment as the needed offset to the current fracturings and pseudo-realistic strivings of *X*, *Y*, and *Z*.

8 Not to be overlooked is the express statement in the text that these postulations *B* are for research through namings, and are *not* set up for all types of search and formulation whatsoever. We cultivate our present gardens, leaving plenty of room for other gardens for future workers.

9 "Varies. . .in accordance with" might be profitably replaced by "is in function with," if we could be sure that the word "function" would be understood as indicating a *kind* of problem, and not as having some positive explanatory value for the particular case. Unfortunately too many of the uses of "function" in psychological and sociological inquiry are of the pontifical type. The problem is to indicate the aspectual status, despite the poverty of available language (Introduction, and Chapter II, n. 1). For discussion of the content of postulation *B*2, see Chapter II, Section IV, No. 4, and compare also the postulation of continuity, *Logic, the Theory of Inquiry*, p. 19, *et al*.

10 Postulations *A* have this characteristic in contrast with postulations *Z*. The free development of subjectmatters in *B*2 and *B*3 coincides in effect with the express rejection of "reals" in *B*9, *C*7, and *H*1. It also removes the incentive to the romantic types of "emergence" which often enter when "substantive reals" depart.

requiring an independent type of observation of its own. What is observed is linked with what is not then and there observed.

6. The subjectmatters of observation are durational and extensional.

7. Technical treatments of extensions and durations developed in one department of subjectmatter are to be accepted as aids for other subjectmatters, but never as controls beyond their direct value in operation.[1]

8. "Objects" in practical everyday identifications and namings prior to organized inquiry hold no permanent priority in such inquiry.[2] Inquiry is free and all "objects" are subject to examination whether as they thus practically come or with respect to components they may be found to contain, or under widened observation as transactional—in all cases retaining their extensional and durational status.[3]

9. Durationally and extensionally observable events suffice for inquiry. Nothing "more real" than the observable is secured by using the word "real," or by peering for something behind or beyond the observable to which to apply the name.[4]

Having focused postulations B upon the aspect of inquiry, we now return to the aspect of event in A.[5] Our declared purpose is to examine naming behaviors as knowings, and to hold the naming behaviors as events in contact with the signaling behaviors on one side and with the symboling behaviors on the other.[6] In expansion from A as events we shall therefore next present postulations C for knowings and D for namings, and shall follow these with indications of what will later be necessary as E for signalings and F for symbolings.

Postulations C are looser than the others, as will be evident at once by our permitting the vague word "knowledge" to creep in. There is sound reason for this. We secure an introductory background in the rough along the lines of ordinary discussion, against which to study namings as knowings. From future study of namings a better postulation for knowings should develop. A comment on the possible outcomes for C will follow.

C. Postulations for Knowings and Knowns as Behavioral Events[7]

1. Knowings and knowns (knowledge, knowledges, instances of knowledge) are natural events. A knowing is to be regarded as the same kind of an event *with respect to its being known* (i.e., just as much "extant") as an eclipse, a fossil, an earthquake, or any other subjectmatter of research.

2. Knowings and knowns are to be investigated by methods that have been elsewhere successful in the natural sciences.

3. Sufficient approach has already been made to knowledge about knowledge through cultural, psychological, and physiological investigations to make it practicable to begin today to use this program.[8]

4. As natural events, knowings and knowns are observable; as observable, they are enduring and extensive within enduring and extensive situations.[9]

5. Knowings and knowns are to be taken together as aspects of one event.[10] The outstanding need for inquiry into knowledge in its present stage is that the knowings and knowns be thus given transactional (as contrasted with interactional) observations.

6. The observable extensions of knowings and knowns run across the inhabited surface of the earth; the observable durations run across cultures,[11] backward into pre-history, forward into futures—all as subjectmatters of inquiry. Persistence (permanence and impermanence) characterize the knowings and the knowns alike.[12]

7. All actualities dealt with by knowledge have aspects of the knowing as well as of the known, with the knowings themselves among such actualities known.

Inspection of postulations C shows that the first two of the group provide for the development of A in accord with $B2$, while the third serves to make emphatic—against the denial everywhere prevalent—our assertion that inquiry *can* proceed on these lines. The fourth is in accord with $B5$, $B6$, and $B7$ as to observation, while the fifth states the type, and the sixth the range, of the observation needed. The seventh, in accord with $B9$, keeps in place the ever-needed bulwark against the traditional totalitarian hypostatizations.

---

1 Bentley, "The Factual Space and Time of Behavior," *The Journal of Philosophy*, XXXVIII (1941), pp. 477-485. No interference is intended with the practical pre-scientific attitudes towards space and time so far as their everyday practical expression is concerned. Although long since deprived of dominance in the physical sciences, these attitudes remain dominant in psychological and sociological inquiry, and it is this dominance in this region that is rejected under our present postulation. See also the footnote to $D3$, postulation $H4$, and comment in the text following postulations D.

2 Chapter II, Section IV, No. 7. Bentley: "The Human Skin: Philosophy's Last Line of Defense," *Philosophy of Science*, VIII (1941), 1-19.

3 Chapter II, Section IV, No. 6. Compare postulation $A4$.

4 $B9$ restates what results if $B2$ is accepted and put to work thoroughly—"the addition of the adjective 'real' to the substantive 'facts' being only for rhetorical emphasis" (Dewey, "Context and Thought," *University of California Publications in Philosophy*, XII, [1931], 203-224). Compare also the statement by Stephen C. Pepper, *The Journal of Philosophy*, XLII (1945), 102: "There is no criterion for the reliability of evidence. . .but evidence of that reliability—that is corroboration." Professor Pepper's discussion of what happens "under the attitude of expecting an unquestionable criterion of truth and factuality to be at hand" runs strongly along our present line.

5 Both "focus" and "aspect" are double-barrelled words, in William James's sense. One cannot focus without something to focus with (such as a lens in or out of his eye) or without something to focus on. As for the word "aspect" (see also p. 108, footnote 1), this word originally stressed the viewing; an archaic meaning was a gaze, glance, or look, and a transitive verb still is usable, "to aspect." In more recent English "aspect" has been transferred in large measure to "object," but there are many mixed uses, even some that introduce locations and directions of action as between observer and observed. In any case the word applies to the "object," not absolutely but with reference to an observer, present or remote.

6 Chapter II, Section IV, No. 8.

7 Two of our earlier experimental formulations may be helpful in their variation of phrasing. Thus: *(a)* knowings are natural events; *(b)* they are known by standard methods; *(c)* enough is known about knowings and knowns to make the use of such methods practicable. Again: *(a)* knowers are in the cosmos along with what is known and to be known; *(b)* knowings are there too, and are to be studied (observed) in the same way as are other subjectmatters.

8 Dewey: "How is Mind to be Known?" *The Journal of Philosophy*, XXXIX (1942), pp. 29-35.

9 Chapter II, Section I.

10 Chapter II, Section I: "We proceed upon the postulate that *knowings* are always and everywhere inseparable from *the knowns*—that the two are twin aspects of common fact."

11 The word "social" is not used, primarily because of its confused status. It is sometimes opposed to "individual," sometimes built up out of "individuals," and, as it stands, it fails to hint at the transactional approach we express. "Culture" is comparatively non-committal, and can be understood much more closely as "behavioral," in the sense we have specified for that word.

12 In contrast to the usual program of concentrating the impermanence (or the fear of it) in the knowing, and assigning the permanence, in measure exceeding that of its being known, to the knowns.

D. Postulations for Namings and the Named as Specimens of Knowings and Knowns[1]

1. Namings may be segregated for special investigation within knowings much as any special region within scientific subjectmatter may be segregated for special consideration.

2. The namings thus segregated are taken as themselves the knowings to be investigated.[2]

3. The namings are directly observable in full behavioral durations and extensions.[3]

4. No instances of naming are observed that are not themselves directly knowings; and no instances of knowings within the range of naming-behaviors (we are not here postulating for signal or symbol behaviors) that are not themselves namings.[4]

5. The namings and the named are one transaction. No instance of either is observable without the other.[5]

6. Namings and named develop and decline together, even though to myopic or close-up observation certain instances of either may appear to be established apart from the participation of the other.[6]

7. Warranted assertion, both in growth and in decline, both as to the warranty and the warranted, exhibits itself as a phase of situations in all degrees of development from indeterminate to determinate. The strongest warranted assertion is the hardest of hard fact, but with neither the determinacy, nor the warranty, nor the hardness, nor even the factuality itself ranging beyond the reach of inquiry—for what is "hard fact" at "one" time is not assuredly "hard" for "all" time.

8. The study of either naming or named in provisional severance as a phase of *the transaction* under the control of postulations *D*4 and *D*5, is always legitimate and useful—often an outstanding need. Apart from such controls it falsifies.[7]

9. The study of written texts (or their spoken equivalents) in provisional severance from the particular organisms engaged, but nevertheless as durational and extensional behaviors under cultural description, is legitimate and valuable.[8] The examination is comparable to that of species in life, of a slide under a microscope, or of a cadaver on the dissection table—directed strictly at what is present to observation, and not in search for non-observables presumed to underlie observation, though always in search for more observables ahead and beyond.[9]

10. Behavioral investigation of namings is to be correlated with the physiology of organism-in-environment rather than with the intra-dermal formulations which physiologists initially employed in reporting their earlier inquiries.[10]

---

[1] An earlier formulation, combining something of both the present postulations *D* and *E*, and perhaps of interest for that reason, ran as follows: *(a)* knowledge is a sign system; *(b)* names are a kind of naturally developed sign; *(c)* naming and "specifying existence" are one process. These statements, however, must all be taken transactionally, if they are to represent our approach properly.

[2] Chapter II, Section III. In other words, under our postulation names do not enter as physical objects, nor as tools or instruments used by a psychical being or object, nor as being constructively separate from behavior in some such form as "products," nor as any other manner of externalization dependent on some supersubtle internalization. Under our postulation all such dismemberments are rejected as superfluous. The procedure, therefore, includes no such nostalgic plaint as that of the legendary egg to the hen: "Now that you have laid me, do you still love me?"

[3] Full duration and extension is not represented adequately and exclusively by such specialized devices as clock-ticks and foot-rules (see *B*7). Though these have developed into magnificent approximations for physics, they lack necessary pasts and futures across continents such as are involved in histories, purposes, and plans. They are therefore inadequate for inquiry into knowings, namings, and other behaviors.

[4] Compare the requirements set up in our appraisal of the logics (Chapter I, Sections I and X) that talkings be treated as "the men themselves in action."

[5] Cf. Chapter I, Section X; Chapter II, Section IV. A full behavioral spacetime form must be employed, comprising (but not limited by) physical and physiological spaces and times. The application of physical and physiological techniques is of course highly desirable, so far as they reach. Objectionable only are claims to dominate beyond the regions where they apply.

[6] Our own experience in the present inquiry is evidence of this, although the postulation ought to be acceptable at sight throughout its whole range of application. Starting out to find careful namings for phases of the subjectmatter discussed in the literature, we were quickly drawn into much closer attention to the named; this phase of the inquiry in turn depended for success on improvement in the namings. The two phases of the inquiry must proceed together. Rigidity of fixation for the one leads to wreckage for the other.

[7] An illustration that casts light on the status of naming and named with respect to each other may be taken from the earlier economics, which tried to hold consumption and production apart but failed miserably. Again, one may study the schemes of debtors and the protective devices of creditors, but unless this is done in a full transactional presentation of credit-activity one gains little more than melodrama or moralizing—equally worthless for understanding.

[8] This procedure was followed, so far as was practicable, in our examination of the logics, where the intention was never criticism of individuals, but always exhibition of the characteristics of the logical-linguistic mechanisms at work at present in America. As a technique of inquiry this is in sharp contrast with the ordinary practice. Through it we secured various exhibits of subjectmatters admitted—indeed even boasted—by their investigators to be neither fish, nor flesh, nor good red herring—neither physical, nor mental, nor linguistic; aliens in the land of science, denizens of never-never land; and likewise of various procedures in the name of consistency, tolerating the abandonment of the simplest standards of accuracy in naming at every other step. Unfortunately, specimens being few, we cannot carry on discussion under the anonymity which an entomologist can grant his bugs when he handles them by the tens and hundreds of thousands. To refer to a writer by name is much the same *sort* of thing as to mention a date, or as to name a periodical with its volume and page numbers. So far as inquiry into "knowledge" is concerned, the "you" and the "I" have their ethical and juridical valuations but offer little definiteness as to the activity under way; and this is certainly as true of the epistemologist's variety of "subject," as it is of any other. Recall the famous observation of William James, which has thus far been everywhere neglected actually in psychological and sociological research, that "the word 'I'...is primarily a noun of position like 'this' and 'here.'" (*A Pluralistic Universe*, New York, 1909, p. 380: *Essays in Radical Empiricism*, New York, 1912, p. 170.)

[9] The classical illustration of the sanctification of the reduplicative nonobservable as an explanation of the observable is, of course, to be found in the third interlude Molière provided for *Le malade imaginaire*, in which the candidate for a medical degree explains the effect of opium as due to its *virtus dormitiva*. Its words should be graven on the breastbone of every investigator into knowledge. The candidate's answer was:

Mihi a docto Doctore
Domandatur causam et rationem quare
Opium facit dormire.
A quoi respondeo,
Quia est in eo
Virtus dormitiva,
Cujus est natura
Sensus assoupire.

Peirce, in quoting this (5.534), remarks that at least the learned doctor noticed that there was *some* peculiarity about the opium which, he implies, is better than not noticing anything at all.

[10] As one stage in dealing with environments physiologists found it necessary to take account of "internal" environments, as in Claude Bernard's "milieu." Since then they have passed to direct consideration of transdermal processes, which is to say: their adequate complete statements could not be held *within* the skin but required descriptions and interpretations running *across* it in physiological analogue of what behaviorally we style "transactional."

Inspection of postulations *D* shows that the first four present definite subjectmatters for inquiry within the mistily presented regions of *C*. The fifth, sixth and seventh give further specifications to *C*5 and *C*6. The eighth provides for legitimate interactional inquiry within the transactional presentation, in sharp contrast with disruption of system, pseudo-interactions of mind-matter, and the total default in results offered by the older procedures for which *X*, *Y*, and *Z* stand as types. The ninth and tenth present supplementary techniques of practical importance.

IV

It is evident from these comments, as well as from the comments on postulations *C*, that although we are doing our best to phrase each separate postulation as definitely as the language available to us will permit, we are nevertheless allowing the selection and arrangement of the postulations within each group to proceed informally, since forced formality would be an artifice of little worth.

Two further comments are of special interest.

The first is that while we felt a strong need in our earlier assemblage of *B* and of *C* for the protective postulations *B*9 and *C*7, and while we shall later find it desirable to re-enforce this protection with postulation *H*1, the program of inquiry into namings as knowings represented by postulations *D*, in accord with *B*2, has already positively occupied the field, to fill which in older days the "reals" were conjured from the depths.

The second comment is that the greatest requirement for progressive observation in this field is freedom from the limitations of the Newtonian space and time grille, and the development of the more complete behavioral space and time frame, for which indications have been given in *B*7 and *D*3 and in the accompanying footnotes, and upon which stress will be placed again in *H*4.

V

In the case of signalings and symbolings which, along with namings, make up the broadest differentiation of behaviors, both as evolutionary stages and as contemporary levels,[1] it would be a waste of time to attempt postulatory elaboration until much further preliminary description had been given. This will be developed elsewhere. For the present the following indications of the need must suffice.

E. Indicated Postulation For Signaling Behaviors

1. Signaling behaviors—the regions of perception-manipulation,[2] ranging from the earliest indirect cues for food-ingestion among protozoa and all the varied conditionings of animal life, to the most delicate human perceptual activities—require transactional observation.

2. The settings for such words as "stimulus" "reaction," and "response," furnished under postulations of the types *X*, *Y*, and *Z*, have resulted in such chaos as to show that this or some other alternative development is urgently required.

F. Indicated Postulation for Symboling Behaviors

1. Symboling behaviors—the regions of mathematical and syntactical consistency—require transactional observation.

2. In current inquiries "foundations" are sought for mathematics by the aid of logic which—if "foundations" are what is needed—is itself notoriously foundation-less.[3]

3. Differentiation of the naming procedures from the symboling procedures as to status (function), methods, and type of results secured—and always under progressive observation—is the indicated step.[4]

We have now postulations *C* and *D* and preparatory comments *E* and *F* focussed upon behaviors in their aspect as Event in expansion from postulations *A*. Over against all of these, but in accord with them, we have postulations *B* focussed upon the aspect of inquiry through designation—the region in which science develops. Postulations *C*, as has already been said, are of lower grade than *D*, as is marked by their employment of the vague word "knowledge," their purpose having been to furnish a rough background for the attempt in *D*, to present namings as knowings direct. Postulations *C* are in further danger of being misinterpreted by some, perhaps many, readers in the sense rather of *B* than of *A*, of designation rather than of event, of the knowing rather than of the known. With postulations *C* thus insecurely seated, what may we say of their probable future?

Of the three types of vagueness in the word "knowledge,"[5] those of localization,[6] distribution, and range of application, the first two have been dealt with in preceding postulations. As for the third, the word bundles together such broadly different (or differently appearing) activities as "knowing how to say" and "knowing how to do"; and, further, from these as a center, has spreading applications or implications running as far down the scale as protozoan sign-behavior, and as far up as the most abstruse mathematical construction. Should future inquiry find it best to hold the word "knowledge" to a central range correlated with, or identified as, language-behaviors, then postulations *C* would merge with *D*. Should it be found preferable to extend the word, accompanied perhaps by the word "sign," over the entire behavioral range, then postulation *C* would return into *A* to find their home. We have no interest in sharp classification under rigid names—observable nature is not found yielding profitable results in that particular form. We do not expect to offer any prescription as to how the word "knowledge" should be used, being quite willing to have it either rehabilitate itself or, as the case may be, fall back into storage among the tattered blanket-wordings of the past. Whatever the future determination, narrow, wide, or medium, for the word "knowledge," postulations *C* keep the action provisionally open.

In the opening paragraphs of this chapter we held that man's knowings should be treated as natural, and should be studied through observation, under express recognition

---

[1] Chapter II, Sections IV and V.

[2] The word "manipulation" is used in its standard widened application and not in limitation to the "manual."

[3] Theorists such as Russell and other logicists are found who in their prideful panoply demand (at least when occasion seems ripe) that no science be recognized as such until it has been dubbed Sir Science and thus legitimatized by Logic.

[4] For introductory considerations, see Bentley, *Linguistic Analysis of Mathematics*, (Bloomington, Indiana, 1932).

[5] Chapter II, Section I.

[6] The old plan of dumping "knowledge" into a "mind" as its peculiar variety of "nature" and thus evading the labor of research, has long since ceased to be attractive to us.

of the postulatory status of observation itself in the transitions of both observations and postulations out of pasts and into futures. We believe that we have not failed throughout in proceeding in accordance therewith. These opening attitudes might perhaps have been themselves set forth as general postulations for the whole inquiry. The objection to this procedure is that the three main words, "nature," "observation," and "postulation," have such varied possible readings that, put together, they make a kite to which too many tails can be attached. From them, however, may be extracted certain statements concerning procedure with namings and things-named which may be offered in postulational form. They present—still from the designational approach as in postulations *B*—the cosmos as in action, the inquirers within it as themselves in action, and the whole process as advancing through time and across space. They are applicable to physical and physiological subjectmatters as well as to behavioral. Whether the aspect of inquiry *B*, as well as that of events *D*, *E* and *F*, will permit broadening in the future is a question that may be left for future discussion.

G. Postulational Orientation[1]

1. Subjectmatters of inquiry are to be taken in full durational spread as present through durations of time, comparable to that direct extensional observation they receive across extensions of space.[2]

2. Namings of subjectmatters are to be taken as durational, both as names and with respect to all that they name. Neither instantaneities nor infinitesimalities, if taken as lacking durational or extensional spread, are to be set forth as within the range of named Fact.

3. Secondary namings falling short of these requirements are imperfections, often useful, but to be employed safely only under express recognition at all critical stages of report that they do not designate subjectmatters in full factuality.[3]

Still lacking in our development and not to be secured until we have gained further knowledge of Signal and Symbol is an efficient postulational organization of Symbol with Designation within modern research. Under Symbol the region of linguistic "consistency" is to be

---

[1] This particular orientation does not preclude recognition of differences between namings that designate subjectmatters across indefinitely extended durations and expanses and those designating subjectmatters definitely limited in these respects. It is suggested, though not here postulated, that such differences may present the grounds for the rigid separations alleged in various traditional theories of knowledge to exist between theoretical and practical, and between rational and empirical, components; likewise for those alleged as between subjectmatters of sense-perception and of scientific knowledge, in ways that constitute radical obstacles to interpretation.

[2] Impressionistically one could say that duration is of the "very nature" of the event, of its "essence," of its "body and texture," though these are types of phrasing to avoid in formal statement, no matter how helpful they may seem for the moment. To illustrate: consider the "texture" of the "situations" in Dewey's *Logic, the Theory of Inquiry* as compared with the usual discussions of his viewpoint. These "situations," both "indeterminate" and "determinate," are cultural. Any report, discussion, or criticism that does not recognize this is waste effort, so far as the issues involved are concerned.

[3] Non-durational applications of such words as "sensation" and "faculty" in psychology have resulted in making these words useless to advanced systematic inquiry. Current words requiring continual watchfulness in this respect are such as "concept," "relation," "abstract," "percept," "individual," "social." In contrast our use of Fact, Event, and Designation is designed for full durational form, however faulty some of the phrasings in provisional report may remain.

presented. Under Designation we consider, as repeatedly stressed, not some "real existence" in a corruptly ultrahuman extension of the words "real" and "exist," but instead an "existency" under thoroughgoing behavioral formulation. It is, we hope, not forcing words too far for the impressionistic statement of the moment if we say that this may be in a "persistency" of durations and extensions such as postulations *G* require.

It is practicable to postulate rejections as well as acceptances. Under postulations *G* we have in effect rejected all non-extensionals, non-durationals, and non-observables of whatever types, including all purported ultimate "isolates." To emphasize this for the present issues of inquiry into knowledge, we now set down the following cases as among the most harmful. Let it be understood that these rejections, like all the other postulations, are offered, not as matters of belief or disbelief, but for the aid they may give research.

H. Postulational Rejects

1. All "reals" beyond knowledge.

2. All "minds" as bearers of knowledge.

3. All assignments of behaviors to locations "within" an organism in disregard of the transactional phases of "outside" participation (and, of course, all similar assignments to "outsides" in similar disruption of transactional event).

4. All forcible applications of Newtonian space and time forms (or of the practical forms underlying the Newtonian) to behavioral events as frameworks or grilles of the checkerboard type, which are either (1) insisted upon as adequate for behavioral description, or (2) considered as so repugnant that behavior is divorced from them and expelled into some separate "realm" or "realms" of its own.

VI

One faces often the temptation to exhibit certain of the postulations as derived from others. We would advise against it, even when the durational postulation is used as source. We are impressed with the needlessness, under our approach, of "deriving" anything from anything else (except, of course, as may be convenient in propaedeutic display, where such a procedure perhaps properly belongs). The postulations present different stresses and offer different types of mutual aid, but no authoritarianism such as logics of ancient ancestry demand, including even (and sometimes peculiarly) those which strive to make their logicism look most positive. Many lines of ordering will suggest themselves as one works. If behaviors are durational, and knowings are behaviors, then the knowings become observable. If knowings and knowns are taken as in system, then one quickly arrives at a durational postulation in trying to report what one has observed; and from the durational one passes to the transactional. On the other hand, from this last, if arrived at first, one passes to the durational. This is, indeed, but a final reiteration of what was stressed in the opening paragraphs. Observation and postulation go hand in hand. The postulations hang together, not by grace of any one, or any two, or any three of them, but by organization in respect to the direction of approach, the points of entry, and the status of the audience—the status, that is, of the group of interested workers at the given time and place in history, and of that whole society-in-cosmos of which they themselves are components.

IV.
INTERACTION AND TRANSACTION

I

OUR preliminary sketch of the requirements for a firm terminology for knowings and knowns placed special stress on two procedures of knowledge[1] called Transaction and Specification. Specification was distinguished from Definition and the immediate development of Transaction was connected with Specification rather than with Definition.

We propose in succeeding chapters to discuss Transaction and Specification at some length, each on its own account, and to show how important it is for any theory of knowledge that their characteristics as natural processes of knowing-men and things-known should be fully understood. Before undertaking this, however, it will be well to display in the present chapter, the extent to which the transactional presentation of objects, and the determination of objects as themselves transactional, has been entering recent physical research. In so doing, the transactional presentation will be brought into contrast with the antique view of self-actions and with the presentation of classical mechanics in terms of interactions. The discussion will not be widened, however, beyond what is needed for the immediate report.

The reader will recall that in our general procedure of inquiry no radical separation is made between that which is observed and the observer in the way which is common in the epistemologies and in standard psychologies and psychological constructions. Instead, observer and observed are held in close organization. Nor is there any radical separation between that which is named and the naming. Comparably knowings and knowns, as inclusive of namings and observings, and of much else as well, are themselves taken in a common system of inquiry, and not as if they were the precarious products of a struggle between severed realms of "being." It is this common system of the knowing and the known which we call "natural," without either preference or prejudice with respect to "nature," such as now often attends the use of that word. Our position is simply that since man as an organism has evolved among other organisms in an evolution called "natural," we are willing under hypothesis to treat all of his behavings, including his most advanced knowings, as activities not of himself alone, nor even as primarily his, but as processes of the full situation of organism-environment; and to take this full situation as one which is before us within the knowings, as well as being the situation in which the knowings themselves arise.[2]

What we call "transaction," and what we wish to show as appearing more and more prominently in the recent growth of physics, is, therefore, in technical expression, neither to be understood as if it "existed" apart from any observation, nor as if it were a manner of observing "existing in a man's head" in presumed independence of what is observed. The "transaction," as an object among and along with other objects, is to be understood as unfractured observation—just as it stands, at this era of the world's history, with respect to the observer, the observing, and the observed—and as it is affected by whatever merits or defects it may prove to have when it is judged, as it surely will be in later times, by later manners.

II

When Comte cast a sweeping eye over the growth of knowledge as far as he could appraise it, he suggested three stages or levels which he called the theological, the metaphysical, and the positive. One would not want to accept these stages today, any more than one would want to adopt Comte's premature scheme for the organization of the sciences. Nevertheless, his general sketch has entered substantially into everyone's comprehension. Roughly speaking, the animistic personifications and personalizations of the world and its phenomena were prevalent in the early days; hypostatizations such as physical "forces" and "substances" followed them; only in recent centuries have we been gaining slowly and often painfully, that manner of statement called positive,[3] objective, or scientific. How the future may view even our best present opinions is still far from clear.

Let us consider a set of opposed tendencies which, for the moment, in everyday English we may call the narrowing and widening of the scope of scientific observation with respect to whatever problem is on hand. By way of introduction, we may trace such an alternation of viewpoints for the most general problems of physics from Newton to Maxwell.

For many generations, beginning with Galileo after his break with the Aristotelian tradition, and continuing until past the days of Comte, the stress in physical inquiry lay upon locating units or elements of action, and determining their interactions. Newton firmly established the system under which particles could be chosen and arrayed for inquiry with respect to motion, and so brought under definite report. But not all discovery resulted in the establishment or use of new particles. In the case of heat, for example, it did not come to pass that heat-particles were identified. "The progress of science," say Einstein and Infeld, "has destroyed the older concept of heat as a substance."[4] Parti-

---

[1] The word "knowledge" has the value here of a rough preliminary description, loosely indicating the field to be examined, and little more.

[2] For formal recognition and adoption of the "circularity" involved in the statement in the text, see Chapter II, Section IV, No. 2.

[3] Comte's "positive" retained something from his "metaphysics," just as his "metaphysics" retained something from his "theological." He substitutes "laws" for "forces," but gives them no extensive factual construction. "Logical positivism" has anachronistically accepted this Comtean type of law, emptied it of what factuality it had, and further formalized it. Such a "positive" does not get beyond short-span, relatively isolated, temporal sequences and spatial coexistences. Its background of expression, combined with a confused notion of the part mathematics plays in inquiry, is what often leads scientists to regard "laws" as the essential constituents of science, instead of stressing directly the factual constructions of science in space-time.

[4] Albert Einstein and Leopold Infeld, *The Evolution of Physics*, (New York, 1938), p. 51. We shall use the Einstein-Infeld book for repeated citation, not at all for confirmation of our views or for support of our development, but in order to have before the reader's eyes in the plainest English, authoritative statements of certain features of physics which everyone ought to know, but which in the fields of knowledge-theory are put to use by few. Since we shall have a good deal to do (although little expressly to say) with the way in which rigidly established views block needed progress—a point to which Max Wertheimer, whom we shall later quote, has recently given vivid illustration—a further comment by Einstein and Infeld is significant: "It is a strange coincidence that nearly all the fundamental work concerned with the nature of heat was done by non-professional physicists who regarded physics merely as their great hobby" (*Ibid.*, p. 51).

cles of a definitely Newtonian type were, it is true, retained in the work of Rumford and Joule, and later of Gibbs; and energy was advocated for a long time as a new substance with heat as one of its forms. But the particle fell upon statistical days (evil, indeed, from the point of view of its older assuredness), and what heat became in the end was a configuration in molecular ranges rather than a particulate presence. Faraday's brilliant observation found that all which happened electrically could not be held within the condenser box nor confined to the conducting wire. Clerk Maxwell took Faraday's observations and produced the mathematical formulation through which they could be expressed.[1] Maxwell's work furnished the structure for the developments of Roentgen, Lorentz, Planck, and Einstein, and their compeers, and for the more recent intra-atomic exploration. His posthumous book, *Matter and Motion*, has a lucidity which makes it a treasure to preserve and a model that all inquirers, especially those in newly opening fields, can well afford to study. The following is from the Preface to this book, dated 1877, and included in the British edition of 1920, edited by Sir Joseph Larmor:

"Physical science, which up to the end of the eighteenth century had been fully occupied in forming a conception of natural phenomena as the result of forces acting between one body and another, has now fairly entered on the next stage of progress—that in which the energy of a material system is conceived as determined by the configuration and motion of that system, and in which the ideas of configuration, motion, and force are generalized to the utmost extent warranted by their physical definitions."

Although Maxwell himself appreciated what was taking place, almost two generations were needed before physicists generally began to admit it: *teste*, their long hunting for that greatest of all victims of the Snark that was Boojum, the ether: the process of re-envisionment is far from completed in physics even yet. The very word "transaction," which we are to stress, was, indeed, used by Maxwell himself in describing physical events; he even speaks of "aspects" of physical transactions in much the sense that we shall employ that word.[2] Thus:

"If we confine our attention to one of the portions of matter, we see, as it were, only one side of the *transaction*—namely, that which affects the portion of matter under our consideration—and we call this aspect of the phenomenon, with reference to its effect, an External Force acting on that portion of matter, and with reference to its cause we call it the Action of the other portion of matter. The opposite aspect of the stress is called the Reaction on the other portion of matter."

Here we see the envisionment that Maxwell had gained in the electromagnetic field actually remodeling his manner of statement for mechanical systems generally. Maxwell was opening up new vistas from a footing in the firmest organization of inquiry the world had ever possessed—that of the Newtonian mechanics. Though our own position is one in which

the best we can hope for is to be able to introduce a small degree of order into an existing chaos, we can use his work, and the results that came from it, in our support, believing as we do that, as progress is made, the full system of human inquiry may be studied as if substantially one.

III

With this much of introductory display let us now set down in broad outlines three levels of the organization and presentation of inquiry in the order of their historical appearance, understanding, however, as is the way with evolutions generally, that something of the old, and often much of it, survives within or alongside the new. We name these three levels, those of Self-Action, Interaction, and Transaction. These levels are all human behaviors in and with respect to the world, and they are all presentations of the world itself as men report it. We shall permit ourselves as a temporary convenience the irregular use of hyphenization in these names as a means of emphasizing the issues involved in their various applications. This is comparable to a free use of capitalization or of quotation marks, and to the ordinary use of italics for stress. It has the particular value that it enables us to stress the inner confusions in the names as currently used.[3]

Self-action: where things are viewed as acting under their own powers.

Inter-action: where thing is balanced against thing in causal interconnection.

Trans-action:[4] where systems of description and naming are employed to deal with aspects and phases of action, without final attribution to "elements" or other presumptively detachable or independent "enti-

---

1 "The most important event in physics since Newton's time," say Einstein and Infeld of Maxwell's equations, "not only because of their wealth of content, but also because they form a pattern for a new type of law" *(Ibid.,* p. 148).

2 *Matter and Motion*, Article XXXVIII. The italics for the word "transaction" are supplied by us.

3 Our problem here is to systematize the three manners of naming and knowing, named and known. Self-action can hardly be written, as writing and reading proceed today, without its hyphen. Transaction, we shall in the end argue, should be established in such a way that hyphenization would be intolerable for it except, perhaps, in purely grammatical or etymological examination. Inter-action, in contrast, within the range of our present specialized field of study, will appear to be the verbal thief-of-the-world in its commoner uses, stealing away "men's minds," mutilating their judgments, and corrupting the very operation of their eyesight. The word "thing" as used in the characterizations in the text is deliberately chosen because it retains its idiomatic uses, and is almost wholly free from the more serious of the philosophers' distortions which commonly go with the whole flock of words of the tribe of "entity." For our future use, where a definite *outcome* of inquiry in its full behavioral setting is involved, the word "object" will be employed.

4 "Transaction," in ordinary description, is used for the consideration as detached of a "deal" that has been "put across" by two or more actors. Such a verbal shortcut is rarely objectionable from the practical point of view, but that is about all that can be said for it. For use in research adequate report of the full event is necessary, and for this again adequate behavioral description must be secured. Dewey's early employment of the word "transaction" was to stress system more emphatically than could be done by "interaction." (See his paper "Conduct and Experience" in *Psychologies of 1930*. [Worcester, Mass.] Compare also his use of "integration" in *Logic, the Theory of Inquiry*.) The beginnings of this attitude may be found in his paper "The Reflex Arc Concept in Psychology" (1896). Bentley's treatment of political events was of the transactional type in his *The Process of Government* (Chicago, 1908), though, of course, without the use of that name. John R. Commons has used the word comparably in his *Legal Foundations of Capitalism* (New York, 1924) to describe that type of economic inquiry in which attention centers on the working rules of association rather than on material goods or human feelings. George H. Mead's "situational" is often set forth in transactional form, though his development is more frequently interactional rather than transactional.

ties," "essences," or "realities," and without isolation of presumptively detachable "relations"[1] from such detachable "elements."

These provisional characterizations will be followed in a later chapter by alternatives showing the variety of points of view from which the issues that are involved must be approached. The reader will note that, while names are given as if for the events observed, the characterizations are in terms of selective observation, under the use of phrasings such as "are viewed," "is balanced against," and "are employed." These are the two aspects of the naming-named transaction, for which a running exhibit is thus given, pending clarification as the discussion advances.

The character of the primitive stage of Self-action can be established easily and clearly by a thousand illustrations, past and present—all confident in themselves as factual report in their times, without suspicion of the way in which later generations would reduce them to the status of naive and simple-minded guesswork.

For Trans-action at the latest end of the development we can show a clean status, not as assertion of its existence, but as a growing manner of observation of high efficiency at the proper time and place, now rapidly advancing to prominence in the growth of knowledge.

As for Inter-action, it furnished the dominant pattern of scientific procedure up to the beginning of the last generation. However, as a natural result of its successes, there grew up alongside it a large crop of imitations and debasements—weeds now ripe for the hoe. To avoid very possible misunderstandings, it is desirable to give a subclassification of the main types of procedure that may from time to time present themselves as, or be appraised as, interactions. We find:

(a) Independently formulated systems working efficiently, such as Newtonian mechanics.

(b) Provisionally separated segments of inquiry given an inter-actional form for convenience of study, though with underlying recognition that their results are subject to reinterpretation in wider systems of description; such, for example, as the investigation of certain inter-actions of tissues and organs within the skin of an organism, while remembering, nevertheless, that the "organism-as-a-whole" transactionally viewed (with perhaps also along with it a still wider transactional observation of the "organism-in-environment-as-a-whole") must come into account before final reports are reached.

(c) Abuses of *(a)* such as often occurred when, before the Einstein development, efforts were made to force all knowledge under the mechanistic control of the Newtonian system.[2]

(d) Grosser abuses much too common today, in which mixtures of self-actional "entities" and inter-actional "particles" are used to produce inter-actional explanations and interpretations *ad lib.:* as when selves are said to inter-act with each other or with environmental objects; when small portions of organisms are said to inter-act with environmental objects as in the traditional theories of sensation; when minds and portions of matter in separate realms are brought by the epistemologies into pseudo-interactional forms; or, probably worst of all, when a word's meaning is severed from the word's actual presence in man's behavior, like a sort of word-soul from a word-body.

IV

Returning now to physics for a further examination of its increasing use of transaction, we may preface discussion with a few general words on self-action. We need not go far back into cultural history to find the era of its dominance. It took Jupiter Pluvius to produce a rainstorm for the early Romans, whereas modern science takes its *pluvius* free from Jupiter. The *Lares* and *Penates* which "did" all that happened in the household multiplied so excessively in Rome that in time they became jokes to their own alleged beneficiaries. The Druid had, no doubt, much tree lore useful for his times, but to handle it he wanted a spirit in his tree. Most magic has this type of background. It took Robin Goodfellow, or one of his kind, a Brownie perhaps, to make cream turn sour. In modern times we have flocks of words of respectable appearance that spring from this source: such words as "substance," "entity," "essence," "reality," "actor," "creator," or "cause," and thus, indeed, the major part of the vocabulary of metaphysics.[3]

Aristotle's physics was a great achievement in its time, but it was built around "substances." Down to Galileo men of learning almost universally held, following Aristotle, that there exist things which completely, inherently, and hence necessarily, possess Being; that these continue eternally in action (movement) under their

---

[1] It should be fairly well evident that when "things" are too sharply crystallized as "elements," then certain leftovers, namely, the "relations," present themselves as additional "things," and from that pass on to becoming a variety of "elements" themselves, as in many current logics. This phase of the general problem we do not intend here to discuss, but we may stop long enough to quote the very instructive comment Max Wertheimer recently made on this sort of thing. He had made a careful study of the way in which a girl who was secretary to the manager of an office described to him the character of her office setup, and he devoted a chapter to her in his book *Productive Thinking* (New York, 1945). His analysis of her account showed it defective in that it "was blind to the structure of the situation" (p. 137), and he was led to the further comment that her procedure was "quite similar to the way a logistician would write a list of relations in a relational network" (p. 138). Compare also Wertheimer's paper "On Truth," *Social Research* (1934), 135-146.

[2] The positions we shall take are in several important respects close to those taken by Richard C. Tolman in his address prepared for the symposium in commemoration of the 300th anniversary of Newton's birth ("Physical Science and Philosophy," *The Scientific Monthly*, LVII (1943), 166-174). Professor Tolman uses a vocabulary of a different type from that which we employ—one relying on such words as "subjective," "objective," "abstraction," "conceptual" etc.—but these wordings are not the significant matter we have in mind. The essential points are that he treats distinctions between the sciences as resting in the techniques of inquiry that are available (pp. 171-172), and that he strongly opposes as a "fallacious assumption" the view that "phenomena at one level of abstraction can necessarily be completely treated at a lower level of abstraction." (p. 174). (Compare our procedure, in Chapter II, Section IV, No. 4). We insert this note not to involve Professor Tolman in our construction, but to provide an alternative form of expression for views comparable in these respects to our own that may better suit the needs of persons who find our own manner unfamiliar or undesirable.

[3] The distinction between ancient rigidities of naming and scientific names of the firm (but not rigid) type, such as we desire to attain in our own inquiry, stands out clearly here. The ancient substances needed rigidity, fixation of names to things in final one-to-one correspondence. Pre-Darwinian classification of living forms showed the rigid trend as opposed to modern freedom of development. We have surviving today in obscure corners numerologies and other superstitions under which things are controlled by the use of the right names. We even find remnants of the ancient view in many of our modern logics which seek domination by verbal development. Bertrand Russell's logical atomism with its never-ceasing striving after minutely named "realities" may be mentioned in this connection.

own power—continue, indeed, in some particular action essential to them in which they are engaged. The fixed stars, under this view, with their eternal circular movements, were instances. What did not, under the older pattern, thus act through its inherent power, was looked upon as defective Being, and the gradations ran down to "matter" on its lowest level, passive and inert.

Galileo's work is generally recognized as marking the overthrow of Self-action in physical doctrine, and it was just this feature which aroused so much hatred among the men of the ancient tradition. An excellent account—probably the best yet given—well be found in Max Wertheimer's book *Productive Thinking*.[1] Departing from the Aristotelian view that eternal force had to be applied to any inert body to put it in motion and to keep it in motion, Galileo made use of an inclined plane in substitution for a falling weight, as a direct aid to observation. Here he identified acceleration as the most significant feature for his purposes. He then considered the opposed case of a weight tossed upwards, using similarly an ascending inclined plane for his guide, and identified negative acceleration. Together, these yielded him the limiting case of the horizontal plane constructively lying between the descending and the ascending planes. He thus identified the fact (more pretentiously spoken of as a "principle" or "law") of inertia in its modern form: a mass once in motion continues in motion in a straight line, if not interfered with by other moving masses. Its motion, in other words, is no longer supposed to be dependent on the continued push applied to it by an "actor." This discovery was the needed foundation for the interactional development to come. Moreover, the new view itself was transactional with respect to the situation of its appearance: what, namely, had been an incident or result of something else was now taken up into direct report as event.[2] Hobbes quickly anticipated what Newton was later to establish, and Descartes made it his prime law of nature. For Newton it became the first law of motion, leading, through a second law concerning direction and proportionality of force, to the third law, namely, that action and reaction are equal and opposed—in other words, to the establishment of the full inter-actional system of mechanics.

The Newtonian construction—unexcelled for its efficiency within its sphere—viewed the world as a process of "simple forces between unalterable particles."[3] Given a closed system of this kind the inter-actional presentation had now been perfected. This, however, had been achieved at the cost of certain great omissions. Space and time were treated as the absolute, fixed, or formal framework within which the mechanics proceeded—in other words, they were omitted from the process itself.

The failure to inquire into the unalterability of the particle was similarly an "omission," though one could freely select whatever "unalterables" one wished for experimental introduction as different problems arose. One immediate effect of Newton's success *within* his accepted restrictions was to hold him to the corpuscular theory of light and make him hostile to the competing wave theory of Huygens.

Einstein's treatment, arising from new observations and new problems, brought space and time into the investigation as among the events investigated. It did more than that: it prepared the scene for the particle itself to go the way of space and time.[4] These steps were all definitely in the line of the transactional approach: the seeing together, when research requires it, of what before had been seen in separations and held severally apart. They provide what is necessary at times and places to break down the old rigidities: what is necessary when the time has come for new systems.

The new foundation that has been given physics on a transactional basis, replacing the old inter-actional extremism, has not yet been made complete. Rival treatments and interpretations have their special places, and what the outcome will be is not wholly clear. Einstein himself devotes his efforts to securing a general field theory, but singularities remain in the field, with which he has not as yet been able successfully to deal. Whether "field" *in physics* is to represent the full situation, or whether it is to be used for an environment to other components is not *our* problem, and is not essential to a general consideration of the transactional phase of inquiry. Our assertion is the right to see in union what it becomes important to see in union; together with the right to see in separation what it is important to see in separation—each in its own time and place; and it is this right, when we judge that we require it for our own needs, for which we find strong support in the recent history of physics. The physicist can readily find illustrations of the two-fold need in his daily work. The changes in stress across the generations, from force as a center to the *vis viva* of Leibnitz, and then on to energy as a special kind of thing in addition to material things, and to the development of the de Broglie equation connecting mass and energy, are in point. Energy now enters more and more in the guise of a described situation rather than in that of an asserted "thing." Long ago some significance, apart from mere puzzlement, was found in the facts that an electric current was not present without a circuit and that all that happened was not "inside" the wire. Twenty years ago physicists began to ask whether light could "start" from a light source, near or distant, if it did not have its place of arrival waiting for it. Today, as indicative of the status of physics, we get discussions strewn with sentences such as the following: "'The path of a light ray,' without including the environment of the light ray in the description, is an incomplete expression and has no operational meaning"; "The term 'path of a particle' has no more operational meaning than 'path of a photon' in ordinary optics"; "Speaking exactly, a particle by itself without the description of the whole experimental set-up is not a physical reality"; "We can not describe the state of a photon on its way from the sun"; "The law [of

---

[1] Quotations of pertinent phrases from both Aristotle and Galileo are given by Einstein and Infeld in the opening chapter of their book previously cited. Wertheimer concentrates attention on the "structure" or "Gestalt" which governed Galileo's search. Seen as a stage of development in understanding and presentation in the cultural setting in which it was produced, this is in the line of our treatment. Seen, however, as Wertheimer has continued to see it, as a mental activity of self-actional parentage applied to an outer world of objects, it falls far short of the manner of statement which we believe to be necessary. The "mind" Wertheimer relies on is far too reminiscent of the older days in which the "physical" opposed to it was an all-too-solid fixture. Wertheimer, in his last book, has nevertheless dropped much of the traditional mentalistic phraseology; and this with no loss to his presentation.
[2] In his early study of perception Max Wertheimer made the comparable demonstration that motion could be *directly* perceived. "Experimentelle Studien über das Sehen von Bewegung," *Zeitschrift für Psychologie*, LXI (1912), 161-265.
[3] Einstein and Infeld, *op. cit.*, pp. 58, 255.

[4] "In so far as wave mechanics has recognized two words that used to be associated with electrons—*position* and *momentum*—and has provided mathematical expressions as sort of tombstones to correspond to these words, it has done so with the least invocation of trouble to itself," W.F.G. Swann, "The Relation of Theory to Experiment in Physics," *Reviews of Modern Physics*, XIII (1941), 193.

causality] in its whole generality cannot be stated exactly if the state variables by which the world is described are not mentioned specifically."[1]

Our aim in this examination of the transformation of viewpoints in physics has been solely to make clear how largely the manner of approach we propose to employ for our own inquiry into knowings and knowns has been already developed by the most potent of all existing sciences. We may supplement what has been said, for the benefit of any still reluctant philosophical, epistemological, or logical reader, by a few citations from the Einstein and Infeld work previously quoted. "The earth and the sun, though so far apart," were, under Newton's laws, "both actors in the play of forces. . . . In Maxwell's theory there are no material actors" (p. 152); "We remember the [mechanical] picture of the particle changing its position with time. . . . But we can picture the same motion in a different way. . . . Now the motion is represented as something which *is*, which exists. . .and not as something which changes. . ." (pp. 216-217); "Science did not succeed in carrying out the mechanical program convincingly, and today no physicist believes in the possibility of its fulfillment" (p. 125); "The concepts of substances, so essential to the mechanical point of view, were more and more suppressed" (p. 148); "The properties of the field alone appear to be essential for the description of phenomena; the differences in source do not matter" (p. 138); "The electromagnetic field is, in Maxwell's theory, something real" (p. 151).

So far as the question of what is called "physical reality" arises in this connection, a reference to a well-known discussion between Einstein and Niels Bohr about ten years ago is pertinent. In contrast with his transactional (i.e., free and open) treatment of *physical* phenomena, Einstein has remained strongly self-actional (i.e., traditionally constrained) in his attitude towards man's activity in scientific enterprise. His positon is that

"physical concepts are free creations of the human mind" *(op. cit.*, p. 33), and that "the concepts of the pure numbers. . .are creations of the thinking mind which describe the reality of our world" *(ibid.*, p. 311).[2] Bohr, in contrast, appears to have a much freer view of a world that has man as an active component within it, rather than one with man by fixed dogma set over against it. In the discussion in question, which involved the issues of momentum in wave theory, Einstein and his associates, Podolsky and Rosen, chose a criterion of reality based upon prediction to the effect that "if" (without disturbance) "one can predict with certainty the value of a physical quantity," then "there exists an element of physical reality corresponding to this physical quantity." In order to have a complete theory (and not merely a "correct" one), they held that "every element of the physical reality must have a counterpart in the physical theory"; and further they offered their proof that either "the quantum-mechanical description of reality given by the wave function is not complete," or "when the operators corresponding to two physical quantities do not commute, the two quantities cannot have simultaneous reality." In reply Bohr, employing his "notion of complementarity," held that the Einstein-Podolsky-Rosen "criteria of physical reality" contained "an essential ambiguity" when applied to quantum phenomena. He asserted further that while relativity had brought about "a modification of all ideas regarding the absolute character of physical phenomena," the still newer features of physics will require "a radical revision of our attitude as regards physical reality."[3] What is involved here is an underlying, though not explicitly developed, conflict as to the manner in which mathematics (as symbolic) applies to physics (as fact-seeking). This in turn involves the organization of symbol with respect to name among the linguistic behaviors of men.

---

[1] These phrasings are all from Philipp Frank's excellent monograph, *Foundations of Physics*, the most recent publication of the *International Encyclopedia of Unified Science* (I, No. 7, [1946], 39, 45, 48, 53).

[2] Various significant comments on Einstein's attitude in this respect, which Wertheimer largely shares, will be found scattered through the latter's book, *Productive Thinking*, previously cited.

[3] "Can Quantum-Mechanical Description of Physical Reality be Considered Complete?" A. Einstein, B. Podolsky, and N. Rosen, *Physical Review*, XLVII (1935); Niels Bohr, *ibid.*, XLVIII (1935).

V.
TRANSACTIONS AS KNOWN AND NAMED

I

FOLLOWING an exhibit in the preceding chapter of the extent to which the manner of observing we call "transactional" is being employed in recent physics, we wish now to show something of its entry into physiology. On this basis we shall discuss its importance for behavioral inquiry and we shall especially stress the outstanding need for its employment in inquiries into knowings and knowns as human behaviors, if such inquiries are to achieve success.

A brief reminder of the terminology provisionally employed is desirable. In a natural factual cosmos in course of knowing by men who are themselves among its constituents, naming processes are examined as the most readily observable and the most easily and practically studied of all processes of knowing.[1] The name "Fact" is applied to such a cosmos with respect both to its naming-knowing and its named-known aspects. The naming aspect of Fact is styled Designation; the named aspect is styled Event. The problem as to whether knowings-knowns of other forms[2] than namings-nameds should be brought into such inquiry prior to its development is postponed on about the same basis that a biologist proceeds with inquiry into either plant or animal life prior to securing a sharp differentiation between the two or a sharp separation of both of them together from physical event. In general, it is to be observed that the range of the known which we have thus far been developing under the name "event" is, later in this book, to be presented as the full range which the word "existence" can cover in coherent application.

The name "Object" is applied to Event well established as the outcome of inquiry. The name "Specification" is applied to that most efficient form of Designation which has developed in the growth of modern science.[3] Transaction is, then, that form of object-presentation in improved Specification, which is becoming more and more importantly employed in the most advanced scientific inquiry, though still grossly disregarded in backward enterprises, and which is wholly neglected in present-day inquiries into knowledge as the knowing-known procedures of men. Transaction will be discussed in the present chapter and Specification in the next.

To reduce the occasion for some of the ordinary forms of misunderstanding, and to avoid frequent reminder of them in the text, attention is now called to certain positions common in whole or in large degree to current epistemologies, psychologies, and sociologies. These are positions which are *not* shared by us, and which may *in no case* be read into our work whether pro or con by persons who wish properly to appraise it.

1. We employ no basic differentiation of subject *vs.* object, any more than of soul *vs.* body, of mind *vs.* matter, or of self *vs.* not-self.

2. We introduce no knower to confront what is known as if in a different, or superior, realm of being or action; nor any known or knowable as of a different realm to stand over against the knower.

3. We tolerate no "entities" or "realities" of any kind, intruding as if from behind or beyond the knowing-known events, with power to interfere, whether to distort or to correct.

4. We introduce no "faculties" or other operators (however disguised) of an organism's behaviors, but require for all investigation direct observation and usable reports of events, without which, or without the effort to obtain which, all proposed procedure is to be rejected as profitless for the type of enterprise we here undertake.

5. In especial we recognize no names that pretend to be expressions of "inner" thoughts, any more than we recognize names that pretend to be compulsions exercised upon us by "outer" objects.

6. We reject the "no man's land" of words imagined to lie between the organism and its environmental objects in the fashion of most current logics, and require, instead, definite locations for all naming behaviors as organic-environmental transactions under observation.

7. We tolerate no finalities of meaning parading as "ultimate" truth or "absolute" knowledge, and give such purported finalities no recognition whatever under our postulation of natural system for man in the world.

8. To sum up: Since we are concerned with what is inquired into and is in process of knowing as cosmic event, we have no interest in any form of hypostatized under-pinning. Any statement that is or can be made about a knower, self, mind, or subject—or about a known thing, an object, or a cosmos—must, so far as we are concerned, be made on the basis, and in terms, of aspects of event which inquiry, as itself a cosmic event, finds taking place.

II

It was said of Transaction in Chapter IV that it represents that late level in inquiry in which observation and presentation could be carried on without attribution of the aspects and phases of action to independent self-actors, or to independently inter-acting elements or relations. We may now offer several additional characterizations[4] correlated with the preliminary one and indicating the wide range of considerations involved. We may take the ancient, indeed, largely archaic, stages of self-action for granted on the basis of what has already been said of them and subject to the illustrations that will be given hereafter, and we may economize space by confining immediate attention to a comparison of transaction with interaction.

---

[1] See Chapter II, Section III. Chapter III, Sections I and III.

[2] These other forms include not only the full range of the perceptive-manipulative (Signal), but also those of non-naming linguistic processes such as mathematics (Symbol). For the words "event" and "existence" see p. 112, footnote 1, and the characterization given the words in Chapter XI.

[3] The word "science" in our use stands for free experimental observation and naming, with the understanding that the advanced branches of scientific inquiry are necessary aids to the backward branches, but never their dictators.

[4] The reader will recall that in the present treatment we do not hope to get beyond characterization, but must leave the greater accuracy of specification for future development, when additional phases of the issue have been examined. The use of hyphenization as a device for emphasizing interior confusions in words continues now and then in the text. The following from the British weekly *Notes and Queries* a hundred years ago may be profitably examined by the muddled victims of unhyphenized "interaction" today: "A neglect of mental hyphenization often leads to mistake as to an author's meaning, particularly in this age of morbid implication."

Consider the distinction between the two as drawn in terms of description. If inter-action is inquiry of a type into which events enter under the presumption that they have been adequately described prior to the formulation of inquiry into their connections, then—

Transaction is inquiry of a type in which existing descriptions of events are accepted only as tentative and preliminary, so that new descriptions of the aspects and phases of events, whether in widened or narrowed form, may freely be made at any and all stages of the inquiry.

Or consider the distinction in terms of names and naming. If inter-action is found where the various objects inquired into enter as if adequately named and known prior to the start of inquiry, so that further procedure concerns what results from the action and reaction of the given objects upon one another, rather than from the reorganization of the status of the presumptive objects themselves, then—

Transaction is inquiry which ranges under primary observation across all subjectmatters that present themselves, and proceeds with freedom toward the re-determination and re-naming of the objects comprised in the system.

Or in terms of Fact. If inter-action is procedure such that its inter-acting constituents are set up in inquiry as separate "facts," each in independence of the presence of the others, then—

Transaction is Fact such that no one of its constituents can be *adequately* specified as fact apart from the specification of other constituents of the full subjectmatter.

Or with respect to Elements. If inter-action develops the particularizing phase of modern knowledge, then—

Transaction develops the widening phases of knowledge, the broadening of system within the limits of observation and report.

Or in terms of Activity. If inter-action views things as primarily static, and studies the phenomena under their attribution to such static "things" taken as bases underlying them, then—

Transaction regards extension in time to be as indispensable as is extension in space (if observation is to be properly made), so that "thing" is in action, and "action" is observable as thing, while all the distinctions between things and actions are taken as marking provisional stages of subjectmatter to be established through further inquiry.

Or with special attention to the case of organism and environment. If inter-action assumes the organism and its environmental objects to be present as substantially separate existences or forms of existence, prior to their entry into joint investigation, then—

Transaction assumes no pre-knowledge of either organism or environment alone as adequate, not even as respects the basic nature of the current conventional distinctions between them, but requires their primary acceptance in common system, with full freedom reserved for their developing examination.[1]

Or more particularly with specialized attention to knowings and knowns. If, in replacement of the older self-action by a knower in person, inter-action assumes little "reals" interacting with or upon portions of the flesh of an organism to produce all knowings up to and including both

the most mechanistic and the most unmechanistic theories of knowledge,[2] then—

Transaction is the procedure which observes men talking and writing, with their word-behaviors and other representational activities connected with their thing-perceivings and manipulations, and which permits a full treatment, descriptive and functional, of the whole process, inclusive of all its "contents," whether called "inners" or "outers," in whatever way the advancing techniques of inquiry require.

And finally, with respect to inquiry in general. Wherever inter-actional presentation, on the basis of its special successes in special fields, asserts itself dogmatically, or insists on establishing its procedure as authoritative to the overthrow of all rivals, then—

Transactional Observation is the fruit of an insistence upon the right to proceed in freedom to select and view all subjectmatters in whatever way seems desirable under reasonable hypothesis, and regardless of ancient claims on behalf of either minds or material mechanisms, or any of the surrogates of either.[3]

Thoroughly legitimate interactional procedures, it will be recalled from our previous discussion,[4] are all those which, like classical mechanics, are held adequately within their frameworks of hypothesis; and also those others which represent provisional partial selections of subjectmatters with recognition of the need for later statement in wider system. Abuses of interactional procedure are found, on the other hand, in the endeavors now happily fast disappearing, to force classical mechanistic control upon other enterprises of inquiry; and in the many quasi-interactional mixtures of diluted self-actors and pseudo-particles which remain largely in control of inquiry in the psychologies, sociologies, and epistemologies.

III

If we turn now to consideration of the biological fields of inquiry we find that much, but not all, of the old-fashioned self-actional has been discarded. The "vital principle" is an outstanding illustration. Employed until recent decades to mark the distinction of "life" from "mechanism," it proved in the end to amount to nothing more than a sort of honorific naming. What is left of it, when it is not a mere appendage to some irrelevant creed, is mostly found lurking in obscure corners, or entering by way of incidental implication. Today the marvelous descriptions we possess of

---

1 How much need there is for precision in these respects is well indicated by a paragraph in a recent book on the general characteristics of evolution by one of America's most distinguished biologists. His phrasings were first that "the organism develops. . .structures and functions," next that "the organism becomes adapted to. . .conditions," and finally that "evolution produces. . .etc." First the organism is actor, next the environment is actor, and lastly "evolution" is hypostatized to do the work. And all in a single paragraph. Such phrasings indicate, of course, inattention to the main issues involved.

2 Descartes, in his discussion of vision in the first five or six chapters of his *Dioptrique*, gives a fascinating account of sensation as mechanistically produced. It should be specially valuable to modern laboratory workers in the field since it lacks the ordinary protective jargon of professional life, and gets down to the verbal bone of the matter. Descartes was far from liking it in its full application, but in the case of vision, he did not see how he could avoid its apparatus of tubes, rods, and animal spirits.

3 The reader of philosophical specialization may be interested in comparing Kant's substance, causation, and reciprocity. Cassirer's substance and function has interest so far as he develops it. The words "analysis" and "synthesis" suggest themselves, but a cursory survey of discussions in that form has shown little of interest. More suggestive, perhaps, for the philosophical specialist, is the now almost wholly discarded "objective idealism" of men like Green, Bradley, and Caird. The basic terminology of this group of men, using "absolute mind" as a starting point, may be stripped off so as to open the way to see more clearly what they were practically seeking. They show us a full system of activity, a dislike for crude dualisms, and a desire to get rid of such breakages as those the epistemologies capitalize. Along with this went a tolerance for, and even an interest in, the growth, and in that sense the "life," of the system itself. Our own development, of course, in contrast, is of the earth earthy, representing strictly an interest in improved methods of research, for whatever they are worth here and now.

4 Chapter IV, Section II.

living processes provide adequate differentiation from the very different, even if themselves equally marvelous, descriptions of physical processes. The orthogenesis of Henry Fairfield Osborn sought to read "direction" into evolutionary lines with the implication of "control," but more and more today, despite his elaborate exhibits, biologists hold that developed description by itself is a far more useful "interpretation" than any appeal to "directives."[1]

Today we find transactional as well as interactional procedures used in the *details* of physiological and biological inquiry; but for *general* formulations we find little more than preliminary approaches to the transactional. This is seen on the large scale in the heavily theoretical separation that is maintained between the organism and the environment and the attribution of many activities to the former as if in independence.[2] As over against the vitalisms the "cell theory" in its radical form stands as a representative of interactional treatment. Views of the type called "organismic," "organismal," etc., except where they contain reminiscences of the old *self*-actional forms, stand for the transactional approach intra-dermally. Such special names as "organismic" were felt to be needed largely because the word "organic," which could serve as an adjective either for "organism" or for "organ," had been too strongly stressed in the latter usage. Transactional treatment, if dominant, would certainly desire to allot the leading adjective rather to the full living procedure of the organism than to minor specialized processes within it; and if ancillary adjectives were needed as practical conveniences, then it would adapt them to the ancillary inquiries in interactional form.[3] The anticipated future development of transdermally transactional treatment has, of course, been forecast by the descriptive spade-work of the ecologies, which have already gone far enough to speak freely of the evolution of the habitat of an organism as well as of the evolution of the organism itself.

The history of the cell in physiology is of great significance for our purposes. For almost a hundred years after Schleiden and Schwann had systematized the earlier scattered discoveries,[4] the cell was hailed as the basic life unit. Today there are only limited regions of physiological report in which the cell retains any such status. What the physiologist sees in it is not what it is, or is supposed to be "in itself," but what it is within its actual environment of tissues. Some types of inquiry are readily carried on in the form of interactions between one cell and other cells. So far as this type of treatment proves adequate for the work that is in hand, well and good. But other types of inquiry require attention in which the interactional presentation is not adequate, and in which broader statements must be obtained in full transactional form in order to secure that wider conveyance of information which is required. One can, in other words, work with cells independently, or with cells as components of tissues and organs; one can put organs into interaction, or one can study the organs as phases of organisms. Biographical treatment of the "organism as a whole" may or may not be profitable. If it is not, this is usually not so much because it fails to go deeply enough into cellular and organic details, as because it fails to broaden sufficiently the organic-environmental setting and system of report. Its defect is precisely that it centers much too crudely in the "individual" so that whether from the more minute or from the more extended viewpoint, the "individual" is precariously placed in knowledge, except as some reminiscence of an ancient self-actional status is slipped in to fortify it for those who accept that kind of fortification.

The gene, when it was first identified by name and given experimental study "on suspicion," seemed almost as if it held the "secret of life" packed into its recesses. Laboratory routine in genetics has become stylized, and is today easy to carry on in standard forms. The routine experimenter who emerges from its interesting specialties and lifts his voice as a radio pundit is apt to tell us all in a single breath unabashed, that many a gene lives a thousand generations unchanged, and that each new-born organism has precisely two genes of each and every kind in each and every cell, one from each parent. One wonders and hunts his textbooks on grammar and arithmetic. But under wider observation and broader viewpoints we find little of that sort of thing. With gene-position and gene-complex steadily gaining increased importance for interpretative statement, the gene, like many a predecessor that has been a claimant for the rank of element or particle in the universe, recedes from its claims to independence *per se*, and becomes configurational within its setting. The genetic facts develop, but the status of self-actor attributed to the gene at the start proves to be a "fifth-wheel" characteristic: the physiological wagon runs just as well without the little genetic selves—indeed, all the better for being freed from their needless encumbrance.[5] In much the way that in the pre-

---

[1] Osborn's use of the word "interaction" is characteristically in contrast with ours. In developing his "energy" theory in his book *The Origin and Evolution of Life* (London, 1917) he considered action and reaction as usually taking place simultaneously between the parts of the organism, and then added interaction as an additional something connecting nonsimultaneous actions and reactions. Interactions therefore appeared as a new product controlling the others, illustrated by such forms as instincts, functions of co-ordination, balance, compensation, co-operation, retardations, accelerations, etc. The "directing power of heredity" was thus set forth as "an elaboration of the principle of interaction" (pp. 4-6, 15-16).

[2] A prevailing type of logical reflection of this older attitude towards the organism will be found in Carnap's assertion that "It is obvious that the distinction between these two branches [physics and biology] has to be based on the distinction between *two kinds of things* which we find in nature: organisms and nonorganisms. Let us take this latter distinction as granted" (*Logical Foundations of the Unity of Science, International Encyclopedia of Unified Science*, I, No. 1, 45; italics not in the original). As against this rigidified manner of approach, compare the discussion of the organism and behavior in John Dewey's *Logic, the Theory of Inquiry*, (New York, 1938), pp. 31-34.

[3] For the intra-organic transactional observation, with occasional still wider envisionments see the works of J. v. Uexkull, W.E. Ritter, and Kurt Goldstein. Lawrence J. Henderson's book *The Fitness of the Environment*, (New York, 1913) should also be examined. Ritter lists among the most forceful of the earlier American advocates of the "organismal theory" as against the extreme forms of the "cell theory" C.O. Whitman, E.B. Wilson, and F.R. Lillie. Goldstein refers in biology to Child, Coghill, Herrick, and Lashley; in psychiatry to Adolf Meyer and Trigant Burrow; in psychology to the *Gestalt* school; and adds references in philosophy to Dilthey, Bergson, Whitehead, and Dewey. Henderson, with reference to Darwin's "fitness," says that it is a "mutual or reciprocal relationship between the organism and the environment," and again that "the fitness of the environment is both real and unique" (*op. cit.*, p. xi, pp. 267-271). To rate as more fundamental than any of these is the discussion by J.H. Woodger in his *Biological Principles* (London, 1929), a book which is far from having received the attention it deserves. Especially to examine are Chapters V on the theory of biological explanation, VII on structure-function, and VIII on the antithesis between organism and environment.

[4] For the slow process of identifying the cell as distinctive structure, see the discussion by E.B. Wilson in *The Cell in Development and Heredity*, 3d ed. (New York, 1937), pp. 2-4.

[5] Such an entitative superfluity exemplifies the position we are taking throughout our entire discussion: Why retain for the purpose of general interpretation "entities" (i.e., supposititious things-named) that no longer figure in actual inquiry, nor in adequate formulation of its results? Why not get rid of such items when worn out and dying, instead of retaining their sepulchral odor till the passing generations cause even the latter to die away? The split of "nature" into two "realms"—two superfluities—is the instance of such entitative survival to which we elsewhere find it necessary to give ever-renewed consideration.

ceding chapter we employed a recent interpretive book in the physical range, for the significance of its wordings rather than for fixation of authority, we may here cite from Julian Huxley's *Evolution, the Modern Synthesis* (New York, 1942). We are told: "Genes, all or many of them, have somewhat different actions according to what neighbors they possess" (p. 48); "The effect produced by any gene depends on other genes with which it happens to be co-operating". . . . "The environment of the gene must include many, perhaps all other genes, in all the chromosomes" (p. 65); "The discreteness of the genes may prove to be nothing more than the presence of predetermined zones of breakage at small and more or less regular distances along the chromosomes" (p. 48); "Dominance and recessiveness must be regarded as modifiable characters, not as unalterable inherent properties of genes" (p. 83); "To say that rose comb is inherited as a dominant, even if we know that we mean the genetic factor for rose comb, is likely to lead to what I may call the one-to-one or billiard-ball view of genetics". . . . "This crude particulate view. . .of unanalyzed but inevitable correspondence. . .is a mere restatement of the preformation theory of development" (p. 19). We have here a clear illustration of the newer feeling and newer expression for physiology comparable to that of other advanced sciences.[1]

Organisms do not live without air and water, nor without food ingestion and radiation. They live, that is, as much in processes across and "through" skins as in processes "within" skins. One might as well study an organism in complete detachment from its environment as try to study an electric clock on the wall in disregard of the wire leading to it. Reproduction, in the course of human history, has been viewed in large measure self-actionally (as fiction still views it) and then interactionally. Knowledge of asexual reproduction was an influence leading to re-interpretation on a fully racial basis, and recent dairy practices for insemination make the transdermally transactional appearance almost the simple, natural one.

Ecology is full of illustrations of the interactional (where the observer views the organism and the environmental objects as if in struggle with each other); and it is still fuller of illustrations of the transactional (where the observer lessens the stress on separated participants, and sees more sympathetically the full system of growth or change). The issue is not baldly that of one *or* the other approach. It is not even an issue as to which shall be the basic underlying construction—since foundations in general in such questions are much less secure than the structures built upon them.[2] It is, in view of the past dominance of the interactional procedure in most scientific enterprise, rather an issue of securing freedom for wider envisionment.

The development of taxonomy since Linnaeus throws much light on the lines of change. He brought system and order among presumed separates. The schematism of taxonomy has at times sought rigidity, and even today still shows such tendencies among certain diminishing

types of specialists. The very wording of Darwin's title, *The Origin of Species*, was a challenge, however, to the entire procedure of inquiry as it had been carried on for untold years. Its acceptance produced a radical change in taxonomic understanding—a method which rendered imperative observation across extended spatio-temporal ranges of events previously ignored. Taxonomy now tends to flexibility on the basis of the widened and enriched descriptions of advancing knowledge.[3]

The distinction of transactional treatment from interactional—the latter often with surviving traces of the self-actional—may be seen in the way the word "emergence" is often used. At a stage at which an inquirer wants to keep "life," let us say, within "nature," at the same time not "degrading" it to what he fears some other workers may think of "nature"—or perhaps similarly, if he wants to treat "mind" within organic life—he may say that life or mind "emerges," calling it thereby "natural" in origin, yet still holding that it is all that it was held to be in its earlier "non-natural" envisionment. The transactional view of emergence, in contrast, will not expect merely to report the advent out of the womb of nature of something that still retains an old non-natural independence and isolation. It will be positively interested in fresh direct study in the new form. It will seek enriched descriptions of primary life processes in their environments and of the more complex behavioral processes in theirs. It is, indeed, already on the way to gain them. The advances in the transactional direction that we can note in biological inquiries, while, of course, not as yet so striking as those in physical sciences, are nevertheless already extensive and important.[4]

IV

We have considered physiological inquiry in transactional forms and we have mentioned, in passing, other biological inquiries such as those concerning trends of evolution, adaptations, and ecologies. We turn now to the wide ranges of adaptive living called behaviors, including thereunder everything psychological and everything sociological in human beings, and embracing particularly all of their knowings and all of their knowns. If physiology cannot successfully limit itself to the interactions between one component of living process within the skin and other components within it, but must first take a transactional view within the skin,

[1] The results secured by R. Goldschmidt and Sewall Wright should also be compared. For the former, see his *Physiological Genetics* (New York, 1938). For the latter see "The Physiology of the Gene," *Physiological Review*, XXI (1941). T. Dobzhansky and M.F. Ashley Montagu write (*Science*, CV, June 6, 1947, p. 588): "It is well known that heredity determines in its possessor not the presence or absence of certain traits, but, rather, the responses of the organism to its environments."

[2] Georg Simmel, *Soziologie: Untersuchungen über die Formen der Vergesellschaftung*, Zweite Auflage. (Leipzig, 1922), p. 13.

[3] E. Mayr, *Systematics and the Origin of Species*, (New York, 1942). The author (pp. 113-122) offers a highly informative account of the learning and naming issues in biological nomenclature, ranging from the "practical devices" of the systematist to the "dynamic concepts" of the evolutionist, and compares a variety of treatments including the morphological, the genetic, the biological-species, and the criterion of sterility. His discussion moves back and forth between the natural processes of naming and the facts-in-nature to be named. When we come later to discuss characterization, description, and specification, it will be evident (1) that the account can be given from the point of view of either aspect, and (2) that the recognition of this very complementarity is basic to our whole procedure. The twenty-two essays in the volume *The New Systematics* (Oxford, 1940) edited by Julian Huxley also furnish much material for profitable examination in this connection.

[4] For a discussion of the entry of the fundamental field theory of physics into biology, see "A Biophysics Symposium," (papers by R.E. Zirkle, H.S. Burr and Henry Margenau) *The Scientific Monthly*, LXIV (1947), 213-231. In contrast, for typical instances of the abuse of field and other mathematical terminology in psychology, *see* Ivan D. London, "Psychologist's Misuse of the Auxiliary Concepts of Physics and Mathematics," *The Psychological Review*, LI, (1944), pp. 226-291.

following this with further allowance for transdermal process, then very much more strongly may behavioral inquiries be expected to show themselves as transdermally transactional.[1] Manifestly[2] the subjectmatter of behavioral inquiries involves organism and environmental objects jointly at every instant of their occurrence, and in every portion of space they occupy. The physiological setting of these subjectmatters, though itself always transactionally organic-environmental, submits itself to frequent specialized investigations which, for the time being, lay aside the transactional statement. The behavioral inquiries, in contrast, fall into difficulties the very moment they depart from the transactional, except for the most limited minor purposes; their traditional unsolved puzzles are indeed the outcome of their rejecting the transactional view whenever it has suggested itself, and of their almost complete failure to allow for it in any of their wider constructions. The ancient custom, of course, was to regard all behaviors as initiated within the organism, and at that not by the organism itself, but rather by an actor or resident of some sort—some "mind," or "psyche," or "person" attached to it—or more recently at times by some "neural center" imitative of the older residents in character. The one-sided inadequacy of this view is what, so often, has called out an equally one-sided opposed view, according to which the organism is wholly passive, and is gradually moulded into shapes adapted to living by independent environmental conditions, mechanistically treated. Both of these views, one as much as the other, are alien to us.

Summing up positions previously taken, we regard behaviors as biological in the broad sense of that word just as much as are any other events which biologists more immediately study. We nevertheless make a technical—indeed almost a technological—distinction between physiological and behavioral inquiries comparable to the technological distinction between physical and physiological. This is simply to stress the difference in the procedures one must use in the respective inquiries, and to note that the technical physiological statement, no matter how far it is developed, does not directly achieve a technical behavioral statement. One may, in other words, take into account all known physical procedures about the moon, and likewise all known physiological procedures of the human body, and yet not arrive, through any combination or manipulation whatsoever, at the formulation, "rustic, all agape, sees man in moon." This last needs another type of research, still "natural," but very different in its immediate procedures. The distinction is never one of "inherent materials," nor one of "intellectual powers," but always one of

subjectmatter at the given stage of inquiry.[3]

As for the self-actional treatment of behaviors (much of which still remains as a heritage of the past in the laboratories) it is probably safe to say that after physicists knocked the animism out of physical reports, the effect was not to produce a comparable trend in organic and behavioral fields, but just the reverse. All the spooks, fairies, essences, and entities that once had inhabited portions of matter now took flight to new homes, mostly in or at the human body, and particularly the human brain. It has always been a bit of a mystery as to just how the commonplace "soul" of the Middle Ages, which possessed many of the Aristotelian virtues as well as defects, came to blossom out into the overstrained, tense, and morbid "psyche" of the last century or two. To Descartes, whether rightly or wrongly, has fallen much of the blame. The "mind" as "actor," still in use in present-day psychologies and sociologies, is the old self-acting "soul" with its immortality stripped off, grown dessicated and crotchety. "Mind" or "mental," as a preliminary word in casual phrasing, is a sound word to indicate a region or at least a general locality in need of investigation; as such it is unobjectionable. "Mind," "faculty," "I.Q.," or what not as an actor in charge of behavior is a charlatan, and "brain" as a substitute for such a "mind" is worse. Such words insert a name in place of a problem, and let it go at that; they pull out no plums, and only say, "What a big boy am I!" The old "immortal soul" in its time and in its cultural background roused dispute as to its "immortality," not as to its status as "soul."[4] Its modern derivative, the "mind," is wholly redundant. The living, behaving, knowing organism is present. To add a "mind" to him is to try to double him up. It is double-talk; and double-talk doubles no facts.

Interactional replacements for self-actional views have had minor successes, but have produced no generally usable constructions. This is true regardless of whether they have presented the organic inter-actors, which they set over against physical objects, in the form of minds, brains, ideas, impressions, glands, or images themselves created in the image of Newtonian particles. Despite all the fine physiological work that has been done, *behavioral* discussions of vision in terms of images of one kind or another are in about as primitive a state as they were a hundred years ago.[5] The

---

1 See Bentley, "The Human Skin: Philosophy's Last Line of Defense," *Philosophy of Science*, VIII (1941), 1-19. Compare J.R. Firth, *The Tongues of Men* (London, 1937), pp. 19-20. "The air we talk and hear by, the air we breathe, is not to be regarded as simply outside air. It is inside air as well. We do not just live within a bag of skin, but in a certain amount of space which may be called living space which we continue to disturb with some success. And the living space of man is pretty wide nowadays. Moreover we never live in the present." "In dealing with the voice of man we must not fall into the prevalent habit of separating it from the whole bodily behaviour of man and regarding it merely as a sort of outer symbol of inward private thoughts."

2 This is "manifest," of course, only where observation has begun to be free. It is far from manifest where ancient categories and other standardized forms of naming control both the observation and the report.

3 Chapter II, Section No. 4 to No. 8. We do not undertake to make a comparable distinction between psychological and sociological inquiries. This latter distinction is standard among "self-actional" treatments, where the "individual" enters in the traditional exaggeration customary in most interactional treatments. Transactionally viewed, a widening or narrowing of attention is about all that remains indicated by such words as "social" and "individual." As we have elsewhere said, if one insists on considering individual and social as different in character, then a derivation of the former from the latter would, in our judgment, be much simpler and more natural than an attempt to produce a social by joining or otherwise organizing presumptive individuals. In fact most of the talk about the "individual" is the very finest kind of an illustration of isolation from every form of connection carried to an extreme of absurdity that renders inquiry and intelligent statement impossible.

4 The historical differentiations between spirit, soul, and body throw interesting light on the subject. Any large dictionary will furnish the material.

5 For example, Edwin G. Boring in *A History of Experimental Psychology* (New York, 1929), p. 100, speaking of the work of Johannes Mueller, writes: "In general, Mueller remains good doctrine today, although we know that perceived size is neither entirely relative nor entirely proportional to visual angle." This despite the fact that he had ascribed to Mueller the view that "It is the retina that the sensorium perceives directly," and added that "it is plain that, for Mueller, the theory of vision is merely the theory of the excitation of the retina by the optical image." This is perhaps mainly carelessness in statement, but what a carelessness!

interactional treatment, as everyone is aware, entered psychological inquiry just about the time it was being removed from basic position by the physical sciences from which it was copied.[1]

The transactional point of view for behaviors, difficult as it may be to acquire at the start, gains freedom from the old duplicities and confusions as soon as it is put to firm use. Consider ordinary everyday behaviors, and consider them without subjection to either private mentalities or particulate mechanisms. Consider closely and carefully the reports we make upon them when we get rid of the conversational and other conventional by-passes and short-cuts of expression.

If we watch a hunter with his gun go into a field where he sees a small animal already known to him by name as a rabbit, then, within the framework of half an hour and an acre of land, it is easy—and for immediate purposes satisfactory enough—to report the shooting that follows in an interactional form in which rabbit and hunter and gun enter as separates and come together by way of cause and effect. If, however, we take enough of the earth and enough thousands of years, and watch the identification of rabbit gradually taking place, arising first in the subnaming processes of gesture, cry, and attentive movement, wherein both rabbit and hunter participate, and continuing on various levels of description and naming, we shall soon see the transactional account as the one that best covers the ground. This will hold not only for the naming of hunter, but also for accounts of his history back into the pre-human and for his appliances and techniques. No one would be able successfully to speak of the hun*ter* and the hun*ted* as isolated with respect to hun*ting*. Yet it is just as absurd to set up hun*ting* as an event in isolation from the spatio-temporal connection of all the components.

A somewhat different type of illustration will be found in the comparison of a billiard game with a loan of money, both taken as events. If we confine ourselves to the problem of the balls on the billiard table, they can be profitably presented and studied interactionally. But a cultural account of the game in its full spread of social growth and human adaptations is already transactional. And if one player loses money to another we cannot even find words in which to organize a fully interactional account by assembling together primarily separate items. Borrower can not borrow without lender to lend, nor lender lend without borrower to borrow, the loan being a transaction that is identifiable only in the wider transaction of the full legal-commercial system in which it is present as occurrence.

In ordinary everyday behavior, in what sense can we examine a talking unless we bring a hearing along with it into account? Or a writing without a reading? Or a buying without a selling? Or a supply without a demand? How can we have a principal without an agent or an agent without a principal? We can, of course, detach any portion of a transaction that we wish, and secure provisional descriptions and partial reports. But all this must be subject to the wider observation of the full process. Even if sounds on the moon, assuming the necessary physical and physiological waves, match Yankee Doodle in intensity, pitch, and timbre, they are not Yankee Doodle by "intrinsic nature," in the twentieth century, whatever they might have been thought to be in the Dark Ages, or may perhaps be thought to be today by echoistic survivals of those days;[2] they need action if they are to yankeedoodle at all.

When communicative processes are involved, we find in them something very different from physiological process; the transactional inspection must be made to display what takes place, and neither the particles of physics nor those of physiology will serve. Many a flint chip fools the amateur archaeologist into thinking it is a flint tool; but even the tool in the museum is not a tool in fact except through users of such tools, or with such tool-users brought into the reckoning. It is so also with the writing, the buying, the supplying. What one can investigate a thing *as*, that is what it *is*, in Knowledge and in Fact.

V

When we come to the consideration of the knowings-knowns as behaviors, we find Self-action as the stage of inquiry which establishes a knower "in person," residing in, at, or near the organism to do (i.e., to perform, or have, or be—it is all very vague) the knowing. Given such a "knower," he must have something to know; but *he* is cut off from it by being made to appear as a superior power, and *it* is cut off from *him* by being made to appear just as "real" as he is, but of another "realm."

Interaction, in the interpretation of knowings, is a somewhat later stage which assumes actual "real" things like marbles which impinge on certain organic regions such as nerve endings or perhaps even brain segments. Here we still have two kinds of "reals" even though superficially they are brought somewhat closer together in physical-physiological organization. The type of connection is superficial in this case because it still requires a mysticism similar to that used for self-actions to bridge across from the little real "thing" to the little "real" sensation as organic, psychic, or psychologic—where by the word "mysticism" is meant nothing "mystic" itself, but merely some treatment that does not yield to description, and quite often does not want to.

The transactional presentation is that, we believe, which appears when actual description of the knowledge process is undertaken on a modern basis. At any rate it is the kind of presentation which has resulted from our own attempts at direct observation, description, and naming; it is for aid in appraising our results that we have, in this present chapter and the one immediately preceding it, examined comparable procedures in other scientific fields and upon other scientific subjectmatters. The steps we have taken, it will be recalled, are to say that we can not efficiently name and describe except through observation; that the word "knowledge" is too broadly and vaguely used to provide a single subjectmatter for introductory inquiry; that we can select as a compact subjectmatter within "knowledges" generally the region of knowings-through-namings; that here observation at once reports

---

[1] The recent work of Egon Brunswik goes as far, perhaps, on the transactional line as any. He recently ("Organismic Achievement and Environmental Probability," *Psychological Review*, L (1943), 259n) suggested coupling "psychological ecology" with "ecological psychology" in what seemed a functional manner from both sides. In contrast Kurt Lewin, speaking at the same meeting, suggested the name "ecological psychology" but rather for the purpose of getting rid of factors undesirable in his mentalistically fashioned "life-space" than for improvement of system. Clark Hull, also on the same program, holds that organic need and organic environment must be "somehow jointly and simultaneously brought to bear" upon organic movement (the phrasing from his book *Principles of Behavior*, [New York, 1943], p. 18, where he italicizes it) and he bridges across the gap by a series of intervening variables of a fictional, pseudological character.

[2] Echolatry might be a good name to apply to the attitudes of our most solemn and persevering remembrancers of things past—and done with. "Echoist," by the way, is a good word in the dictionaries, and should not be wholly lost from sight.

that we find no naming apart from a named, and no named apart from a naming in such separation that it can be used as direct subject of *behavioral* inquiry—whatever physical or physiological observations we can incidentally make on the namings and the named in provisional separations; that such observations in fused systems *must* be steadily maintained if we are to attain complete behavioral report; and that, if this procedure requires an envisionment for behavioral purposes of space and time that is more extensive and comprehensive than the earlier physical and physiological reports required, such envisionment is then what we must achieve and learn to employ.

The outcome of self-actional and interactional procedures, so far as any competent theory of knowledge is concerned, has been and still is chaos, and nothing more. One can easily "think of" a world without a knower, or perhaps even of a knower without a world to belong to, and to know. But all that "think of" means in such a statement is "to mention in crude language," or "to speak crudely." The hypostatizing fringes of language are what make this "easy." While "easy," it is nevertheless not "possible," if "possible" covers carrying through to a finish, and if "think" means sustained consideration that faces all difficulties, holds to coherent expression, and discards manifestly faulty experimental formulations wherever and whenever it finds them—in short, if the "thinking" strives to be "scientifically" careful. A "real world" that has no knower to know it, has, so far as human inquiry is concerned (and this is all that concerns us), just about the same "reality" that has the palace that in Xanadu Kubla Khan decreed. (That, indeed, has had its reality, but it was not a reality beyond poetry, but in and of it.) A knower without anything to know has perhaps even less claim to reality than that. This does not deny the geologic and cosmic world prior to the evolution of man within it. It accepts such a world as known to us, as within knowledge, and as with all the conditionings of knowledge; but it does not accept it as something superior to all the knowledge there is of it. The attribute of superiority is one that is, no doubt, "natural" enough in its proper time and place, but it too is "of and in" knowledge, not "out of" or "beyond."[1] In other words, even these knowings are transactions of knowing and known jointly; they themselves as knowings occupy stretches of time and space as much as do the knowns of their report; and they include the knower as himself developed and known within the known cosmos of his knowledge.

How does it come to pass, one may ask, if the naming-named transaction as a single total event is basic as we say it is, that historically our language has not long since developed adequate special naming for just this basic process itself? The answer lies partially in the fact that, so far as ordinary conversational customs are concerned, it frequently happens that the most matter-of-fact and commonplace things are taken for granted and not expressly written down. For the rest of the answer, the part that concerns the professional terminology of knowledge and of epistemology, the sad truth is that it has long been the habit of the professionals to take words of the common vocabulary, stiffen them up somewhat by

purported definition, and then hypostatize "entities" to fit. Once given the "entities" and their "proper names," all factual contact, including carefully managed observation, defaults. The names ride the range (in the west) and rule the roost (in the east). All too often the bad names get crowned while the good names get thumbs down. The regions in which this happens are largely those in which procedure is governed by the grammatical split between the subject and the object of the sentence rather than by observation of living men in living linguistic action. In such theoretical interpretations an unobservable somewhat has been shoved beneath behavioral naming, so that "naming as such" is personified into a ready-made faculty-at-large simply waiting for entities to come along for it to name; though most regrettably without that supernatural prescience in attaching the right name to the right animal which Adam showed in the Garden of Eden. The absurdity is thus standardized; after which not merely epistemology but linguistics, psychology, sociology, and philosophy proceed to walk on artificial legs, and wobbly-creaky legs at that. Turn the subject and object of the sentence into disconnected and unobservable kinds of entities, and this is what happens.

The organism, of course, seems in everyday life and language to stand out strongly apart from the transactions in which it is engaged. This is superficial observation. One reason for it is that the organism is engaged in so many transactions. The higher the organism is in the evolutionary scale, the more complicated are the transactions in which it is involved. Man especially is complex. Suppose a man engaged in but one transaction and that with but one other man, and this all his life long. Would he be viewed in distinction from that transaction or from that other man? Hardly. Much analysis, if an analyzer existed, would at least be necessary to separate him out as a constituent of what went on. A "business man" would not be called a business man at all if he never did any business; yet the very variety of his other transactions is what makes it easy to detach him and specialize him as a "business man." Consider the great variety of his other transactions, and it becomes still easier to make "a man" out of him in the sense of an "essence" or "substance," or "soul" or "mind," after the pattern demanded by the general noun. He comes thus, in the end, to be considered as if he could still be a man without being in *any* transaction. It is precisely modern science which reverses this process by driving through its examinations more thoroughly. When actions were regarded as separate from the actor, with the actor regarded as separate from his actions, the outcome, individually and collectively, was to bring "essence" into authority. The procedures of Galileo, Newton, and Darwin, steadily, bit by bit, have destroyed this manner of observation; and the procedures which must follow hereafter will complete it for the most complex human behavioral activities. They will reverse the old processes and bring the transactions into more complete descriptive organization without the use of either self-actional powers, or interactional "unalterable particles" behind them.[2]

---

1 Many a man is confident in saying that he knows for certain (and often with a very peculiar certainty) what is behind and beyond his personal knowings. We are well aware of this. Nevertheless, we do not regard it as good practice in inquiry when dependable results are sought.

2 A discussion of "The Aim and Progress of Psychology" by Professor J.R. Kantor (*American Scientist*, XXXIV, [1946], 251-263) published after the present paper was written, may be examined with profit. It stresses the modern "integrated-field stage" of science, with special reference to psychology, in contrast with the earlier "substance-property" and "statistical-correlation" stages.

VI.
SPECIFICATION

I

HAVING discussed at length the status of those events of the known and named world which we have styled "transactions," we proceed now to examine that linguistic activity through which Transaction is established: namely, Specification.[1]

Specification, in our provisional terminology, is the most efficient form of Designation, and Designation is that behavioral procedure of naming which comprises the great bulk of linguistic activities, and which, in the line of evolution, lies intermediate between the earlier perceptional activities of Signaling and the later and more intricately specialized activities of Symboling.

It will be recalled that we have inspected Fact most generally as involving and covering at once and together the naming process and the "that" which the naming is about. The choice of the word "fact" to name the most general transaction of "knowledge," was made because in practically all of its many varied uses this word conveys implications of the *being known* along with those of the *what* that is known; moreover, Fact applies to that particular region of the many regions covered by the vague word "knowledge" in which namings are the prominent feature. It is in this region that "knowledge" is most generally considered to be "knowledge of existence" in perhaps the only sense of the word "existence" having practical utility—that, namely, in which the existence is being affirmed with a considerable measure of security as to its details.[2]

Taking Fact as inclusive of both the naming and the named in common process, we adopted Designation for the naming phase of the transaction, and Event for the phase of the named. Events (or "existences," if one is prepared to use the latter word very generally and without specialized partisan stress) were distinguished as Situations, Occurrences, and Objects; and Objects were then examined in their presentations as Self-actions, Inter-actions, and Trans-actions—all of this, of course, not as formal classification, but as preliminary descriptive assemblage of varieties. The "self," "inter," and "trans" characteristics appear in Situations and Occurrences as well as in Objects, but it is in the more determinate form of Objects that the examination can most closely be made.

When we now turn to the examination of the processes of Designation we must on the one hand place designation definitely within the evolutionary range of behaviors; on the other hand we must examine the stages of its own development, leading up to Specification as its most efficient and advanced stage. The first of these tasks is necessary because a disjunction without a conjunction is usually more of a deception than of a contribution; but the pages we give to it furnish no more than a sketch of background, the further and more detailed treatment being reserved for a different connection.[3] In the second of these tasks we shall differentiate Cue, Characterization, and Specification as the three stages of Designation, and shall give an account of Specification freed from the hampering limitations of the symbolic procedures of Definition.[4]

II

Designation, as we have said, covers naming. "Naming" would itself be an adequate name for the processes to be considered under Designation—and it would be our preferred name—if the name "name" itself were not so tangled and confused in ordinary usage that different groups of readers would understand it differently, with the result that our own report would be largely *mis*-understood. For that reason, before going further, we shall insert here a few paragraphs about the common understandings as to "name," and as to their difference from the specialized treatment we introduce as Designation. Some of these positions have been discussed before, and others will be enlarged upon later.

Naming we take as behavior, where behavior is process of organism-in-environment. The naming type of behavior, by general understanding so far as present information goes, is one which is characteristic of *genus homo* in which almost alone it is found. Except as behavior—as living behavioral action—we recognize no name or naming whatever. Commonly, however, in current discussions, name is treated as a third type of "thing" separate both from organism and from environment, and intermediate between them. In colloquial use this makes little difference. But in the logics and epistemologies, a severed realm of phenomena, whether explicit or implicitly introduced, matters a great deal. Such an intervening status for "name," we, by hypothesis, reject.

Name, as a "thing," is commonly spoken of as a tool which man or his "mind" uses for his aid. This split of a "thing" from its function is rejected. Naming is before us not as a tool (however it may be so described from

---

[1] We shall continue, as heretofore, to capitalize some of our main terms where stress on them seems needed, and particularly where what is in view is neither the "word" by itself nor the "object" by itself, but the general presentation of the named-as-in-naming. We shall continue also the occasional use of hyphenization as a device for emphasis.

[2] For naming and knowing see Chapter II, Section III. For comment on "existence" see Chapter XI.

[3] Of psychology today one can sharply say (1) if its field is behavior, and (2) if human behavior includes language, then (3) this behavioral language is factor in all psychology's presentations of assured or suspected fact, and (4) psychological construction today shows little or no sign of taking this linguistic factor into account in its double capacity of being itself psychologic fact and at the same time presenter of psychologic fact. The problem here is, then, the terminological readjustment of psychological presentation to provide for this joint coverage of the naming and the named in one inquiry.

[4] The word "definition" is used throughout the present chapter, as in preceding chapters, to stand for procedures of symboling as distinct from those of designating. This choice was made mainly because recent developments of technique, such as those of Tarski, Carnap, and symbolic logicians generally, have either adopted or stressed the word in this sense. After the present chapter, however, we shall abandon this use. In preparing our succeeding chapter, to appear under the title "Definition," we have found such complex confusions that misunderstanding and misinterpretations seem to be inevitable, no matter how definition is itself "defined." The effect of this change will be to reduce the word "definition" from the status of a "specification" to that of a "characterization" as this distinction is now to be developed. Progress towards specification in the use of the word "definition" is, of course, what is sought, no matter how unattainable it may seem in the existing logical literature.

limited viewpoints), but as behavioral process itself in action, with the understanding, nevertheless, that many forms of behavior, and perhaps all, operate as instrumental to other behavioral processes which, in turn, are instrumental to them.

Treatments of name as thing or tool accompany (or are accompanied by; the point is not here important) the splitting of "word" from "meaning"—"word," whether crudely or obscurely, being taken as "physical," with "meaning" as "mental." The split of a sign-vehicle from a sign, stressed as one of maximum theoretical importance in certain recent efforts at construction in this general field, is merely the old rejected split in a new guise. Under the present approach such a treatment of name, or of any other word, is regarded as deficient and inefficient, and is therefore banished.[1]

Under the above approach naming is seen as itself directly a form of knowing, where knowing is itself directly a form of behavior; it is the naming type of knowing behavior (if one wishes to widen the scope of the word "knowledge"), or it is the distinctive central process of knowledge (if one prefers to narrow the scope of the word "knowledge" thus far). Our hypothesis is that by treating naming as itself directly knowing, we can make better progress than in the older manners.

Naming does things. It states. To state, it must both conjoin and disjoin, identify as distinct and identify as connected. If the animal drinks, there must be liquid to drink. To name the drinking without providing for the drinker and the liquid drunk is unprofitable except as a tentative preliminary stage in search. Naming selects, discriminates, identifies, locates, orders, arranges, systematizes. Such activities as these are attributed to "thought" by older forms of expression, but they are much more properly attributed to language when language is seen as the living behavior of men.[2] The talking, the naming, is here oriented to the full organic (currently "organismic" or "organismal") process rather than to some specialized wording for self, mind, or thinker, at or near, or perhaps even *as*, a brain.

All namings are positive. "Not-cow" is as much positive naming as is "cow"—whatever the cow itself might think about that. The cow's local point of view does not govern all theoretical construction. If the negatives and the positives alike stand for something, this something is as thoroughly "existential" in the one case as in the other.

Written names are behavioral process as much as as spoken names are. Man's diminishment of the time-period, say, to the span of his day or of his life, does not govern decision as to what is behavioral or what is not.

The "what" that is named is no fiction. "Hercules" was a name in its time for something existently cosmic or cultural—not as "reality at large," but always as "specified existence." "Sea-serpents" and "ghosts" have played their parts, however inactive they may be as existential namings today. Trilobites are inactive, but they nevertheless made animal history.

These viewpoints that we have set down are not

---

[1] The issue here is not one of personal "belief," whether pro or con. It is one of attitude, selection, decision, and broader theoretical formulation. Its test is coherence of achievement. *Practical* differentiations of specialized investigation upon half a dozen lines with respect to word, or along half a dozen other lines with respect to word-meaning, are always legitimate, and often of great practical importance. For some account of the abuses of sign and sign-vehicle see Chapter XI.

[2] However, if language is not regarded as life-process by the reader, or if thought is regarded as something other and higher than life-process, then the comparison in the text will not be acceptable to him.

separates fortuitously brought together. They are transactional. They form, for this particular region of inquiry, the substance of what is meant by "transaction" in our use. That they will not "make sense" from the inter-actional point of view, or from the self-actional point of view, is only what is to be expected. They make sufficient sense as fact to be usable by us in hypothesis, and the test of their value will be in the outcome of such use.

III

If we are to examine Designations as behaviors, we must first establish the characteristics of behavior as we see it. That the name "behavior," however elsewhere used, can, in biological studies, be applied without misunderstanding to certain adjustmental types of animal activities, will hardly be disputed. That a behavioral statement in this sense is not itself directly a physiological statement, nor a physiological statement itself directly a behavioral one, will likewise hardly be disputed, as matters stand today, however much one may hope or expect the two forms of statement to coalesce some time in the future, or however valuable and indeed indispensable the primary understanding of the physiological may be for any understanding of the behavioral. Extend either form of statement as far as you wish, holding it closely within its own vocabulary; it will, nevertheless, not directly convert itself into the other. Moreover attempts to limit the application of the word "behavior" to the overt muscular and glandular activities of an organism in the manner of a generation ago have not proved satisfactory. Too much development in terms of the participation of the "whole organism"—or, better said, of "the rest of the organism" has of late been made; and recent attempts to revive the older narrow construction for the interpretation of knowledge have had misfortunes enough to serve as ample warnings against such programs.

We shall take the word "behavior" to cover all of the adjustmental activities of organism-environment, without limiting the word, as is sometimes done, to overt outcomes of physical or physiological processes. This latter treatment involves too crude a disregard of those factual processes which in older days were hypostatized as "mental," and which still fall far short of acquiring "natural" description and reports. In the older psychologies (and in many still with us), whether under "mental" or "physiological" forms of statement, the distinction of the typically human behaviors from non-human and also of behaviors generally from the non-behavioral, was made largely in terms of "faculties" or "capacities" assumed to be inherent in the organism or its running mate, the "mind" or "soul." Thus we find "purposiveness" stressed as the typically "animal" characteristic; or accumulations of complexly-interrelated habits, or certain emotional, or even moral, capacities. In our case, proceeding transactionally, nothing, so far as we know, of this "capacitative" manner of statement remains in stressed use at critical points of research. Regarding behaviors as events of organism-environment in action, we shall find the differentiation of behavioral processes (including the purposive) from physical or physiological to rest upon types of action that are observable directly and easily in the full organic-environmental locus.

Sign: Developing behaviors show indirections of action of types that are not found in physical or physiological processes. This is their characteristic. The word "indirection" may, no doubt, be applicable to many physiological processes as compared with physical, but it is not the word by itself that is important. The particular type of indirection that is to be found in behaviors we shall call *Sign*, and we shall so use the word "sign" that where sign is found we have behavior, and where behavior occurs sign-process is involved. This is an extremely broad usage, but we believe that, if we can make a sound report on the factual case, we are justified in applying the word as we do.[1]

At a point far down in the life-scale Jennings identified sign as a characteristic behavioral process forty years ago. He was studying the sea-urchin, and remarked that while it tends to remain in dark places and light is apparently injurious to it, "yet it responds to a sudden shadow falling upon it by pointing its spines in the direction from which the shadow comes." "This action," Jennings continues, "is defensive, serving to protect it from enemies that in approaching may have cast the shadow. The reaction is produced by the shadow, but it *refers*, in its biological value, to something behind the shadow."[2]

This characteristic of Sign is such that when we have followed it back in protozoan life as far as we can find traces of it, we have reached a level at which we can pass over to the physiological statement proper and find it reasonably adequate for what we observe as happening. This makes the entry of the "indirection" which we call "sign" a fair border-line marker between the physiological and the behavioral. The sign-process characterizes perceptions all the way up the path of behavioral evolution; it serves directly for the expanded discussion of differentiated linguistic representation; it deals competently with the "properties" and the "qualities" that have for so long a time at once fascinated and annoyed philosophers and epistemologists; it can offer interpretation across all varieties of expressive utterance up to even their most subtle forms. All these phases of behavior it can hold together simply and directly.

Having adopted an interchangeability of application for sign and behavior, our position will be as follows: If we fall away from it, that fact will be evidence of defect in our development; if we fall seriously away that will be indication of an insecurity in our basic hypotheses themselves; if we can maintain it throughout—not as *tour de force* but as reasonably adequate factual statement—this will furnish a considerable measure of evidence that the manner of construction is itself sound.

We have indicated that behavior is envisaged transactionally and that sign itself is a transaction. This means that in no event is sign in our development to be regarded as consisting of an "outer" or detached "physical" thing or property; and that in no event is it to be regarded as the kind of an ear-mark that has no ear belonging to it, namely, as a detached "mental thing." Sign, as we see it,

will not fit into a self-actional interpretation at all; nor will it fit into an interactional interpretation.

If this is the case an important question—perhaps the most important we have to face—is the exact location of sign. Precisely *where* is the event that is named when the name "sign" is applied? Sign is process that takes place only when organism and environment are in behavioral transaction. Its locus is the organism and the environment, inclusive of connecting air, electrical and light-wave processes, taken all together. It is these in the duration that is required for the event, and not in any fictive isolation apart from space, or from time, or from both. A physiologist studying breathing requires air in lungs. He can, however, temporarily take for granted the presence of air, and so concentrate his own attention on the "lungs"—on what *they* do—and then make his statement in that form. He can, that is, for the time being, profitably treat the transaction as interactional when the occasion makes this advantageous. The student of the processes of knowings and knowns lacks this convenience. He can not *successfully* make such a separation at any time. Epistemologies that isolate two components, that set them up separately and then endeavor to put them together again, fail; at least such is our report on the status of inquiry, and such our reason for proceeding transactionally as we do.

It is evident that time in the form of clock-ticks and space in the form of foot-rules yield but a poor description of such events as we report signs to be. Treat the events as split into fragments answering to such tests as clocks and rules may give, and you have a surface account, it is true, but one that is poor and inadequate for the full transaction. Even physics has not been able to make the advances it needs on any such basis. The spatial habits of the electrons are bizarre enough, but they are only the prologue. When physicists find it practicable to look upon 92 protons and 142 neutrons as packed into a single nucleus in such a way that the position of each is "spread out" over the entire nuclear region, certainly it should be permissible for an inquirer into man's behavioral sign-processes to employ such pertinent space-forms with pasts and futures functioning in presents, as research plainly shows to be necessary if observation is to be competent at all. At any rate any one who objects to freedom of inquiry in this respect may properly return to his own muttons, for subsequent proceedings will not interest him at all.

Taking Sign, now, as the observable mark of all behavioral process, and maintaining steadily transactional observation in replacement of the antique fixations and rigidities, we shall treat Signal, Designation, and Symbol as genera of signs, and so of behaviors. Similarly within the genus Designation, we may consider Cue, Characterization, and Specification as species. In this we shall use "genus" and "species" not metaphorically, but definitely as natural aids to identification.

Signal: All the earlier stages of sign up to the entry of language we group together under the name "signal." Signal thus covers the full sensori-manipulative-perceptive ranges of behavior, so far as these are unmodified by linguistic behaviors. (Complex problems of linguistic influencings will surely have to be faced at later stages of inquiry, but these need not affect our terminology in its preliminary presentation.) Signals like all signs are transactional. If a dog catches sight of the ear of a rabbit and starts the chase, signal behavior is involved. The signal is not the rabbit's ear for itself, nor is it the identification mechanism in the dog; it is the particular "fitness" of

---

[1] The *Oxford Dictionary* has twelve main dictionary definitions of sign, and a number of subdivisions. The *Century* has eleven. In modern discussion the uses are rapidly increasing, but no one usage is yet fixed for the field we are at work in. Usages range from saying that sign is a form of energy acting as a stimulus, followed by the application of the word for almost any purpose that turns up, to presenting it as a product of mind-proper. No one use can claim the field till it has been tried out against others; and certainly no candidate should even enter itself until it has been tried out in its own backyard and found capable of reasonably coherent usage.

[2] H.S. Jennings, *Behavior of the Lower Organisms* (New York, 1906), p. 297. Jennings has himself never made a development in terms of sign, despite the highly definite description he so early gave it. Karl Bühler, who was one of the first men to attempt a broad use of a sign-process for construction, quoted this passage from Jennings in his *Die Krise der Psychologie*, (Jena, 1927), p. 80, at about the time it began to attract attention among psychologists in the United States.

environment and organism—to use Henderson's word; it is the actual fitting in the performance. Pavlov's conditioned reflex, as distinguished from simple reflexes that can tell their stories directly in terms of physical-physiological excitations and reactions, is typical signaling, and Pavlov's own use of the word "signal" in this connection is the main reason for our adoption of it here.[1] The Pavlov process must, however, be understood, not as an impact from without nor as a production from within, but as a behavioral event in a sense much closer to his own descriptions than to those of many comparable American laboratory inquiries. It must be a feature of the full stimulus-response situation of dog and environmental objects together. If we take bird and berry in place of dog and rabbit, berry is as much a phase of signal as bird is. Divorce the two components—disregard their common system—and there *is* no signal. Described in divorcement the whole picture is distorted. Signaling is always action; it is event; it occurs; and only as occurrence does it enter inquiry as subjectmatter. It is not only transactional as between dog and rabbit and between bird and berry, but each instance of it is involved in the far wider connections of the animal's behaviors. No such fact is ever to be taken as an isolate any more than one animal body is to be taken as an isolate from its genus, species, race, and family. If one takes either the sensory, the motor, or the perceptional as an isolate, one again distorts the picture. Each case of signal, like every other case of sign, is a specific instance of the continued durational sign-activity of life in the organic-environmental locus. The motor phase has its perceptive-habitual aspect, and the perceptive phase has its motor aspect, with training and habit involved.

IV

Designation: Designation develops from a basis in Signaling. Signaling is organic-environmental process that is transactional. Designation in its turn is transactional organic-environmental process, but with further differentiation both with respect to the organism and with respect to the environment. With respect to the organism the "naming" differentiates; with respect to the environment the "named" differentiates. On neither side do we consider detachment as factual. The organism is not taken as a "capacity" apart from its environmental situation. The environment is not taken as "existing" in detachment from the organism. What is "the named" is, in other words, not detached or detachable environmental existence, but environment-as-presented-in-signaling-behavior. In other words, signalings are the "named," even though the namer in naming develops a language-form presumptively presenting an "outer" as detachable. Neither "naming" nor "named" under our procedure is taken as either "inner" or "outer," whether in connections or separations. The process of designation becomes enormously more complex as it proceeds; in it environmental determinations and namings unfold together. We make our approach, however, not in terms of the late complex specializations, but instead in terms of the growth in its

early stages. The *what* that is assumed in the earliest instances is, then, not a *thing* in detachment from men (as most logicians would have us believe); much less is it some "ultimate reality," "provisional reality," "subsistence," or metaphysical "existence" (whatever such "things" may be taken to be). What is "cued" in the earliest forms of naming is some action-requirement within the sign-process, that is, within behavior. When one of a pair of birds gives a warning cry to his mate, or when a man says "woof" to another man as sign of bear-trail or bear-presence, it is behavior that is brought in as named; it is transactionally brought in, and is transaction itself as it comes. One can go so far back along the evolutionary line that the bird-call or the "woof" or some more primitive predecessor of these has, under such observation and report as we can make, not yet reached so much as the simplest differentiating stage with respect to "naming and named." But when the differentiating stage *is* reached, then the "named" that differentiates within the behavior is an impending behavioral event—an event in process—the environing situation included, of course, along with the organism in it. Both bird-call and woof indicate something doing, and something to be done about it.

The transactional locus of a designation in one of its earliest forms is very narrow—just the range of the creatures in communication, and of the sensori-manipulative-perceptive events directly presented in the communication. When and as the designation-event develops more complexity, the locus widens. Intermediate stages of namings intervene. Some of them push themselves temporarily into the foreground of attention, but even so are in fact members of a total inclusive transaction, and are given isolation and independence only in theories that depart from or distort observation. One may name a law, say the price-control act, without ever putting one's "finger" on it. In fact our experts in jurisprudence talk indefinitely about a statutory or other law without being able to specify what any law *is*, in a way equivalent to a direct "fingering." And while, in this talking and writing about the law, limiting intervening namings become temporarily the focus of direct attention, the *what* that is named is the law in its entire reach.

It is in this transition to more and more complex designations that the descriptive accounts are most likely to go astray. The cry "Wolf" is quickly brought to rest through actions that yield a "yes" or "no." The cry "Atomic Bomb" is evidently on a different level. It is in the cases of highly developed designations that it is most necessary to take our clue from the simpler cases so as to be firmly and solidly aware that name can not be identified as a process in an organism's head or "mind," and that the named can not be identified with an object taken as "an entity on its own account"; that the naming-transaction has locus across and through the organisms-environments concerned in all their phases; and that it is subject to continued development of indefinite scope so that it is always in transit, never a fixture.

We shall give attention to the two less complex stages of designation, namely, Cue and Characterization, merely far enough to lead up to the presentation of Specification as the perfected (and ever-perfecting) stage of naming, and so as to provide the ground for its differentiation from symbol and definition. So far as the terminology used is concerned, it may seem strange to group the thing-name, Cue, with the action-names, Characterization and Specification, as we are doing. But since all designations are designations in and of behavioral

---

[1] Allowing for a difference in the forms of expression shown by the use of such a word as "relation," Bartley and Chute in *Fatigue and Impairment in Man* (New York, 1947) plan to differentiate the word "signal" along very much the lines of our text. They write (p. 343): "Neither items in the physical world nor perceived items are themselves signals. A signal merely expresses the relation between the two, as determined by the functional outcome." They, however, still retain the word "stimulus" separately for the "physical items from which the signals arise."

activities, the preliminary noun-form used does not greatly matter. We might, perhaps, set up Cue, Common Noun, Term[1] as one series of names to range the field; or, as an alternative, we might use Ejaculation, Characterization, and Specification. Provided the behavioral transactions are taken as names with respect to developing action, the selection of terminology may well be left open for the present.

Cue: By Cue is to be understood the most primitive language-behavior. Wherever transactional sign on the signal level begins to show differentiation such that out of it will grow a verbal representation of any signal process, we have the beginnings of Cue. It is not of prime importance whether we assert this as first arising on the subhuman animal level, and say that language comes into being there, or whether we place the first appearance of true language among men. The general view is that the regions of cue, in contrast with those of signal, are characteristically communicative, but this issue, again, is not one of prime importance. Such questions lie in the marginal regions which modern science (in distinction from older manners of inquiry) does not feel it necessary to keep in the forefront of attention. Life is life, whether we can put a finger on the line that marks the boundary between it and the non-living, or whether a distinction here is far beyond our immediate powers; and much energy will be saved if we postpone such questions till we have the facts. Biology learned about its marginal problems from the viruses and could have got along just as well or better without the oceans of opinionative disputation over the "vital principle" in older days.

The illustrations of designation above were mainly from the lower levels and will serve for cue. Cue, as primitive naming, is so close to the situation of its origin that at times it enters almost as if a signal itself. Face-to-face perceptive situations are characteristic of its type of locus. It may include cry, expletive, or other single-word sentences, or any onomatopoeic utterance; and in fully developed language it may appear as an interjection, exclamation, abbreviated utterance, or other casually practical communicative convenience. Though primarily name grown out of signal, it may at times have the guise of more complex name reverted to more primitive uses. We may perhaps say that cue is signal with focal localization shifted from organism-object to organism-organism, but with object still plain in reach.

The transition from signal to cue may be indicated in a highly artificial and wholly unromantic way through a scheme which, fortunately, is devoid of all pretense to authority as natural history. On the branches of a tree live three snakes protectively colored to the bark, and enjoying vocal chords producing squeaks. Transients at the tree are squeaking birds: among them, A-birds with A-squeaks that are edible by snakes, and B-birds with B-squeaks that pester snakes. Bird-squeaks heard by snakes enter as signals, not as bird-squeaks alone, nor as snake-heard sounds alone, but strictly as events in and of the full situation of snake-bird-tree activity. Snake-squeaks, onomatopoeically patterned, are cues between snakes—primitive verbalisms we may call them, or pre-verbalisms. The evolutionary transition from bird-squeaks warning snakes, to snake-squeaks warning snakes is not one from external signs to internal signs, nor from the automatic to the mental, but just *a slight shift in the*

stresses of the situation. When cue appears, we have a changed manner of action. When cue is studied transactionally, we change our stress on these subject-matters of inquiry. Our change is slight, and one of growth in understanding, elastic to the full development of inquiry. It is not a breakage such as a self-actional account produces, nor even a set of minor breakages such as interactional treatment involves. The change to transactional treatment permits descriptions such as those on which perfected namings are built up.

The cue-stage of designation was not mentioned in our sketch of terminology in Chapter II, our arrangement there being designed to give preliminary stress to the distinction of definition from specification. Signal was chosen as a name for the perceptive-manipulative stage of sign process largely on the basis of Pavlov's use of it. Cue was chosen for its place because all "dictionary definitions" (except one or two that lack the sign character altogether) make it verbal in nature. It may be, however, especially in view of Egon Brunswik's recent studies,[2] that the words "cue" and "signal" could better be made to shift places. Our purpose here is solely to establish at once the manners of disjunction and of conjunction of cue and signal, and an interchange of names would not be objectionable.

Characterization: Out of cue there develops through clustering of cues—i.e., through the growth of language—that type of naming which makes up almost all of our daily conversation. It is the region of evolving description, which answers well enough for current practical needs, but is limited by them in scope. The wider the claims it makes, the less value it has. It is the region where whale in general is a fish because it lives in the water like any "proper" fish. Words cease to be of the type of "this," or "that," or "look," or "jump quick," and come to offer a considerable degree of connection among and across environmental situations, occurrences, and objects. The cues overlap and a central cue develops into a representative of a variety of cues. The interconnections are practical in the colloquial sense of everyday life. Horse is named with respect to the way one does things with and about horses, and with respect to the way horse does things with and about us. The noun enters as an extension of the pronoun, which is a radically different treatment from that of ordinary grammar. The characterizations move forward beyond the "immediately present" of the cues as they widen their connections, but for the most part they are satisfied with those modes of linguistic solvency which meet the requirements set by an immediately present "practical" communicative situation.

The first great attempt to straighten out the characterizations and bring them under control was perhaps made by the Greek sophists, and this led the way to Aristotle's logic. The logics that have followed Aristotle, even those of today that take pride in calling themselves non-Aristotelian, are still attempting to bring characterizations under the control of rules and definitions—to get logical control of common namings. All theories of linguistics, at least with a rare exception or two, make their developments along these lines. In the region of characterization the

[1] Decision as to the use of the word "terms" is one of the most difficult to make for the purposes of a safe terminology. Mathematics uses the word definitely, but not importantly. Logics, as a rule, use it very loosely, and with much concealed implication.

[2] Egon Brunswik, "Organismic Achievement and Environmental Probability," *Psychological Review*, L (1943), 255. See also Tolman and Brunswik, "The Organism and the Causal Texture of the Environment," *Psychological Review*, XLII (1935), 43. Both cue and signal overlap in ordinary conversational use, a fact of interest here. George H. Mead occasionally used "signal" in much the region where we use "cue." Mead's treatment of the animal-man border regions will be of interest (*Mind, Self, and Society*, [Chicago, 1934], pp. 61-68, 81, *et al.*).

view arises that if naming occurs there must exist a "some one" to do the naming; that such a "some one" must be a distinctive kind of creature, far superior to the observed world—a creature such as a "mind" or personified "actor"; and that for such a "some one" to give a name to "anything," a "real" thing or "essence"[1] must exist somewhere apart and separate from the naming procedure so as to get itself named. (The word "must" in the preceding sentences merely reports that where such practical characterizations are established they think so well of themselves that they allege that every form of knowledge "must" adapt itself to them.) Alien as this is from modern scientific practice, it is, nevertheless, the present basis of most linguistic and logical theory and of what is called "the philosophy of science."[2] It is in this stage that namings and the named get detachable existences assigned to them by reflecting or theorizing agents, their immediate users being, as a rule, protected against this abuse by the controls exercised in conversational exchange by the operative situation directly present to those who participate in the oral transaction. Indeed, one may go so far as to doubt whether the distorted theory would have arisen if it had not been for the development of written documents with their increasing remoteness from determination by a directly observed situation. Given the influence of written, as distinct from spoken, language, it is dubious whether theoretical or philosophical formulations could have taken any form other than the one they now have, until a high degree of the development of the methods of inquiry now in use in advanced subjects had provided the pattern for statement in the form we term specification as complementary with transaction.

Description: Before passing to specification it will be well to attend to the status of names and naming with respect to descriptions. Phrasings develop around namings, and namings arise within phrasings. A name is in effect a truncated description. Somewhat similarly, if we look statically at a stable situation after a name has become well established, a description may be called an expanding naming. The name, in a sense which is useful if one is careful to hold the phrasing under control, may be said to name the description, and this even more properly at times than it is said to name the object. For naming the object does not legitimately, under our approach, name an object unknown to the naming system; what it names is the object-named (with due allowance for the other forms of knowing on the sensori-manipulative-perceptive level of signal); and the object-named is far more fully set forth in description than by the abbreviated single word that stands for the description. Beebe[3] mentions a case in which a single word, Orthoptera, in the Linnaean scheme precisely covered 112 words which Moufet had required for his description a hundred years earlier. The process of description begins early and is continuous while naming proceeds in its own line of growth, whatever arbitrary substitutes for it may at times be sought. Take two yellow cats and one black cat. Some little while afterwards, culturally speaking (a few tens of thousands of years, perhaps) primitive man will mark the color distinction, not as color for itself, but as color in contrast with other color. Put his color-naming in system with

cat-naming, and you have the beginnings of description. "Cat" begins now to stand not merely for anti-scratch reaction, or for cat-stew, but for an organization of words into description. Bertrand Russell and several of his contemporaries have had a great deal of trouble with what they call "descriptions" as compared with what Russell, for instance, calls "logical proper names." Fundamentally Russell's "proper names" are analogues of the cue—reminiscent of primeval yelps and of the essences and entities that descend from them, to which it is that Russell wishes to reduce all knowledge. At the far extreme from his form of statement stands specification as developed out of characterization by expanding descriptions which in the end have attained scientific caliber. It is to Specification rather than to survivals of primitive catch-words that our own procedure directs itself in connection with progress in knowledge. Our most advanced contemporary cases of scientific identification should certainly not be compelled to comply with a demand that they handcuff themselves "logically" to a primitive type of observation and naming, now scientifically discarded.

V

Specification: Specification is the type of naming that develops when inquiry gets down to close hard work, concentrates experimentally on its own subjectmatters, and acquires the combination of firmness and flexibility in naming that consolidates the advances of the past and opens the way to the advances of the future. It is the passage from conversational and other "practical" namings to namings that are likewise practical—indeed, very much more practical—for research. The whale ceases to be a fish on the old ground that it lives in water and swims, and becomes established instead as a mammal because of characterizations which are pertinent to inquiries covering wide ranges of other animals along with the whale, bringing earlier "knowns" into better system, and giving direction to new inquiries. "Fish," as a name, is displaced for whale, not because it fails to conform to "reality," but because in this particular application it had been limited to local knowings which proved in time to be obstructive to the further advance of inquiry in wider ranges. Scientific classificatory naming, as it escapes from the bonds of rigidity, illustrates the point in biology. In physics it is illustrated by the atom which ceases to be a little hard, round, or cubical "object" that no one can make any smaller, harder, or rounder, and has become instead a descriptive name as a kind of expert's shorthand for a region of carefully analyzed events. Incidentally this procedure of specification is marked by notable inattention to the authority so often claimed for ancient syllogistic reasoning carried on in patterns fixed in advance.[4] The surmounting of the formal or absolute space and time of Newton, and the bringing of space and time together under direct physical description, is the outstanding illustration of the work of specification in recent physics, and our account in a preceding chapter[5]

---

[1] The recent revival of the word "essence" in epistemological discussion, as in Santayana's writings, is of itself convincing evidence of this statement.

[2] The difficulties in which the logics find themselves are examined in Chapters I and VIII.

[3] William Beebe, Editor, *The Book of Naturalists*, (New York, 1944), p. 9.

[4] The issues, of course, are of the type so long debated under the various forms of contrast between what is called empiricism and what is called rationalism, these names merely marking the condition surrounding their entry into specialized modern prominence. Such issues are, however, held down by us to what we believe we are able to report under direct observation of the connections between language and event in current scientific enterprise in active operation.

[5] Chapter IV. For a complete account, of course, a full appraisal of the participation of mathematics would be necessary; that is, of the system of organization of symbol with name.

of the advance of transactional presentation of physical phenomena might in large part have been developed as a report upon specification. The developmental process in "science" is still far from complete. In biological work, organism and object still often present themselves in the rough as characterizations without specification, even though much specification has occurred in the case of physiological inquiry. In psychological and societal subjectmatters procedures are even more backward. It is astonishing how many workers in these latter fields relegate all such issues to "metaphysics" and even boast that they are "scientific" when they close their eyes to the directly present (though unfortunately most difficult) phases of their inquiry.

In our preliminary account of naming we have said that it states and connects. Cue states and characterization connects. Specification goes much further. It opens and ranges. By the use of widened descriptions it breaks down old barriers, and it is prepared to break down whatever shows itself as barrier, no matter how strongly the old characterizations insist on retention. What it opens up it retains for permanent range from the furthest past to the best anticipated futures. Also it retains it as open. It looks back on the ancient namings as at least having been designational procedure, no matter how poor that procedure was from man's twentieth-century point of view. It looks upon further specifications as opening a richer and wider world of knowledge. In short it sees the world of knowledge as in growth from its most primitive forms to its most perfected forms. It does not insert any kind of a "still more real" world behind or beneath its world of knowledge and fact.[1] It suspects that any such "real" world it could pretend to insert behind the known world would be a very foolish sort of a guessed-at world; and it is quite content to let full knowledge come in the future under growth instead of being leaped at in this particular instant. It welcomes hypotheses provided they are taken for what they are. Theories which sum up and organize facts in ways which both retain the conclusions of past inquiries and give direction to future research are themselves indispensable specifications of fact.

The word "specification" will be found making occasional appearances in the logics though not, so far as we have observed, with definitely sustained use. A typical showing of contrasted use appears in Quine's *Mathematical Logic*, where a "principle of application or specification" is embodied in Metatheorem *231. The name "specification" itself hardly appears again in his book, but the principle so named—or, rather, its symbolic embodiment—once it has entered, is steadily active thereafter. Non-symbolically expressed, this principle "leads from a general law, a universal quantification, to each special case falling under the general law." In other words, whereas we have chosen the name "specification" to designate the most complete and accurate description that the sustained inquiry of an age has been able to achieve based on all the inquiries of earlier ages, this

alternative use by Quine employs it for the downward swoop of a symbolically general law to fixate a substitute for the name of the thing-named. This is manifestly one more illustration of the extremes between which uses of words in logical discussion may oscillate.

Specification, as we thus present it, *is* science, so far as the word "science" is used for the reporting of the known. This does not mean that out of the word "science" we draw "meanings" for the word "specification," but quite the contrary. Out of a full analysis of the process of specification we give a closer meaning to the word "science" as we find it used in the present era. Scientists when confronting an indeterminacy alien to classical mechanics, may seem as agitated as if on a hot griddle for a month or a year or so; but they adapt themselves quickly and proceed about their business. The old characterizations did not permit this; the new specifications do; this is what is typical of them. There is a sense in which specification yields the veritable object itself that is present to science; specification, that is to say, as one aspect of the process in which the object appears in knowledge, while, at the same time, the object, as event, yields the specification. It is not "we" who are putting them together in this form; this is the way we find them coming. The only object we get is the object that is the result of inquiry, whether that inquiry is of the most primitive animal-hesitation type, or of the most advanced research type. John Dewey has examined this process of inquiry at length on its upper levels—those known as "logic"—and has exhibited the object in the form of that for which warranted assertion is secured.[2]

The scientific object, in this broad sense, is that which *exists*. It reaches as far into existence as the men of today with their most powerful techniques can reach. In our preliminary suggestions for terminology we placed event in contrast with designation as the existential aspect of fact. We should greatly prefer to place the word "existence" where we now provisionally place event, and shall probably do so when we are ready to write down the determinations at which we aim. Exist, the word, is derivative of the Latin *sto* in its emphatic form, *sisto*, and names that which stands forth. What stands forth requires temporal and also spatial spread. Down through the ages the word "existence" has become corrupted from its behavioral uses, and under speculative philosophy has been made to stand for something which is present as "reality" and on the basis of which that which is "known" is rendered as "phenomenon" or otherwise to the knower. Common usage, so far as the dictionaries inform us, leans heavily towards the etymological and common-sense side, though of course, the philosophical conventions get their mention. The common man, not in his practical use, but if asked to speculate about what he means, would probably offer his dogmatic assurance that the very "real" is what exists. *Solvitur ambulando* is a very good practical solution of a practical question, but *solvitur* in the form of a dogmatic assertion of reality is something very different. Dr. Johnson (if it was Dr. Johnson) may kick the rock (if it was a rock), but what he demonstrates is kicked-rock, not rock-reality, and this only within the linguistic form then open to him. We believe we have ample justification for placing existence where we now place event in our terminological scheme—only delaying until we can employ the word without too much risk of misinterpretation by hearer or reader. If, however, we do this, then specification and

1 Philipp Frank, *Between Physics and Philosophy* (Cambridge, 1941), using a terminology and a psychological base very different from ours, writes: "Our modern theoretical physics, which admits progress in all parts of the symbol system, is skeptical only when viewed from the standpoint of school philosophy" (p. 102); "There are no boundaries between science and philosophy" (p. 103); "Even in questions such as those concerning space, time and causality, there is scientific progress, along with the progress in our observations" (p. 102); "The uniqueness of the symbol system can be established within the group of experiences itself without having recourse to an objective reality situated outside, just as the convergence of a sequence can be established without the need of discussing the limit itself" (p. 84).

2 John Dewey, *Logic, the Theory of Inquiry,* (New York, 1938), p. 119.

existence are coupled in one process, and with them science; though again it must be added, not science in a purely "physical" or other narrow rendering of the word, but science as it may hope to be when the best techniques of observation and research advance into the waiting fields.

VI

The passage from characterization to specification is not marked by any critical boundary. Nor is the passage from everyday knowledge to scientific knowledge, nor that from everyday language to scientific language. Our attention is focused on lines of development and growth, not on the so-called "nature" of the subject-matter of inquiry. If we are wrong about observing events in growth, then the very inquiry that we undertake in that form should demonstrate that we are wrong. Such a demonstration will be more valuable than mere say-so in advance that one should, or should not, make such an attempt. The regions of vagueness remain in specification, but they decrease. They are Bridgman's "hazes." Their main implication is, however, transformed. The earlier vagueness appeared as defect of human capacity, since

this latter did not seem to succeed in reaching the infinite or the absolute as it thought it ought to. The newer vagueness, under the operation of specification, is a source of pride. It shows that work to date is well done, and carries with it the assurance of betterment in the future.

It is common for those who favor what is called "naturalism" to accept, with qualifications, many phases of the development above. We are wholly uninterested in the phases of the "ism," and solely concerned with techniques of inquiry. For inquiry into the theory of knowledge, to avoid wastage and make substantial progress, we believe the attitude indicated must be put to work one hundred per cent, and without qualification either as to fields of application or ranges of use. We have, however, not yet discussed the manner in which symbol and definition, which we do not permit to interfere with designation, may be put to work in the service of the latter. Nor have we shown the intimate connection between the techniques of specification and the establishment of transaction as permissible immediate subject-matter and report. These problems are among those remaining for a further inquiry which, we trust, will be continued along the lines we have thus far followed.

VII.
THE CASE OF DEFINITION

I

IT is now time to give close attention to the status of the word "definition" in present-day discussions of knowings and knowns, and especially in the regions called "logic." We began by accepting the word as having soundly determinable specialized application for mathematics and formal logic, but by rejecting it for use with the procedures of naming.[1] Naming procedures were styled "designations," and their most advanced forms, notable especially in modern science, were styled "specifications." Thereby definition and specification were held in terminological contrast for the uses of future inquiry.

Throughout our inquiry we have reserved the privilege of altering our terminological recommendations whenever advancing examination made it seem advisable. This privilege we now exercise in the case of the word "definition." For the purposes of the present discussion we shall return the word to its ordinary loose usage, and permit it to range the wide fields of logic in its current great variety of ways. This step was forced upon us by the extreme difficulty we found in undertaking to examine all that has to be examined under "definition," while we ourselves stood committed to the employment of a specialized use of the word. It is much better to abandon our suggested preference than to let it stand where there is any chance that it may distort the wider inquiry.

Our present treatment in effect deprives the word "definition" of the status we had planned to allot it as a "specification" for procedures in the mathematical and formal logical fields. Since we had previously rejected it as a specification for namings, it will now as a name, for the time being at least, be itself reduced to the status of a characterization.[2]

Regardless of any earlier comments,[3] we shall for the present hold in abeyance any decision as to the best permanent employment of the word. The confusions that we are to show and the difficulties of probing deep enough to eliminate them would seem sufficient justification for rejecting the word permanently from any list of firm names. On the other hand the development of its specialized use in formal logic along the line from Frege and Hilbert down to recent "syntactics" (as this last is taken in severance from its associated "semantics") would perhaps indicate the possibility of a permanent place for it, such as we originally felt should be alloted it.

If we begin by examining the ordinary English dictionaries, the Oxford, Century, Standard, and Webster's, for the definitions they provide for definition itself, we shall find them vague and often a bit shifty in setting forth the nature of their own peculiar type of "definition," about which they might readily be expected to be the most definite: namely, the traditional uses of words. Instead, they are strongly inclined to take over some of the authority of the philosophies and the logics, in an attempt to make the wordings of these latter more intelligible to the general reader. Two directions of attention are manifest, sharply phrased in the earlier editions, and still present, though a bit more vaguely, in the later. First, there is a distinction between definition as an "act" (the presence of an "actor" here being implied) and definition as the "product" of an act (that is, as a statement in verbal form); and then there is a distinction between the defining of a word and the defining of a "thing," with the "thing," apparently, entering the definition just as it stands, as a component directly of its own right, as a word would enter. What is striking here, moreover, is the strong effort to *separate* "act" and "product" as different *kinds* of "meanings" under differently numbered entries, while at the same time *consolidating* "word" and "thing" in close *phrasal union*, not only inside the definition of "act" but also inside that of "product." In the Oxford Dictionary (1897)[4] entry No. 3 is for the "act" and entry No. 4 is for the statement *produced* by the act. The "act," we are told, concerns "what a thing is, or what a word means," while the "product" provides in a single breath both for "the essential nature of a thing," and for the "form of words by which anything is defined." Act and product are thus severed from each other although their own "definitions" are so similar they can hardly be told apart. So also with the Century (1897), in which act and product are presented separately in definitions that cover for each not only "word or phrase," but also what is "essential" of or to a "thing." The Standard has offered continuously for fifty years as conjoined illustrations of definition: "a *definition* of the word 'war'; a *definition* of an apple." The latest edition of Webster (revision 1947) makes "essential nature" now "archaic"; runs acts and processes of explaining and distinguishing together, with formulations of meaning such as "dictionary definitions" added to them for good measure; and then secures a snapshot organization for Logic by a scheme under which "traditional logic" deals with the "kinds of thing" in terms of species, genera, and specific differences, while "later schools of logic" deal with statements "either of equivalences of connotation, or intension, or of the reciprocal implications of terms."

Now, a distinction between words and things other than words along conventional lines is easy to make. So is one between an "act" and the products of acts, especially when an "actor," traditionally hypostatized for the purpose, is at hand ready for use. In the present case of the dictionaries, what apparently happens is that if an actor is once obtained and set to work as a "definer," then all his "definings" are taken to be one *kind* of act, whether concerned with words or with things: whereupon their products are taken as equally of one *kind*, although *as* products they belong to a realm of "being" different from that to which actors as actors belong. In the present inquiry we shall have a continuous eye on the dealings logical definition has with words and things, and on the manner in which these dealings rest on its separation of product from acts. We shall not, however, concern ourselves directly with the underlying issues as to the

---

1 See especially Chapter II, Section IV, No. 5.
2 For this terminology, see Chapter VI, Section IV.
3 See Chapter II, Section IV, No. 5; p. 132, footnote 4; see also Chapter XI.

4 We omit, of course, entries irrelevant to the problem of knowings and knowns.

status of acts and products with respect to each other.[1] As to this it is here only to be remarked that in general any such distinction of product from act is bad form in modern research of the better sort. Fire, as an "actor," expired with phlogiston, and the presence of individually and personally existing "heat" is no longer needed to make things hot.

These remarks on what the ordinary dictionaries accomplish should keep our eyes close to the ground—close to the primary facts of observation—as we advance in our further examinations. Whether a dictionary attempts patternings after technical logical expressions, whether it tries simplified wordings, or whether it turns towards evasiveness, its troubles, under direct attention, are in fairly plain view. Elsewhere the thick undergrowth of verbiage often subserves a concealment.

II

This inclusion of *what a thing is* with *what a word says* goes back to Aristotle. Aristotle was an observer and searcher in the era of the birth of science. With him, as with his contemporaries, all knowledge, to be sound, or, as we should say today, to be "scientific," had to win through to the completely fixed, permanent, and unchangeable: to the "essence" of things, to "Being." Knowledge was present in definition, and *as* definition.[2] Word and thing, in this way, came before him conjoined. The search for essences came to be known classically as "clarification."[3] Clarification required search in two directions at once. Definition must express the essences; it must also be the process of finding them. Species were delimited through the "forms" that make them *what* they are. It was in the form of Speech *(Logos)* that logic and ontology must come into perfect agreement.

Aristotle thus held together the subjectmatters which came together. He did not first split apart, only to find himself later forced to try to fit together again what had thus been split. The further history is well known to all workers in this field. The Middle Ages retained the demand for permanence, but developed in the end a sharp split between the name and thing, with an outcome in "isms." On one side were the nominalists (word-dizzy, the irreverent might say), and on the other side the realists (comparably speaking, thing-dizzy). Between them came to stand the conceptualists, who, through an artificial device which even today still seems plausible to most logicians, inserted a fictitious locus—the "concept"—in which to assemble the various issues of word and thing. The age of Galileo broke down the requirements of immutability, and substituted uniformities or "laws" for the old "essences" in the field of inquiry. Looking back upon that age, one might think that the effect of this change on logic would have been immediate and

profound. Not so. Even today the transformation is incomplete in many respects, and even the need for it is often not yet brought into the clear. John Stuart Mill made a voyage of discovery, and developed much practical procedure, as in the cases of naming and induction, but his logic held to dealings with "laws" as separate space-time connections presented to knowledge, and was essentially pre-Darwinian in its scientific setting, so that many of the procedures it stresses are now antiquated.[4]

The Aristotelian approaches were, however, sufficiently jarred to permit the introduction in recent times of "non-Aristotelian" devices. These were forecast by a new logic of relations. These "relations," though not at all "things" of the ancient types, nevertheless struggled from the start (and still struggle) to appear as new variations of the old, instead of as disruptive departures from it. Logical symbols were introduced profusely after the pattern of the older mathematical symbols, but more as usable notations than as the recognition of a new outlook for logic.

In addition to the greatly changed appearance since Greek days of the "objects" presented as "known," as the so-called "contents" of knowledge,[5] there are certain marked differentiations in techniques of presentation (in the organization of "words" and "sentences" to "facts") which are of high significance for the logic of the future. For one thing, there is the difference between what "naming" has come to be in science since Darwin, and what it was before his time; for another, there is the difference between what a mathematical symbol is in mathematics today and what it was when it was still regarded as a type of naming.[6] Neither of these changes has yet been taken up into logical understanding to any great extent, however widely discussion in the ancient forms of expression has been carried on. The common attempt is to reduce logical, mathematical, and scientific procedures to a joint organization (usually in terms of some sort of single mental activity presumed at work behind them) in such a way as to secure a corresponding forced organization of the presumptive "things," logical, mathematical, and scientific, they are supposed to deal with.

One may illustrate by such a procedure as that of Bertrand Russell's "logical atomism," in which neither Russell nor any of his readers can at any time—so far as the "text" goes—be quite sure whether the "atoms" he proposes are minimal "terms" or minimal "reals." Comparably, in those logical systems which use "syntac-

---

1 For some illustrations of the separation of "acts" and "products" in the logics, under a variety of formulations, see Chapter I, particularly Sections I and X. In Chapter IX product follows product and by-product follows by-product; here Sections I and IV exhibit the range of wordings employed. In Chapter VIII five of the six logicians examined make use of separable products under one form or another as basic to their constructions.

2 "Opinion" was allowed for as dealing with uncertainties, but on a lower level. It was *not* science; it was *not* definition.

3 Felix Kaufmann is one of the comparatively few workers in this field who make deliberate and sustained—and in his case, powerful—efforts to develop "clarification" in the classical sense *(Methodology of the Social Sciences,* [New York, 1944]). He expressly affirms this approach *(Philosophy and Phenomenological Research,* Vol. V, 1945, p. 351) in the sense of *Meno,* 74 ff., and *Theaetetus,* 202 ff.

4 Mill managed to see adjectives as names *(Logic,* I, i, Chapter II, Section 2) but not adverbs. By way of illustration drawn from our immediate range of subjectmatters, consider how much sharper and clearer the adverb "definitely" is in its practical applications than is an adjectival assertion of definite*ness*, or a purported "nounal" determination of what "definition" substantively *is*.

5 Consider, for example, astronomy, with respect to which Greek science found itself inspecting a fixed solar system moving about the fixed earth, with sun and moon moving backward and forward, and the firmament of fixed stars rotating above. Its physics offered four fixed elements, different in essence with earth movements downward toward their proper "end"; fire movements upwards into the heavens; air and gas movements upwards as far as the clouds or moon; and water movements, and those of all liquids, back and forth. Its biology had fixed animal and plant species, which remained fixed until Darwin.

6 The problem in this respect began as far back as the first uses of zero or minus-one, and has only disappeared with the heavy present employment of the wave in mathematical formulation. Professor Nagel has lately given such fine illustrations of this status that, with his permission, we should like to recommend to the reader the examination of his pages as if directly incorporated at this point as a part of our text *(The Journal of Philosophy,* XLII, 1945, pp. 625-627).

tics" and "semantics," as soon as these distinctions have been made, an attempt follows to bring them together again by "interpretations." But the best of these "interpretations" are little more than impressionistic manifestations of wishfulness, gathering within themselves all the confusions and uncertainties which professedly have been expelled from the primary components. Although the "definitions" in such treatments are established primarily with reference to "syntactics" they spawn various sub-varieties or queer imitations on the side of "semantics." We seem to have here exhibits of the conventional isolations of form from content, along with a companionate isolation of things from minds, of a type that "transactional"[1] observation and report overcome.

In summary we find word and thing in Aristotle surveyed together but focused on permanence. In the later Middle Ages they came to be split apart, still with an eye on permanence, but with nothing by way of working organization except the tricky device of the "concept" as a third and separate item. Today logic presents, in this historical setting, many varieties of conflicting accounts of definition, side-slipping across one another, compromising and apologizing, with little coherence, and few signs of so much as a beginning of firm treatment. We shall proceed to show this as of the present. What we may hope for in the future is to have the gap between name and object done away with by the aid of a modern behavioral construction which is Aristotelian in the sense that it is freed from the post-Aristotelian dismemberment of man's naming activities from his named world, but which at the same time frees itself from Aristotle's classical demand for permanence in knowledge, and adapts itself to the modern view of science as in continuing growth.[2] Act and product belong broadly together, with product, as proceeds, always in action, and with action always process. Word and thing belong broadly together, with their provisional severance of high practical importance in its properly limited range, but never as full description nor as adequate theoretical presentation, and always *in* action.

III

Since, as we have said, we are attempting to deal with this situation on the ground level, and in the simplest wordings we can find, it may be well to preface it with a brief account of an inquiry into definition carried on throughout in highly sophisticated professional terminology, which exhibits the confusions in fact, though without denouncing them at their roots. Walter Dubislav's *Die Definition*[3] is the outstanding work in this field. In discussing Kant and Fries he remarks that they do not seem to observe that the names they employ bring together into close relations things that by rights should be most carefully held apart; and in another connection he suggests that one of the important things the logician can do is to warn against confounding definitions with

verbal explications of the meaning of words. His analysis yields five types *(Arten)* of definitions, the third of which, he is inclined to think, is merely a special case of the first. These are: (1) Special rules of substitution within a calculus; (2) Rules for the application of the formulas of a calculus to appropriate situations of factual inquiry; (3) Concept-constructions; (4) Historical and juristic clarifications of words in use; (5) Fact-clarifications, in the sense of the determination of the essentials *(Inbegriff)* of things *(Gebilde)*, these to be arrived at under strictly logical-mathematical procedure out of basic presuppositions and perceptual determinations, within a frame of theory; and from which in a similar way and under similar conditions all further assertions can be deduced, but with the understanding (so far as Dubislav himself is concerned) that things-in-themselves are excluded as chimerical.[4] A comparison of the complexly terminological composition of No. 5 with the simple statements of Nos. 1, 2, and 4, or even with the specialized appearance of simplicity in No. 3, gives a fair idea of the difficulties of even talking about definition from the older viewpoints.

IV

Definition may be—and often is—talked about as an incidental, or even a minor, phase of logical inquiry. This is the case both when logic is seen as a process of "mind" and when the logician's interest in it is primarily technological. In contrast with this view, the processes of definition may be seen as the throbbing heart—both as pump and as circulation—of the whole knowledge system. We shall take this latter view at least to the extent that when we exhibit the confusions in the current treatments of definition, we believe that we are not exhibiting a minor defect but a vital disease. We believe, further, that here lies the very region where inquiry into naming and the named is the primary need, if dependable organization is to be attained. The field for terminological reform in logic generally is much wider, it is true, than the range of the word "definition" alone, and a brief reminder of these wider needs may be in order. In logic a definition may enter as a proposition, or as a linguistic or mental procedure different from a proposition; while, alternatively, perhaps all propositions may be viewed as definitions. A proposition itself may be almost anything;[5] it consists commonly of "terms," but terms, even while being the "insides" of propositions, may be either words or nonverbal "things." The words, if words enter, may either be meanings, or have meanings. The meanings may be the things "themselves" or other words. Concepts may appear either as "entities" inserted between words and things, or as themselves words, or as themselves things. Properties and qualities are perhaps the worst performers of all. They may be anything or everything, providing it is not too definite.

The following exhibits, some of confusion, others of efforts at clarification, are offered much as they have happened to turn up in current reading, though

1 See Chapter V.

2 Compare the development in John Dewey, *Logic, The Theory of Inquiry*, Chap. XXI, on "Scientific Method: Induction and Deduction," especially pages 419, 423, 424.

3 Third edition, (Leipzig, 1931), 160 pages *(Beihefte der "Erkenntnis"*, 1). Citations from pages 17, 131. Historically Dubislav finds four main "theories" of definition: the Aristotelian essence and its successors to date *(Wesensbestimmung);* the determination of concepts *(Begriffsbestimmung);* the fixation of meanings, historical and juristic *(Feststellung der Bedeutung. . . bzw., der Verwendung);* and the establishment of new signs *(Festsetzung über die Bedeutung. . .bzw., der Verwendung).*

4 The characterization of type No. 5 in the text above is assembled from two paragraphs of Dubislav's text (pp. 147, 148) which are apparently similar, but still not alike in content. Dubislav's own view as to what should be regarded as definition makes use of type No. 1 along lines of development from Frege to Hilbert (with a backward glance at Pascal), and in the expectation that its "formal" can be made "useful" through definitions of type No. 2.

5 For illustrations and references as to propositions, see Chapter VIII, Section V.

supplemented by a few earlier memoranda and recollections. We shall display the confusions directly on the body of the texts, but intend thereby nothing invidious to the particular writers. These writers themselves are simply "the facts of the case," and the case itself is one, at this stage, for observation, not for argumentation or debate.

For philosophers' views we may consult the philosophical dictionaries of Baldwin (1901), Eisler (German, 1927), Lalande (French, 1928), and Runes (1942). If definition "clarifies," then the philosophical definitions of definition are far from being themselves definitions. Ancient terminologies are at work, sometimes with a slight modernization, but without much sign of attention to the actual life-processes of living men, as modern sciences tell us about them. Robert Adamson, in Baldwin, proceeds most firmly, but also most closely under the older pattern.[1] Both acts and resultant products are, for him, definitions, but the tops and bottoms of the process—the individual objects, and the *summa genera*—are "indefinables."[2] Nominal definitions concern word-meanings, and real definitions concern the natures of things defined; analytic definitions start with notions as given, and synthetic definitions put the notions together; rational definitions are determined by thought, and empirical definitions by selective processes. This is all very fine in its way, but effectively it says little of present-day interest.

Lalande first identifies "definition" in a "general logic" dealing with the action of *l'esprit*; and then two types in "formal" logic, one assembling known terms to define the concept, the other enunciating equivalences. He also notes the frequent extension of the word "definition" to include all propositions whatsoever. Eisler adds to *Nominaldefinition* and *Realdefinition*, a *Verbaldefinition*, in which one word is merely replaced by others. Besides this main division he reports minor divisions into analytic, synthetic, and implicit, the last representing Hilbert's definition of fundamental geometrical terms.

Alonzo Church, in Runes's dictionary, stands closer to contemporary practice, but shows still no interest in what in ordinary modern inquiry would be considered "the facts of the case," namely, actual uses. He mixes a partial report on contemporary practices with a condensed essay on the proprieties to such an extent that it is difficult to know what is happening.[3] For a first grouping of definitions, he offers *(a)* (in logistic systems proper) nominal or syntactical definitions which, as conventional abbreviations or substitutions, are merely a sort of minor convenience for the logician, though nevertheless, it would seem, they furnish him one hundred per cent of

his assurance; and *(b)* (in "interpreted logistic systems") semantical definitions which introduce new symbols, assign "meanings" to them, and can not appear in *formal* development, although they may be "carried" implicitly by nominal definitions, and are candidates for accomplishing almost anything that may be wished, under a properly adapted type of "intent." As a second grouping of definitions he offers the "so-called real" definitions which are not conventions as are syntactical and semantical definitions, but instead are "propositions of equivalence" (material, formal, etc.) between "abstract entities"; which require that the "essential nature" of the definiendum be "embodied in" the definiens; and which sometimes, from other points of approach, are taken to convey assertions of "existence," or at least of "possibility." He notes an evident "vagueness" in "essential nature" as this controls "real" definition, but apparently sees no source of confusion or any other difficulty arising from the use of the single word "definition" for all these various purposes within an inquiry, nor in the various shadings or mixtures of them, nor in the entry of definienda and definientia, sometimes in "nominal" and sometimes in "real" forms, nor in livening up the whole procedure, wherever it seems desirable, with doses of "interpretative" intent.[4]

Here are other specimens, old and new, of what is said about definition. Carnap expressly and without qualification declares that a definition "means" that a definiendum "is to be an abbreviation for" a definiens, but in the same inquiry introduces definitions that are not explicit but recursive, and provides for definition rules as well as for definition sentences. Elsewhere he employs two "kinds" of definition, those defining respectively logical and empirical "concepts," and makes use of various reduction processes to such an extent that he is spoken of at times as using "conditioned" definitions.[5] Tarski makes definition a stipulation of meaning which "uniquely determines," differentiating it thus from "designation" and "satisfaction."[6] An often quoted definition of definition by W.L. Davidson is that "It is the object of Definition to determine the nature or meaning or signification of a thing: in other words definition is the formal attempt to answer the question 'What is it?' "[7] H.W.B. Joseph writes: "The definition of anything is the statement of its essence: what makes it that, and not something else."[8] J.H. Woodger limits the word for axiom-systems to "one quite

---

[1] Adamson, who was one of the most impartial observers and keenest appraisers of logical theory, himself remarked in *A Short History of Logic* (Edinburgh and London, 1911), pp. 19-20, that "looking to the chaotic state of logical textbooks. . .one would be inclined to say that there does not exist anywhere a recognized, currently received body of speculations to which the title logic can be unambiguously assigned."

[2] The introduction of such an "indefinable," of which we shall later find various examples, is, in effect, to make proclamation of ultimate impotence precisely at the spot in logical inquiry where sound practical guidance is the outstanding need.

[3] It is proper to recall that the contributors to Runes's dictionary had much fault to find with the way their copy was edited, and that, therefore, the dictionary text may not fully represent Professor Church's intention. The difficulty to be stressed is, however, not peculiar to his report. A similar confused mixture of what is "historical" with what is "factual" is general. The Runes classification, as we find it, is in this respect vague at almost every point of differentiation. More broadly for all four philosophical dictionaries, the manifestly unclarified phrasings seem to outnumber the attempted clarifications a dozen to one.

[4] How curiously this sort of thing works out can be seen in the opening pages of Church's *Introduction to Mathematical Logic* Part I (Princeton, 1944), in which he writes (p. 1): "In the formal development we eschew attributing any meanings to the symbols of the Propositional Calculus"; and p. 2): "We shall be guided implicitly by the interpretation we eventually intend to give it." Repeated examinations which several interested inquirers have made into Church's words, "meanings," "natural," "necessary," "language," "implicit," and "interpretation," as he has used them in the context of the sentences just quoted, have given no aid towards reducing the fracture in his constructional bone. As pertinent to the issue it may be recalled that Kurt Gödel, in analyzing Russell's procedure, came to the conclusion that Russell's formalism could not "explain away" the view that "logic and mathematics (just as physics) are built up on axioms with a real content" *(The Philosophy of Bertrand Russell*, P.A. Schilpp, editor [Chicago, 1944] p. 142). Again, Hermann Weyl in the *American Mathematical Monthly*, (1946), p. 12, remarks of Hilbert that "he is guided by an at least vaguely preconceived plan for such a proof."

[5] Carnap, R., *Introduction to Semantics* (Cambridge, 1942), pp. 17, 31, 158. "Testability and Meaning," *Philosophy of Science*, III, (1936), pp. 419-471, especially pp. 431, 439, 448; IV, (1937), pp. 1-40.

[6] *Introduction to Logic* (New York, 1941), pp. 33-36, pp. 118ff. "The Semantic Conception of Truth," *Philosophy and Phenomenological Research*, IV (1944), 345, and notes 20 and 35.

[7] *The Logic of Definition* (London, 1885), p. 32.

[8] *An Introduction to Logic* (Oxford, 1916), p. 72.

definite sense," the "explicit"—understanding thereby substitutability; however differently the word might elsewhere be used.[1] Henry Margenau distinguishes between constitutive and operational.[2] A.J. Ayer has "explicit" or "synonymous" definition in contrast with philosophical "definition in use."[3] Morris Weitz employs the names "real" and "contextual."[4] A.W. Burks proposes to develop a theory of ostensive definition which describes "definition in terms of presented instances, rather than in terms of already defined concepts," and says that counter-instances as well as instances must be pointed at; such definition is thereupon declared to be classification.[5] W.E. Johnson decides that "every definition must end with the indefinable," where the indefinable is "that whose meaning is so directly and universally understood, that it would be mere intellectual dishonesty to ask for further definition"; he also thinks, interestingly enough, that it would "seem legitimate. . .to define a proper name as a name which *means* the same as what it *factually indicates.*"[6] G.E. Moore insists on a sharp separation between defining a word and defining a concept, but leaves the reader wholly uncertain as to what the distinction between word and concept itself might be in his system, or how it might be defined to others.[7] What current philosophizing can accomplish under the aegis of the loose and vagrant use of "definition" by the logics is illustrated by Charles Hartshorne in a definition of "reality" which ideal knowledge is said to "provide" or "give" us by the aid of a preliminary definition of "perfect knowledge." This "definition" of "reality" is: "The real is whatever is content of knowledge ideally clear and certain,"[8] and in it, however innocent and simple the individual words may look separately, there is not a single word that, in its immediate context, is itself clear or certain.

The above specimens look a good deal like extracts from a literature of escape, and some might rate well in a literature of humor. No wonder that Professor Nagel says of Bertrand Russell (who may be regarded as one of the ablest and most active investigators of our day in such matters) that "it is often most puzzling to know just what he is doing when he says that he is 'defining' the various concepts of mathematics";[9] and that Professor Skinner, discussing the problems of operational definition with some of his psychological colleagues, tells them that while definition is used as a "key-term," it itself is not "rigorously defined," and that "modern logic. . .can scarcely be appealed to by the psychologist who

recognizes his own responsibility."[10]

V

Turn next to recent reports of research into the question. A paper by H.H. Dubs, "Definition and its Problems,"[11] gives an excellent view of the difficulties logicians face when they strive to hold these many processes together as one. Present theories of definition are recognized by him as confused, and the time is said to have arrived when it has become necessary to "think through" the whole subject afresh. Throwing dictionary definition[12] aside as irrelevant, "scientific definitions" are studied as alone of logical import, with logic and mathematics included among the sciences. Science itself is described as a linkage of concepts, and the general decision is reached for definition that it must tell us what the "concepts" are which are to be associated with a "term or word" so as to determine when, "in immediate experience or thought," there is present the "entity or event" denoted by that term or word.

Definitions, in Dubs' development, are classified as conceptual or non-conceptual. The former is, he says, what others often call "nominal" and occurs where term is linked only with term. The latter are inevitable and occur where the linkage runs back to "logical ultimates" or "indefinables." Cutting across this classification appears another into "essential" and "nominal" (sometimes styled "real" and "accidental") depending on whether we can or can not so define a term as to denote "all" the properties of the object or other characteristic defined. Practically, he reports, scientific definition consists almost entirely of the conceptual and the nominal, even though "scientific" has been adopted as the name of *all* definition pertinent to logic.

If we examine the non-conceptual indefinables in this presentation, we find that they consist of *(a)* causal operations (necessarily "wordless"), *(b)* direct pointings or denotings (that "cannot be placed in books"), and *(c)* intermediate hermaphroditic specimens, half pointings and half verbalizings. Dubs is not at all pleased to find his scheme of definition falling back upon the "indefinable," but his worry is mainly in the sense that he would prefer to reach "ultimates." He solaces himself slightly by hoping that it is just an affair of nomenclature. Nevertheless, since in certain cases, such as those of logic and mathematics, indefinables are seen entering which, he feels sure, are *not* ultimates, he feels compelled to keep the "indefinable" as an outstanding component of definition (or, shall we say, of the "definable"?) without permitting this peculiarity to detract from the hope that he has secured a new "practical and consistent" theory.

Especially to note is that while his leading statement about definition depends upon the use of such words as "concept," "term or word," "immediate experience," "thought," and "entity or event," no definition or explanation of any sort for these underlying words is given. They rate thus, perhaps, as the indefinables of the definition of definition itself, presumably being taken as so well known in their mental contexts that no question

---

[1] *The Axiomatic Method in Biology* (Cambridge, England, 1937), p. 4.

[2] "On the Frequency Theory of Probability," *Philosophy and Phenomenological Research*, VI (1945), p. 17.

[3] *Language, Truth and Logic* (New York, 1936), pp. 66-69. A comparison of the respective uses of the word "explicit" by Carnap, Woodger, and Ayer might be instructive.

[4] "The Unity of Russell's Philosophy," in *The Philosophy of Bertrand Russell* (P.A. Schilpp, editor) (Chicago, 1944), pp. 120-121. The following pronouncement (p. 121) on the subject would seem worthy of profound pondering: "The value or purpose of real and contextual definitions is that they reduce the vaguenesses of certain complexes by calling attention to their various components."

[5] "Empiricism and Vagueness," *The Journal of Philosophy*, XLIII (1946), p. 479.

[6] *Logic* (Cambridge, England, 1921), Vol. I, pp. 105-106; p. 93.

[7] "A Reply to My Critics," in *The Philosophy of G.E. Moore* (P.A. Schilpp, editor) (Chicago, 1942), pp. 663-666.

[8] "Ideal Knowledge Defines Reality: What was True in 'Idealism,'" *The Journal of Philosophy*, XLIII (1946), p. 573.

[9] In his contribution to *The Philosophy of Bertrand Russell*, edited by P.A. Schilpp, p. 326.

[10] "Symposium on Operationism," *Psychological Review*, LII (1945), pp. 270-271.

[11] *Philosophical Review*, LII (1943), pp. 566 ff.

[12] It is interesting to note that one of his requirements for a dictionary definition is that it "must be capable of being written down." This is not demanded for the "scientific definition" for which there are "two and only two" specific requirements: it must be commensurate with what it defines, and it must define a term only in terms that have been previously defined.

about them will be raised. We have already seen the indefinable mentioned by Adamson and Johnson (and the related "ostensive" by Burks), and we shall see more of the peculiar problem they raise as we go along.

Consider another type of examination such as that offered by John R. Reid under the title: "What are Definitions?"[1] Here the effort is to solve the problem of definition not by classifying, but by the building up, or "integration," of a "system of ideas" into an "articulated unity." Taking for consideration what he calls a "definitional situation," he distinguishes within it the following "factors" or "components": a "definitional relation," a "definitional operation," and a "definitional rule." While distinct, these factors are not to be taken as "isolated." The "rule" seemingly is given top status, being itself three-dimensional, and thus involving sets of symbols, sets of cognitive interpretants, and particular cases to which the rule can apply. He holds that we can not *think* at all without this definitional equipment, and stresses at the same time that the symbols, as part of the equipment, would not exist at all except in mental activity. This makes "mental activity" both cause and product, and apparently much the same is true for "symbol." The above points are elaborately developed, but all that is offered us in the way of information about definition itself is a recognition, in currently conventional form, that definitions may be either "syntactical" or "semantical," and that the word "definition" remains ambiguous unless it has an accompanying adjective thus to qualify it. This difference is assumed, but the differentiation is not studied; nor, apparently, is it considered to be of much significance in theory.

Not by pronouncement, but in their actual procedures, both Dubs and Reid bring definition under examination as a facultative activity in the man who does the defining and who is the "definer," and as having, despite its many varieties, a single "essential" type of output. Where Dubs undertakes through cross-classifications to make the flagrant conflicts in the output appear harmonious, Reid strives to establish unity through a multiplication of "entities," arriving thus only at a point at which the reader, according to his likings, will decide either that far too many entities are present, or nowhere near enough.[2]

VI

Let us turn next to the deliberation of a group of six scientists and logicians in a recent symposium from which we have already quoted one speaker's pungent comment.[3] On the list of questions offered by the American Psychological Association for especial examination was one (No. 10), which asked: "What is a Definition, operational or otherwise?" Two of the contributors, so far as they used the word "definition" at all, used it in conventional ways, and gave no direct attention to its problems, so we can pass them by. Two others, Feigl and

Boring, used stylized phrasings, one in a slightly sophisticated, the other in a currently glib form, and have interest here merely to note the kind of tunes that can be played.

Feigl regards definition as useful in minor ways in helping to specify meanings for terms or symbols. Since it deals with terms or symbols it is always "nominal," and the "so-called real" definitions rate as mere descriptions or identifications. Nevertheless, although all definition is "nominal," it always terminates in something "not nominal," namely, observation. This might leave the reader uncertain whether a definition is still a definition after it has got beyond the nominal stage, but Feigl gets rid of this difficulty by calling it "a mere question of terminology." Boring defines definition as a "statement of equivalence," and says it can apply between a term and other terms or between a term and "events"—blandly inattentive to the question as to just in what sense a "term" can be *equivalent* to an "event." He further distinguishes definitions as either operational or non-operational, without, apparently, any curiosity as to the nature of the difference between the operational and the non-operational, nor as to what might happen to an assured equivalence on the operation table with or without benefit of anesthetic.

This leaves two contributors to the symposium, Bridgman, the physicist, and Skinner, the psychologist, to give useful practical attention to the business in hand, operational and definitional.[4] Bridgman, standing on the firm practical ground he has long held with respect to physical procedure, treats definitions as statements applying to terms. Such definitions, he says, presuppose checking or verifying operations, and are thus not only operational, but so completely so that to call them "operational" is a tautology. Skinner, making the same kind of hard, direct operational analysis of the psychological terms that Bridgman made twenty years ago for physical terms, tosses out by the handful the current evasive and slippery phrasings wherever he finds them, and lays the difficulties, both in the appraisals of operation and in the appraisals of definition, to the fact that underlying observation and report upon human behaviors are still far too incomplete to give dependable results. He rejects dualisms of word and meaning, and then settles down to the application of plain, practical, common sense to the terminology in use—to the problem, namely, of what words can properly stand for in observation and experimentation in progress, and to the tentative generalizations that can be made from the facts so established. Skinner agrees with Bridgman that operational analysis applies to all definitions. For him a good answer to the question "What is a definition?" would require, first of all, "a satisfactory formulation of the effective verbal behavior of the scientist." His own undertaking is to contribute to the answer "by example." The others in this symposium might well be asked to become definite as to the status of whatever it is they mean when they say "*non*-operational."

The examinations, both by Bridgman and by Skinner, are held to the regions we ourselves have styled "specification" in distinction from those of "symbolization." For us they lead the way from the antiquated manners of approach used by the other workers we have just examined to three papers, published in *The Journal*

[1] *Philosophy of Science*, XIII (1946), pp. 170-175.

[2] Reid's asserted background for his inquiry into definition may be illustrated by the further citations: "A. . .relation is the symbolic product of an. . .operation. . .according to. . .rule." "These distinctions are. . .not only 'real'. . .but. . .fundamental for. . .understanding." We must not deny "the irreducible complexity of the relevant facts." "Definitions. . .are. . .not assertable statements." Reid is here frank and plain about what ought to be in the open, but which in most discussions of definition is left tacit.

[3] "Symposium on Operationism," *Psychological Review*, LII (1945), pp. 241-294. Introduction by the editor, Herbert S. Langfeld. Contributions by Edwin G. Boring, P.W. Bridgman, Herbert Feigl, Harold E. Israel, Carroll C. Pratt, and B.F. Skinner.

[4] Bridgman, however, is still deeply concerned with his old query as to the "public" *vs.* the "private" in knowledge; and the other contributors were so bemused by it that in the seventeen pages of "rejoinders and second thoughts" following the primary papers, one-third of all the paragraphs directly, and possibly another third indirectly, had to do with this wholly fictitious, time-wasting issue.

of Philosophy in 1945 and 1946, in which much definite progress has been made. They are by Abraham Kaplan, Ernest Nagel, and Stephen C. Pepper. We shall note the advances made and the openings they indicate for future work. We regret it if other equally advanced investigations have failed to attract our attention.

VII

Kaplan's paper is styled "Definition and Specification of Meaning."[1] In it he examines "specification of meaning" as a process for improving the applicability of terms, and as thus leading the way towards definition in the older logical sense of a "logical equivalence between the term defined and an expression whose meaning has already been specified." He sets up the connection between specification and equivalence of meaning as a goal, but does not undertake to deal with its problems, limiting himself here instead to an examination of some of the matters to take account of in such a theory. He acutely observes that much of the best work of science is done with "concepts" such as that of species for which all the long efforts of the biologists have failed to secure any "definition" whatever, and proceeds to ask how this can be possible if definition is so potent and so essential as logicians make it out to be. Treating "specification of meaning" as "hypothetical throughout," he leads up to the question: How does such a development of meaning come to approximate, and in the end to attain, the character of logical definition in which the meanings are no longer held within the limits of hypothesis (though, nevertheless, under the reservation that in the end, definition may possibly come to appear as "only a special form of specification")?

A great field of inquiry is thus opened, but certain difficulties at once demand attention. Kaplan employs, as if well understood, certain key-words in connection with which he gives no hint of specification on their own account. These include such words as "concept," "term," and "meaning." In what sense, for example, in his own development, does "specification of meaning" tell more than "specification" alone? What additional "meaning" is added by the word "meaning"? Is this "meaning" in any way present apart from or in addition to its "specification"? If "meaning" adds anything, should its particular contribution not be made clear at an early point in the treatment—a point earlier, indeed, than Kaplan's present discussion? What is the difference between "meaning" by itself and "meaning of a term"—a phrase often alternatively used? If "term" has to have a "meaning" and its "meaning" has to be separately "specified," is the inquiry not being carried on at a stage twice removed—and unnecessarily removed at that—from direct observation? If three stages of "fact" are thus employed, should not their differentiation be clearly established, or their manner of entry at least be indicated? These questions are asked, not to discourage, but as encouragement for, the further examination which Kaplan proposes, if the relations of equivalence and specification are to be understood.

VIII

Nagel, in his paper "Some Reflections on the Use of Language in the Natural Sciences,"[2] understands by "definition" very much what Kaplan understands by specification. He does not, for the immediate purposes of his discussion of language, generalize the problem as

Kaplan does, but he shows brilliantly the continuous revision and reconstruction of uses which an active process of definition involves. We have already cited, in passing, his valuable pages on the growing abandonment by mathematicians of their older expectations that their symbols should have efficient status as names. He provides an illuminating illustration of the underlying situation in the case of "constant instantial velocity" which by its very manner of phrasing makes it operationally the "name" of nothing at all, while nevertheless it steadily maintains its utility as a phrase.[3] He eliminates the claims of those types of expression which applied in a variety of meanings through a variety of contexts, manage to fascinate or hypnotize the men who use them into believing that, as expressions, they possess "a generic meaning common to all" their varieties of applications. He could even have used his own word "definition" as an excellent example of such a form of expression. In the background of his vividly developed naturalistic appraisal of the processes of knowledge, all of these steps have high significance for further progress.

IX

Pepper's account of "The Descriptive Definition"[4] undertakes a form of construction in the very region which Kaplan indicates as locus of the great problem. His "descriptive definition" is so close to what we ourselves have described as "specification" that, so far as his introductory characterization goes, we could gladly accept his account of it, perhaps free from any reservation, as meeting our needs. It offers at the least a fair alternative report to that which we seek. His framework of interpretation is, however, another matter, and his "descriptive definition," as he sees it, is so intricately built into that framework that it must be appraised as it there comes.

Like Kaplan, with his specification of meaning, Pepper envisions the old logical definition as basic to the display and justification of a descriptive definition. He notes that "in empirical enquiry observers desire expressions which ascribe meaning to symbols with the definite proviso that these meanings shall be as nearly true to fact as the available evidence makes possible." To this end the reference of description is "practically never" to be taken as "unalterable." He does not offer a fully positive statement, but at least, as he puts it in one phrasing, the "reference" of the symbol under the description is "intended" to be altered whenever "the description can be made more nearly true." The descriptive definition thus becomes "a convenient tool constantly responsible to the facts," rather than "prescriptive in empirical enquiry."[5]

---

1 *The Journal of Philosophy*, XLIII (1946), pp. 281-288.
2 *The Journal of Philosophy*, XLII (1945), pp. 617-630.

3 His statement, of course, is not in terms of "naming." As he puts it, such expressions "are *prima facie inapplicable* to anything on land or sea"; they "*apparently* have no pertinent use in connection with empirical subject-matter," as "no overtly identifiable motions of bodies can be characterized" by them. *Ibid.*, pp. 622-623.
4 *The Journal of Philosophy*, XLIII (1946), pp. 29-36.
5 Merely as a curiosity, showing the way in which words used in logic can shift back and forth, Peirce once undertook (2.330) to suggest much the same thing as Pepper now develops. Pepper's language is that "a nominal definition is by definition prescriptive." Peirce's wording was: "this definition—or rather this precept that is more serviceable than a definition." Precept and prescription are not the same etymologically, but their uses are close. Peirce's "precepts" would be a close companion for Pepper's descriptive definition; while Pepper's "prescription," in its none too complimentary use, matches closely what Peirce felt compelled to understand by definition.

So far so good. Men are shown seeking knowledge of fact, elaborating descriptions, and changing names to fit improved descriptions. But Pepper finds himself facing the query whether this is "definition" at all? Perhaps, he reflects, the nominal and the real definitions of logical theory, the equations, the substitutions, and the ostensive definitions have exclusive rights to the field and will reject "descriptive" definitions as intruders? At any rate he feels it necessary to organize descriptive definition with respect to these others, and with a continuing eye on the question whether he can get from or through them authority for what he wishes to accomplish. Though still committing himself to the wearing of a coat of many definitional colors, the organization he seeks is in a much more modern spirit than the classifications of Dubs or the "articulated unity" of Reid.

He begins by making descriptive definition one of two great branches of definition. Against it he sets nominal definition. The former is known by its being "responsible to facts meant by the symbol." The latter is not thus responsible, but either assumes or ignores facts, or else is irrelevant to them. Nominal definition in turn has two species: the equational and the ostensive.

The equational species is said to be strictly and solely a matter of the substitution of symbols. Whether this is an adequate expression for all that goes on in the equational processes of mathematics and in the development of equivalences, and whether it is really adequate for his own needs, Pepper does not discuss. What he is doing is to adopt a manner of treatment that is conventional among logicians who deal with logical "products" displayed on logical shelves instead of with the logical activity of living men, even though his own procedure is, culturally speaking, much further advanced than is theirs. For our present purposes we need to note that in *his* program equational definition is substitution—it is *this, and nothing more.*

The ostensive species of definition is (or "involves" or "refers to") a non-symbolic meaning or source of meaning for a symbol. It is primarily and typically a "pointing at." Pepper here examines the "facts" before him with excellent results. He suggests the interpretation of such "pointing" as "indication," and then of "indication" as "operation." The indicative "operation" at which he arrives in place of ostensive definition along the older lines is, however, still allotted status as itself a "definition." In this, there is a survival of influences of the "word and thing" mixture, even at the very moment when important steps are being taken to get rid of such conventional congealment. Oldtimers could talk readily of "ostensive *definition*" because they lived reasonably misty lives and avoided analysis in uncomfortable quarters. Pepper makes a pertinent analysis, and we are at once startled into asking: Why and how can such an ostensive or indicative operation be itself called a "definition" in any careful use of either of the words, "ostensive" or "definition"? Is there something logically in common between finger-pointing without name-using, and name-using without finger or other pointing action? If so, just *what* is it? And above all, just *where* does it "existentially" have specific location?

Pepper's problem, now, is to organize the descriptive to the ostensive and to the equational. The problem is of such great importance, and every fresh exploration of it is of such great interest, that we shall take the space to show what is apt to happen when "substitution" in the guise of "equation" is employed as organizer for a presumptively less dependable "description." We are given three diagrams, and these with their accompanying texts should be carefully studied. The diagram for equational definition shows that the symbol, "*S*," is "equated with" and is "*equationally* defined by" other "symbols." The diagram for ostensive definition shows that such an "*S*" "indicates" and is "*ostensibly* defined by" an "empirical fact," "*O*." The diagram for descriptive definition expands to triadic form. "*S*" is symbol as before. "*O*" is advanced from "fact" to "field," while remaining empirical. "*D*" is added to stand for description. Further distinctive of this diagram is the entry into its formulations of the words "tentative," "hypothetical," and "verifies," along with "describes." The scheme of the diagram then develops in two parallel manners of expression, or formulations:

Formulation A	Formulation B

1 *(D-O):* D hypothetically describes (or) is verified by........*O.*
2 *(S-D):* S is tentatively equated with (or) is descriptively defined by *D.*
3 *(S-O):* S tentatively indicates.... (or) is ostensively defined by *O.*[1]

Under Formulation A, it would seem easy or "natural" to condense the statements to read that, given a hypothetical description of "something," we can take a word to stand tentatively for that description, and this word will then also serve as a tentative indication of the "something." Under Formulation B, we might similarly say (though various other renderings are possible) that, given a verified description of something, that description descriptively defines a word which in turn is (or is taken as) ostensively defined by the something; or we might perhaps more readily think of an ostensive hint leading through naming and description to verification or imagined verification followed by a thumping return to "ostensive *definition.*" (In a discussion in another place Pepper, to some extent, simplifies his report by saying of the descriptive definition that "strictly speaking, it is an arbitrary determination of the meaning of a symbol in terms of a symbol group, subject, however, to the verifiability of the symbol group in terms of certain indicated facts.")[2] But would it not be still more informative if one said that what substantially this all comes to is a report that men possess language in which they describe events (facts, fields)—that they can substitute single words for groups of words—that the groups of words may be called descriptive definitions—and that the words so substituted serve to indicate the facts described, and, when regarded as "verified" are linked with "indicative operations" styled "ostensive"?

Thus simplified (if he will permit it) Pepper's development may be regarded as an excellent piece of work towards the obliteration of ancient logical pretenses,

[1] Certain features of the diagram should be mentioned in connection with the above transcription. There is a bare possibility, so far as diagrammatic position goes, that the "tentatively indicates" and the "ostensively defined" on the *S-O* line should change places; our choice was made to hold the "tentatives" and "hypothetical" together in one set. The *S-D* and *S-O* statements under Formulation A (the latter transformed to read "is...taken to indicate") are noted as "at the same time," which possibly indicates orders of successsion elsewhere which we have overlooked. (We are far from wishing to force any such successions into the treatment.) The shiftings from active to passive verb forms may have some significance which we overlook, and the specific subject indicated for the passive verbs would be interesting to know. These difficulties are slight, and we trust none of them has interfered with a proper rendering by us of Professor Pepper's position.

[2] *The Basis of Criticism in the Arts* (Cambridge, 1945), p. 31. The structural diagram in the book for the *D-O* of Formulation A uses the word "describes," and has not yet made the limitation "hypothetically describes."

and it might well be made required reading in preliminary academic study for every budding logician for years to come. Our only question is as to the effectiveness of his procedure, for he carries it on as if it were compelled to subject itself to the antique tests supplied by the traditional logical scheme. We may ask: if equation is substitution or substitutability—precisely this, and nothing else—why not call it substitution in place of equation? Would not such an unequivocal naming rid us of a bit of verbal trumpery, and greatly heighten definiteness of understanding? If equation runs pure and true from symbol to symbol, what possibly can "tentative" equating be as a type of equating itself, and not merely as a preliminary stage in learning? If one turns the phrase "is equated with" into the active form "equals," how pleasing is it to find oneself saying that "*S* tentatively equals *D*"? Or, if we are told, as in one passage, that the description is "not flatly equated" with the symbol, but only equated "to the best of our abilities," does this add clarity? Descriptions are meant to be altered, but equations not; again, the question arises, how is it that descriptive definitions can be equated? How can "verifies" become an alternative phrasing for "hypothetically describes"? Should not the alternative form in the diagram be perhaps, "hypothetically verifies," but then what difference would there be between it and "description" itself? In introducing the "ostensive," Pepper views it in the older manner as dealing with "facts of immediacy," despite his own later reduction of it to indicative operation, and his retention of it as definition.[1] But just what could a tentative immediacy be?

The main question, further, remains: Why employ one single name, even under the differentiations of genus and species, for such varied situations in human behavior? If we take "*e*" for equationally, "*o*" for ostensively, and "*d*" for descriptively, Pepper's varieties of definition may be set down as follows:

e-defined is where a word *S* is substitutable for other words.

o-defined is where a word is used operationally to indicate an object, *O*.

d-defined is where a word tentatively indicates an *O*, by being tentatively substitutable for a description, *D*, which latter hypothetically describes the *O*; or alternatively (and perhaps at some different phase of inquiry), *d*-defined is where a word is *o*-defined *via*, or in connection with, the verification of the *D* by the *O*.

In preliminary, conventional forms of statement, the report would seem to be that the *e*-definition, *as he offers it* (though we do not mean to commit ourselves to such a view), seems to be primarily a matter of language-organization; the *o*-definition seems to lie in a region commonly, though very loosely, called "experience"; and the *d*-definition, so far as any one can yet see, does not seem to lie comfortably anywhere as a species of a common genus.

X

We have seen many conflicting renderings of the word "definition" offered us by acute investigators who are currently engaged in a common enterprise in a common field. Recalling these conflicts, may one not properly say that this display by itself, and just as it stands, provides sufficient reason for a thorough terminological overhaul; and that, without such an overhaul and reorganization, the normal practical needs for intercommunication in research will fail to be properly met? The one word "definition" is ex-

pected to cover acts and products, words and things, accurate descriptions and tentative descriptions, mathematical equivalences and exact formulations, ostensive definitions, sensations and perceptions in logical report, "ultimates," and finally even "indefinables."[2] No one word, anywhere in careful technical research, should be required to handle so many tasks. Where, outside of logic—except, perhaps, in ancient theology or modern stump-speaking—would such an assertion be tolerated as that of the logicians when, among the "definables" of definition, they push "indefinables" boldly to the front? Here seems to be a witches' dance, albeit of pachydermally clumsy logical will-o'-the-wisps. What more propriety is there in making definition cover such diversities than there would be in letting some schoolboy, poring perversely over the pages of a dictionary, report that the Bengal tiger, the tiger-lily, the tiger on the box, and the tiger that one on occasion bucks, are all species of one common genus: *Tiger?*

The types of definition we have inspected appear to fall roughly into three groups: namely, equivalence as in mathematics, specification as in science, and a traditionally derived mixed logical form which hopes and maneuvers to establish specifications ultimately under a perfected logical pattern of equivalences. The worst of the affair is that present-day logic not only accepts these different activities and all their varieties as evident phases of a common process, but actually sees the great goal of all its labors to lie in their fusion into a single process, or unit in the logical system. The outcome is just the chaos we have seen.

The problem here to be solved is not one for a debating society, nor is it one for a formal calculating machine. It requires to be faced in its full historical linguistic-cultural setting. The great phases of this setting to consider are: the *logos* and the Aristotelian essences; the late medieval fracture of namings from the named as separate "things" in the exaggerated forms of "nominals" and "reals"; the artificially devised "concept" inserted to organize them; the survival into modern times of this procedure by conceptual proxy under a common, though nowhere clarified, substitution of "word" (or "sentence") for "concept"; the resultant confusedly "independent" or "semi-independent" status of "words," "terms," and "propositions" as components of subjectmatter; and finally the unending logical discussion of the connections of science and mathematics carried on in the inherited terminology, or in slight modifications of it, with no adequate factual examination at any stage, of the modern developments of scientific designation and mathematical symbolization in their own rights.

In this setting, and in the illustrations we have given, one feature appears that has great significance for our present consideration as showing the excesses to which the existing terminological pretenses may lead. This is the entry, of which we have repeatedly taken note,[3] and which will now receive a little closer attention, of the "ultimates" and

---

[1] *Ibid.*, pp. 27ff.

[2] As we have already indicated, this confusion of a great variety of things under a single name is most probably maintained under some primitive form of reference of them all to a purportedly common source in a single human "capacity," "action," or "act," derivative of the medieval soul.

[3] In the preceding reports we have had mention of indefinables by Adamson, Johnson, and Dubs. Ostensive definition has been variously treated by Pepper and Burks. Feigl also considers the ostensive definition, saying that it is "rather fashionable nowadays." He believes there should be no trouble with it, as it is either "a designation rule formulated in a semantical metalanguage" or "a piece of practical drill in the learning of the 'right use' of words." Dubs, for his part, finds it "not quite clear" enough to use.

"indefinables" as components at once of "definition" and of "reality." What we have called "specifications" and "symbolizations" can surely rate as current coin of the logical fields, no matter who does the investigating, and no matter how thorough or how precarious today's understanding of them is. The "indefinables" and the "ultimates," in contrast, are counterfeit. In them, as they enter logical discussion, we have neither good working names, nor intelligible equivalences, nor verifiable factual references, but instead pretenses of being, or of having more or less the values of, all three at once. They enter through "ostensive" definitions, or through some verbal alternative for the ostensive, but in such a way that we are unable to tell whether the "definition" itself is "about" something, or "is" the something which it is "about," or how such phrasings as "is" and "is about" are used, or just what meanings they convey. John Stuart Mill remarked a hundred years ago that, however "unambiguously" one can make known who the particular man is to whom a name belongs "by pointing to him," such pointing "has not been esteemed one of the modes of definition."[1] "Pointing" on the basis of previous mutual understanding is one thing, but the kind of understanding (or definition) that might be developed from pointing alone in a communicational vacuum, offers a very different sort of problem. Nevertheless, regardless of all such absurdities the ostensive definition, since Mill's time, has gained very considerable repute, and is, indeed, a sort of benchmark for much modern logic, to which, apparently, the possession of such a name as "ostensive definition" is guarantee enough that somewhere in the cosmos there must exist a good, hard fact to go with the name. The ostensives, and their indefinables and ultimates, seem, indeed, to be a type of outcome that is unavoidable when logic is developed as a manipulation of old terminologies using "definers," "realities," and "names" as separate components, instead of being undertaken as an inquiry into a highly specialized form of the behaviors of men in the world around them. Ostensive behavior can be found. Definitional behavior can be found. But the mere use of the words "ostensive definition" is not enough to solve the problem of their organization.

If we try, we may take these procedures of defining the indefinables apart so as to see, in a preliminary way at least, what they are made of and how they work. What are these indefinables and ultimates assumed to be (or, verbally, to stand for) as they enter definition? Are they regarded as either "physical" or "mental"? Usually not. Instead they are spoken of as "logical" entities or existences, a manner of phrasing as to which the less it is inquired into, the better for it. Certainly there are words involved, because, without its linguistic presentation as "ostensive," there is no way, apparently, in which such definition would be before our attention at all. Certainly, also, there are "things" involved—"things" in the sense of whatever it is which is beyond the finger in the direction the finger points. Certainly also there is a great background of habit and orientation, of behavior and environment, involved in every such pointing. More than this, in any community using language—and it is a bit difficult to see how definition in any form would be under close scrutiny except in a language-using community—a large part of the background of such pointing is linguistic. Suppose we consider as sample exhibits in the general region of pointings a masterless wild dog alert and tense with nose turned towards scent of game; a trained pointer in field with hunting master, the master pointing with hand for benefit of comrade towards sign of motion in brush; a savage hunter pointing or following with his eye another hunter's pointing; a tropic savage as guest in arctic watching Eskimo's finger pointed towards never-before-seen snow, and finally, the traditional Patagonian getting first sight of locomotive as Londoner points it out. Traditionally the Patagonian sees nothing of what a locomotive "is," and certainly it would be stretching matters to assume that the immediate case of "pointing" *defined* the locomotive to him.[2] Hardly anywhere would a theorist speak of the wild dog as engaged in definition. In the intervening cases there are various gradations of information established or imparted *via* sign. Where does distinctive "definition" begin, and why? Where does it cease, and why? These questions concern varieties of events happening, and names needed in their study.

It is our most emphatically expressed belief that such a jumble of references as the word "definition" in the logics has today to carry can not be brought into order until a fair construction of human behaviors across the field is set up, nor until within that construction a general theory of language on a full behavioral basis has been secured.[3] We have sketched tentatively in preceding chapters some of the characteristics which we believe such construction will have. Identifying behavior in general with organic-environmental sign-process, transactionally viewed, we have noted the perceptive-manipulative activities at one end of the range, and then three stages of the designatory use of language, followed by another type of use in symbolization. Given such a map of the behavioral territory, the various sorts of human procedure insistently lumped together under the name of "definition" could be allocated their proper operating regions. Among them the "ostensive," now so absurdly present, should be able, under a much needed transmutation, to find a proper home.

---

[1] *Op. cit.* Book I, Chapter VIII, Section 1. For a discussion in a wider background than the present of the whole problem of "demonstratives" including both the "pointings" and the "objects" pointed-at, see J. Dewey, *Logic, The Theory of Inquiry*, pages 53-54, 124, and 241-242.

[2] We have already cited Skinner's view that without a developed behavioral base modern logic is undependable, and we repeat it because such a view so rarely reaches logicians' ears. Their common custom is to discard into the "pragmatic" all uncomfortable questions about logic as living process, forgetting that the "pragmatic" of Peirce and James, and of historical status, is quite the opposite, since it interprets meanings, concepts, and ideas in life. Skinner's conclusion is that eventually the verbal behavior of the scientists themselves must be interpreted, and that if it turns out that this "invalidates our scientific structure from the point of view of logic and truth-value, then so much the worse for logic, which will also have been embraced by our analysis."

[3] Cf. the discussion of "demonstratives" in Dewey's *Logic, The Theory of Inquiry*, pp. 125-127.

VIII.
LOGIC IN AN AGE OF SCIENCE[1]

I

AMONG the subjectmatters which logicians like at times to investigate are the forms of postulation that other branches of inquiry employ. Rarely, however, do they examine the postulates under which they themselves proceed.[2] They were long contented to offer something they called a "definition" for logic, and let it go at that. They might announce that logic dealt with the "laws of thought," or with "judgment," or that it was "the general science of order." More recently they are apt to connect it in one or another obscure way with linguistic ordering.

We may best characterize the situation by saying that while logicians have spent much time discussing how to apply their logic *to* the world, they have given almost no examination to their own position, as logicians, *within* the world which modern science has opened. We may take Darwin's great demonstration of the "natural" origin of organisms as marking the start of the new era in which man himself is treated as a natural member of a universe under discovery rather than as a superior being endowed with "faculties" from above and beyond, which enable him to "oversee" it. If we do this, we find that almost all logical enterprises are still carried on in pre-Darwinian patterns. The present writer is, indeed, aware of only two systems (and one of these a suggested project rather than a developed construction) which definitely undertake an approach in a modern manner. The rest are almost wholly operated under the blessing, if not formally in the name, of "thinkers," "selves," or superior realms of "meanings." The present memorandum will sketch the new form of approach and contrast it with typical specimens of the old.

Two great lines of distinction between pre-Darwinian and post-Darwinian types of program and goals for logic may readily be set down.

While the former are found to center their attention basically upon *decisions* made by individual human beings (as "minds," "deciders," or otherwise "actors"), the latter describe broadly, and appraise directly, the presence and growth of knowings in the world, with "decisions" entering as passing phases of process, but not as *the* critical acts.

While enterprises of pre-Darwinian types require certainties, and require these to be achieved with perfection, absoluteness, or finality, the post-Darwinian logic is content to hold its results within present human reach, and not strive to grasp too far beyond.

Examined under these tests the recent logics of the non-Aristotelian, multivalued, and probability types all still remain in the pre-Darwinian or "non-natural" group, however they may dilute their wordings with respect to the certainties. A century ago Boole undertook to improve logic by mathematical aid, and there was great promise in that; but Russell, following the mind-steeped Frege, and himself already thoroughly indoctrinated to understanding and interpretation by way of "thought" or "judgment," reversed this, and has steadily led the fight to make logic master and guardian[3] in the ancient manner, with never a moment's attention to the underlying problem: *Quis custodiet ipsos custodes?*

The lines of distinction we have mentioned above might, perhaps, be made the basis for two contrasting sets of postulations. In some respects such postulations could be developed as sharply as those which geometers set up with respect to parallels. But such a course would be practicable only on the condition that the key words employed in them could be held to sharply established meanings. However, as logic and its surrounding disciplines now stand, this necessary linguistic precision cannot be attained. A single man might allege it, but his fellows would not agree, and at his first steps into the linguistic market-place he would find each logician he approached demanding the right to vary the word-meanings, and to shape them, here subtly, there crudely, out of all semblance to the proposed postulational use.[4]

Since such a course will not avail, we may try another. We may hunt down in several logics the most specific statements each of them has made in regard to the issues in question. We may then assemble these as best we can. We shall not in this way obtain postulations[5] in the sense in which more securely established inquiries can obtain them, but we may at times secure fair approximations. Where we can not get even this far forward, we can at least present skeletons of the construction of logical systems, such that they contain the materials out of which postulations may perhaps be derived if in the end the logicians involved will ever attend closely enough to what they are doing. If careful appraisals are to be secured, work of this kind is essential, even though it as yet falls far below the standards we could wish.

We shall consider six logical procedures, half of them in books published in 1944 and 1945, and all now under active discussion. We shall take them in three groups: first

---

1 This chapter is written by Bentley.

2 Sub-postulations within a wider, tacitly accepted (i.e., unanalysed) postulatory background are common enough. The present viewpoint is that of Morris R. Cohen when he writes: "The philosophic significance of the new logic, the character of its presuppositions, and the directions of its possible application are problems which have attracted relatively little reflective thought." *A Preface to Logic* (New York, 1944), p. ix.

3 In his very latest publication Bertrand Russell still writes: "From Frege's work it followed that arithmetic, and pure mathematics generally, is nothing but a prolongation of deductive logic." *A History of Western Philosophy*, (New York, 1945), p. 830.

4 Samples of logicians' linguistic libertinism can be found anywhere, anytime, in current periodicals. Thus, for instance, in a paper just now at hand, we find "principles" of deduction referred to "intuition" for their justification, and this along with the suggestion that intuition should be "reinforced by such considerations as. . .ingenuity may suggest." A few paragraphs later a set of "principles" containing wholly naive uses of the word "true" are declared to be "intuitively obvious." Lack of humor here goes hand in hand with inattention to the simpler responsibilities of speech; Max Black, "A New Method of Presentation of the Theory of the Syllogism," *The Journal of Philosophy*, XLII (1945), 450-51.

5 Compare Chapter III where groups of postulations are presented looking towards a natural theory of knowings and knowns.

the "natural"[1] constructions of John Dewey and J.R. Kantor;[2] next, the sustained efforts of Morris R. Cohen and Felix Kaufmann to adapt old mentalistic-individua-listic forms of control to modern uses; finally the desperate struggles of two outstanding logician-philo-sophers, Bertrand Russell and G.E. Moore, to secure victory with their ancient banners still waving. Our purpose is not so much to debate the rights and wrongs of these procedures, as it is to exhibit the differences in materials and workmanship, and to indicate the types of results thus far offered.

II

John Dewey's wide professional and public following derives more from his philosophical, educational, and social studies than from his logic. Nevertheless for over forty years he has made logic the backbone of his inquiry. His preliminary essays on the subject go back, indeed, to the early nineties. The *Studies of Logical Theory* appeared in 1903. The *Essays in Experimental Logic* in 1916, and *Logic, the Theory of Inquiry* in 1938, all in a single steadily maintained line of growth which stresses inquiry directly as the great subject-matter of logic along a line of development foreseen and tentatively employed by Charles Sanders Peirce.[3] With Dewey the method and outcome of inquiry becomes warranted assertion. "Proof," which the older logics endeavored to establish under validities of its own for the control of knowledge, is here to be developed within, and as a phase of, inquiry; all certainty becomes subject to inquiry including the certainties of these very canons of logic which older logics had treated as the powerful possessors of certainty in their own right. Man is thus seen to advance in his logical action as well as in all his other affairs *within* his cosmos, so that the dicta and ultimacies of the older *super*-natural rationalities, presumptively possessed by men, fall forfeit. The basic attitudes adopted in Dewey's *Logic*, the makings of a postulation for it, will be found in his first chapter. We list six section headings from this chapter and supplement them with two other characteristic atti-tudes. These are numbered 1 to 8, and are followed by a dozen more specialized determinations, here numbered 9 to 20.

[1] In characterizing these logics as "natural," it is to be understood that the word "natural," as here used, is not to be taken as implying something specifically "material" as contrasted with something specifically "mental." It stands for a single system of inquiry for all knowledge with logic as free to develop in accordance with its own needs as is physics or physiology, and to develop in system with either or both of these as freely as they develop in system with each other. Many logicians rated by us as non-natural would label them-selves "naturalistic." Thus Russell declares that he "regards knowl-edge as a natural fact like any other" (*Sceptical Essays* [New York, 1928] page 70), though our examination of his materials and pro-cedures will give him quite the contrary rating.

[2] If Otto Neurath had lived to develop his position further than he did, we could doubtless list him also on the "natural" side. He was from the beginning much further advanced in this respect than others of his more active associates in the projected *International Encyclopedia of Unified Science*, of which he was editor-in-chief. His most recent publication is "Foundations of the Social Sciences," a monograph contributed to the Encyclopedia.

[3] "As far as I am aware, he (Peirce) was the first writer on logic to make inquiry and its methods the primary and ultimate source of logical subjectmatter," John Dewey, *Logic, the Theory of Inquiry*, (New York, 1938) p. 9n. The fourth of the postulates for Dewey in the text is frequently called "the postulate of continuity," and perhaps offers the straightest and widest route from Darwin through Peirce to Dewey.

DEWEY[4]

1. "Logic is a progressive discipline" (p. 14);
2. "The subjectmatter of logic is determined operation-ally" (p. 14);
3. "Logical forms are postulational" (p. 16);
4. "Logic is a naturalistic theory;. . .rational operations *grow out of* organic activities" (pp. 18-19);
5. "Logic is a social discipline" (p. 19);
6. "Logic is autonomous; inquiry into inquiry. . .does not depend upon anything extraneous to inquiry" (p. 20);
7. "Every special case of knowledge is constituted as the outcome of some special inquiry" (p. 8);
8. "Logical theory is the systematic formulation of controlled inquiry" (p. 22) with "the word 'con-trolled'. . .standing for the methods of inquiry that are developed and perfected in the processes of continuous inquiry" (p. 11).
9. Inquiry, through linguistic development of terms and propositions, arrives in judgment at warranted assertions upon existence (Chapter VII);
10. Propositions and propositional reasonings are inter-mediate and instrumental in inquiry (pp. 113, 166, 181, 310, *et al.*); propositions are not found in independence or as isolates, but only as members of sets or series (p. 311);
11. Terms enter as constituents of propositions and as conditioned by them, never in independence or as isolates (p. 328);
12. Singular and generic propositions are conjugates, the former specifying "kinds," the latter organizations of "kinds" (pp. 358-9);
13. The development of propositions in "generic deriva-tion or descent where differentiation into kinds is conjoined with differentiation of environing conditions" is an equiva-lent in logic to the biological advance which established the origin of species (pp. 295-6);
14. Singular propositions (and with them "particulars") appear as incomplete or imperfect, rather than as "simple," "atomic," or otherwise primordial (p. 336n, p. 342);[5]
15. The propositions called "universal" are intermediate stages of inquiry like all others, and are to be examined on various levels of instrumental differentiation (Chapters XIV, XV, XX);
16. The adequate development of the theory of inquiry must await the development of a general theory of language in which form and matter are not separated (p. iv);
17. Mathematical forms, and logical forms generally, are properly to be studied in severance from their subjectmatters only when it is recognized that the severance is provisional, and that their full setting in determinate human action is to be taken into account in the final construction (Chapter XVII);
18. The canons of the old logic (including non-contradic-tion), now entering as forms attained in and with respect to inquiry, lose all their older pretenses to authority as inherent controls or as intuitively evident (pp. 345-6), and, when detached from their place in "the progressive conduct of inquiry," show themselves as "mechanical and arbitrary" survivals (p. 195).

[4] All page references are to *Logic, the Theory of Inquiry*. Professor Dewey has made further development since the *Logic* was published, particularly with respect to the organization of language, logical forms, and mathematics. Such advances are intimated, but not expressly set forth, in the numbered paragraphs of the text, since it is desirable, for all logics discussed, to hold the presentation to what can be directly verified by the reader in the pages of the works cited.

[5] The radical nature of the advance in postulate 14 over older treatments will be plain when the postulations for Russell are considered. For the equally radical postulate 19 see postulate B-8, and its context, in Chapter III, Section III.

19. "Objects" as determined through inquiry are not determined as existences antecedent to all inquiry, nor as detached products; instead they enter knowledge as conditioned by the processes of their determination (p. 119);

20. No judgments are to be held as *super*-human or as final; organisms and environments alike are known to us in process of transformation; so also are the outcomes in judgment of the logical activities they develop (Chapters I to V; p. 345n).

The other natural approach to logic to be considered is that of J.R. Kantor in his book *Psychology and Logic* (Bloomington, Indiana, 1945). He makes his development upon the basis of his interbehavioral psychology which rates as one of the most important advances in psychological construction since William James. The "natural" characteristic of this psychology is that it undertakes not merely to bring organism and environmental object into juxtaposition, but to investigate their behavioral activities under a form of functional interpretation throughout. Applying his approach to the field of logic, Kantor offers eight postulates for a "specificity logic." These follow in his own wordings, and in the main from the first chapter of his book. Two ancillary statements, 2.1 and 4.1, and a few other phrasings, have been supplied from other chapters to compensate compactly for the detachment of the leading postulates from their full contextual exposition.

KANTOR[1]

1. "Logic constitutes primarily a series of operations."
2. "Logical theory is continuous with practice"; the "theory...constitutes...the *study* of operations"; the "practice...consists of these *operations* themselves."
2.1 "Interbehavioral psychology assumes that organisms and objects exist before they become the subjectmatter of the various natural sciences."
3. "Logical operations constitute interbehavioral fields." "The materials must be regarded as...performing operations co-ordinate with those of the logician."
4. "Logical operations have evolved as techniques for achieving systems as products." "No other generalization is presupposed than that system-building goes on."
4.1 "Not only can the work be separated from the product, but each can be given its proper emphasis." "We may interbehave with...the objects of our own creation."
5. "Logic is essentially concerned with specific events."
6. "Since logic consists of actual interbehavior it sustains unique relations with the human situations in which it occurs." "As a human enterprise logic cannot escape certain cultural influences."
7. "Logic is inseverably interrelated with psychology." "Logic...entails a psychological dimension."
8. "Logic is distinct from language." It "is not...primarily concerned with...linguistic things." "Contrary to logical tradition, for the most part, symbols, sentences, or statements are only means for referring to...or for recording..."

The two procedures so outlined resemble each other in their insistence upon finding their subjectmatter in *concretely* observable instances of logical behavior; in their stress upon *operational* treatment of their subjectmatter; in their establishment of *natural* and *cultural* settings for the inquiry; and in their insistence that organism and environment be viewed together as *one* system. Kantor's 1, 2, 4, 5 and 6 follow in close correspondence with Dewey's 2, 8, 4, 7 and 5, while Dewey's 1 and 3 should easily be acceptable to him.

Within so similar a framework, however, marked differences of treatment appear. This, of course, is as it should be when a live field of fresh research is being developed. Dewey's 6 and Kantor's 3 and 7 might perhaps raise problems of interpretation in their respective contexts. The marked difference, however, is to be found in Kantor's 2.1 and 4.1 as compared with Dewey's 19, and in Kantor's 8 as compared with Dewey's 9 and 16. Kantor treats the system of organism and environment "interactionally," while Dewey makes the "transactional" approach basic. Kantor introduces "pre-logical materials" as requisite for logical activity, distinguishes logical activity sharply from linguistic activity, and offers as outcomes logical products akin in pattern to physical products and serving as stimulants to men in the same manner these latter do. Dewey, in contrast, exhibits inquiry as advancing from indeterminate to determinate situations in full activity throughout, and requires the "objects" determined by inquiry to be held within its system, future as well as past. Where Kantor holds himself to what can be accomplished from a start in which human organisms and environmental objects are presented to logic ready-formed as the base of its research, Dewey brings within his inquiry those very processes of knowledge under which organisms and objects are themselves identified and differentiated in ordinary life and in specific research as components of such a natural world.[2] We have thus within a very similar "natural" background two contrasting routes already indicated for further development.

III

The four other logics which we shall consider retain as presumptively basic various materials and activities derived from the vocabularies and beliefs of pre-Darwinian days. Such items of construction include, among many others, "sense-data," "concepts," "propositions," "intuitions," "apprehensions," "meanings," and a variety of "rationals" and "empiricals" taken either as individually separable "mental existences," as directly present "objects of mind," or as philosophical offspring of terminological interbreeding between them. The question which concerns us is as to how such materials enter into construction and how they behave.[3]

---

[1] All wordings are those of the section-headings of the postulates or of the immediately following text, except as follows: The sentence in 2.1 is from page 140, lines 11-12; the second sentence in 4 is from page 168, lines 13-14; the sentences in 4.1 are from page 294, lines 9-10, and page 7, lines 3-4; the sentence in 5 is from page xiii, lines 2-3; the second sentence in 7 is from page xiii, line 6.

[2] This difference is well brought out by a remark of Bridgman's which Kantor quotes in order to sharpen his statement of his own position. Bridgman holds that "from the operational point of view it is meaningless to separate 'nature' from 'knowledge of nature.'" Kantor finds Bridgman's view a departure from correct operational procedure. Dewey, on the contrary, would be in full agreement with Bridgman in this particular respect. P.W. Bridgman, *The Logic of Modern Physics* (New York, 1927), p. 62; Kantor, "The Operational Principle in the Physical and Psychological Sciences," *The Psychological Record*, II (1938), p. 6. For an appraisal of Kantor's work under a point of view sharply contrasted with that taken in the present text see the review by Ernest Nagel, *The Journal of Philosophy*, XLII ([1945], 578-80).

[3] Typical confusions of logical discussion have been examined from a different point of view in Chapter I. Certain characteristics of the work of Carnap, Cohen and Nagel, Ducasse, Lewis, Morris, and Tarski are there displayed. A thorough overhauling has long been needed of the procedures of Carnap and other logical positivists, both with respect to their logic and their positiveness, and this is now promised us by C.W. Churchman and T.A. Cowan (*Philosophy of Science*, XII [1945], 219). One device many logicians employ to justify them in maintaining the antiquated materials is their insistence that logic and psychology are so sharply different that they must leave each other alone—in other words, that while psychology may be allowed to "go natural," logic may not be so allowed. This argument of the logicians may be all very well as against an overly narrowed psychology of the non-natural type; but by the same token an overly narrowed logic results. The problem is one of full human behavior—how human beings have evolved with all their behaviors—no matter how convenient it has been found in the popular speech of the past to scatter the behaviors among separate departments of inquiry.

In the work of Cohen and Kaufmann we shall find earnest endeavor to smooth them into place in a modern world of knowledge. Thereupon in the light of what these men give us—or, rather, of what they have failed to give us thus far—we shall be able to get a clearer view of the violent struggles of Russell, and the intricate word-searchings of Moore, as they strive to establish and organize logical controls under their ancient forms of presentation.[1]

Professor Cohen's desire to strengthen logical construction had stimulation from Peirce on one side and from the later blossoming of symbolic logic on the other. He has not, however, taken a path which permits him to find Dewey's manner of construction adequate in the direct line from Peirce.[2] The citations we assemble are taken verbatim from his latest book. While they have been removed from their immediate contexts and rearranged, it is hoped that no one of them has been warped from its accurate significance in his construction.

COHEN[3]

1. "Symbols. . .represent. . .only. . .general properties" (p. 8, line 6);

2. "Science studies the. . .determinate properties of things" (p. 17, line 18). Physics, e.g., "starts with material assumptions, i.e., assumptions true only of certain objects, namely entities occupying time and space" (p. 16, line 3);

3. In the manipulation of symbols. . ."the meaning of our final result follows from our initial assumptions" (p. 8, line 13);

4. "The assumption that the objects of physics. . .must conform to logic is necessary in the sense that without it no science at all can be constructed" (p. 16, lines 15-16);

5. "The rules according to which. . .symbols can be combined are by hypothesis precisely those according to which the entities they denote can be combined" (p. 8, lines 8-10);

6. "Logically. . .existence and validity are strictly correlative" (p. 15, line 26).

This reads smoothly enough but it makes science, and apparently logic also, depend for foundations upon a "necessary assumption"—where "necessary" is what we cannot avoid, and "assumption" what we have to guess at. It sepa-

rates two great ranges of human attention, one called that of "symbols," the other that of the "determinate properties of things." However modernized their garb, these are little other than the ancient "reason" and "sense." Their organization is by *fiat*, by the flat assertion that they *must* be correlated. Such *fiat* is employed precisely because "system" has *not* been established. If the "entities occupying time and space" make up "nature," then the "symbols" remain "non-natural" in the sense in which we have employed the word, so long as they are not brought within a common system of interpretation, but enter merely by decree.

Professor Kaufmann, as we shall report him, works under a similar severance of certain of man's activities from the environing "nature" upon, within, or with respect to which, these activities are carried on. He develops a far-reaching, intricate, and in his own way powerful, analysis to establish organization for them conjointly. He accepts and admires Dewey's "theory of inquiry" as an outstanding contribution to knowledge but not by itself as an adequate logic. He holds that the theory of deduction must be grounded in intuitional meanings, and that with it must be correlated a theory for the empirical procedures of science in terms of the scientists' *decisions*. Far from regarding himself as severing the logical process and its canons from nature, he holds that what he is doing is to "define inquiry in terms of the canons."[4] This, however, still leaves the contrast with Dewey striking since the latter's undertaking has been to describe the canons along with all other logical activity as inquiry going on, rather than to use canons as criteria of its definition. The following are Professor Kaufmann's "tenets":

KAUFMANN[5]

1. The work of the logic of science is to clarify the rules of scientific procedure.

---

[1] For an extreme "mentalistic" and hopefully "solipsistic" base for logic, the procedures of C.I. Lewis may be brought into comparison by anyone sufficiently interested. Lewis is represented by the following "postulates," which, from any "natural" point of view, rate as disintegrating and unworkable traditions: (1) Knowledge involves three components, the activity of thought, the concepts which are produced by thought, and the sensuously given; (2) The pure concept and the content of the given are mutually independent; neither limits the other; (3) The concept gives rise to the *a priori* which is definitive or explicative of concepts; (4) Empirical knowledge arises through conceptual interpretation. *See Mind and the World Order* (New York, 1929), pp. 36ff.; "The Pragmatic Element in Knowledge," *Univ. of California Publications in Philosophy*, VI (1926); *A Survey of Symbolic Logic* (Berkeley, Calif., 1918). A characteristic determination arising in such a background is that if "analytic facts" can "function propositionally," then "they are called propositions"; so that "the proposition 'Men exist' is literally one and the same with the fact that men might exist." Lewis and Langford, *Symbolic Logic* (New York, 1932) p. 472. For other illustrations of what happens to ordinary integrity of expression under such a construction see my notes on Lewis' vocabulary, *The Journal of Philosophy* (1941), pp. 634-5. See also Chapter I, Section VI.

[2] See his discussion of Dewey's *Experimental Logic* (1916) reprinted as an appendix to his book *A Preface to Logic* (New York, 1944).

[3] All citations in the text are from *A Preface to Logic*. Compare the following from Cohen's essay, "The Faith of a Logician," in *Contemporary American Philosophy* (New York, 1930) p. 228: "Logical laws are thus neither physical nor mental, but the laws of all possible significant being."

[4] From private correspondence.

[5] The book here characterized is Felix Kaufmann's *Methodology of the Social Sciences* (New York, 1944). Page references are not given as the presentation in the text has Professor Kaufmann's endorsement as it stands with the proviso that "he does not maintain that scientists always consciously apply the rules in their inquiries" but that "he does maintain the reference to the rules is logically implied when the correctness of scientific decisions or the appropriateness of the methods applied is judged." "Formulations of such judgments which do not contain reference to the rules," he regards as "elliptical." The following citations, which Professor Kaufmann quite properly insists should be understood in the full context of the book, are assembled by the present writer who, properly also, he hopes, believes they are essential to show the manner in which expression under this procedure develops: "The contrast between deductive reasoning (in the strict sense) and empirical procedure. . .will be the guiding principle of our analysis and. . .the key to the solution of. . .problems" (*op. cit.* p. 3); "The most general characterization of scientific thinking" is "that it is *a process of classifying and reclassifying propositions by placing them into either of two disjunctive classes in accordance with presupposed rules*" (p. 40); "The distinction between the logical order of meanings and the temporal order of inquiry" is "all important" (p. 39); The "temporal aspect of inquiry does not enter into the timeless logical relations among propositions" (p. 30); "The fundamental properties of the system of rules are invariable" (p. 232); The "genuine logical theory of empirical procedure" is "*a theory of correct scientific decisions in given scientific situations*" (p. 65); Language requires "a system of rules that gives to particular acoustical phenomena the function of symbols for particular thoughts" (p. 17); "Lack of distinction in language is, in most cases, the consequence of unclear thought." (p. 27); "Concepts and propositions *are* meanings" (p. 18); In "problems of empirical science" and "logical analysis". . ."we have to presuppose (elementary) meanings" (p. 19). Kaufmann reiterates and emphasizes his difference from Dewey in a late paper (*Philosophy and Phenomenological Research*, VI [1945] 63n.) when he states that he cannot follow Dewey when the latter dismisses "intuitive knowledge of meanings" along with "intuitive knowledge of sense-data."

2. The corpus of science consists of propositions that have been accepted in accordance with such rules.

3. Changes in the corpus of a science, either by acceptance of a proposition into it, or by the elimination of a proposition from it, are called scientific decisions.

4. Scientific decisions are distinguished as correct or incorrect in terms of rules of procedure called *basic*. (Other rules called *preference* rules concern appropriateness of approach.) Basic rules as well as preference rules may be changed. Standards for the correctness of such changes are called rules of the second (or higher) order.

5. Principles governing the scientific acceptance or elimination of propositions, and placing limitation upon the changes of basic rules, are the reversibility of all decisions, the recognition of observational tests, the exclusion of contradictions, and the decidability in principle of all propositions.

6. The two last mentioned principles are called procedural correlates of the principles of contradiction and of excluded middle respectively. The former states that the basic rules of procedure must be such as to foreclose the emergence of contradiction in science. The latter states that the basic rules must be such as not to foreclose the verifiability of any given statement.

7. The correctness of scientific decisions in terms of basic rules depends solely on the knowledge established at the time, i.e., on previously accepted propositions which now serve as *grounds* for the acceptance of new ones.

8. Identifiable propositional meanings are presupposed in scientific decisions.

9. The presupposition of identifiable propositional meanings implies that we take it for granted that for any two given propositions it is determined whether or not one is entailed in the other, and whether or not one contradicts the other.

10. Entailment and contradiction are recognizable either directly in immediate apprehension of meanings or indirectly by deductive process.

11. Deductive processes are autonomous within scientific inquiry and can be described without reference to verification or invalidation.

In his construction Professor Kaufmann rejects the demand for the logical determination of ontological certainty in its older and more brazen form.[1] He is, indeed, unfriendly in many respects to its newer and more insidious forms. However, although he can content himself without the ontological specialized search, he can not content himself without the ontological searcher. He retains the non-natural "mental"—the "ego," "person," "decider," or basic "knower"—if not as existential possessor, then at least as subsistential vehicle or conveyor, of meanings. He sees science as composed of propositions, the propositions as being meanings (i.e., generally as "thoughts," or "concepts"), and the meanings as enjoying some sort of logically superordinate[2] existence over, above, or beyond, their physical, physiological, or behavioral occurrence. He requires decisions to

get the propositions, rules to get the decisions, and higher rules to get changes in the lower rules; behind all of which he puts a backlog of invariant (i.e., unchangeable) properties which the rules possess. Underlying all logical procedure he requires the presupposed meanings and the invariant properties; and underlying these he requires the intuition or immediate apprehension which operates them. Deduction is intuition indirectly at work. We are in effect asked to adopt a sort of indirectly immediate apprehension. It feels uncomfortable.

If we compare this with Dewey's "natural" procedure,[3] we find Dewey offering us science, not as a corpus of "propositions" embalmed in "decisions" made in accordance with prescribed "rules," but as the actual observable ongoing process of human inquiry in the cosmos in which it takes place. "Propositions," for him, are instruments, not exhibits. What happens, happens, and no need is found to insert "intuition" behind it to *make* it happen, or to be the happening. Meaningful utterance is taken as it comes, and not as separated from life and language. "Decision" is the long process of appraisal, often requiring cultural description; it is never some intermediate act-under-rule. The outcome in judgment is not a "conception" nor even a "pronouncement," but the full activity that rounds out inquiry. Finally the canons are to Dewey outgrowths of inquiry, not its presuppositions; their high value, when in active inquiry is fully recognized, but when set off by themselves they are found mechanical, arbitrary, and often grievously deceptive. This difference is not one of creed or opinion—Dewey's work is not to be taken on any such basis as that; it is a difference of practical workmanship, with the "credal" aspects trailing behind, and with the report we here give furnishing merely the clues to the practice.

IV

We come now to the struggle of Russell and the subtleties of Moore in their efforts to secure a logic under these ancient patterns of speech—logical, ontological, psychological, and metaphysical—in which sensings and conceivings, world and man, body and mind, empiricals and rationals, enter in opposing camps. Russell offers rich complexes of such materials, and their kaleidoscopic shiftings are so rapid that it is most of the time difficult to center one's eye on the spot where the issues are clearest. His great and everywhere recognized early achievements in symbolic logic and in planning its organization with mathematics have ended with his efforts in the last half of his life to find out what actually he has been dealing with. His view today seems as strong as in his earlier years—perhaps even stronger—that unless a man adopts some metaphysics and puts it to work, he cannot even make a common statement in everyday language.[4]

To represent Russell we shall establish a base in his Logical Atomism of 1918-1919 and 1924, and supplement this, where it seems desirable, by

---

[1] See the two typical marks of distinction between pre-Darwinian and post-Darwinian programs and goals suggested in the opening paragraphs of this chapter. Kaufmann's tenets No. 5 and No. 7 mark steps of his advance.

[2] The word "superordinate" is here employed by me as an evasive compromise. Kaufmann would say that "the meanings" are "presupposed in," "essential to," "logically implied by," or "necessary for the definition of" the "inquiry." I would say that what his development actually accomplishes is to retain them as "prior to," "superior to," "independent of," or "in a realm apart from" the "inquiry." He fully satisfies me that my wording is not what he intends, but without affecting my view that I am nevertheless describing what he in effect does.

[3] Direct comparison of particular phrases is not simple, because the whole method of expression—the "linguistic atmosphere"—varies so greatly. However, K2 may be compared with D10 and D15; K4 with D5; K7 with D19; and K10 with D18. In addition the citations about language on p. 153, footnote 5, taken from Kaufmann's pages 17 and 27 are at the extreme opposite pole, so far as present issues go, from D16.

[4] Or at least this seems to be the purport of such a conclusion as that "the goal of all our discussions" is "that complete metaphysical agnosticism is not compatible with the maintenance of linguistic propositions" (*An Inquiry into Meaning and Truth* [New York, 1940] p. 437).

citations from earlier and later papers. The clumsiness of our report is regrettable, but it is due to overlapping and ever-shifting applications by Russell of such words as "simple," "particular," "entity," and "symbol," which make plain, direct citation often risky, and at times altogether impracticable.[1]

RUSSELL[2]

1. "Ultimate simples," (in theory if not in practical research) are entities "out of which the world is built" (M, 1919, 365). They "have a kind of reality not belonging to anything else" (M, 1919, 365). "Simple" objects are "those objects which it is impossible to symbolize otherwise than by simple symbols" (M, 515).

2. Propositions and facts are complexes. "I do not believe that there are complexes. . .in the same sense in which there are simples" (LA, 374).

3. Complexes are to be dealt with through their component simple entities or simple symbols. "It seems obvious to me that what is complex must be composed of simples" (LA, 375). "Components of a proposition are the symbols. . . ; components of the fact. . .are the meanings[3] of the symbols" (M, 518).

4. Simple symbols are those "not having any parts that are symbols, or any significant structure" (LA, 375. *Cf.* M, 515).

5. Knowledge is attained through the fixation of the right simples by the right logical proper names, i.e., symbols (the argument of M and LA throughout). "An atomic proposition is one which does. . .actually name. . .actual particulars" (M, 523).[4]

6. As a controlling principle: "Wherever possible, substitute constructions out of known entities for inferences to unknown entities" (LA, 363).[5]

7. Among the simples consider the particulars (M, 497).[6] These are "the *terms* of the relation" in atomic facts (M, 522). Proper names properly apply to them and to them alone (M, 508, 523, 524). "Particulars have this peculiarity. . .in an inventory of the world, that each of them stands entirely alone and is completely self-subsistent" (M, 525).

8a. Particulars are known by direct acquaintance. "A name. . .can only be applied to a particular with which the speaker is acquainted" (M, 524).[7] "The word 'red' is a simple symbol. . .and can only be understood through acquaintance with the object" (M, 517).

8b. "Simples" are not "experienced as such"; they are "known only inferentially as the limit of analysis" (LA, 375).[8]

9. For success in attaining knowledge it becomes necessary to sort propositions into types. "The doctrine of types leads. . .to a more complete and radical atomism than any that I conceived to be possible twenty years ago" (LA, 371).

10. In "The Type's Progress," the stages thus far (1945) have been:

a) The *entities*[9] exist in a variety of types;

b) "The theory of types is really a theory of *symbols*, not of things" (M, 1919, 362);

c) Words (symbols) are all of the *same* type (LA, 369);

d) The meanings of the symbols may be of *any* type (I, 44; *see also* LA, 369);

e) (when the going seems hard) "Difference of type means difference of syntactical function" (RC, 692);[10]

f) (when the going seems easy) "There is not one relation of meaning between words and what they stand for, but as many relations of meaning, each of a different logical type, as there are logical types among the objects for which there are words" (LA, 370);

---

1 A specimen of Russell's conflicting phrasings from his book *What I Believe*, is quoted by Cassius J. Keyser in *Scripta Mathematica*, III (1935), 28-29 as follows: (page 1) "Man is a part of nature, not something contrasted with nature"; (p. 16) "We are ourselves the ultimate and irrefutable arbiters of value, and in the world of value Nature is only a part. Thus in this world we are greater than Nature."

2 The sources of the citations from Russell are indicated as follows:
M. "The Philosophy of Logical Atomism," *Monist*, (1918), 495-527; (1919), 32-63, 190-222, 345-380. Page references are to the 1918 volume unless otherwise indicated.
LA. "Logical Atomism," *Contemporary British Philosophy*, New York, First Series, 1924, pp. 359-383.
RC. "Reply to Criticisms," *The Philosophy of Bertrand Russell*, P.A. Schilpp, editor, (Chicago, 1944), pp. 681-741.
I. *An Inquiry into Meaning and Truth.*

3 What Russell intends by meaning is, in general, very difficult to determine. It is not that no light is thrown on the question but entirely too many kinds of light from too many points of view, without sifting. Most profitable is an examination of all the passages, a dozen or more, indexed in the *Inquiry*. See also M, 506-8; LA, 369, and Bertrand Russell, *Mysticism and Logic*, (New York, 1918), pp. 223-4.

4 For a discussion in terms of "basic propositions" see I, p. 172, p. 362, p. 414. Here the contrast between Russell and Dewey is so sharp (see Dewey, No. 14, preceding) that the extensive discussions between the two men could be reduced to a one-sentence affirmation on this point and a one-page exhibit of the context of discussion, historical and contemporary.

5 An alternative form will be found in a paper in *Scientia*, 1914, reprinted in *Mysticism and Logic*, p. 155: "Wherever possible, logical constructions are to be substituted for inferred entities."

6 These are variously called logical atoms, ultimate constituents, simple entities, etc. "Such ultimate simples I call 'particulars'" *An Analysis of Mind* (New York, 1921), p. 193. They are the hardest of hard facts, and the most resistant to "the solvent influence of critical reflection." They may be sense-data, or entities called "events" (LA, 381) or sometimes point-instants or event-particles. Mathematical-physical expressions sometimes join them among the ultra-safe. If Russell would establish definite usage for at least one or two of these words, his reader might have an easier time doing justice to him. It is particularly disconcerting to find the particulars turning out to be themselves just words, as where (I, 21) he speaks of "egocentric particulars, i.e., words such as 'this,' 'I,' 'now,' which have a meaning relative to the speaker." If "terms" are "words" for Russell (I would not presume to say) then the second sentence in point No. 7 in the text also makes particulars out to be symbols rather than entities. For indication of Russell's logical atoms and proper names as of the nature of "cues" and similar primitive behaviors, see p. 122, footnote 3, Chapter VI, Section IV and Chapter VII, Section II.

7 Compare *Problems of Philosophy* (New York, 1912), p. 91; "On the Nature of Acquaintance," *Monist*, (1914).

8 If there has been any systematic progress in Russell's work as the years pass by with respect to attitudes 8A and 8B, I have failed to detect it. The difference seems rather one of stress at different stages of argumentation. If the clash as here reported seems incredible, I suggest an examination of a particularly illuminating passage in Professor Nagel's contribution to *The Philosophy of Bertrand Russell*, p. 341, in which, though without directly mentioning the incoherence, he notes (a) that Russell holds that some particulars are perceived, and at least some of their qualities and relations are immediately apprehended; (b) that Russell believes his particulars are simples; and (c) that Russell admits that simples are not directly perceived, but are known only inferentially as the limit of analysis. Further light on the situation may be gained from Nagel's penetrating analysis of *An Inquiry into Meaning and Truth. The Journal of Philosophy*, XXXVIII (1941), 253-270.

9 RC, 691; *Principles of Mathematics*, (Cambridge, England, 1903); *Introduction to Mathematical Philosophy*, (New York, 1919); also, off and on, at any stage of his writings. Note the similar difficulty for "particulars," point No. 7 and footnote 6 above.

10 I, p. 438: "Partly by means of the study of syntax, we can arrive at considerable knowledge concerning the structure of the world."

g) (and at any rate) "Some sort of hierarchy is necessary. I hope that in time, some theory will be developed which will be simple and adequate, and at the same time be satisfactory from the point of view of what might be called logical common sense." (RC, 692).[1]

Probably the sharpest criticism to be made of Russell's workmanship is to point out his continual confounding of "symbol" and "entity." We have had illustrations of this in the cases, both of the "particular" and of the "type." Fusion of "symbol" and "entity" is what Russell demands, and confusion is what he gets. With an exhibit as prominent as this in the world, it is no wonder that Korzybski has felt it necessary to devote so much of his writings to the insistent declaration that the word is *not* the thing.[2] His continued insistence upon this point will remain a useful public service until, at length, the day comes when a thorough theory of the organization of behavioral word and cosmic fact has been constructed.

Turning now to Moore, we find him using much the same line of materials as does Russell, but he concentrates on the ultimate accuracies of expression in dealing with them. Whether primarily classified as a logician or not, he outlogics the logics in his standards of logical perfection. Where Russell proposes to force the ultimate simples of the world to reveal themselves, Moore takes a frank and open base in the most common-sense, matter-of-fact, personal experiences he can locate, accepting them in the form of "simplest" propositions. He then takes account of sense-data, linguistic expressions, the conceptual and propositional meaningfulness of these latter, man's belief in them, his feelings of certainty about his beliefs, and his assertions of known truth; plus, of course, the presumptive "realities" he takes to be involved. Where Russell finds himself compelled continually to assert that his critics fail to understand him, Moore is frank in his avowal that he is never quite sure that he understands himself.[3] He is as willing to reverse himself as he is to overthrow others. His virtues of acuity and integrity applied to his presuppositions have yielded the following development:

MOORE[4]

1. Start with common sense statements, such as "this is my body," "it was born," "it has grown," "this is a chair," "I am sitting in this chair," "here is my hand," "here is another hand." Examine these as propositions, and in all cases under reduction to the simplest expression that can be reached—such, that is to say, as is most secure

of ordinary acceptance, and is least liable to arouse conflict (CS, 193-195).[5]

2. Accept these common-sense propositions as "wholly true" (CS, 197); as what "I know with certainty to be true" (CS, 195).[6]

3. In such a proposition, if it is thus held true, "there is always some *sense-datum* about which the proposition in question is a proposition," i.e., its "subject" is always a sense-datum (CS, 217). Moreover such a proposition is "unambiguous," so that "we understand its meaning" in a way not to be challenged, whether we do or do not "know what it means" in the sense of being able *"to give a correct analysis"* of its meaning" (CS, 198).[7]

4. On the basis of such simplified common sense propositions having sense-data for subjects, very many[8] other instances of knowledge in propositional form can be tested and appraised through Analysis.

5. In Analysis a "concept, or idea, or proposition" is dealt with, "and *not* a verbal expression" (RC, 663, 666). The word "means" should not be used since that implies a "verbal expression," and therefore gives a "false impression" (RC, 664; this passage is seventeen years later than that in No. 3 above, where the word "means" is still employed.)

6. To "give an analysis" of a concept you must come across (or, at least, you "must mention") another concept which, like the first, "applies to" an object (though it neither "means" nor "expresses" it) under circumstances such that (a) "nobody can know" that the first concept "applies" without "knowing" that the second applies; (b) "nobody can verify" that the first applies without "verifying" that the second applies; and (c) "any expression which expresses" the first "must be synonymous with any expression which expresses" the second (RC, 663).[9]

7. Otherwise put, a "correct" analysis in the case of concepts is one which results in showing that two concepts expressed by different expressions "must, in some sense" be the same concept[10] (RC, 666).

8. To establish itself in firm status for the future, Analysis has two primary tasks, (a) it must make a successful analysis of sense-data (CS, p. 216-222); (b) it must make a successful analysis of Analysis itself (Compare RC, 660-667).

9. Analysis of sense-data has thus far been unsatisfactory. Its present status is best exhibited in a particular case of analysis. Take, for example, "the back of my hand" as "something seen," and seek to establish what,

---

1 For the latest illustration of Russell's confusion of statement, pages 829-834 of his *A History of Western Philosophy*, (New York, 1945) may profitably be examined. A passing glance will not suffice since the main characteristic of philosophical language is to make a good appearance. A cold eye, close dissection, and often much hard work is necessary to find out what kind of a skeleton is beneath the outer clothing.

2 Alfred Korzybski: *Science and Sanity: an Introduction to Non-Aristotelian Systems and General Semantics,* (New York, 1933).

3 Russell remarked to Professor Schilpp, the editor of the volume *The Philosophy of Bertrand Russell,* that "his greatest surprise, in the reading of the twenty-one contributed essays, had come from the discovery that 'over half of their authors had *not* understood' him (i.e. Russell)." *(Op. cit., p.* xiii). For Moore see No. 9 and No. 10 of the skeleton construction of his logical procedure, which follows.

4 The sources of the citations from Moore are indicated as follows:
    CS. "A Defense of Common Sense," *Contemporary British Philosophy.* Second Series, (New York and London, 1925), pp. 193-223;
    RC. "A Reply to my Critics" *The Philosophy of G.E. Moore,* P.A. Schilpp, editor, (Chicago, 1942), pp. 535-677.

5 See also "Proof of an External World," *Proceedings of the British Academy,* XXV (1939), pp. 273-299. Professor Nagel's comment in his review of *The Philosophy of G.E. Moore* in *Mind,* [1944], 60-75 will be found of interest.

6 Included are physical objects, perceptive experiences taken as mental, remembered things, and other men's bodies and experiencings. "I think I have nothing better to say than that it seems to me that I *do* know them, with certainty. It is, indeed, obvious that, in the case of most of them, I do not know them *directly"*. . .,but. . ."In the past I have known to be true *other* propositions which were evidence for them" (CS, 206).

7 "I think I have always both used, and intended to use, 'sense-datum' in such a sense that the mere fact that an object is *directly apprehended* is a *sufficient* condition for saying that it is a sense-datum" (RC, 639). A remarkable illustration of his careful expression may be found in the passage on page 181 of his paper, "The Nature of Sensible Appearances," *Aristotelian Society,* Supplementary Vol. VI (1926).

8 "Very many" is to be understood in the sense in which Moore uses the words (CS, 195), with a trend towards, but not immediate assertion of, "all."

9 Note the confidently reiterated "nobody can" and the "must."

10 In the typical case, however, one concept is opposed to two or more concepts, these latter being accompanied in their consideration by explicit mention of their method of combination (RC, 666).

precisely, *is* the sense-datum that enters as subject of the "common sense" and "wholly true" proposition: "This is the back of my hand." In 1925 Moore reported that "no philosopher has hitherto suggested an answer that comes anywhere near to being *certainly* true" (CS, 219). In 1942 his report was "The most fundamental puzzle about the relations of sense-data to physical objects is that there does seem to be some reason to assert. . .paradoxes"(RC, 637).[1]

10. Analyses of Analysis itself also arrives at paradox, so that in the outcome it may be said: "I do not know, at all clearly, *what* I mean by saying that '*x* is a brother' is identical with '*x* is a male sibling,' and that '*x* is a cube' is *not identical* with '*x* is a cube with twelve edges' "(RC, 667).[2]

It may be suggested on the basis of the above display of Moore's techniques and results that his Analysis could reasonably be carried still further. Analysis of "concept" and of "proposition," of "expression" and of "meaning," and of "datum" as well as of "sense," might lead towards solutions. This, however, would involve untrammelled inquiry into "man's analyzing procedures" for whatever such procedures might show themselves operationally to be, in a full naturalistic setting. And this, again, would require throwing off the limitations imposed by the old vocabularies that place "man the analyzer" outside of, or over against the world of his analysis.[3] The differences in spatial and temporal location are huge between what is "sense-datum" and what is "wholly true." The range of the one is a bit of an organism's living in a bit of environment. The range the other seems to claim is all, or even more than all, of space and time. Analysis will surely need to be super-jet, if it is to make this transit, fueled as it is proposed it be by "concept, idea, and proposition," and these alone.

V

The reader who wishes to appraise for himself the situations we have exhibited—and especially the reader who has been accustomed to the use of his hands and eyes on materials such as enter any of the natural sciences—may be interested in an experiment. Let him look on logicians' writings as events taking place in the world. Let him pick out some phase of these events for study. Let him be willing to examine it at least as carefully as he might the markings of a butterfly's wings, remembering also that the present level of inquiry into logic is not much further advanced than that into butterflies was when they were still just museum curiosities, and modern physiology undreamt of. This will mean clearing his work bench of all superfluous terminologies, and "getting down to cases," with the cases under examination, whatever they are, pinned down on the bench and not allowed to squirm themselves out of all decent recognition. Suppose such an inquirer has noticed the word "proposition" frequently present in the text. On the assumption, however rash, that logical terms are supposed to denote, name, designate, point at, or refer to something factually determinate, let him then select the presumptive fact "proposition" for his examination. By way of preliminary orientation, if he should examine the six logicians we have considered he would find that for Dewey a proposition is an instrumental use of language (D9, D10, D15, D16); for Kantor it is a "product" of logical interbehavior;[4] for Cohen "propositions are linguistic forms with meanings that are objective relations between such forms and certain states of fact";[5] for Kaufmann a proposition is a "meaning" developed from a base in intuition, fundamentally presupposed as such by logic, and not therein to be investigated.[6] For Russell it may be a class of sentences, or the meaning or significance of a sentence, or even at times something he doubts the existence of;[7] for Moore, it is a dweller in a land of thoughts, companion of concepts and ideas, and to be found midway (or perhaps some other way) between words and objects.[8] Here is surely not merely "food for thought" but much incentive to matter-of-fact research. A few further trails for searchers to follow are mentioned in the footnote.[9]

In the preceding examination I have done my best to be accurate and fair. I hope I have at least in part succeeded. Certainly I have squandered time and effort triple and quadruple what I would have agreed to at the start. I find myself unwilling to close without expressing my personal opinion more definitely than I have heretofore. The procedures of Russell and Moore seem so simple-minded it is remarkable they have survived at all in a modern world. Those of Cohen and Kaufmann are heroic efforts to escape from the old confusions, yet futile because they fail to pick up the adequate weapons. In what may grow from the two other enterprises I have, of course, great hopes.

---

[1] Moore has written: "I define the term (sense-datum) in such a way that it is an open question whether the sense-datum which I now see in looking at my hand and which is a sense-datum of my hand is or is not identical with that part of its surface which I am now actually seeing" (CS, 218). In simplified report his analysis in the case of "the back of my hand" discriminates "a physical object," "a physical surface," and a certain "directly seen" (such as one has in the case of an after-image or double-image). Moore's analysis with respect to the second and third of these has results which indicate to him that at the very time at which he not only feels sure but *knows* that he is seeing the second, he is in a state of doubt whether the third, which also he is seeing (and that *directly* in the indicated sense), is identical with the second or not; recognizing that it may be identical, in which case he is in a position of both "feeling sure of and doubting the very same proposition at the same time" (paradox I); or "so far as I can see," at any rate, "I don't *know* that I'm not (Paradox II). It is to the second form of the paradox that the comment cited in the text above refers (RC, 627-653, and particularly 636-637, also CS, 217-219).

[2] The analysis of Analysis which Moore offers (RC, 664-665) declares equivalence as to concepts between expressions of the form: this "concept" is "identical" with that, this "propositional function" is "identical" with that, and "to say this" is "the same thing" as "to say that." But if we proceed to another form which also we feel we must accept, such as "to be this" is "the same thing" as "to be that," we have, we are told, reached a paradox, which, as between expressions and concepts, remains unresolved.

[3] It is significant in this connection that Moore tells us that it is always "things which other philosophers have said" that suggest philosophical problems to him. "I do not think," he remarks, "that the world or the sciences would ever have suggested to me any philosophical problems." *The Philosophy of G.E. Moore*, p. 14.

[4] J.R. Kantor, *op. cit.*, p. 223, pp. 282-3; also "An Interbehavioral Analysis of Propositions," *Psychological Record*, 5 (1943) p. 328.

[5] M.R. Cohen, *op. cit.*, p. 30. Also: "Acts of judgment, however, are involved in the apprehension of those relations that are called meanings." See also M.R. Cohen and E. Nagel, *An Introduction to Logic and Scientific Method*, (New York, 1934), pp. 27, 28, 392, where facts are made of propositions, and propositions are specifically declared to be neither physical, mental, linguistic, nor communication, and to be identifiable by the sole characteristic that whatever else they are they are "true or false."

[6] Felix Kaufmann, *op. cit.*, pp. 18, 19.

[7] B. Russell: *An Inquiry into Meaning and Truth*, pp. 208, 210, 217, 237 *et. al. Proceedings of the Aristotelian Society*, XI, (1911), 117. *Mysticism and Logic*, p. 219; *Monist* (1918) p. 504.

[8] See phrasings in Moore, No. 1, No. 3, No. 5 *et al.* To Moore all such items are as familiar as the tongues of angels, and it is difficult, perhaps even impossible to find a direct cite.

[9] Kaplan and Copilowish, *Mind*, (1939), 478-484; Lewis and Langford, *Symbolic Logic*, p. 472; A.P. Ushenko, *The Problems of Logic* (1941) pp. 171, 175, 219; Roy W. Sellars in *Philosophy and Phenomenological Research*, V, (1944) 99-100; G. Ryle, *Proceedings of the Aristotelian Society*, Supplementary Vol. IX (1929) pp. 80-96. An excellent start, and perhaps even a despairing finish, may be made with the Oxford Dictionary, or some other larger dictionary.

IX.

A CONFUSED "SEMIOTIC" [1]

I

CHARLES MORRIS, in *Signs, Language, and Behavior* (New York, 1946) declares himself a semiotician (p. 354) operating in harmony with "behavioristicians" (pp. 182, 250). "Semiotic," he tells us, is "the science of signs," and "semiosis" is that sign-process which semioticians investigate (p. 353). If he is to "lay the foundation for a comprehensive and fruitful science of signs," his task is, he says, "to develop a language in which to talk about signs" (pp. v, 4, 17, 19), and for this, he believes, "the basic terms. . .are statable in terms drawn from the biological and physical sciences" (p. 19). It is possible in this way, he believes, to "suggest connections between signs and the behavior of animals and men in which they occur." (p. 2)

Here is a most laudable enterprise. I wish to examine carefully the technical language Professor Morris develops, find out whether it contains the makings of dependable expression such as we commonly call "scientific," and appraise his own opinion that the terms he adopts are "more precise, interpersonal, and unambiguous" than those favored by previous workers in this field (p. 28). The numerous special features of this book, often of high interest and value, I shall leave to others to discuss.[2]

We are greatly aided in our task by the glossary the author furnishes us. In it he "defines"[3] or otherwise characterizes the main "terms" of semiotic, and stars those which he deems "most important" as "the basis" for the rest. We shall center our attention on a central group of these starred terms, and upon the linguistic material out of which they are constructed. The reader is asked to keep in mind that the problem here is not whether, impressionistically, we can secure a fair idea as to what Professor Morris is talking about and as to what his opinions are, but rather whether his own assertion that he is building a scientific language, and thus creating a science, can be sustained through a close study of his own formulations. The issue will be found to be one of maximum importance for all future research and appraisal of knowings and knowns. Our conclusion will be that his attempt is a failure.

We are somewhat hampered by the fact that, although he builds throughout with respect to behavior, he does not "define" the terms he takes over from "general behavior theory," but says that these "really operate as undefined terms in this system" (p. 345). It is evident that this manner of being "undefined" is not at all the same as the manner we find in a geometrician's postulated "elements." Instead of freeing us from irrelevant questions, it burdens us at almost every step with serious problems as to just how we are to understand the writer's words.

There are other difficulties such as those that arise when we find a term heavily stressed with respect to what it presents, but with no correlated name or names to make clear just what it excludes. The very important term "preparatory-stimulus" is a case in point; the set of variations on the word "disposition," later listed, is another. The difficulty here is that in such instances one is compelled to interpolate other names to make the pattern a bit clearer to oneself, and this always invokes a risk of injustice which one would wish to avoid.

From this point on I shall use the word semiotic to name, and to name only, the contents of the book before us. I shall use the word semiosis to name, and to name only, those ranges of sign-process[4] which semiotic identifies and portrays. It is evident that, so proceeding, the word "semiotician" will name Professor Morris in his characteristic activity in person, and nothing else.

Four none too sharply maintained characteristics of the point of view that underlies semiotical procedure may now be set down for the reader's preliminary guidance:

1. Semiotic "officially"[5] declares the word "behavior" as in use to name, and to name only, the muscular and glandular actions of organisms in goal-seeking (i.e., "purposive") process.[6]

2. Semiosis is expressly envisaged, and semiotic is expressly constructed, with reference to behavior as thus purposive in the muscular-glandular sense. If there exists anywhere any sign-process not immediately thus oriented, it is *technically* excluded from the semiotic which we have before us. (One form of behavioral process which most psychologists regard as involving sign, but which Morris' formulation excludes, is noted in footnote 1, p. 165 following.) The assurance the semiotician gives us that semiotic provides us with a universal sign-theory does not alter this basic determination; neither does the weft of "sign-signify-significatum" and "sign-denote-denotatum" woven upon this muscular-purposive warp to make a total web.[7]

3. The two other main "factors" of semiotical inquiry—namely, stimulus and disposition to respond—are *not* behavior in the strict sense of the term in semiotic (even though now and then referred to nontechnically as behavioristic or behavioral).

4. With a very few, wholly incidental, exceptions all

---

1 This chapter is written by Bentley.

2 A discussion by Max Black under the title "The Limitations of a Behavioristic Semiotic" in *The Philosophical Review*, LVI (1947), 258-272, confirms the attitude of the present examination towards several of Morris' most emphasized names such as "preparatory", "disposition" and "signification." Its discussion is, however, on the conventional lines of yes, no, and maybe so, and does not trace back the difficulty into traditional linguistic fixations as is here attempted. See also reviews by A.F. Smullyan in *The Journal of Symbolic Logic*, XII (1947), 49-51; by Daniel J. Bronstein in *Philosophy and Phenomenological Research*, VII (1947), 643-649; and by George Gentry and Virgil C. Aldrich in *The Journal of Philosophy*, XLIV (1947), 318-329.

3 I shall permit myself in this chapter to use the words "define" and "term" casually and loosely as the author does. This is not as a rule, safe practice, but in the present case it eliminates much incidental qualification of statement, and is, I believe, fairer than would be a continual quibbling as to the rating of his assertions in this respect.

4 "Sign-process" is used by Morris in a very general and very loose sense. See Assertion No. 25 following.

5 I shall use the word "official" occasionally to indicate the express affirmations of the glossary as to terminology; this in the main only where contrasts suggest themselves between the "official" use and other scattered uses.

6 The word "behavioristics" is used loosely for wider ranges of inquiry. The compound "sign-behavior" is sometimes loosely, sometimes narrowly used, so far as the component "behavior" is concerned.

7 This statement applies to semiotic as it is now before us and to the range it covers. Professor Morris leaves the way open for other "phenomena" to be entered as "signs" in the future (p. 154, *et passim*). These passages refer in the main, however, to minor, marginal, increments of report and do not seem to allow for possible variations disruptive of his behavioral construction.

"official" reports in semiotic are made through the use of such key words as "produce," "direct," "control," "cause," "initiate," "motivate," "seek," and "determine."[1] Semiotic works thus in terms of putative "actors" rather than through direct description and report upon occurrences. This characteristic is so pronounced as to definitely establish the status of the book with respect to the general level of scientific inquiry.

Recall of the above characteristics will be desirable to avoid occasional misunderstandings.

Our primary materials of inquiry are, as has been said, to be found in a central group of the terms that are starred as basic. In fabricating them, the semiotician uses many other words not starred in the glossary, and behind and beyond these certain other words, critical for understanding, though neither starred nor listed. Among the starred terms that we shall examine as most important for our purposes are *sign, *preparatory-stimulus, *response-sequence, *response-disposition, and *significatum. Among unstarred words conveying key materials are *behavior, response, stimulus* and *stimulus-object*. Among key words neither starred nor listed are 'reaction,' 'cause,' 'occasion,' 'produce,' 'source,' and 'motivate.' It is interesting to note that *preparatory-stimulus is starred, but *stimulus* and *stimulus-object* are not (while "object" is neither indexed nor discussed in any pertinent sense); that *behavior-family is starred but *behavior* is not; that *response-disposition and *response-sequence are starred but *response* is not; that *sign and *sign-family are starred but *sign-behavior* is not. We have thus the "basic" terms deliberately presented in nonbasic settings.

II

Before taking up the terminological organization of semiotic, it will be well to consider two illustrations of the types of statement and interpretation that frequently appear. They serve to illuminate the problems that confront us and the reasons that make necessary the minuteness of our further examination.

Consider the following: "For something to be a sign to an organism. . .does not require that the organism signify that the something in question is a sign, for a sign can exist without there being a sign that it is a sign. There can, of course, be signs that something is a sign, and it is possible to signify by some signs what another sign signifies." (p. 16).

The general purport of this statement is easy to gather and some addicts of Gertrude Stein would feel at home with it, but precision of expression is a different matter. The word "sign" is used in semiotic in the main to indicate either a "stimulus" or an "object,"[2] but if we try to substitute either of these in the statement we find difficulty in understanding and may lose comprehension altogether. Moreover, the verb "signify," closely bound with "sign" and vital in all semiotical construction, is found strangely entering with three types of subjects: an "organism" can signify; a "sign" can signify; and indefinitely "it is possible" to signify.

Try, next, what happens in the development of the following short sentence: "Signs in the different modes of signifying signify differently, have different significata."

(p. 64).

We have here a single bit of linguistic expression (centering in the word "sign") differentiated with respect to participations as subject, verb, or object, and with the three phases or aspects, or whatever they are, put back together again into a sentence. What we have before us looks a bit like a quasi-mathematical organization of sign, signify, and significatum, the handling of which would require the firm maintenance of high standards; or else like a pseudo-physical construction of the general form of "Heat is what makes something hot." We shall not concern ourself with the possible difficulties under these respective interpretations, but solely with what happens to the words in the text.

The sentence in question opens a passage dealing with criteria for differentiating modes of signifying (pp. 64-67). I have analyzed the elusive phrasings of its development half a dozen times, and offer my results for what they may be worth as a mere matter of report on the text, but with no great assurance that I have reached the linguistic bottom of the matter. It appears that the semiotician starts out prepared to group the "modes of signifying" into four types: those answering respectively to queries about "where," "what," "why," and "how" (p. 65, lines 9, 10, 11; p. 72, lines 6-7 from bottom of page). To establish this grouping semiotically, he employs an extensive process of phrase-alternation. He first gives us a rough sketch of a dog seeking food, thereby to "provide us with denotata of the signs which we wish to introduce" (p. 65). Here he lists four types of "stimuli," presents them as "signs," and calls them identifiors, designators, appraisors, and prescriptors. He tells us (p. 65, bottom) that these stimuli "influence" behaviors, "and so" dispositions (although, in his official definition for sign,[3] behaviors do not influence dispositions but instead these latter must be built up independently prior to the behaviors). Next he shifts his phrasings in successive paragraphs from disposition to interpretant, and then from interpretant to significatum, saying what appears to be the same thing over again, but each time under a different name. Finally he revamps his phrasing again into a form in which it is not the stimulus that "disposes" but the interpreter who "is disposed."[4] He then suggests that a new set of names be introduced for four major kinds of significata: namely, *locatum*, *discriminatum*, *valuatum*, and *obligatum*. Since there is no official difference between significatum and signification (p. 354) he now has acquired names indicative of the four "modes of signifying" which is what was desiderated.

If the reader will now take these two sets of names and seek to discover what progress in inquiry they achieve, he will at once find himself involved in what I believe to be a typical semiotic uncertainty. This is the problem of verbal and nonverbal signs, their analysis and organi-

---

[1] A longer list of such words with illustration of their application will be given later in this chapter.

[2] A variety of other ranges of use for the word will be noted later: *see* Assertions No. 1, 2, 3, 19, 29, and 32; also p. 162, footnote 5.

[3] See Assertion No. 19 later in this paper, and the accompanying comment.

[4] Such a shift as this from an assertion that the stimulus (or sign or denotatum) "disposes the dog" to do something to the assertion that "the interpreter (i.e., the dog) is disposed" to do it, is common in semiotic. The trouble is that the "is disposed to" does not enter as a proper passive form of the verb "disposes," but is used practically (even if not categorically) to assert power in an actor; and this produces a radical shift in the gravamen of construction and expression. As a personal opinion, perhaps prematurely expressed, I find shifts of this type to be a major fault in semiotic. They can be successfully put over, I believe, only with verbs carefully selected *ad hoc,* and their employment amounts to something very much like semiotical (or, perhaps more broadly, philosophical) punning.

zation.[1] Taking the case of identifior and *locatum* as developed on pages 64-69, (I am following here the typographical pattern of the text) one finds that both of these words enter without addition of the single quotation marks which are added when it is the word, as a *word*, that is under examination. Now in the case of identifior the lack of single quotation marks corresponds with the use of the word in the text where certain nonverbal facts of life, such as dog, thirst, water, and pond, are introduced. In the case of *locatum*, however, the word enters directly as sign, with indirect reference to it as a term. The italics here are apparently used to stress the status of locatum and its three italicised companions as "special terms for the special kinds of significata involved in signs in the various modes of signifying." (p. 66). The textual introduction of *locatum* in extension from identifior is as follows: "We will use *locatum, discriminatum, valuatum,* and *obligatum* as signs signifying the significata of identifiors, designators, appraisors, and prescriptors." (p. 67). Under this treatment semiotic yields the following exhibits:

a) The identifior has for its significatum location in space and time.[2]

b) *Locatum* is a sign signifying the significatum of identifior.

c) *Locatum* therefore has for its significatum a location in space and time.

d) The significatum of *locatum* thus differentiates one of the great "modes of signifying" which are the subject of investigation—the one, namely, concerning locations.

Here we have an army of words that march up the hill, and then march down again. What is the difference between "location" at the beginning and "location" at the end? How great is the net advance? This can perhaps best be appraised by simplifying the wording. If we drop the word 'significatum' as unproductively reduplicative with respect to 'sign' and 'signify,' we get something as follows:

a) That which a sign of location signifies is location.

b) *Locatum* is a sign used to signify that which a sign of location signifies.

c) *Locatum* thus signifies location.

d) *Locatum* now becomes a special term to name this particular "mode of signifying."

A second approximation to understanding may be gained by substituting the word 'indicate' for 'signify,' under a promise that no loss of precision will thereby be involved. We get:

a) Signs of location indicate locations (and now my story's begun).

b) *Locatum*, the word, is "used" to indicate what signs of location indicate.

c) *Locatum* thus indicates location.

d) Location, thus indicated by *locatum*, enables the isolation behaviorally (p. 69) of that "mode of signifying" in which signs of location are found to indicate location. (and now my story is done).

In other words the progress made in the development from terms in *or* to terms in *um* is next to nothing.

The semiotician seems himself to have doubts about his terms in "um," for he assures us that he is not "peopling the world with questionable 'entities' " and that the "um" terms "refer only to the properties something must have to be denoted by a sign" (p. 67). But " 'property' is a very general term used to embrace. . .the denotata of signs" (p. 81), and the locatum and its compeers have been before us as significata, not denotata; and signifying and denoting are strikingly different procedures in semiosis, if semiotic is to be believed (pp. 347, 354).[3] The degree of salvation thus achieved for the terms in "um" does not seem adequate.

These and other similar illustrations of semiotical procedure put us on our guard as to wordings. The second of them is important, not only because it provides the foundation for an elaborate descriptive classification of significations which is one of the main developments of semiotic,[4] but further, because it displays the attitude prominent throughout semiotic whereunder subjects, verbs, and objects are arbitrarily severed and made into distinct "things" after which their mechanistic manipulation over against one another is undertaken as the solution of the semiotical problem.

III

With this much of a glimpse at the intricacy of the terminological inquiry ahead of us, we may proceed to examine the semiotician's basic construction line upon line. We shall take his main terminological fixations, dissect their words (roughly "lansigns" in semiotic, p. 350),[5] and see if, after what microscopic attention we can give them, they will feel able to nest down comfortably together again. We shall consider thirty-three such assertions, numbering them consecutively for ease of reference. Only a few of them will be complete as given, but all of them, we hope, will be true-to-assertion, so far as they go, whether they remain in the original wordings

---

[1] The words sign, signify, and significatum are employed, often indiscriminately, for both language and nonlanguage events. Available typographical marks for differentiation are often omitted, as with the cited matter in the text. Distinction of interpretation in terms of interpreters and their powers to "produce" seems here wholly irrelevant. This situation is high-lighted by almost any page in the Glossary. The glossary entries are at times technically offered as "definitions," at times not, and they are frequently uncertain in this respect. The reports on these entries may begin "A sign. . ."; "A term. . ."; or "A possible term. . . ." But also they may begin: "An object. . ."; "An organism. . ."; "A significatum. . ."; or "The time and place. . . ." Thus the entry for *locatum* reads: "Locatum. . . .A significatum of an identifior." To correspond with the treatment in the text, this should perhaps have been put: Locatum. . . .A sign (word?, name?, term?) for the significatum of an identifior. This form of differentiation is usually unimportant in nontechnical cases, and I do not want to be understood as recommending it or adopting it in any case; it is only for the comprehension of semiotic that it here is mentioned. My report on the cases of identifior and *locatum* as first presented in magazine publication was defective in phrasing in this respect. Reexamination has shown this blind spot in semiotic to be much more serious than I had originally made it out to be.

[2] Elsewhere expressed: "Identifiors *may be said* to signify location in space and time" (p. 66, italics supplied).

[3] The status of denotation with respect to signification is throughout obscure in semiotic. The practical as distinguished from the theoretical procedure is expressed by the following sentence from p. 18: "Usually we start with signs which denote and then attempt to formulate the significatum of a sign by observing the properties of denotata." Unfortunately before we are finished "properties" will not only have appeared as the source of signs but also as the last refuge of some of the significata. As concerns Morris' "where," "what," "why," and "how" modes of "signifying," comparison with J.S. Mill's five groups (existence, place, time, causation, and resemblance), *A System of Logic* (I, i, Chap. VI, Sec. 1) may have interest, as also the more elaborate classification by Ogden and Richards in connection with their treatment of definition *(The Meaning of Meaning,* [New York, 1923], pp. 217 ff.).

[4] Not examined in the present chapter, which is confined to the problem of underlying coherence. See p. 166, footnote 5.

[5] However, "the term 'word'. . .corresponds to no single semiotical term" (p. 222).

or are paraphrased. Paraphrases are employed only where the phrasings of the text involve so much correlated terminology that they are not clear directly and immediately as they stand.

Where first introduced, or where specially stressed, typographical variations will be employed to indicate to the reader whether the term in question is stressed as basic by the semiotician in person, or is selected for special attention by his present student. Stars and italics are used for the basic starred word of the glossary; italics without stars are used for words which the glossary lists unstarred; single quotes are used for unlisted words which semiotic apparently takes for granted as commonly well enough understood for its purposes. Where no page reference is given, the citation or paraphrase will be from the glossary definition for the term in question. Practical use in this way of italics, asterisks, and single quotation marks has already been made in the last paragraph of Section I of this chapter.

We first consider the materials for prospective scientific precision that are offered by the general linguistic approach to the word "sign."

1. Sign (preliminary formulation): "Something" that "controls behavior towards a goal" (p. 7).

2. Sign (roughly): "Something[1] that directs behavior with respect to something that is not at the moment a stimulus" (p. 354).

3. *Sign* (officially): A kind of "stimulus." [2]

4. *Stimulus*:[3] A "physical energy."

5. *Stimulus-object:* "The 'source' of a stimulus."

6. 'Stimulus-properties': "The 'properties' of the 'object' that produce stimuli" (p. 355).

We have here the presentation of sign on one side as an object or property, and on the other side as an energy or stimulus. We have the unexplained use of such possibly critical words as "source of," "produce," "direct," "control." We are given no definite information as to what organization the words of this latter group have in terms of one another, and so far as one can discover the problems of their organization are of no concern to semiotic. The way is prepared for the semiotician to use the word "sign" for either object or stimulus, when and as convenient, and if and as equivalent.

A second group of words involved in the presentation of the basic "preparatory-stimulus" has to do with impacts upon the organism.

7. 'Reaction': Something that "a stimulus 'causes'. . .in an organism" (p. 355).

8. *Response:* "Any action of a muscle or gland."

9. *Preparatory-stimulus:* "A stimulus that 'influences' a response to some other stimulus." It "necessarily 'causes'. . .a reaction. . .but this reaction need not be a response." [4]

10. Evocative Stimulus (at a guess)[5]: a presumptively primary or standard form of stimulus which is *not* "preparatory", i.e., which, although a stimulus, is not in the semiotic sense a "sign."

To its primarily established "object" or "stimulus" semiotic has now added the effect that the object or stimulus has—that which it (or energy, or property) causes (or produces, or is the source of)—namely, the reaction. One form of reaction it declares to be "any action of a muscle or gland," and it names this form response. Another form (or kind, or variant, or differentiation) of stimulus is one which "influences" some other response by necessarily causing a reaction which is not a response; this form is called "preparatory."

It is important to know what is happening here.[6] Names widely used, but thus far not established in firm dependable construction by the psychologies, are being taken over "as is," with no offer of evidence as to their fitness for semiotical use.[7] "Stimulus" is, of course, the characteristic word of this type. The word "response," although it is much more definitely presented as presumptively a form of reaction, is almost always (I could perhaps venture to say, always) called "action" rather than "reaction"—an attitude which has the effect of pushing it off to a distance and presenting it rather "on its own" than as a phase of semiosical process.[8]

We shall next see that the part of reaction which is *not* response (or, at least, some part of that part) is made into a kind of independent or semi-independent factor or component, viz., disposition; and that a part of that part which *is* response is made into another such factor, viz., behavior. Dispositions and behaviors are thus set over against each other as well as over against stimuli; and the attempt is made to organize all three through various unidentified types of causation without any apparent inquiry into the processes involved.

11. *Response-disposition:*[9] "The state of an organism at a given time such that" (under certain additional conditions) "a given response takes place." "Every preparatory-stimulus causes a disposition to respond" but "there may be dispositions to respond which are not caused by preparatory-stimuli" (p. 9).

12. 'Disposition': Apparently itself a "state of an organism." Described as like being "angry" before "behaving angrily"; or like having typhoid fever before showing the grosser symptoms (p. 15).[10]

13. 'State of an organism': Illustrated by a 'need' (p. 352) or by a brain wave (p. 13). It is a something that can be 'removed' by a goal-object (p. 349), and something "such that" in certain circumstances "a response takes place" (p. 348). (Semiotic rests heavily upon it, but as

---

[1] For the use of "thing" in "something" compare: "The buzzer is the sign" (p. 17); "The words spoken are signs" (p. 18).

[2] For type of stimulus and conditions see Assertion No. 19 following and compare Nos. 29 and 32.

[3] "*Stimulus:* Any physical energy that acts upon a receptor of a living organism" (p. 355).

[4] "If something is a preparatory stimulus of the kind specified in our. . .formulation it is a sign" (p. 17).

[5] "In a sign-process something becomes an evocative stimulus only because of the existence of something else as a preparatory-stimulus" (p. 308). This name does not appear, so far as I have noted, except in this one passage. I insert it here because something of the kind seems necessary to keep open the question as to whether, or in what sense, psychological stimuli are found (as distinct from physiological excitations) which are *not* signs. I do not want to take issue here on either the factual or terminological phases of the question, but merely to keep it from being overlooked. (See p. 252, note D.) The words quoted may, of course, be variously read. They might, perhaps, be intended to indicate, not a kind of stimulus genetically prior to or more general than "preparatory stimulus," but instead a kind that did not come into "existence" at all except following, and as the "product" of, preparatory stimulation.

[6] A little attention to such reports as that of the committee of the British Association for the Advancement of Science which spent seven years considering the possibility of "quantitative estimates of sensory events" would be of value to all free adaptors of psychological experiment and terminology. See S.S. Stevens, "On the Theory of Scales of Measurement," *Science,* CIII (1946), 677-80.

[7] See, however, Morris' appendices No. 6 and 7, and remarks on his relation to Tolman toward the end of the present chapter.

[8] John Dewey's "Reflex Arc" paper of 1896 should have ended this sort of thing forever for persons engaged in the broader tasks of construction. The point of view of recent physiology seems already well in advance of that of semiotic in this respect.

[9] The same as *disposition to respond* (pp. 348, 353). The "additional conditions" are "conditions of need" and "of supporting stimulus-objects" (p. 11). "*Need*" is itself an 'organic state' (p. 352), but no attempt to "probe" it is made (p. 250).

[10] I have noticed nothing more definite in the way of observation or description. Discussion of dispositions and needs (and of producers and interpreters) with respect to expression, emotion, and usage, will be found pp. 67-69

with 'disposition' there is little it tells us about it.)

14. *Interpretant:* "The disposition in an interpreter to respond because of a sign." "A readiness to act" (p. 304). Perhaps "synonymous" with "idea" (pp. 30, 49).[1]

15. *Interpreter:* "An organism for which something is a sign."

We now have needs, states of the organism, and dispositions, all brought loosely into the formulation. Beyond this some dispositions are response-dispositions, and some response-dispositions are caused by signs. Also as we shall next find (No. 16) some sign-caused responses are purposive, and under the general scheme there must certainly be a special group of sign-caused, purposive dispositions to mediate the procedure, though I have not succeeded in putting a finger clearly upon it. What for the moment is to be observed is that the sign-caused, purposive-or-not, response-disposition gets rebaptized as "interpretant." Now a sharp name-changing *may* be an excellent aid to clarity, but this one needs its clarity examined. Along with being an interpretant, it demands an "interpreter," not professedly in place of the "organism," but still with a considerable air of being promoted to a higher class. While dispositions are mostly "caused," interpretants tend to be "produced" by interpreters and, indeed, the radical differentiation between signals and symbols (Nos. 20 and 21) turns on just this difference. Dispositions have not been listed as "ideas," but interpretants are inclined to be "synonymous" with ideas, while still remaining dispositions. There is also a complex matter of "signification" which runs along plausibly, as we shall later see, in terms of interpretants, but is far from being at home among dispositions directly arising out of stimulant energy. These are matters, not of complaint at the moment, but merely to be kept in mind.

Having developed this much of semiotic—the disposition factor—so as to show, at least partially, its troublesome unclarity, we may now take a look at "response" in semiosis as distinct from stimulus and from disposition; in other words, at behavior, remembering always that the problem that concerns us is one of precision of terminology and of hoped-for accuracy of statement.

16. *Behavior:* "Sequences of. . .actions of muscles and glands" (i.e., of "responses") "by which an organism seeks goal-objects." "Behavior is therefore 'purposive.' "

17. *Behavior-family:* A set of such sequences similar in initiation and termination with respect to objects and needs.[2]

18. *Sign-behavior:* "Behavior in which signs occur." Behavior "in which signs exercise control" (p. 7).

Here we have behavior as strictly muscle-gland action to put alongside of sign as stimulus-energy and of interpretant as nonmuscular, nonglandular reaction. Despite this distinctive status of behavior, it appears that sign-behavior is a kind of behavior that has signs occurring

in it, or, alternatively, a kind *in which* signs exercise control. In such a rendering sign-behavior becomes approximately equivalent to the very loosely used "sign-process" (No. 25, q.v.).[3] This is no trifling lapse but is a confusion of expression lying at the very heart of the semiotical treatment of semiosical process.

We know fairly well where to look, not only when we want to find physical "objects" in the environment, but also when we seek the "muscles and glands" that make up "behavior," being in this respect much better off than when comparably we seek to find a locus for a disposition or an interpretant. Nevertheless a variety of problems arise concerning the technical status of behavior which may be left to the reader to answer for himself, reminding him only that precision of statement is what is at stake. Such problems are whether (1) muscle-gland action, set off independently or semi-independently for itself is intelligibly to be considered as itself "purposive"; (2) what muscle-gland action would be as theoretically "purposive," apart from stimulation; and (3) what part the "glands" play in this purposive semiotical construction. Probably only after the semiotical plan of locations for stimuli, signs, and purposes in terms of receptors, muscles, and glands has been worked out, can one face the further problem as to what locations are left over for dispositions and interpretants. On this last point the semiotician is especially cagey.[4]

We are now, perhaps, in a position to consider more precisely what a sign may be in semiotic:

19. *Sign* (officially)[5]: a preparatory-stimulus which,

    (a) in the absence of certain evocative stimulation,[6]

    (b) secures a reinvocation of, or replacement for, it, by

    (c) "causing" in the organism a response-disposition,[7] which is

    (d) capable of achieving[8] a response-sequence such as the evocative stimulus would have 'caused.'

All this takes place under a general construction that semiosis has its outcomes in purposive behavior, where the words "purposive" and "behavioral" are co-applicable,

---

[1] Semiotic, while not using "mentalist" terms at present, retains mentalist facts and suggests the possibility that "all mentalist terms" may be "incorporable" within semiotic at some later time (p. 30).

[2] This is a very useful verbal device, but not one, so far as I have observed, of any significance in the construction, though it is listed (pp. 8-11) as one of the four prominent "concepts" in semiotic along with stimulus, disposition and response. What it accomplishes is to save much complicated phrasing with respect to similarities absent and present. The typically pleonastic phrasing of the "definition" is as follows: "Any set of response-sequences which are initiated by similar stimulus-objects and which terminate in these objects as similar goal-objects for similar needs."

[3] For loose uses of "sign-behavior" see pp. 15 and 19.

[4] Professor George V. Gentry, in a paper "Some Comments on Morris' 'Class' Conception of the Designatum," *The Journal of Philosophy*, XLI (1944), 383-384, examined the possible status of the interpretant and concluded that a neurocortical locus was indicated, and that Morris did not so much reject this view as show himself to be unaware that any problem was involved. This discussion concerned an earlier monograph by Morris *(Foundations of the Theory of Signs, International Encyclopedia of Unified Science,* I, 2, 1938) and is well worth examining both for the points it makes and for the manner in which Morris has disregarded these points in his later development.

[5] Many other manners of using the word "sign" appear besides those in Assertions No. 1, 2, 3, 29, and 32. A sign may be an activity or product (p. 35). It may be "any feature of any stimulus-object" (p. 15). "An action or state of the interpreter itself becomes (or produces) a sign" (p. 25). "Actions and states and products of the organism. . .may operate as signs" (p. 27). Strictly "a sign is not always a means-object" (p. 305). Thus despite the definitions, formal and informal alike, a sign may be an action, an act, a thing, a feature, a function, an energy, a property, a quality, or a situation; and this whether it is produced by an object (as in the opening statements) or is produced by an organism in its quality as interpreter (as in much later development).

[6] Officially: "in the absence of stimulus-objects initiating response-sequences of a certain behavior-family."

[7] "Causes in some organism a disposition to respond by response-sequences of this behavior-family."

[8] I have found no verb used at this point, or at least do not recall any and so introduce the word "achieve" just by way of carrying on. A form of "delayed causation" is implied but not definitely expressed.

and where behavior proper in the semiotic sense is an affair of muscles and glands.[1]

It should now be sufficiently well established on the basis of the body of the text that a sign in semiotic is officially a kind of stimulus, produced by an object, which "causes" a disposition (perhaps one named "George") to appear, and which then proceeds to "let George do it," the "it" in question being behavior, that is, muscle-gland action of the "purposive" type. Under this *official formulation*, thunder, apparently, would not semiotically be a sign of storm unless it "caused" a disposition to put muscles and glands into purposive action.[2] Sign, as stimulus, belongs strictly under the first of the three basic, major, operative, relatively independent or semi-independent (as they are variously described) factors: stimulus, disposition, and overt body-action. Not until this is plainly understood will one get the full force and effect of the dominant division of signs in semiotical construction, viz., that signs are divided officially into two groups: those produced by interpreters and all others.

20. *Symbol:* "A sign that is produced by its interpreter and that acts as a substitute for some other sign with which it is synonymous."

21. *Signal:* "A sign that is not a symbol."

22. *Use of a Sign:* "A sign is used. . .if it is produced by an interpreter as a means. . . ."[3] "A sign that is used is thus a means-object."

Certain questions force themselves upon our attention.

If a sign is by official definition a "stimulus" produced by a "property" of an "object" which is its "source," in what sense can the leading branch of signs be said to be produced by "interpreters," rather than by "properties of objects"?

Assuming factual distinctions along the general line indicated by signal and symbol, and especially when such distinctions are presented as of maximum importance, ought not semiotic, as a science stressing the need of terminological strength, be able to give these distinctions plain and clear statement?[4]

What sense, precisely, has the word "use" in semiotic when one compares the definition for "symbol" with that for "use"?[5]

Three other definitions, two of them of starred terms, next need a glance:

23. *Sign-vehicle:* "A particular event or object. . .that functions as a sign." "A particular physical event—such as a given sound or mark or movement—which *is* a sign will be *called* a *sign-vehicle*" (p. 20; italics for "is" and "called" not used in Morris' text).

24. *Sign-family:* "A set of similar sign-vehicles that for a given interpreter have the same signification."

25. 'Sign-process': "the status of being a sign, the interpretant, the fact of denoting, the significatum" (p. 19).[6]

The peculiarities of expression are great. How is an object that "functions" as a sign different from another object that "stimulates" us as a sign or from one that "is" a sign? Is the word "particular" which modifies "event" the most important feature of the definition, and what is its sense? We are told (p. 20) that the distinction of sign-vehicle and sign-family is often not relevant, but nevertheless is of theoretical importance. Just what can this mean? We hear talk (p. 21) of sign-vehicles that have "significata"; but is not signification the most important characteristic of sign itself rather than of vehicle? If sign is energy is there some sense in which its vehicle is *not* energetic?

On the whole we are left with the impression that the distinction between "sign" and "sign-vehicle," so far as linguistic signs go, is nothing more than the ancient difference between "meaning" and "word," rechristened but still before us in all its ancient unexplored crudity. What this distinction may amount to with respect to non-linguistic signs remains still more in need of clarification.[7]

Our attention has thus far been largely concerned with the semiosis of goal-seeking animals by way of the semiotical vocabulary of object, stimulus, disposition, need, muscle, and gland. We are now to see how there is embroidered upon it the phraseology of the epistemological logics of the past in a hoped-for crystallization of structure for the future.

26. *Signify:* "To act as a sign." "To have signification." "To have a significatum." (The three statements are said to be "synonymous.")

27. 'Signification': "No attempt has been made to differentiate 'signification' and 'significatum' " (p. 354).

28. *Significatum:* "The conditions" for "a denotatum."[8]

29. Sign (on suspicion): The "x" in "x *signifies its significatum.*"[9]

---

[1] For this background of construction see the nondefinitional statement for 'behavior' in the glossary, as this is factually (though not by explicit naming) carried over into the formal definition of *Behavior-family.*

[2] If a discussion of this arrangement were undertaken, it would need to be stressed that the causation found in semiotic is of the close-up, short-term type, such as is commonly called mechanistic. No provision seems to be made for long-term intricate interconnection. See also footnote 1, p. 164.

[3] The omitted words in the definition for "use of a sign" cited above are "with respect to some goal." Insert them and the definition seems plausible; remove them and it is not. But they add nothing whatever to the import of the definition, since sign itself by the top definition of all exists only with respect to some goal.

[4] The section on signal and symbol (Chapter I, Sec. 8) has impressed me as one of the most obscure in the book, quite comparable in this respect with the section on modes of significance used earlier for illustration.

[5] The probable explanation of the separation of use from mode can be found by examining the first pages of Chapter IV. *Cf.* also pp. 92, 96, 97, 104, 125.

[6] The text rejects the word "meaning" as signifying "any and all phases of sign-process" and specifies for "sign-processes" by the wording above. Apparently the ground for rejection of "meaning" would also apply to "sign-process." "Sign-behavior" (No. 18 above) is often used as loosely as is "sign-process." The phrasing cited above is extremely interesting for its implicit differentiation of "status" and "fact" in the cases of sign and denotatum from what would appear by comparison to be an implied actuality for interpretant and significatum.

[7] By way of showing the extreme looseness of expression the following phrasings of types not included in the preceding text may be cited. Although signs are not interpretants or behaviors but stimuli, they "involve behavior, for a sign must have an interpretant" (p. 187), they are "identified within goal-seeking behavior" (p. 7), they are "described and differentiated in terms of dispositions. . ." (p. v). Interpretants, although dispositions, are "sign-produced behavior" (pp. 95, 104) or even "sign-behavior" (p. 166). A fair climax is reached in the blurb on the cover of the book (it is a good blurb in showing, as many others do, which way the book-wind blows), where all the ingredients are mixed together again in a common kettle by the assertion that this "theory of signs" (incidentally here known as semantics rather than as semiotic) "defines signs in terms of 'dispositions to respond'—that is, in terms of behavior." Along with these one may recall one phrasing already cited in which signs were spoken of as influencing behaviors first and dispositions later on in the process.

[8] Significatum: "The conditions such that whatever meets these conditions is a denotatum of a given sign" (p. 354).

[9] "A sign is said to signify its significatum" (p. 354). "Signs in the different modes of signifying signify differently" (p. 64). "Signs signifying the significata of. . ." (p. 67).

30. *Denote:* "A sign that has a denotatum. . .is said to denote its denotatum."

31. *Denotatum:* "Anything that would permit the completion of the response-sequences to which an interpreter is disposed because of a sign." "Food in the place sought. . .is a denotatum" (p. 18). "A poet. . .is a denotatum of 'poet' " (p. 106).

32. Sign (on suspicion:) the "y" in "y *denotes its denotatum.*"

33. *Goal-object:* "An object that partially or completely removes the state of an organism (the need) which motivates response-sequences."

The above is obviously a set of skeletons of assertions, but skeletonization or some other form of simplification is necessary if any trail is to be blazed through this region of semiotic. If we could be sure whether denotata and goal-objects were, or were not, "the same thing" for semiotic we might have an easier time deciphering the organization.[1] The characterizations of the two are verbally fairly close: "anything" for denotata is much like "an object" for goal-objects; "permitting completion" is much the same as "removing the need"; "is disposed" is akin to "motivates." But I have nowhere come across a definite statement of the status of the two with respect to each other, though, of course, I may have easily overlooked it. The first semiotical requirement for a denotatum is that it be "actual," or "existent" (pp. 17-20, 23, 107, 168; disregarding, perhaps, the case [p. 106] in which the denotatum of a certain ascriptor is "simply a situation such that. . ."). As "actual" the denotatum is that which the significatum is "conditions for." The significatum may remain "conditions" in the form of an "um" component of semiotic even if no denotatum "actually" exists,[2] so that the goal-object would then apparently be neither "actual" nor "existent" (except, perhaps, as present in "the mind of the interpreter" or in some terminological representative of such a "mind"). If goal-object and denotatum could be organized in a common form we might, perhaps, be able to deal more definitely with them. We are in even worse shape when we find, as we do occasionally, that significata may be "properties" as is the case with "formators" (pp. 157-158), or in their coverage of "utilitanda properties" (p. 304; see also p. 67); and that "property" itself is "a very general term used to embrace the denotata *(sic)* of signs. . ." (p. 81). Perhaps all that we can say descriptively as the case stands is that "denotatum" and "goal-object" are two different ways of talking about a situation not very well clarified with respect to either.

IV

I have endeavored to limit myself thus far to an attempt to give what may be called "the facts of the text." I hope the comments that I have interspersed between the numbered assertions have not gone much beyond what has been needed for primary report. In what follows I shall call attention to some of the issues involved, but even now not so much to debate them as to show their presence, their complexities, and the lack of attention given them.

1 There is also a very interesting question as to means-objects: whether they enter as sign-produced denotata or as directly acting objects which are not denotata at all. But we must pass this one over entirely. Compare Assertion No. 22, and footnote 2, p. 163.

2 "All signs signify, but not all signs denote." "A sign is said to signify (but not denote) its significatum, that is, the conditions under which it denotes" (pp. 347, 354).

In our preliminary statement of the leading characteristics of semiotic it was noted that the interpretation was largely in terms of causation and control. What this type of statement and of terminology does to the subject-matter at the hands of the semiotician may be interestingly seen if we focus attention upon the verbs made use of in the official accounts of "sign." What we are informed is (1) that if we are provided with a "stimulus-object" possessing "properties," then (2) these properties *produce* a kind of stimulus which (3) *influences* by (4) *causing* a disposition to appear, so that if (5) a state of the organism (a need) *motivates*, and if (6) the right means-objects are in place, then (7) it will come to pass that that which was produced at stage No. 4 proves to *be such that* (8) a response-sequence *takes place* wherein or whereby (9) the stimulus object of stage No. 1 or some other object is *responded* to as a goal-object which (10) in its turn *removes* the state of the organism (the need) that was present in stage No. 5.

What these shifting verbs accomplish is clear enough. Whichever one fits most smoothly, and thus most inconspicuously, into a sentence is the one that is most apt to be used. A certain fluency is gained, but no precision. I have not attempted to make a full list of such wordings but have a few memoranda. "Produce," for example, can be used either for what the organism does, for what a property does, for what an interpreter does, or for what a sign does (pp. 25, 34, 38, 353, 355). It may be voluntary or involuntary (p. 27), though non-humans[3] are said seldom to produce (p. 198). In the use of a comparable verb, "to signify," either organisms or signs may be the actors (p. 16). Among other specimens of such linguistic insecurity are 'because of' (p. 252), 'occasion' (pp. 13, 155), 'substitute for' (p. 34), 'act as' (p. 354), 'determine' (p. 67), 'determine by decision' (p. 18), 'function as' (p. 354), 'be disposed to' (p. 66), 'connects with' (p. 18), 'answers to' (p. 18), 'initiates' (p. 346), 'affects in some way' (p. 9), 'affects or causes' (p. 8), 'controls' (p. 7), 'directs' (p. 354), 'becomes or produces' (p. 25), 'seeks' (p. 346), and 'uses' (p. 356). One can find sentences (as on p. 25) which actually seem to tell us that interpreters produce signs as substitutes for other signs which are synonymous with them and which originally made the interpreter do what they indicated, such that the substitutes which the interpreter himself has produced now make him do what the signs from without originally made him do.[4] The "fact" in question is one of familiar everyday knowledge. Not this fact, but rather the peculiarities of statement introduced by semiotical terminology are what here cause our concern.

V

Though vital to any thorough effort at research and construction, two great problems are left untouched by semiotic. These problems are, first, the factual organization of what men commonly call "stimulus" with that which they commonly call "object"; and, secondly, the

3 Another interesting remark about animals, considering that semiotic is universal sign-theory, is that "even at the level of animal behavior organisms tend to follow the lead of more reliable signs" (p. 121).

4 No wonder that a bit later when the semiotician asks, "Are such words, however, substitutes for other synonymous signs?" he finds himself answering, "This is a complicated issue which would involve a study of the genesis of the signs produced" (p. 34). The "such words" in question are the kind that "are symbols to both communicator and communicatee at least with respect to the criterion of producibility."

corresponding organization of what the semiotician calls "interpreter" with what he calls "interpretant," or, more generally, of the factual status with respect to each other of "actor" and "action." The interpreter-interpretant problem is manifestly a special case of the ancient grammatical-historical program of separating a do-er from his things-done, on the assumption that the do-er is theoretically independent of his things-done, and that the things-done have status in some fairy realm of perfected being in independence of the doing-do. The case of stimulus-object on close inspection involves a quite similar issue. In semiotic the interior organization of disposition, interpretant, and significatum offers a special complexity. We can best show the status of these problems by appraising some of the remarks which the semiotician himself makes about the stepping stones he finds himself using as he passes through the swamps of his inquiry. No systematic treatment will be attempted since the material we have before us simply will not permit it without an enormous amount of complicated linguistic dissection far greater than the present occasion will tolerate.

Semiotic stresses for its development three main components: sign, disposition, and behavior; the first as what comes in; the second as a sort of intervening storage warehouse; the third as what goes out. For none of these, however, despite the semiotician's confidence that he is providing us "with words that are sharpened arrows" (p. 19), can their semiotical operations be definitely set down. Sign, as we have seen, is officially stimulus, practically for the most part object or property, and in the end a glisteningly transmogrified denoter or signifier. Behavior parades itself like a simple fellow, just muscles and glands in action; but while it is evidently a compartment of the organism it doesn't fit in as a compartment of the more highly specialized interpreter, although this interpreter is declared to be the very organism itself in sign action, no more, no less; moreover, behavior is purposive in its own right, though what purposive muscles and glands all on their own may be is difficult to decipher. As for disposition (or rather response-disposition, since this is the particular case of disposition with which semiotic deals), it is, I shall at least allege, a monstrosity in the form of Siamese triplets, joined at the butts, hard to carve apart, and still harder to keep alive in union. One of these triplets is disposition physiologically speaking, which is just common habit or readiness to act. Another is interpretant which is disposition-in-signing (though why such double naming is needed is not clearly made evident). The third member of the triplet family is significatum, a fellow who rarely refers to his low-life sib but who, since he is not himself either incoming stimulus or outgoing muscle or gland action, has nowhere else to be at home other than as a member of the disposition-triplets—unless, indeed, as suspicion sometimes suggests, he hopes to float forever, aura-like or soul-like, around and above the other two.

The semiotician offers us several phrasings for his tripartite organization of "factors," (of which the central core is, as we have just seen, itself tripartite). "The factors operative in sign-processes are all either stimulus-objects or organic dispositions or actual responses" (p. 19). "Analysis," we are told, yields "the stimulus, response, and organic state terminology of behavioristics" (p. 251). The "three major factors" correspond to the "nature" of the environment, its "relevance" for needs, and the "ways in which the organism must act" (p. 62). The "relative independence of environment, need, and response" is mentioned (pp. 63-64).

Despite this stressed threeness in its various forms, the practical operation of semiotic involves five factors, even if the "disposition-triplets" are seen as fused into one. The two needed additions are object as differentiated from stimulus at one end, and interpreter (or personified organism) as differentiated from interpretant at the other. (This does not mean that the present narrator wishes to introduce such items. He does not. It merely means that he finds them present and at work in the text, however furbished.) Object and stimulus we have seen all along popping in and out alternatively. "Interpreter" enters in place of interpretant whenever the semiotician wishes to stress the organism as itself the performer, producer, or begetter of what goes on. What this means is that, at both ends, the vital problems of human adaptational living in environments are entirely ignored—the problems, namely, of stimulus-object[1] and of actor-action.

What evidence does the semiotician offer for the presence of a disposition? He feels the need of evidence and makes some suggestions as to how it may be found (pp. 13-15). Each of his remarks exhibits an event of sign-process such that, if one *already believes* in dispositions as particulate existences, then, where sign-process is under way, it will be quite the thing to call a disposition in to help out. None of his exhibits, however, serves to make clear the factual presence of a disposition, whether for itself or as interpretant or as significatum, in any respect whatever as a separate factor located between the stimulation and the action. The only manifest "need" that the introduction of such a disposition seems to satisfy is the need of conforming to verbal tradition.

The issue here is not whether organisms have habits, but whether it is proper semiotically (or any other way) to set up a habit as a thing caused by some other thing and in turn causing a third thing, and use it as a basic factor in construction. Three passages of semiotic let the cat neatly out of the bag. The first says that even though a preparatory stimulus is the 'cause' of a disposition, "logically. . . .'disposition to respond' is the more basic notion" (p. 9). The second tells us that sign-processes "within the general class of processes involving mediation" are "those in which the factor of mediation is an interpretant" (p. 288). The third citation is possibly even more revealing, for we are told that "the merit of this formulation" (i.e., the use of a conventional, naively interpolated "disposition") "is that it does not require

---

[1] A few references occur in semiotic to modern work on perception (pp. 34, 191, 252, 274), but without showing any significant influence. The phenomenal constancy studies of Katz, Gelb, Bühler, Brunswik, and others on foundations running back to Helmholtz would, if given attention, make a great difference in the probable construction. (For a simple statement in a form directly applicable to the present issues see V.J. McGill, "Subjective and Objective Methods in Philosophy," *The Journal of Philosophy*, XLI [1944], 424-427.) There is little evidence that the developments of Gestalt studies even in the simpler matters of figure and ground have influenced the treatment. The great question is whether "property," as semiotic introduces it, is not itself sign, to start out with. Semiotic holds, for example, that sign-process has nothing to do with a man reaching for a glass of water to drink, unless the glass of water is a sign of something else. The reaching is "simply acting in a certain way to an object which is a source of stimulation," (p. 252) from which it would appear that in semiotic no "response-disposition" is involved in getting water to drink—a position which seems strange enough to that manner of envisionment known as common sense, but which nevertheless will not be objected to in principle by the present writer in the present chapter, if consistently maintained and successfully developed.

that the dog or the driver respond to the sign itself" (p. 10);[1] this being very close to saying that the merit of semiotic is that it can evade the study of facts and operate with puppet inserts.

There is another very interesting employment of disposition which should not be overlooked even though it can barely be mentioned here. Semiotic employs a highly specialized sign about signs called a "formator." The signs corresponding to the "modes of signification," at which we took an illustrative glance early in this chapter, are called "lexicators." The formator, however, is not a lexicator. Nevertheless it has to be a "sign," in order to fill out the construction; while to be a sign it has to have a "disposition" (interpretant). This, in the ordinary procedure, it could not attain in ordinary form. It is therefore allotted a "second-order disposition" (p. 157); and this,—since "interpretant" *via* "interpreter" represents the ancient "mind" in semiotic,—is about equivalent to introducing a two-story "mind" for the new "science" to operate with.

As concerns disposition-to-respond and interpretant in joint inspection, all that needs to be said is that if interpretant is simply one species of disposition and can so be dealt with, there is no objection whatever to naming it as a particular species. But, as we have seen in repeated instances, disposition shows itself primarily as a thing seemingly 'caused' from 'without,' while interpretant is very apt to be a thing, or property, or characteristic 'produced' from 'within.' Evading the words 'within' and 'without,' and switching names around does not seem to yield sufficient "science" to cope with this problem.

Consider next the significatum in its status in respect to the interpretant. Remarks upon this topic are rare, except in such a casual form as "a significatum. . .always involves an interpretant (a disposition. . .)" (p. 64-65). At only one point that I have noted is there a definite attempt at explication. We are told (p. 18) that "the relation between interpretant and significatum is worth noting." Here we find the significatum as a sort of interpretant turned inside out. The situation will be well remembered by many past sufferers from the ambiguity of the word "meaning." In effect, if the interpretant is a disposition with a certain amount of more or less high-grade "meaning" injected into it, then a significatum is this meaning more or less referable to the environment rather than to the interpreter. "The interpretant," we are told, "*answers to* the behavioral side of the behavior-environment complex"; as against this, "the significatum. . .*connects with* the environmental side of the complex" (p. 18, italics supplied). Here the interpretant enters "as a disposition," and the significatum enters "as a set of terminal[2] conditions under which the response-sequence can be completed," i.e., under which the "disposition" can make good. What this whole phase of semiotic most needs is the application to itself of some of its specialized ascriptors with designators dominant.

As for the organization of significatum with denotatum, and of both with ordinary muscle action and goal objects, there seems little that can be said beyond the few problems of fact that were raised following Assertions No.

26 to 33 in the text above. These comments had to be held to a minimum because the interior organization lies somewhere behind a blank wall. To be noted is that while to be "actual" or "existent" is the great duty imposed on the denotatum, the significatum is alloted its own type of actuality[3] and thingness, which is manifestly not of the denotatum type, but yet is never clearly differentiated from the other. Here is one of the greatest issues of semiotic—one which may be put in the form of the question "how comes that conditions are *ums?*" The semiotician could well afford to keep this question written on his every cuff.[4] The other great question as to the significata is, of course: How does it come about that the sign (stimulus) of No. 19 in any of its crude forms, "object," "property," "thing," or "energy," mushrooms into the stratosphere of "the good," "the beautiful," and "the true," with or without the occasional accompanying "denotation" of a few actual goodies, pretties, or verities?

VI

At the start of this chapter it was said that our examination would be expressly limited to an appraisal of the efficiency of the technical terminology which semiotic announced it was establishing as the basis for a future science; we left to others the discussion of the many interesting and valuable contributions which might be offered along specialized lines. The range of our inquiry has thus been approximately that which Professor Morris in a summary and appraisal of his own work (p. 185) styles "the behavioral analysis of signs." The specialized developments which he there further reports as "basic to his argument" are the "modes" of signifying, the "uses" of signs, and the "mode-use" classification of types of discourse, with these all together leading the way to a treatment of logic and mathematics as discourse in the "formative mode" and the "informative use" (pp. 169 ff., 178 ff.)[5] Reminder is made of these specialized developments at this point in order to maintain a proper sense of proportion as to what has here been undertaken. It is, of course, practicable for a reader primarily interested in mode, use, and type to confine himself to these subjects, without concern over the behavioral analysis underlying their treatment.

With respect to the materials which semiotical terminology identifies, we may now summarize. The organism's activities with respect to environments are divided into stimulations, dispositions, and responses. Sign-processes are similarly divided: a certain manner of indirect stimulation is called sign; the sign produces, not a

---

[1] The probable reason why the semiotician is so fearful of getting objects and organisms into direct contact (and he repeatedly touches on it) is that his view of "causation" is of the billiard-ball type, under the rule "once happen, always happen." His "intervening third" is a sort of safety valve for the cases in which his rule does not work. Which is again to say that he makes no direct observation of or report upon behavioral process itself.

[2] "Terminal" in this use seems much more suggestive of goal-object or denotatum than it is of significatum.

[3] See also the paper by Professor Gentry, previously mentioned, which very competently (and from the philosophical point of view far more broadly than is attempted here) discusses this and various other deficiencies in Morris' sign theory.

[4] Semiotic offers, however, a set of working rules under which it believes difficulties such as those of the theory of types can be readily solved (p. 279). These are: that a sign as sign-vehicle can denote itself; that a sign cannot denote its own significatum; that a sign can neither denote nor signify its interpretant (pp. 19, 220, 279). Herein lies an excellent opening for further inquiries into the fixations of "um."

[5] Something of the manner in which "modes of signifying" were identified was presented in an illustrative way in the earlier part of the present paper. The distinction, and at the same time close relation, of uses and modes is discussed in the book (pp. 96-97). The combination of use and mode for the classification of types of discourse is displayed in tabular form on p. 125. As for "everything else" in the book, Morris composedly writes (p. 185) that "our contention has been merely that it is possible to deal with all sign phenomena in terms of the basic terminology of semiotic, and hence to define any other term signifying sign phenomena in these terms."

response in muscle-gland action, but a kind of disposition called interpretant; the interpretant, in turn, under proper conditions, produces a particular kind of muscle-gland action—the "purposive" kind—which is called behavior.[1] Sign must always be a stimulus; disposition (so far as sign-process is concerned) always the result of a sign;[2] and behavior always a purposive muscular or glandular action; if semiotic is to achieve its dependable terminological goal.[3]

With respect to the actualization of this program, we quickly discover that semiotic presents a leading class of signs (symbols) which are *not* stimuli in the declared sense, but instead are "produced by interpreters" (all other signs being signals). We learn also that many interpretants are commonly produced by interpreters (by way of symbols) although they are themselves dispositions, and dispositions (so far as sign-process goes) are caused by properties of objects. We discover that significata have been introduced into the system without any developed connection with the terminology of goal-objects, purposive behaviors, dispositions, interpretants, or even with that of sign, save as the word "sign" enters into the declaration that "signs signify significata." We find also certain interstitial semiotical appellations called denotata and identified only in the sense of the declaration that "signs" (sometimes) "denote denotata." We have the "use" of a sign made distinct from its behavioral presence; we have denotata declared to be actual existences in contrast with significata which are "the conditions" for them; we then have significata gaining a form of actuality while denotata shrink back at times into something "situational." As a special case of such terminological confusion we have significata showing themselves up in an emergency as "properties," although "property" is in general the producer of a stimulus (p. 355) and although it is in particular described as "a very general term used to embrace...denotata" (p. 81); so that the full life-history of the process property-sign-signify-significatum-denote-denotatum-property ought to be well worth inquiry as an approach to a theory of sign-behavior.[4]

A glance at some of the avowed sources of semiotic may throw some light on the way in which its confusions

arise. Its use of the word "interpretant" is taken from Charles Sanders Peirce,[5] and its treatment in terms of "purposive" response is from what Professor Morris calls "behavioristics," more particularly from the work of Edward C. Tolman.[6] The difficulty in semiotic may be fairly well covered by saying that these two sources have been brought into a verbal combination, with Tolman providing the basement and ground floor while Peirce provides the penthouse and the attics, but with the intervening stories nowhere built up through factual inquiry and organization.

Peirce very early in life[7] came to the conclusion that all thought was in signs and required a time. He was under the influence of the then fresh Darwinian discoveries and was striving to see the intellectual processes of men as taking place in this new natural field. His pragmaticism, his theory of signs, and his search for a functional logic all lay in this one line of growth. Peirce introduced the word "interpretant," not in order to maintain the old mentalistic view of thought, but for quite the opposite purpose, as a device, in organization with other terminological devices, to show how "thoughts" or "ideas" as subjects of inquiry were not to be viewed as psychic substances or as psychically substantial, but were actually processes under way in human living. In contrast with this, semiotic uses Peirce's term in accordance with its own notions as an aid to bring back *sub rosa*,[8] the very thing that Peirce—and James and Dewey as well—spent a good part of their lives trying to get rid of.[9]

Tolman has done his work in a specialized field of recognized importance. Along with other psychologists of similar bent he took animals with highly developed yet

---

1 The fact that some of these names are starred as basic and others not, and that those not starred are the underlying behavioral names, was noted earlier in this chapter. The attempt is thus made to treat sign authoritatively without establishing preliminary definiteness about the behavior of which sign is a component. It should now, perhaps, be clear that the confusion of terminology is the direct outgrowth of this procedure, as is also the continual uncertainty the reader feels as to what precisely it is that he is being told.

2 "There may be dispositions to respond which are not caused by preparatory-stimuli" (p. 9).

3 It is to be understood, of course, that semiotic presents itself as open to future growth. The open question is whether the present terminology will permit such a future growth by further refinement, or whether the primary condition for growth is the eradication of the terminology from the ground up.

4 The position of the writer of this report is that defects such as we have shown are not to be regarded, in the usual case, as due to the incompetence of the workman, but that they are inherent in the manner of observation and nomenclature employed. Generations of endeavor seem to him to reveal that such components when split apart as "factors" will not remain split. The only way to exhibit the defects of the old approach is upon the actual work of the actual workman. If Professor Morris or any one else can make good upon the lines he is following, the credit to him will be all the greater.

---

5 See Morris, *op cit.*, p. v, and Appendix 2. On page 27 of his text, his analysis of semiotic is "characterized as an attempt to carry out resolutely the insight of Charles Peirce that a sign gives rise to an interpretant and that an interpretant is in the last analysis a 'modification of a person's tendencies toward action.'"

6 In addition to a citation in the opening paragraph of this chapter see *op. cit.*, p. 2: "A science of signs can be most profitably developed on a biological basis and specifically within the framework of the science of behavior." For Tolman see Appendix 6.

7 "Questions Concerning Certain Faculties Claimed for Man," *Journal of Speculative Philosophy*, II (1868); Collected Papers, 5.253.

8 This assertion is made categorically despite Morris' sentence (p. 289) in which he assures us that "The present treatment follows Peirce's emphasis upon behavior rather than his more mentalistic formulations." A typical expression by Peirce (2.666, *circa* 1910) is "I really know no other way of defining a habit than by describing the kind of behavior in which the habit becomes actualized." Dewey's comment (in correspondence) is that it is a complete inversion of Peirce to identify an interpretant with an interpreter. Excellent illustrations of the creation of fictitious "existences" in Morris' manner have recently been displayed by Ernest Nagel *(The Journal of Philosophy*, XLII [1945], 628-630) and by Stephen C. Pepper *(Ibid.*, XLIII [1946], 36).

9 John Dewey in a recent paper "Peirce's Theory of Linguistic Signs, Thought, and Meaning" *(The Journal of Philosophy*, XLIII [1946], 85-95) analysed this and other of Morris' terminological adaptations of Peirce, including especially the issues of pragmatism, and suggested that "'users' of Peirce's writings should either stick to his basic pattern or leave him alone." In a short reply Morris evaded the issue and again Dewey stressed that Morris' treatment of Peirce offered a "radically new version of the subjectmatter, intent, and method of pragmatic doctrine," for which Peirce should not be called a forerunner. Again replying, Morris again evaded the issue *(ibid.*, pp. 196, 280, 363). Thus, so far as this discussion is concerned, the issue as to the propriety of Morris' statement that he offers "an attempt to carry out resolutely the insight of Charles Peirce" remains still unresolved. In still another way Morris differs radically from Dewey. This is in regarding his development of semiotic as made "in a way compatible with the framework of Dewey's thought." *(Signs, Language and Behavior*, p. 273.)

restricted ranges of behavior, and channelized them
as to stock, environment, and activities. He then,
after many years, developed a terminology to cover
what he had observed. I keep his work close to
my table though I may not use it, perhaps, as
often as I should. The fact that the results which
Tolman and his fellow workers have secured may
be usefully reported in terms of stimulus, need,
and response does not, however, suggest to me that
this report can be straightway adopted as a basic
formulation for all procedures of human knowledge.
When Tolman, for example, recognizes "utilitanda"
one can know very definitely what he intends; but
when Morris takes up Tolman's "utilitanda properties" and includes them, "when signified, under the
term 'significatum'" (p. 304) just as they stand,
intelligibility drops to a much lower level.

Semiotic thus takes goal-seeking psychology at the
rat level, sets it up with little change, and then
attempts to spread the cobwebs of the older logics
and philosophies across it. The failure of Morris'
attempt does not mean, of course, that future
extensions of positive research may not bring the
two points of approach together.

Broadening the above orientation from immediate
sources to the wider trends in the development of
modern knowledge, we may report that much of
the difficulty which semiotic has with its terminology lies in its endeavor to conciliate two
warring points of view. One point of view
represents the ancient lineage of selves as actors, in
the series souls, minds, persons, brains. The other
derives from Newtonian mechanics in which particles
are seen as in causal interaction. The former is
today so much under suspicion that it makes its
entries largely under camouflage. The latter is no
longer dominant even in the physics of its greatest
successes. Harnessing together these two survivors
from the past does not seem to yield a live
system which enables sound descriptions of observations in the manner that modern sciences strive
always to attain.

VII

So great are the possibilities of misinterpretation in
such an analysis as the above that I summarize anew as to
its objectives. I have aimed to make plain the "factors"
(as purported "facts") which Professor Morris' "terms"
introduce, but to reject neither his "factors" nor his
"terms" because of my own personal views. I admit them
both freely *under hypothesis* which is as far as I care to
go with any alternatives which I myself propose. This,
manifestly, is not easy to achieve with this subject and in
this day, but one may at least do his best at trying.
Under this approach his "terms" are required to make
good both as between themselves and with respect to the
"facts" for which they are introduced to stand. To test
their success I take the body of his text for my material
and endeavor to ascertain how well his terms achieve their
appointed tasks. What standards we adopt and how high
we place them depends on the importance of the theory
and on the claims made for it. When in his preface
Professor Morris names an associate as having done "the
editing of the various rewritings," although in the
immediately preceding paragraph this same associate had
been listed among advisers none of whom "saw the final
text," we recognize a very trifling slip. When slips of
this kind in which one statement belies another appear
in the body of a work in such an intricate field as
the present one, we recognize them as unfortunate but
as something our poor flesh is heir to. But when such
defects are scattered everywhere—in every chapter
and almost on every page of a book purported to
establish a new science to serve as a guide to
many sciences, and when they affect each and
every one of the leading terms the book declares
"basic" for its construction, then it is time to cry
a sharp halt and to ask for a redeployment of the
terminological forces. This is the state of the new
"semiotic" and the reason for our analysis. Only
the radical importance of the inquiry for many
branches of knowledge can justify the amount of
space and effort that have been expended.

X.

COMMON SENSE AND SCIENCE[1]

THE discussion that follows is appropriately introduced by saying that both common sense and science are to be treated as transactions.[2] The use of this name has negative and positive implications. It indicates, negatively, that neither common sense nor science is regarded as an entity—as something set apart, complete and self-enclosed; this implication rules out two ways of viewing them that have been more or less current. One of these ways treats them as names for mental faculties or processes, while the other way regards them as "realistic" in the epistemological sense in which that word is employed to designate subjects alleged to be knowable entirely apart from human participation. Positively, it points to the fact that both are treated as being marked by the traits and properties which are found in whatever is recognized to be a transaction:—a trade, or commercial transaction, for example. This transaction determines one participant to be a buyer and the other a seller. No one exists as buyer or seller save *in and because of* a transaction in which each is engaged. Nor is that all; specific things *become* goods or commodities because they are engaged in the transaction. There is no commercial transaction without things which only are goods, utilities, commodities, in and because of a transaction. Moreover, because of the exchange or transfer, both *parties* (the idiomatic name for *participants)* undergo change; and the goods undergo at the very least a change of *locus* by which they gain and lose certain connective relations or "capacities" previously possessed.

Furthermore, no given transaction of trade stands alone. It is enmeshed in a body of activities in which are included those of *production*, whether in farming, mining, fishing, or manufacture. And this body of transactions (which may be called industrial) is itself enmeshed in transactions that are neither industrial, commercial, nor financial; to which the name "intangible" is often given, but which can be more safely named by means of specifying rules and regulations that proceed from the system of customs in which other transactions exist and operate.

These remarks are introductory. A trade is cited as a transaction in order to call attention to the traits to be found in common sense and science *as* transactions, extending to the fact that human life itself, both severally and collectively, consists of transactions in which human beings partake together with non-human things of the milieu along with other human beings, so that without this togetherness of human and non-human partakers we could not even stay alive, to say nothing of accomplishing anything. From birth to death every human being is a *Party*, so that neither he nor anything done or suffered can possibly be understood when it is separated from the fact of participation in an extensive body of transactions—to which a given human being may contribute and which he modifies, but only in virtue of being a partaker in them.[3] Considering the dependence of life in even its

physical and physiological aspects upon being parties in transactions in which other human beings and "things" are also parties, and considering the dependence of intellectual and moral growth upon being a party in transactions in which cultural conditions partake—of which language is a sufficient instance,—the surprising thing is that any other idea has ever been entertained. But, aside from the matters noted in the last footnote (as in the part played by religion as a cultural institution in formation and spread of the view that soul, mind, consciousness are isolated independent entities), there is the fact that what is necessarily involved in that process of living gets passed over without special attention on account of its familiarity. As we do not notice the air in the physiological transaction of breathing till some obstruction occurs, so with the multitude of cultural and non-human factors that take part in all we do, say, and think, even in soliloquies and dreams. What is called *environment* is that in which the conditions called physical are enmeshed in cultural conditions and thereby are more than "physical" in its technical sense. "Environment" is not something around and about human activities in an external sense; it is their *medium*, or *milieu*, in the sense in which a *medium* is *inter*mediate in the execution or carrying *out* of human activities, as well as being the channel *through* which they move and the vehicle *by* which they go on. Narrowing of the medium is the direct source of all unnecessary impoverishment in human living; the only sense in which "social" is an honorific term is that in which the medium in which human living goes on is one by which human life is enriched.

I

I come now to consideration of the bearing of the previous remarks upon the special theme of this paper, beginning with common sense. Only by direct active participation in the transactions of living does anyone become *familiarly acquainted* with other human beings and with "things" which make up the world. While "common sense" includes more than knowledge, this acquaintance knowledge is its distinguishing trait; it demarcates the frame of reference of common sense by identifying it with the life actually carried on as it is enjoyed or suffered. I shall then first state why the expression "common sense" is a usable and useful name for a body of facts that are so basic that without systematic attention to them "science" cannot exist, while philosophy is idly speculative apart from them because it is then deprived of footing to stand on and of a field of significant application.

Turning to the dictionary we find that the expression "common sense" is used as a name for "the general sense, feeling, judgment, of mankind or of a community." It is highly doubtful whether anything but matters with which actual living is directly concerned could command the attention, and control the speech usage of "mankind," or of an entire community. And we may also be reasonably sure that some features of life are so exigent that they impinge upon the feeling and wit of all mankind—such as need for food and means of acquiring it, the capacity of

---

[1] This chapter is written by Dewey.
[2] See Chapters IV and V of this volume.
[3] No better illustration of this fact can be found than the fact that it was a pretty extensive set of religious, economic, and political transactions which led (in the movement named individualism) to the psychological and philosophical theories that set up human beings as "individuals" doing business on their own account.

fire to give warmth and to burn, of weapons for hunting or war, and the need for common customs and rules if a group is to be kept in existence against threats from within and without. As for a community, what can it be but a number of persons having certain beliefs in common and moved by widely shared habits of feeling and judgment? So we need not be surprised to find in the dictionary under the caption "common sense" the following: "Good sound practical sense. . .in dealing with everyday affairs." Put these two usages together and we have an expression that admirably fits the case.[1]

The everyday affairs of a community constitute the *life* characteristic of that community, and only these common life-activities can engage the general or common wits and feelings of its members. And as for the word "sense" joined to "common," we note that the dictionary gives as one usage of that word "intelligence in its bearing on action." This account of sense differs pretty radically from the accounts of "sensation" usually given in books on psychology but nevertheless it tells how colors, sounds, contacts actually function in giving direction to the course of human activity. We may summarize the matters which fall within the common sense frame of reference as those of the uses and enjoyments common to mankind, or to a given community. How, for example, should the *water* of direct and familiar acquaintance (as distinct from H₂O of the scientific frame) be described save as that which quenches thirst, cleanses the body and soiled articles, in which one swims, which may drown us, which supports boats, which as rain furthers growth of crops, which in contemporary community life runs machinery, including locomotives, etc., etc.? One has only to take account of the water of common use and enjoyment to note the absurdity of reducing water to an assemblage of "sensations," even if motor-muscular elements are admitted. Both sensory qualities and motor responses are without place and significance save as they are enmeshed in uses and enjoyments. And it is *the latter* (whether in terms of water or any substance) which is a *thing* for common sense. We have only to pay attention to cases of which this case of water is representative, to learn respect for the way in which children uniformly describe things,—"It's what you do so-and-so with." The dictionary statement in which a thing is specified as "*that* with which one is occupied, engaged, concerned, busied," replaces a particular "*so-and-so*" by the generalized "*that*," and a particular *you* by the generalized *one*. But it retains of necessity the children's union of self-and-thing.

II

The words "occupied, engaged, concerned, busied," etc., repay consideration in connection with the distinctive subjectmatter of common sense. *Matter* is one of the and-so-forth expressions. Here is what the dictionary says of it:—"A thing, affair, concern, corresponding to the Latin *res*, which it is often used to render." A further statement about the word brings out most definitely the point made about children's way of telling about anything as something in which a human being and environmental conditions co-operate:—"An event, circumstance, state or course of things which is the object of consideration or of practical concern." I do not see how anything could be more inclusive on the side of what philosophers have regarded as "outer or external" than the words found in the first part of the statement quoted; while "consideration and practical concern" are equally inclusive on the side of the "inner" and "private" component of philosophical dualisms.[2]

Since, "subject, affair, business" are mentioned as synonyms of matter, we may turn to them to see what the dictionary says, noting particularly the identification of a "*subject*" with "*object of consideration.*" *Concern* passed from an earlier usage (in which it was virtually a synonym of *dis-cern)* over into an object of care, solicitude, even anxiety; and then into that "with which one is busied, occupied," and *about* which one is called upon to act. And in view of the present tendency to restrict *business* to financial concern, it is worth while to note that its original sense of force was *care, trouble. Care* is highly suggestive in the usage. It ranges from solicitude, through caring *for* in the sense of fondness, and through being deeply stirred, over to caring *for* in the sense of *taking* care, looking after, paying attention systematically, or *minding. Affair* is derived from the French *faire*. Its usage has developed through love-intrigues and through business affairs into "that one has to do with or has ado with;" a statement which is peculiarly significant in that *ado* has changed from its original sense of that which is *a doing* over into a doing "that is forced on one, a difficulty, trouble." *Do* and *ado* taken together pretty well cover the conjoint under*takings* and under*goings* which constitute that "state and course of things which is the object of consideration or practical concern." Finally we come to *thing*. It is so far from being the metaphysical substance or logical entity of philosophy that is external and presumably physical, that it is "that with which one is concerned in action, speech, or thought":—three words whose scope not only places *things* in the setting of transactions having human beings as partners, but which so cover the whole range of human activity that we may leave matters here for the present.[3] I can not refrain, however, from adding that the words dealt with convey in idiomatic terms of common sense all that is intended to be conveyed by the technical term *Gestalt* without the rigid fixity of the latter and with the important addition of emphasis on the human partner.

It does not seem as if comment by way of interpretation were needed to enforce the significance of what has been pointed out. I invite, however, specific attention to two points, both of which have been mentioned in passing. The words "concern," "affair," "care," "matter," "thing," etc., fuse in indissoluble unity senses which when discriminated are called *emotional, intellectual, practical*, the first two being moreover marked traits of the last named. Apart from a given context, it is not even possible to tell which one is uppermost; and when a context of use is present, it is always a question of emphasis, never of separation. The supremacy of subjectmatters of concern, etc., over distinctions usually made in psychology and philosophy, cannot be denied by anyone who attends to the facts. The other consideration is even more significant. What has been completely divided in philosophical discourse into man *and* the world, inner *and* outer, self *and* not-self, subject *and* object, individual *and* social, private *and* public, etc., are in actuality parties in life-transactions. The philosophical "problem" of how to get them together is artificial. On the

---

[1] Both passages are quoted from the Oxford Dictionary. The first and more general one dates in the illustrative passage cited over one hundred years earlier than the more limited personal usage of the second use. Together they cover what are sometimes spoken of as "objective" and "subjective" uses, thus anticipating in a way the point to be made next.

[2] This case, reinforced by others to follow, is perhaps a sufficient indication of the need philosophy has to pay heed to words that focus attention upon human activities as transactions in living.
[3] All passages in quotation marks are from the Oxford Dictionary.

basis of fact, it needs to be replaced by consideration of the conditions under which they occur as *distinctions*, and of the special uses served by the distinctions.[1]

Distinctions are more than legitimate *in their place*. The trouble is not with making distinctions; life—behavior develops by making two distinctions grow where one—or rather none—grew before. Their place lies in cases of uncertainty with respect to *what* is to be done and *how* to do it. The prevalence of "wishful thinking," of the danger of allowing the emotional to determine what is taken to be a cognitive reference, suffices to prove the need for distinction-making in this respect. And when uncertainty acts to inhibit (suspend) immediate activity so that what otherwise would be *overt* action is converted into an *examination* in which motor energy is channeled through muscles connected with organs of looking, handling, etc., a distinction of the factors which are obstacles from those that are available as resources is decidedly in place. For when the obstacles and the resources are referred, on the one hand, to the self as a factor and, on the other hand, to conditions of the medium-of-action as factors, a distinction between "inner" and "outer," "self" and "world" *with respect to cases of this kind* finds a legitimate place within "the state and course" of life-concerns. Petrifaction of distinctions of this kind, that are pertinent and recurrent in specific conditions of action, into inherent (and hence absolute) separations is the "vicious" affair.

Philosophical discourse is the chief wrong-doer in this matter. Either directly or through psychology as an ally it has torn the intellectual, the emotional, and the practical asunder, erecting each into an entity, and thereby creating the artificial problem of getting them back into working terms with one another. Especially has this taken place in philosophy since the scientific revolution of a few centuries ago. For the assumption that it constituted natural science an entity complete in and of itself necessarily set man and the world, mind and nature as mindless, subject and object, inner and outer, the moral and the physical, fact and value, over against one another as inherent, essential, and therefore *absolute* separations. Thereby, with supreme irony, it renders the very existence of extensive and ever-growing knowledge the source of the "problem" of how knowledge is possible anyway.

This splitting up of things that exist together has brought with it, among other matters, the dissevering of philosophy from human life, relieving it from concern with administration of its affairs and of responsibility for dealing with its troubles. It may seem incredible that human beings as *living* creatures should so deny themselves as alive. In and of itself it is incredible; it has to be accounted for in terms of historic-cultural conditions that made heaven, not the earth; eternity, not the temporal; the supernatural, not the natural, the ultimate worthy concern of mankind.

It is for such reasons as these that what has been said about the affairs and concerns of common sense is a significant matter (in itself as well as in the matter of connections with science to be discussed later) of concern. The attention that has been given to idiomatic, even colloquial, speech accordingly has a bearing upon

philosophy. For such speech is closest to the affairs of everyday life; that is, of common (or shared) living. The intellectual enterprise which turns its back upon the matters of common sense, in the connection of the latter with the concerns of living, does so at its peril. It is fatal for an intellectual enterprise to despise the issues reflected in this speech; the more ambitious or pretentious its claims, the *more* fatal the outcome. It is, I submit, the growing tendency of "philosophy" to get so far away from vital issues which render its problems not only technical (to some extent a necessity) but such that the more they are discussed the more controversial are they and the further apart are philosophers among themselves:—a pretty sure sign that somewhere on the route a compass has been lost and a chart thrown away.

III

I come now to consideration of the frame of reference that demarcates the method and subjectmatter of science from that of common sense; and to the questions which issue from this difference. I begin by saying that however the case stands, they are *not* to be distinguished from one another on the ground that science is *not* a human concern, affair, occupation. For that is what it decidedly is. The issue to be discussed is that of the *kind* of concern or care that marks off scientific activity from those forms of human behavior that fall within the scope of common sense; a part of the problem involved (an important part) being how it happened that the scientific revolution which began a few short centuries ago has had as one outcome a general failure to recognize science as itself an important human concern, so that, as already remarked, it is often treated as a peculiar sort of entity on its own account—a fact that has played a central role in determining the course taken by epistemology in setting the themes of distinctively *modern* philosophy.

This fact renders it pertinent, virtually necessary in fact, to go to the otherwise useless pains of calling attention to the various features that identify and demarcate science as a concern. In the first place, it is a *work* and a work carried on by a distinct group or set of human beings constituting a profession having a special vocation, exactly as is the case with those engaged in law or medicine, although its distinction from the latter is becoming more and more shadowy as an increasing number of physicians engage in researches of practically the same kind as those engaged in by the men who rank as scientists; and as the latter increasingly derive their special problems from circumstances brought to the fore in issues arising in connection with the source and treatment of disease. Moreover, scientific inquiry as a particular kind of work is engaged in by a group of persons who have undergone a highly specialized training to fit them for doing that particular kind of work—"job" it would be called were it not for the peculiar aura that clings to pursuits labeled "intellectual." Moreover, the work is done in a special kind of workshop, specifically known as *labor*atories and observatories, fitted out with a particular kind of apparatus for the carrying on of a special kind of occupation—which from the standpoint of the amount of monetary capital invested in it (although not from the side of its distinctive returns) is a business. Just here is a fitting place, probably *the* fitting place to note that not merely the physical equipment of scientific workshops is the net outcome of long centuries of prior *cultural* transformation of physiological processes (themselves developed throughout no one knows how many

---

[1] The list given can be much extended. It includes "pursuit, report, issue, involvement, complication, entanglement, embarrassment; enterprise, undertaking, undergoing," and "experience" as a double-barreled word. As a general thing it would be well to use such words as *concern, affairs*, etc., where now the word *experience* is used. They are specific where the latter word is general in the sense of vague. Also they are free from the ambiguity that attends *experience* on account of the controversies that have gathered about it. However, when a name is wanted to emphasize the interconnectedness of all concerns, affairs, pursuits, etc., and it is made clear that *experience* is used in that way, it may serve the purpose better than any word that is as yet available.

millions of years), but that the *intellectual* resources with which the work is done indeed, the very problems involved, are but an aspect of a continuing cultural activity: an aspect which, if one wishes to call attention to it *emphatically*, may be called a *passing* phase in view of what the work done *there and then* amounts to in its intimate and indispensable connection with all that has gone before and that is to go on afterwards. For what is done on a given date in a given observatory, laboratory, study (say of a mathematician) is after all but a re-survey of what *has* been going on for a long time and which *will* be incorporated, absorbed, along with it into an activity that will continue as long as the earth harbors man.

The work done could no more be carried out without its special equipment of apparatus and technical operations than could the production of glass or electricity or any one of the great number of industrial enterprises that have taken over as integral parts of *their* especial work processes originating in the laboratory. Lag of popular opinion and repute behind actual practice is perhaps nowhere greater than in the current ignoring of—too often ignorance of—the facts adduced; one of which is the supposition that scientific knowing is something done by the "mind," when in fact science as practiced today began only when the work done (i.e., life activities) by sense and movement was refined and extended by adoption of material devices and technological operations.

I may have overdone the task of indicating how and why "science" is a concern, a care, and an occupation, not a self-enclosed entity. Even if such is the case, what has been said leads directly up to the question:—What is the distinctive concern of science as a concern and occupation by which it is marked off from those of common sense that grow directly out of the conduct of living? In principle the answer is simple. Doing and knowing are both involved in common sense and science—involved so intimately as to be necessary conditions of their existence. Nor does the difference between common sense and science consist in the fact that knowing is the *important* consideration in science but not in common sense. It consists of the position occupied by each member in relation to the other. In the concerns of common sense knowing is as necessary, as important, as in those of science. But knowing there is for the sake of *agenda*, the *what* and the *how* of which have to be studied and to be learned—in short, *known* in order that the necessary affairs of everyday life be carried on. The relation is reversed in science as a concern. As already emphasized, doing and making are as necessarily involved as in any industrial technology. But they are carried on for the sake of advancing the system of knowings and knowns. In each case doing remains doing and knowing continues to be knowing. But the concern or care that is distinctively characteristic of common sense concern and of scientific concern, with respect to *what* is done and known, and *why* it is done and known, renders the subjectmatters that are proper, necessary, in the doings and knowings of the two concerns as different as is H_2O from the water we drink and wash with.

Nevertheless, the first named is *about* the last named, although what one consists of is sharply different from what the other consists *of*. The fact that what science is *of* is *about* what common sense subjectmatter is *of*, is disguised from ready recognition when science becomes so highly developed that the *immediate* subject of inquiry consists of what has *previously* been found out. But careful examination promptly discloses that unless the materials involved can be traced back to the material of

common sense concerns there is nothing whatever for scientific concern to be concerned with. What is pertinent here is that science is the example, *par excellence*, of the liberative effect of abstraction. Science is *about* in the sense in which "about" is *away* from; which is *of* in the sense in which "of" is *off* from:—how far off is shown in the case repeatedly used, water as H_2O where use and enjoyment are sweepingly different from the uses and enjoyments which attend laboratory inquiry into the makeup of water. The liberative outcome of the abstraction that is supremely manifested in scientific activity is the transformation of the affairs of common sense concern which has come about through the vast return wave of the methods and conclusions of scientific concern into the uses and enjoyments (and sufferings) of everyday affairs; together with an accompanying transformation of judgment and of the emotional affections, preferences, and aversions of everyday human beings.

The concern of common sense knowing is "practical," that of scientific doing is "theoretical." But *practical* in the first case is not limited to the "utilitarian" in the sense in which that word is disparagingly used. It includes all matters of direct enjoyment that occur in the course of living because of transformation wrought by the fine arts, by friendship, by recreation, by civic affairs, etc. And "theoretical" in the second instance is far away from the *theoria* of pure contemplation of the Aristotelian tradition, and from any sense of the word that excludes elaborate and extensive doings and makings. Scientific knowing is that particular form of *practical* human activity which is concerned with the advancement of *knowing* apart from concern with *other* practical affairs. The adjective often affixed to knowing of this kind is "pure." The adjective is understandable on historic grounds, since it demanded a struggle—often called *warfare*—to free natural inquiry from subordination to institutional concerns that were irrelevant and indeed hostile to the business of inquiry. But the idea that exemption from subjection to considerations extraneous and alien to inquiry as such is inherent in the essence or nature of science *as an entity* is sheer hypostatization. The exemption has itself a *practical* ground. The actual course of scientific inquiry has shown that the best interests of human living in general, as well as those of scientific inquiry in particular, are best served by keeping such inquiry "pure," that is free, from interests that would bend the conduct of inquiry to serve concerns alien (and practically sure to be hostile) to the conduct of knowing as its own end and proper terminus. This end may be called the *ideal* of scientific knowing in the *moral* sense of that word—a guide in conduct. Like other directive moral aims, it is far as yet from having attained complete supremacy:—any more than its present degree of "purity" was attained without a hard struggle against adverse institutional interests which tried to control the methods used and conclusions reached in which was asserted to be science:—as in the well-known instance when an ecclesiastical institution dictated to "science" in the name of particular religious and moral customs. In any case, it is harmful as well as stupid to refuse to note that "purity" of inquiry is something to be striven for and to be sustained by the scrupulous attention that depends upon noting that scientific knowing is one human concern growing out of and returning into other more primary human concerns. For though the existing state of science is *one* of the interests and cares that determine the selection of things to be investigated, it is not the only one. Problems are not self-selecting, and the *direction* taken by inquiry is determined by the human factors of dominant interest and concern that affect the choice of the matters to be specifically inquired into.

The position here taken, namely that science is a matter of concern for the conduct of inquiry *as inquiry* sharply counters such statements as that "science is the means of obtaining practical mastery over nature through understanding it," especially when this view is expressly placed in contrast with the view that the business of scientific knowing is to find out, to *know* in short. There can be no doubt that an important, a very important *consequence* of science is to obtain human mastery over nature. That fact is identical with the "return wave" that is emphasized. The trouble is that the view back of the quotation ignores entirely the kind of human *uses* to which "mastery" is put. It needs little discernment to see that this ignoring is in the interest of a preconceived dogma—in this particular case—a Marxist one—of what genuine mastery consists of. What *"understanding"* nature means is dogmatically assumed to be already known, while in fact anything that legitimately can be termed *understanding* nature is the outcome of scientific inquiry, not something established independent of inquiry and determining the course of "science." That science is itself a form of doing, of practice, and that it inevitably has reflex consequences upon other forms of practices, is fully recognized in the account here given. But this fact is the very reason why scientific knowing should be conducted without pre-determination of the practical consequences that are to ensue from it. That is a question to be considered on its *own* account.

There is, then a problem of high importance in this matter of the relation of the concerns of science and common sense with each other. It is not that which was taken up by historic epistemologies in attempting to determine which of the two is the "truer" representative of "reality." While a study of the various human interests, religious, economic, political-military, which have at times determined the direction pursued by scientific inquiry, contributes to clear vision of the problem, that study is itself historical rather than philosophical. The problem of concern may be introduced (as I see it) by pointing out that a reference to the *return* of scientific method and conclusions into the concerns of daily life is purely factual, descriptive. It contains no implication of anything honorific or *intrinsically* desirable. There is plenty of evidence that the outcome of the return (which is now going on at an ever-increasing speed and in ever-extending range) is a mixture of things approvable and to be condemned; of the desirable and the undesirable. The *problem*, then, concerns the possibility of giving direction to this return wave so as to minimize evil consequences and to intensify and extend good consequences, and, if it is possible, to find out how such return is to be accomplished.

Whether the problem is called that of philosophy or not is in some respects a matter of names. But the problem is *here* whatever name be given. And for the future of philosophy the matter of names may prove vital. If philosophy surrenders concern with pursuit of Reality (which it does not seem to be successful in catching), it is hard to see what concern it can take for its distinctive care and occupation save that of an attempt to meet the need just indicated. Meantime, it is in line with the material of the present paper to recur to a suggestion already made: namely, that perhaps the simplest way of getting rid of the isolations, splits, divisions, that now trouble human living, is to take seriously the concerns, cares, affairs, etc., of common sense, as far as they are transactions which (i) are constituted by the indissoluble active union of human and non-human factors; in which (ii) traits and features called intellectual and emotional are so far from being independent of and isolated from practical concerns, things done and to be done, *facta* and *facienda*, that they belong to and are possessed by the one final practical affair—the state and course of life as a body of transactions.[1]

1 In the course of consulting the Oxford Dictionary (s.v. Organism) I found the following passage (cited from Tucker, 1705-1774): "When an artist has finished a fiddle to give all the notes in the gamut, but not without a hand to play upon it, this is an organism." Were the word *organism* widely understood as an organization in which a living body and environing conditions cooperate as fiddle and player work together, it would not have been necessary to repeat so often the expression "organic-environmental." The passage may also stand as a typical reminder of what a transaction is. The words "not without" are golden words, whether they are applied to the human or to the environmental partners in a transaction.

XI.

A TRIAL GROUP OF NAMES

UNDERTAKING to find a few firm names for use in connection with the theory of knowledge—hoping thereby to promote co-operation among inquirers and lessen their frequent misinterpretations of one another—we at once found it essential to safeguard ourselves by presenting in explicit postulation the main characteristics of our procedure.[1]

The first aspect of this postulatory procedure to stress is that the firm namings sought are of that type of firmness attained by modern science when it aims at ever-increasing accuracy of specification rather than at exactness (q.v.) of formulation, thus rejecting the old verbal rigidities and leaving the paths of inquiry freely open to progress.

An observation which, we believe, any one can make when the actual procedures of knowledge theorists are examined is that these procedures deal with knowings in terms of knowns, and with knowns in terms of knowings, and with neither in itself alone. The epistemologist often comments casually on this fact, and sometimes discusses it at length, but rarely makes any deliberate effort to act upon it. No attempt at all, so far as we are aware, has been made to concentrate upon it as a dependable base for operations. We accept this observation and report as a sound basis for an inquiry under which the attainment of firm names may be anticipated, and we adopt it as our guiding postulation.

Such a postulation, wherever the inquiry is not limited to some particular activity of the passing moment but is viewed broadly in its full scope, will at once bring into the knowing and the known as joint subjectmatter all of their positings of "existence," inclusive of whatever under contrasting manners of approach might be presumed to be "reality" of action or of "being" underlying them. Taking this subjectmatter of inquiry as one single system, the factual support for any theory of knowings is then found to lie within the spatial and temporal operations and conclusions of accredited science. The alternative to this—and the sole alternative—is to make decision as to what is and what is not knowledge rest on dicta taken to be available independent of and prior to these scientific subjectmatters, but such a course is not for us.

Under this postulation we limit our immediate inquiry to knowings through namings, with the further postulation that the namings (as active behaviors of men) are themselves before us as the very knowings under examination. *If* the namings alone are *flatus vocis*, the named alone and apart from naming is *ens fatuum*.

The vague word "knowledge" (q.v.) in its scattered uses covers in an unorganized way much territory besides that of naming-knowing.[2] Especially to remark are the regions of perception-manipulation on the one hand, and the regions of mathematically symbolic knowledge on the other. These remain as recognized fields of specialized study for all inquiry into knowledge. Whether or not the word "knowledge" is to be retained for all of these fields as well as for namings-knowings is not a question of much importance at the present imperfect stage of observation and report.

Some of the words here appraised may be taken as key-names for the postulation employed, and hence as touchstones for the other names. *Fact* is thus used for knowings-knowns in system in that particular range of knowings-knowns, namely, the namings-nameds, which is studied. *Designation* is used as a most general name for the naming phases of the process, and *Existence* as a most general name for the named phases. Attention is called to the distinction between *inter* and *trans* (the former the verbal locus of much serious contemporary confusion), and to the increasingly firm employment of the words "aspect" and "phase" within the transactional framework of inspection.

Certain changes are made from our earlier recommendations.[3] "Existence" replaces "event," since we have come to hope that it may now be safely used. "Event," then, replaces "occurrence." "Definition" has been demoted from its preliminary assignment, since continued studies of its uses in the present literature show it so confused as to rate no higher than a crude characterization. "Symbolization" has been given the duty of covering the territory which, it was earlier hoped, "definition" could cover. "Exact," for symbolization has been substituted for "precise," in correlation with "accurate" for specification. The names "behavior-object" and "behavior-agent" have been dropped, as not needed at the present stage of inquiry, where object and organism suffice.

The reader will understand that what is sought here is clarification rather than insistent recommendation of particular names; that even the most essential postulatory namings serve the purpose of "openers," rather than of "determiners"; that if the distinctions herein made prove to be sound, then the names best to be used to mark them may be expected to adjust themselves in the course of time under attrition of the older verbal abuses; and that every division of subjectmatters through disjunction of names must be taken in terms of the underlying conjunctions that alone make the disjunctions soundly practicable by providing safety against absolutist applications.

Accurate: When specification is held separate from symbolization (q.v.), then separate adjectives are desirable to characterize degrees of achievement in the separate ranges. Accurate is recommended in the case of specification. See *Exact.*

Action, Activity: These words are used by us in characterizations of durational-extensional subjectmatters only. Where a stressed substantive use of them is made, careful specification should be given; otherwise they retain and promote vagueness.

Actor: A confused and confusing word; offering a primitive and usually deceptive organization for the complex behavioral transactions the organism is engaged in. Under present postulation Actor should always be

1 See Chapter III.

2 How much territory the word "knowledge" is made to cover may be seen from what is reported of it in Runes' *The Dictionary of Philosophy* (1942). Knowledge appears as: "Relations known. Apprehended truth. Opposite of opinion. Certain knowledge is more than opinion, less than truth. Theory of knowledge, or epistemology (q.v.), is the systematic investigation and exposition of the principles of the possibility of knowledge. In epistemology: the relation between object and subject."

3 Compare especially the tentative list of words suggested at the close of Chapter II.

taken as postulationally transactional, and thus as a Trans-actor.

Application: The application of a name to an object may often be spoken of advantageously where other phrasings mislead. See *Reference*.

Aspect: The components of a full transactional situation, being not independents, are aspects. The word is etymologically correct; the verb "aspect" is "to look out." See *Phase*.

Behavior: A behavior is always to be taken transactionally: i.e., never as *of* the organism alone, any more than *of* the environment alone, but always as of the organic-environmental situation, with organisms and environmental objects taken as equally its aspects. Studies of these aspects in provisional separation are essential at many stages of inquiry, and are always legitimate when carried on under the transactional framework, and through an inquiry which is itself recognized as transactional. Transactionally employed, the word "behavior" should do the work that "experience" has sought to do in the past, and should do it free from the shifting, vague, and confused applications which have in the end come to make the latter word so often unserviceable. The phrase "human behavior" would then be short for "behavior with the understanding that is human."

Behavioral: Behavioral inquiry is that level of biological inquiry in which the processes examined are not currently explorable by physical or physiological (q.v.) techniques. To be understood in freedom equally from behavioristic and from mentalistic allusions. Covers equally the ranges called "social" and those called "individual."

Biological: Inquiry in which organic life is the subjectmatter, and in which the processes examined are not currently explorable by physical (q.v.) techniques; covers both physiological and behavioral inquiry.

Characterization: The intermediate stage of designation in the evolutionary scale, with cue (q.v.) preceding, and specification (q.v.) following; includes the greater part of the everyday use of words; reasonably adequate for the commoner practical purposes.

Circularity: Its appearance is regarded as a radical defect by non-transactional epistemological inquiries that undertake to organize "independents" as "reals." Normal for inquiry into knowings and knowns in system.

Coherence: Suggests itself for connection (q.v.) as established under specification, in distinction from consistency attained in symbolic process.

Concept, Conception: Conception has two opposed uses: on one side as a "mentalistic entity"; on the other as a current phrasing for subjectmatters designed to be held under steady inspection in inquiry. Only the latter is legitimate under our form of postulation. In any event the hypostatization set up by the word "concept" is to be avoided; and this applies to its appearance in formal logic even more than elsewhere.

Connection: To apply between objects under naming. See *Reference* and *Relation*.

Consciousness: The word has disappeared from nearly all research, but survives under various disguises in knowledge theory. Where substantively used as something other than a synonym of a comparable word, "awareness," we can find under our postulation no value whatever in it, or in its disguises, or in the attitudes of inquiry it implies.

Consistency: To be used exclusively in symbolic ranges. See *Coherence*.

Context: A common word in recent decades carrying many suggestions of transactional treatment. However, where it obscures the issues of naming and the named, i.e., when it swings obscurely between verbal and physical environments, it is more apt to do harm than good.

Cosmos: Commonly presents "universe as system." If the speaking-knowing organism is included in the cosmos, and if inquiry proceeds on that basis, cosmos appears as an alternative name for Fact (q.v.).

Cue: The earliest stage of designation in the evolutionary scale. Some recent psychological construction employs cue where the present study employs signal. Firm expression is needed in some agreed form. If a settled psychologist's use develops, then it, undoubtedly, should govern.

Definition: Most commonly employed for specification (q.v.), though with varied accompanying suggestions of dictionary, syllogistic, or mathematical adaptation. These latter, taken in a group, provide a startling exhibit of epistemological chaos. In recent years a specialized technical application has been under development for the word in formal logic. Establishment in this last use seems desirable, but the confusion is now so great that it is here deemed essential to deprive the word of all terminological status above that of a characterization (q.v.) until a sufficiently large number of experts in the fields of its technical employment can establish and maintain a specific use.[1]

Designation: The knowing-naming phase of fact. To be viewed always transactionally as behavior. The word "name" (q.v.) as a naming may advantageously be substituted wherever one can safely expect to hold it to behavioral understanding. Extends over three levels: cue, characterization, and specification.

Description: Cues organizing characterizations; characterizations developing into specifications. Not to be narrowed as is done when brought too sharply into contrast with narration as temporal. A name is, in effect, a truncated description. Somewhat similarly, with respect to an established name, a description may be called an expanded naming.

Entity: Assumed or implied physical, mental, or logical independence or semi-independence (the "semi" always vague or evasive) in some part of a subjectmatter under inquiry; thus, a tricky word, even when not positively harmful, which should be rejected in all serious inquiry. See *Thing* that, in its idiomatic use, is free from the misleading pretentiousness of entity.

Environment: Situations, events, or objects in connection (q.v.) with organism as object. Subject to inquiry physically, physiologically, and, in full transactional treatment, behaviorally.

Epistemological: As far as this word directly or indirectly assumes separate knowers and knowns (inclusive of to-be-knowns) all epistemological words are ruled out under transactional procedure.

Event: That range of differentiation of the named which is better specified than situation, but less well specified than object. Most commonly employed with respect to durational transition. (In earlier sketches employed where we now employ Existence.)

Exact: The requirement for symbolic procedure as distinguished from the requirement of accuracy (q.v.) for specification.

Excitation: A word suggested for specific use where *physiological* process of environment and organism is concerned and where distinction from behavioral stimulus (q.v.) in the latter's specific use is required.

---

[1] Chapter II, p. 110, footnote 1, p. 112, footnote 2; Chapter VI, p. 132, footnote 4; and Chapter VIII, Section I.

Existence: The known-named phase of fact, transactionally inspected. Established through designation under an ever-increasing requirement of accuracy in specification. Hence for a given era in man's advance, it covers the established objects in the evolving knowing of that era. Not permitted entry as if at the same time both a "something known" and a "something else" supporting the known. Physical, physiological, and behavioral subject-matters are here taken as equally existential, however different the technical levels of their treatment in inquiry at a given time may be. Both etymologically and in practical daily uses this application of the word is far better justified than is an extra-behavioral or absolutist rendering (whether physicalist or mentalist) under some form of speculative linguistic manipulation.

Experience: This word has two radically opposed uses in current discussion. These overlap and shift so as to cause continual confusion and unintentional misrepresentation. One stands for short extensive-durational process, an extreme form of which is identification of an isolated sensory event or "sensation" as an ultimate unit of inquiry. The other covers the entire spatially extensive, temporally durational application; and here it is a counterpart for the word "cosmos." The word "experience" should be dropped entirely from discussion unless held strictly to a single definite use: that, namely, of calling attention to the fact that *Existence* has organism and environment as its aspects, and can not be identified with either as an independent isolate.

Fact: The cosmos in course of being known through naming by organisms, themselves among its phases. It is knowings-knowns, durationally and extensionally spread; not what is known to and named by any one organism in any passing moment, nor to any one organism in its lifetime. Fact is under way among organisms advancing in a cosmos, itself under advance as known. The word "fact," etymologically from *factum*, something done, with its temporal implications, is much better fitted for the broad use here suggested than for either of its extreme and less common, though more pretentious applications: on the one hand for an independent "real"; on the other for a "mentally" endorsed report. Whether the word may properly apply to the cosmic presentation of inferior non-communicating animals, or to that of a superior realm of non-naming symbols, is for others to develop at other times and places. *See* Chapter II, Section IV.

Field: On physical analogies this word should have important application in behavioral inquiry. The physicist's uses, however, are still undergoing reconstructions, and the definite correspondence needed for behavioral application can not be established. Too many current projects for the use of the word have been parasitic. Thorough transactional studies of behaviors on their own account are needed to establish behavioral field in its own right.

Firm: As applied to a proposed terminology for knowings and knowns this word indicates the need of accuracy (q.v.) of specification, never that of exactness of symbolization. For the most firm, one is to take that which is least vague, and which at the same time is most free from assumed finality—where professed finality itself, perhaps, is the last word in vagueness.

Idea, Ideal: Underlying differences of employment are so many and wide that, where these words are used, it should be made clear whether they are used behaviorally or as names of presumed existences taken to be strictly mental.

Individual: Abandonment of this word and of all substitutes for it seems essential wherever a positive *general theory* is undertaken or planned. Minor specialized studies in individualized phrasing should expressly name the limits of the application of the word, and beyond that should hold themselves firmly within such limits. The word "behavior" (q.v.) as presented in this vocabulary covers both individual and social (q.v.) on a transactional basis in which the distinction between them is aspectual.

Inquiry: A strictly transactional name. It is an equivalent of knowing, but preferable as a name because of its freedom from "mentalistic" associations.

Inter: This prefix has two sets of applications (see Oxford Dictionary). One is for "between," "in-between," or "between the parts of." The other is for "mutually," "reciprocally." The result of this shifting use as it enters philosophy, logic, and psychology, no matter how inadvertent, is ambiguity and undependability. The habit easily establishes itself of mingling without clarification the two sets of implications. It is here proposed to eliminate ambiguity by confining the prefix *inter* to cases in which "in between" is dominant, and to employ the prefix *trans* where the mutual and reciprocal are intended.

Interaction: This word, because of its prefix, is undoubtedly the source of much of the more serious difficulty in discussion at the present time. Legitimate and illegitimate uses in various branches of inquiry have been discussed in chapters IV and V. When transactional and interactional treatments come to be explicitly distinguished,[1] progress in construction should be more easily made. For the general theory of knowings and knowns, the interactional approach is entirely rejected under our procedure.

Knowledge: In current employment this word is too wide and vague to be a *name* of anything in particular. The butterfly "knows" how to mate, presumably without learning; the dog "knows" its master through learning; man "knows" through learning how to do an immense number of things in the way of arts or abilities; he also "knows" physics, and "knows" mathematics; he knows *that, what,* and *how*. It should require only a moderate acquaintance with philosophical literature to observe that the vagueness and ambiguity of the word "knowledge" accounts for a large number of the traditional "problems" called *the problem of knowledge*. The issues that must be faced before firm use is gained are: Does the word "knowledge" indicate something the organism possesses or produces? Or does it indicate something the organism confronts or with which it comes into contact? Can either of these viewpoints be coherently maintained? If not, what change in preliminary description must be sought?

Knowings: Organic phases of transactionally observed behaviors. Here considered in the familiar central range of namings-knowings. The correlated organic aspects of signalings and symbolings are in need of transactional systematization with respect to namings-knowings.

Knowns: Environmental phases of transactionally observed behaviors. In the case of namings-knowings the range of the knowns is that of existence within fact or cosmos, not in a limitation to the recognized affirmations of the moment, but in process of advance in long durations.

Language: To be taken as behavior of men (with extensions such as the progress of factual inquiry may show to be advisable into the behaviors of other organisms). Not to be viewed as composed of word-bodies

---

1 Transactions: doings, proceedings, dealings. Interaction: reciprocal action or influence of persons or things on each other (Oxford Dictionary).

apart from word-meanings, nor as word-meanings apart from word-embodiment. As behavior, it is a region of knowings. Its terminological status with respect to symbolings or other expressive behaviors of men is open for future determination.

Manipulation: See *Perception-manipulation.*

Matter, Material: See *Physical* and *Nature.* If the word "mental" is dropped, the word "material" (in the sense of matter as opposed to mind) falls out also. In every-day use, both "mental" and "material" rate at the best as "characterizations." In philosophy and psychology the words are often degraded to "cues."

Mathematics: A behavior developing out of earlier naming activities, which, as it advances, more and more gains independence of namings and specializes on symboling. See *Symbol.*

Meaning: A word so confused that it is best never used at all. More direct expressions can always be found. (Try for example, speaking in terms of "is," or "involves.") The transactional approach does away with that split between disembodied meanings and meaningless bodies for meanings which still enters flagrantly into much discussion.

Mental: This word not used by us. Usually indicates an hypostatization arising from a primitively imperfect view of behavior, and not safe until the splitting of existence into two independent isolates has been generally abandoned. Even in this latter case the word should be limited to service as emphasizing an aspect of existence. See *Behavior* and *Transaction.*

Name, Naming, Named: Language behavior in its central ranges. Itself a form of knowing. Here, at times, temporarily and technically replaced by the word "designation," because of the many traditional, speculatively evolved, applications of the word "name," closely corresponding to the difficulties with the word "concept" (q.v.), many of them still redolent of ancient magic. The word "name" will be preferred to the word "designation," as soon as its use can be assumed in fully transactional form and free from conventional distortions.

Nature: See *Cosmos* and *Fact.* Here used to represent a single system of subjectmatters of inquiry, without implication of predetermined authoritative value such as is usually intended when the word "naturalism" is used.

Object: Within fact, and within its existential phase, object is that which acquires firmest specification, and is thus distinguished from situation and event. This holds to the determination of Dewey *(Logic,* p. 119); also pp. 129, 520, *et al.)* that in inquiry object "emerges as a definite constituent of a resolved situation, and is confirmed in the continuity of inquiry," and is "subjectmatter, so far as it has been produced and ordered in settled form."

Objective: A crude characterization which seems easily enough intelligible until one observes that in the behavioral sciences almost every investigator calls his own program objective, regardless of its differences from the many self-styled objective offerings that have gone before. As often employed the word has merely the import of impartial, which might advantageously replace it. Objective is used so frequently to characterize aspects of "subject" rather than of "object," that its own status with respect both to subject and to object should be carefully established before use.

Observation: To be taken as durationally and extensionally transactional, and thus neither separately in terms of the observing, nor separately in terms of the observed. Always to be viewed in the concrete instance but never as substantively stressed "act," nor in any other way as isolated or independent. Always to be postulationally guarded in current technical employment, and always to remain tentative with respect to future observing and knowing. See *Experience.*

Operational: The word "operation" as applied to behavior in recent methodological discussions should be thoroughly overhauled and given the full transactional status that such words as "process" and "activity" (q.v.) require. The military use of the word is suggestive of the way to deal with it.

Organism: Taken as transactionally existent in cosmos. Presentations of it in detachment or quasi-detachment are to be viewed as tentative or partial.

Organization: See *System.*

Percept: To be taken transactionally as phase of signaling behavior. Never to be hypostatized as if itself independently "existing."

Perception-Manipulation: Taken jointly and inseparably as the range of signal behaviors. Differences between perception and manipulation seemed striking in the earlier stages of the development of psychology, but today's specialization of inquiry should not lose sight of their common behavioral status.

Phase: Aspect of fact in sufficiently developed statement to exhibit definite spatial and temporal localizations.

Phenomenon: A word that still has possibilities of convenient use, if deprived of all of those implications commonly called subjective, and used for provisional identifications of situation with no presumptive "phenomenine" behind it for further reference.

Physical: One of the three, at present, outstanding divisions of the subjectmatters of inquiry. Identifiable through technical methods of investigation and report, not through purported differences in material or other forms of purported substance.

Physiological: That portion of biological inquiry which forms the second outstanding division of the subjectmatter of all inquiry as at present in process; differentiated from the physical by the techniques of inquiry employed more significantly than by mention of its specialized organic locus. See *Behavioral.*

Pragmatic: This word is included here (but no other of its kind except epistemological) solely to permit a warning against its current degradation in making it stand for what is practical to a single organism in limited durational spread—this being a use remote from that of its origin.

Process: To be used aspectually or phasally. See *Activity.*

Proposition: Closely allied to proposal both etymologically and in practical daily use. Widely divorced from this, and greatly confused in its current appearances in the logics. Many efforts in the last two decades to distinguish it clearly from assertion, statement, sentence, and other words of this type upon the basis of the older self-oriented logics, have only served to increase the difficulties. Sufficient light is thrown upon its status by its demand, concealed or open, that its component terms be independent fixities while at the same time it hypostatizes itself into an ultimate fixity. Treated in Dewey's *Logic, the Theory of Inquiry* under radically different construction as an intermediate and instrumental stage in inquiry.

Reaction: To be coupled with excitation in *physiological* reference (q.v.).

Real: Its use to be completely avoided when not as a recognized synonym for genuine as opposed to sham or counterfeit.

Reality: As commonly used, it may rank as the most metaphysical of all words in the most obnoxious sense of metaphysics, since it is supposed to name something which lies underneath and behind all knowing, and yet, as Reality, something incapable of being known in fact and as fact.

Reference: Behavioral application of naming to named. See *Connection* and *Relation.*

Relation: various current uses, ranging from casual to ostentatious; rarely with any sustained effort at localization of the "named," as is shown by ever-recurrent discussions (and, what is worse, evasions) as to whether relation (assumed to have a certain existence somewhere as itself factual) is "internal" or "external." Suggested by us to name system among words, in correlation with reference and connection (q.v.).[1]

Response: To be coupled with stimulus in the signal range of behavior.

Science, Scientific: Our use of this word is to designate the most advanced stage of specification of our times—the "best knowledge" by the tests of employment and indicated growth.

Self: Open to aspectual examination under transactional construction. Where substantively stressed as itself an object, self should not be permitted also an aura of transactional values, tacitly, and apart from express development.[2]

Self-action: Used to indicate various primitive treatments of the known, prior in historical development to interactional and transactional treatments. Rarely found today except in philosophical, logical, epistemological, and a few limited psychological regions of inquiry.

Sentence: No basic distinction of sentence from word nor of meaning of sentence from verbal embodiment of sentence remains when language is viewed as transactionally behavioral.

Sign: This name applied transactionally to organic-environmental behavior. To be understood always as sign-process; never with localization of sign either in organism or in environment separately taken. Hence never as if signs were of two kinds: the natural and the artificial. Coterminous with behavioral process, and thus technically characteristic of all behaviors viewed in their knowing-known aspects. Distinctive as technical mark of separation of behavioral from physiological process, with the disjointure of research in the present day on this borderline more marked than that on the borderline between physics and physiology, where biophysics is making strong advance. Evolutionary stages and contemporary levels differentiated into signal, name, and symbol.

Sign-Process: Synonym for *Sign.*

Signal: The perceptive-manipulative level and stage of sign in transactional presentation. Border-regions between signaling and naming still imperfectly explored, and concise characterizations not yet available.

Situation: The more general, and less clearly specified, range of the named phase of fact. In our transactional development, the word is not used in the sense of environment; if so used, it should not be allowed to introduce transactional implications tacitly.

Social: The word in its current uses is defective for all *general* inquiry and theory. See *Individual.*

Space-Time: Space and time alike to be taken transactionally and behaviorally—never as fixed or given frames (formal, absolute, or Newtonian) nor exclusively as physical specializations of the types known since relativity.[3]

Specification: The most highly perfected naming behavior. Best exhibited in modern science. Requires freedom from the defectively realistic application of the form of syllogism commonly known as Aristotelian.

Stimulus: An unclarified word, even for most of its key-word uses in psychology. The possibility of an adequate transactional specification for it will be a critical test of transactional construction. The indicated method of procedure will be through the thorough-going substitution of nouns of action such as "stimulation" in place of substantive nouns such as "stimulus" is usually taken to be.

Subject: This word can profitably be dropped, so long as subjects are presented as in themselves objects. Subject was object in Greece and remains unclarified today. Might be properly used, perhaps, in the sense of "topic" as "subjectmatter undergoing inquiry," in differentiation from "object" as "subjectmatter determined by inquiry."

Subjective: Even less dependable as a word than objective (q.v.).

Subjectmatter: Whatever is before inquiry where inquiry has the range of namings-named. The main divisions in present-day research are into physical, physiological, and behavioral.

Substance: No word of this type has place in the present system of formulation. See *Entity.*

Symbol: A non-naming component of symboling behavior. To be taken transactionally, and not in hypostatization. Thus comparable with name and signal.

Symboling, Symbolization: An advance of sign beyond naming, accompanied by disappearance of specific reference (q.v.) such as naming develops.

System: Perhaps a usable word where transactional inquiry is under way. Thus distinguished from organization which would represent interaction. "Full system" has occasionally been used to direct attention to deliberately comprehensive transactional procedure.

Term: This word has today accurate use as a name only in mathematical formulation where, even permitting it several different applications, no confusion results. The phrase "in terms of" is often convenient and, simply used, is harmless. In the older syllogism term long retained a surface appearance of exactness (q.v.) which it lost when the language-existence issues involved became too prominent. For the most part in current writing it seems to be used loosely for "word carefully employed." It is, however, frequently entangled in the difficulties of concept. Given sufficient agreement among workers, term could perhaps be safely used for the range of specification, and this without complications arising from its mathematical uses. It might, then, be characterized as follows: Term: a firm name as established through inquiry; particularly, a name for the group of all those names that name whatever has acquired technically assured standing as object.

Thing: Most generally used for anything named. This very generality gives it frequent advantage over its pretentious substitutes, Entity and Substance, and more particularly over Object in the common case in which the type of objectivity involved is not specified. Though sometimes facilitating epistemological or logical evasion,

---

1 See Dewey, *Logic, the Theory of Inquiry* (New York, 1938), p. 55, for such a presentation.

2 In illustration: Mead's wide-ranging transactional inquiries are still taken by most of his followers in the sense of interesting comments on an object, namely the "self," in independence.

3 See Bentley, "The Factual Space and Time of Behavior," *The Journal of Philosophy,* XXXVIII (1941), 477-485.

its very looseness of application is safer than the insufficiently analyzed rigidities of the other words mentioned. *See* p. 105, footnote 1; p. 121, footnote 3; and pp. 170-171.

Time: See *Space-time.*

Trans: This prefix has, in older usage, the sense of beyond, but in much recent development it stands for across, from side to side, etc. To be stressed is the radical importance at the present time of a clear differentiation between trans and inter (q.v.).

Transaction: The knowing-known taken as one process in cases in which in older discussions the knowings and knowns are separated and viewed as in interaction. The knowns and the named in their turn taken as phases of a common process in cases in which otherwise they have been viewed as separated components, allotted irregular degrees of independence, and examined in the form of interactions. See *Interaction.*

Transactor: See *Actor.*

True, Truth: These words lack accuracy in modern professedly technical uses, in that the closer they are examined, it frequently happens, the more inaccurate they appear. "Warranted assertion" (Dewey) is one form of replacement. Confinement to "semantic" instances is helpful, so far as "semantic" itself gains accuracy of use. A subjectmatter now in great need of empirical inquiry, with such inquiry apparently wholly futile under traditional approaches.

Vague: This word is itself vaguely used, and this as well in our preceding inquiries as elsewhere. It should be replaced by names specifying the kind and degree of inaccuracy or inexactness implied.

Word: To be used without presumptive separation of its "meaning" as "mental" from its "embodiment" (airwaves, marks on paper, vocal utterances, etc.) as "physical"; in other words, to be taken always as behavioral transaction, and thus as a subjectmatter examined whole as it comes, rather than in clumsily fractured bits.

Some of the above words enter our trial group of names as representative of the postulation we have adopted. The remainder fall into two sub-groups: namely, that may probably be clarified and salvaged, and others that show themselves so confused and debased that we unqualifiedly urge their rejection from all technical discourse at the present time. This is as far as we have been able to proceed in terminological systematization under the chaotic state of current discussion.

With respect to our central postulations: first, that knowings-knowns are to be transactionally studied, and secondly, that namings, when transactionally studied, show themselves as directly existential knowings, we renew our repeated reminder and caution. We are all aware that knowings, as behaviors, lie within, or among, wider ranges of behaviors. We are also all aware that the word "knowing" is itself variously applied to phenomena at perhaps every scattered stage of behavior from the earliest and simplest organic orientations to the most complex displays of putatively extrapolated supra-organic pseudocertainties. The range of our own inquiry—the central range of technically transactional fact-determination—will be declared by some readers to demand its own "interpretation" on the basis of behavioral activities taken as antecedent and "causal" to it. By others all inquiry in our range will be declared to be under the control of powers detached from, and presumptively "higher" than, any such behavioral activity. Our own assertion is that, no matter how dogmatically either of these declarations may be made, the passage of time will more and more require an ever broadening and deepening inquiry into the characteristic processes of organization and system they involve. It is our hope that the more naive fiats will some day cease to be satisfactory even to their most ardent pronouncers. Progress from stylized cue or loose characterization to careful specification becomes thus a compelling need, and it is with the possibilities of such progress under postulation that we have here experimented. Detachable empiricals and detachable rationals are alike rejected.

Finally, both with regard to postulation and to terminology, we are *seeking* the firm (q.v.) and not trying to decree it.

XII.
SUMMARY OF PROGRESS MADE

THE research upon which we have made report has exhibited itself in three main phases: at the beginning, an endeavor to secure dependable namings in the chosen field; next, a display of the current linguistic insecurity in activities in those fields; thirdly, an initial development of the transactional approach which becomes necessary, in our view, if reliable namings are to be secured. The first of these phases is presented in Chapters II, III, and XI, and has been allowed to rest with such terminological suggestions as the last of these chapters offers. The second, seen in Chapters I, VII, VIII, and IX, was expanded far beyond preliminary expectation, as it became clear to us that, without increased recognition of the extent of the underlying linguistic incoherence, little attention would be paid to the need for reform. The third was sketched in Chapters IV, V, VI, and X; its further development remains for later presentation in psychological, linguistic, and mathematical ranges corresponding to the levels of Signal, Designation, and Symbol within Behavior.

In most general statement our chosen postulatory approach presents the human organism as a phase of cosmic process along with all of his activities including his knowings and even his own inquiries into his knowings as themselves knowns. The knowings are examined within the ranges of the known, and the knowns within the ranges of the knowing; the word "ranges" being here understood, not as limiting the research in any way, but as vouching for its full freedom and openness. This approach does not imply an absorption of knowing activities into a physical cosmos, any more than it implies an absorption of the physical cosmos into a structure of knowings. It implies neither. This must be most emphatically asserted. Emphasis is all the more necessary because the position of the present writers whether in their separate inquiries or in the present joint undertaking, is so frequently mis-stated. In illustration, two recent notices of our procedure in the technical journal that we regard as standing closest to our field of inquiry[1] have described us as neglecting a difference, radical in nature, taken to exist between psychological and logical facts: a difference which, they appear to hold, ought to be known to everyone as crucial for all inquiries in this field. One reviewer goes even further, in disregard of our most explicit expression at other points, when from a detached preliminary phrase he infers that we reject "abstraction" from both mathematical and logical operations. This latter opinion will, we feel sure, be dissipated upon even the most hasty survey of our texts. The former is likewise a misunderstanding that cannot maintain itself under study. We may assure all such critics that from early youth we have been aware of an *academic* and *pedagogical* distinction of logical from psychological. We certainly make no attempt to deny it, and we do not disregard it. Quite the contrary. Facing this distinction in the presence of the actual life processes and behaviors of human beings, we deny any rigid *factual* difference such as the academic treatment implies. Moreover, it has been our sustained effort throughout all our inquiry to show the practicability of theoretical construction upon a new basis by offering the beginnings of its development. We have as strong an objection to the assumption of a science of psychology severed from a logic and yet held basic to that logic, as we have to a logic severed from a psychology and proclaimed as if it existed in a realm of its own where it regards itself as basic to the psychology. We regard knowings and reasonings and mathematical and scientific adventurings even up to their highest abstractions, as activities of men—as veritably men's behaviors— and we regard the study of these particular knowing behaviors as lying within the general field of behavioral inquiry; while at the same time we regard psychological inquiry itself and all its facts and conclusions as being presented to us under the limitations and qualifications of their being known. None of this involves any interference with the practical differentiations of inquiry as between logic and psychology, any more than it interferes with differentiations within either of these fields taken separately. Specializations of attention and effort based on methods and on subjectmatters methodologically differentiated remain as valid and usable as ever.

The difficulty in mutual understanding in such cases as the above lies, we believe, in the various conventionally frozen sets of implications which many of the crucial words that are employed carry over from the past, and which have not yet been resolved under factual examination. They are like the different focussings of different linguistic spectacles which yield strangely different pictures of presumptive fact. It is this deficiency in communication that calls for the extended examination we have given several of the leading current texts. It is this deficiency also that explains the often clumsy and labored expression we have permitted ourselves to retain in the endeavor to keep the right emphasis upon the intended subjects of our statements. Striking illustration of the dangers of ordinary rhetorical formulation have been provided several times in the course of preliminary publication through the effects that have followed some of the kindly efforts of proofreaders, copyeditors or other good friends to improve our diction by the use of conventional phrasings.

It is often claimed that work in our field of research should be confined to specific problems in limited regions, and that in this way alone can be found safety and escape from metaphysical traps. However we cannot accept this claim. For any reader who regards our procedures and postulations as more general than the present state of inquiry justifies, we suggest consideration of the closing words of Clerk Maxwell in his treatise, *Matter and Motion*, from which we have made earlier citations.[2] Maxwell was discussing the development of material systems, while we are interested in the development of knowledge systems. We cite him strictly upon an issue as to *methods of inquiry* useful in their proper times and places to man, the irrepressible inquirer, and without any implication whatever of preference for

---

[1] Alonzo Church, Review of three papers by John Dewey and Arthur F. Bentley, *The Journal of Symbolic Logic*, X (1945), pp. 132-133; Arthur Francis Smullyan, Review of the paper, "Definition," by John Dewey and Arthur F. Bentley, *Ibid*, XII (1947), p. 99.

[2] J. Clerk Maxwell, *Matter and Motion*, (London, 1894), Articles CXLVIII and CXLIX.

material systems over knowledge systems, or *vice versa*. His attention became concentrated upon the use of hypothesis in "molecular science," and he declared that the degree of success in its use "depends on the generality of the hypothesis we begin with." Given the widest generality, then we may safely apply the results we hypothetically secure. But if we frame our hypothesis too specifically and too narrowly then, even if we get resulting constructs agreeable to the phenomena, our chosen hypothesis may still be wrong, unless we can prove that no other hypothesis would account for the phenomena. And finally:

"It is therefore of the greatest importance in all physical inquiries that we should be thoroughly acquainted with the most general properties of material systems, and it is for this reason that in this book I have rather dwelt on these general properties than entered on the more varied and interesting field of the special properties of particular forms of matter."

With the word "behavioral" inserted for the words "physical" and "material," this well expresses our attitude towards our own inquiry. Since it was the mathematics of Clerk Maxwell, dealing with the unparalleled observations of Faraday, that led in the end to the Einsteinian transformation of Newtonian physics, upon one of the highest levels of the use of hypothesis that the world has known, there is much justification for citing Maxwell authoritatively upon this issue. The citation, of course, is not in any way used to give support to our own form of generalization. It applies, instead, to whatever wide-ranging treatment in this field may in the course of time succeed in establishing itself, whosesoever it may be. For the moment the argument is used solely against men of epistemological despair.

We stress once more what has been our theme throughout: namely, that Specification and Transaction, the one on the side of the knowings, the other on the side of the knowns, make common advance. Once under way, once free of the negations and suppressions of ancient verbal lineage, they will be able to make ever more rapidly their joint advances. They make possible at once full spatial-temporal localization, and reference within it to the concrete and specific instance.

Since we have repeatedly said that the recognition of underlying problems and the opening of paths for further construction seems more important to us than the pronouncement of conclusions, we add a memorandum of the places of original publication of our reports for the possible use of anyone desirous of appraising the changes of procedure that came about in the course of the development. The original of Chapter VIII appeared in *Philosophy of Science*, XIII (1946); that of Chapter IX in *Philosophy and Phenomenological Research*, VIII (1947). The publication of the material of the other chapters was in *The Journal of Philosophy*, XLII, XLIII, XLIV, XLV (1945, 1946, 1947, 1948) and, except for that of Chapter X, in the order in which the chapters appear in this volume. The preface and the summary in Chapter XII were later added. The present Introduction accompanied Chapter I in the original publication.

APPENDIX

A Letter from John Dewey

The following letter was written by John Dewey to a philosopher friend after the chapters of this volume were in type. The friend's questionings that elicited this reply will be found in *The Journal of Philosophy*, XLVI, (1949), pp. 329-342.

I

Discovery Bay,
Jamaica

My dear A____ :

In sending you this letter I can not do otherwise than begin with expressing my appreciation of the spirit in which you have written. I also wish to express my gratitude to you for affording me this opportunity to restate the position which, as you suggest, has occasioned difficulties to others as well as to yourself.

When, however, I began to write to you in reply, I found myself in a quandary; in fact, on the horns of a dilemma. On the one hand it seemed obligatory for me to take up each one of your difficulties one by one, and do what I could to clarify each point. The more, however, I contemplated that course, the more I became doubtful of its success in attaining the desired end of clarification. If, I thought, I had not been able to make my position clear in the course of several hundred pages, how can I expect to accomplish that end in the course of a small number of pages devoted to a variety of themes? The other horn of the dilemma was that failure to take up all your points might seem to show a disrespect for your queries and criticism which I am very far from feeling. While I was pondering this matter, I received a letter from a younger fellow student of philosophy. In this letter, written naturally in ignorance of our proposed discussion, he quoted some words written by me some thirty years or more ago. The passage reads: "As philosophers, our disagreements with one another as to conclusions are trivial in comparison with our disagreements as to problems; to see the problem another sees, in the same perspective and at the same angle—that amounts to something. Agreement as to conclusions is in comparison perfunctory."

When I read this sentence it was as if a light dawned. It then occurred to me that I should proceed by trying to show that what is said by me in the book which is the source of your intellectual difficulties, is set forth in a context which is determined, entirely and exclusively, by problems that arise in connection with a development of a Theory of Inquiry; that is, in the context of problems that arise in undertaking an inquiry into the facts of inquiry. Accordingly, I concluded that I might best accede to your request for clarification of the difficulties you have experienced by means of a fresh statement of some of the fundamentals of my position. Since your difficulties and questions hang together, I am sure you will find no disrespect in my treating them as a systematic whole instead of as if they were scattered,

independent, and fragmentary. There is also no disrespect in the belief that their systematic nature is due to the fact that you read what was actually written in the context of connection with the conduct of *inquiry* as if it were written in an *ontological* context—especially as this latter context is classic, in comparison with which that set forth in my *Theory of Inquiry* is an upstart.

I hope, accordingly, dear A___, that you will understand why what is here said delays in coming to a direct answer to specific questions you raise. In order to make my position clear as a whole I have to begin at the beginning, which in the present case lies far back of your questions. I think, for example, that the importance in my writings of what is designated by the words "situation" and "problematic" must have escaped you. Whether this be so or not, we have right here at hand what seems to be an excellent example of their meaning. "Situation" stands for something inclusive of a large number of diverse elements existing across wide areas of space and long periods of time, but which, nevertheless, have their own unity. This discussion which we are here and now carrying on is precisely part of a situation. Your letter to me and what I am writing in response are evidently parts of that to which I have given the name "situation"; while these items are conspicuous features of the situation they are far from being the only or even the chief ones. In each case there is prolonged prior study: into this study have entered teachers, books, articles, and all the contacts which have shaped the views that now find themselves in disagreement with each other. It is this complex of fact that determines also the applicability of "problematic" to the present situation. That word stands for the existence of something questionable, and hence provocative of investigation, examination, discussion—in short, inquiry. However, the word "problematic" covers such a great variety of occasions for inquiry that it may be helpful to specify a number of them. It covers the features that are designated by such adjectives as confusing, perplexing, disturbed, unsettled, indecisive; and by such nouns as jars, hitches, breaks, blocks—in short, all incidents occasioning an interruption of the smooth, straightforward course of behavior and that deflect it into the kind of behavior constituting inquiry.

The foregoing, as I quite recognize, my dear friend, is an indirect approach to the questions you raise. Perhaps I can render it somewhat more direct by calling attention to the fact that the unsettled, indecisive character of the situation with which inquiry is compelled to deal affects all of the subjectmatters that enter into all inquiry. It affects, on the one hand, the observed existing facts that are taken to locate and delimit the problem; on the other side, it affects all of the suggestions, surmises, ideas that are entertained as possible solutions of the problem. There is, of course, nothing at all sacred in employing the words "potentiality" and "possibility" to designate the subjectmatters in inquiry that stand for progress made in determining, respectively, the problem and its solution. What is important, and from the standpoint of my position, all important, is that the tentative, on-trial nature of the subjectmatters involved in each case be recognized; while that recognition can hardly be attained unless some names are given. The indecisive and tentative nature of the subjectmatters involved might have been expressed by using either the word "potentiality" or the word "possibility" for the subjectmatters of both the problem and solution. But in that case, it would have been at once necessary to find sub-terms to designate the distinctive places held and the specific offices or functions performed by subjectmatters constituting what is taken during the conduct of inquiry, as on the one hand the problem to be dealt with and on the other hand the solution suggested: both of them, let it be recalled, being tentative on-trial since both are equally implicated in doubt and inquiry.

From the standpoint of conduct of inquiry it directly follows that the nature of the problem as well as of the solution to be reached is *under* inquiry; failure in solution is sure to result if the problem has not been properly located and described. While this fact is not offered as a justification of the use of the particular words "potentiality" and "possibility," given the standpoint of connection with inquiry, it does imperatively demand the use of two different words as *names* and as names for two disparate but complementary uses.

In any case, dear friend, what has been said has a much wider application than simply to the meaning to be assigned to these two words. For it indicates how and why meaning assigned to *any* phase or aspect of my position which puts what is said in an ontological context instead of that of inquiry is sure to go amiss in respect to understanding. And when I say this, I say it in full recognition of the fact that exclusion of the need of ontological backing and reference of any kind may quite readily convert your difficulty and doubt into outright rejection. But, after all, rejection based upon understanding is better than apparent agreement based on misunderstanding. I should be happy indeed, dear A——, to obtain your assent to my view, but failing that, I shall be quite content if I can obtain an understanding of what it is that my theory of inquiry is trying to do if and when it is taken to be, wholly and exclusively, a theory of knowledge.

II

I hardly need remind you that there is nothing new in recognizing that both observed facts and ideas, theories, rational principles, have entered in fundamental ways into historic discussion of philosophical theories of knowledge. There is nothing new to be found in the fact that I have made them the subjectmatter of a problem. Whatever relative novelty may be found in my position consists in regarding the *problem* as belonging in the context of the conduct of inquiry and not in either the traditional ontological or the traditional epistemological context. I shall, accordingly, in the interest of elucidation attempt another line of approach: one in terms of familiar historical materials.

One outstanding problem of modern philosophy of knowledge is found in its long preoccupation with the controversy between empiricism and rationalism. Even today, when the controversy has receded at least temporarily into the background, it can not be denied by one who surveys the course of the historical discussion that important statements were made with respect both to what was called experience and what was called reason,

and this in spite of the fact that the controversy never reached the satisfactory conclusion constituted by the two parties arriving at agreement. It is not a mere biographical fact, accordingly, if I call attention to the fact that I am in no way an inventor of the problem in a theory of knowledge of the relation to each other of observed factual material on one side and ideational or theoretical material on the other side. The failure of the controversy to arrive at solution through agreement is an important ground of the idea that it is worth while to take these constituents of controversy out of an ontological context, and note how they look when they are placed in the context of the use they perform and the service they render in the context of *inquiry*. The net product of this way of viewing the two factors in the old controversy is expressed in the phrase "The Autonomy of Inquiry." That phrase does more than merely occur in the book that is the source of the discussion in which we are now engaged, since its use there was intended to serve as a key to understanding its contents. The elimination of ontological reference that at first sight may seem portentous actually amounts to the simple matter of saying that whatever claims to be or to convey knowledge has to be found in the context of inquiry; and that this thesis applies to *every* statement which is put forth in the theory of knowledge, whether the latter deals with its origin, its nature, or its possibility.

III

In approaching the special topic of mathematical subjectmatter and mathematical inquiry, I find it necessary, as well as advisable, to begin with the topic of Abstraction. According to the standpoint taken in *The Theory of Inquiry*, something of the nature of abstraction is found in the case of *all* ideas and of all theories. Abstraction from assured and certain existential reference belongs to *every* suggestion of a possible solution; otherwise inquiry comes to an end and positive assertion takes its place. But subjectmatters constituting during the course of inquiry what is taken to be the *problem* are also held in suspense. If they are not so maintained, then, to repeat, inquiry comes automatically to an end. It *terminates* even though the termination is not, with respect to inquiry, a *conclusion*. A flight away from what there and then exists does not of itself accomplish anything. It may take the form of day-dreaming or building castles in the air. But when the flight lands upon what for the purpose of inquiry is an idea, it at once becomes the point of departure for instigating and directing new observations serving to bring to light facts the use of which will develop further use and which thereby develop awareness of the problem to be dealt with, and consequently serve to indicate an improved mode of solution; which in turn instigates and directs new observation of existential material, and so on and on till both problem and solution take on a determinate form. In short, unless it is clearly recognized that in *every* case of obstructed ongoing behavior "*ideas*" are temporary deviations and escapes, what I have called their functional and operational standing will not be understood. Every *idea* is an *escape*, but escapes are saved from being *evasions* so far as they are put to use in evoking and directing observations of further factual material.

I am reasonably confident, dear A——, that in this one point at least we shall find ourselves in agreement. I do not believe that either of us is in sympathy with the wholesale attacks upon abstractions that are now being made in some quarters. Theories as they are used in scientific inquiry are themselves matters of systematic abstraction. Like ideas, they get away from what may be called the immediately given facts in order to be applicable to a much fuller range of relevant facts. A scientific theory differs from the ideas which, as we say, "pop into our heads," only in its vast and systematic range of applicability. The peculiarity of *scientific* abstraction lies in the degree of its freedom from *particular* existential adhesions.

It follows as a matter of course that abstraction is carried on indefinitely further in scientific inquiry than there is occasion for carrying it on in connection with the affairs of everyday life. For, in the latter case, an abstraction loses its serviceability if it is carried beyond applicability to the *specific* difficulty then and there encountered. In the case of scientific inquiry, theory is carried to a point of abstraction which renders it available in dealing with a maximum variety of possible uses. What we call *comprehensiveness* in the case of a theory is not a matter of its own content, but of the serviceability in range of application of that content. It is perhaps worth while to notice that the Newtonian theory was, for a long time, believed to be completely comprehensive in respect to all astronomical subjectmatter; not merely that which had already been observed but to all that ever could possibly be observed. Finally, there occurred what in the case of an everyday affair of life would be called a *hitch* or *block*. Instead of the discrepancy being accepted as a finality it was, however, at once *put to use* in suggesting further development upon the side of theory as abstraction. The outcome constitutes what is known as "The Relativity Theory." Newton had carried *his* abstraction to a point which was shocking to many of his contemporaries. They felt that it took away the reality which gave point and zest to the affairs of life, moral and esthetic as well as practical in a utilitarian sense. In so doing they made the same mistake that professional philosophers made after them. They treated a use, function, and service rendered in conduct of inquiry as if it had ontological reference apart from inquiry.

When viewed from the standpoint of its position in the conduct of inquiry, the relativity theory rendered space and time themselves subjectmatters of inquiry instead of its fixed limits. In the Newtonian theory they had been treated as an *Ultima Thule* beyond which scientific inquiry could not possibly go. These considerations may be used, dear A ___ , as an example of how submitting inquiry to ontological reference obstructs it. But here they are mentioned on account of their bearing on the question of mathematical subjectmatter. No matter how far physical theory carries its abstractions, it would contradict the very intent of the latter if they went beyond possibility of application to every kind of *observable* existential materials. The privilege of *that* use and office is reserved for mathematical inquiry. The story of the development of mathematical inquiry shows that its advances have usually been occasioned by something which struck some inquirer as a hitch or block in the previous state of its subjectmatter. But in the course of the last one or two generations, mathematicians have arrived at the point at which they see that the heart of the work they are engaged in is the method of free postulation. It is hardly necessary to note how the constructions in which the interior angles of a triangle are, as the case may be, either less or more than two right angles, have removed the ontological obstructions that inhered in Euclidean geometry. While in most respects I

am compelled to admit that important features of my position are incompatible with philosophical theories that have received authoritative and, so to say, official formulations, in this matter of mathematics, I believe, Mr. A___, that I am "on the side of the angels." At all events, I did not invent the position that I have taken in the foregoing statements. I took it over almost bodily from what the mathematicians have said who have brought about the recent immense advances in that subject. It is the progress of mathematical inquiry *as* mathematical which has profoundly shaken the ontological rigidity once belonging to the circle and the triangle as their own immutable "essences." I can not, accordingly, refrain from mentioning the role that considerations similar to those just mentioned have played in inducing me to undertake an attempt to convert all the *ontological*, as prior to inquiry, into the *logical* as occupied wholly and solely with what takes place in the conduct of inquiry as an evergoing concern.

IV

In the hope that it may further a clarified understanding of my position, I shall now take up another outstanding problem of modern epistemological philosophy. It is a familiar fact that the historical systems of epistemological philosophy did their best to make ontological conclusions depend upon prior investigation of the conditions and nature of knowledge. A fact which is not so familiar, which indeed is often ignored, is that this attempt was itself based upon an ontological assumption of literally tremendous import; for it was assumed that whatever else knowledge is or is not, it is dependent upon the independent existence of a *knower* and of something *to be known;* occurring, that is, between mind and the world; between self and not-self; or, in words made familiar by use, between subject and object. The assumption consisted in holding that the subjectmatters designated by these antithetical terms are separate and independent; hence the problem of problems was to determine some method of harmonizing the status of one with the status of the other with respect to the possibility and nature of knowledge. Controversy on this topic, as is the case with the other historic problem already mentioned, has now receded into the background. It can not be affirmed, however, that the problem is settled by means of reaching an agreed-upon solution. It is rather as if it had been discovered that the competing theories of the various kinds of realism, idealism, and dualism had finally so covered the ground that nothing more could be found to say.

In this matter also it accordingly occurred to me that it might be a good idea to try the experiment of placing in the context of inquiry whatever matters were of moment and weight in what was urged by the various parties to the controversy. For observed and observable facts of inquiry are readily available: there is a mass of fact extending throughout the whole recorded intellectual history of man, in which are manifest for study and investigation both failures and successes—much as is the case in the story of any important human art. In this transfer of matters at issue from their prior ontological setting into a context that is set *wholly and only* by conditions of the conduct of inquiry, what had been taken to be inherent ontological demands were seen to be but arbitrary assumptions from their own standpoint, but important distinctions of use and office in the progressive carrying on of inquiry.

In pursuing this line of inquiry, it proved to be a

natural affair to take as a point of departure the physiological connection and distinction of organism and environment as the *most readily observable* instance of the *principle* involved in the matter of the connection and distinction of "subject and object." Consideration of the simpler physiological activities which significantly enough already bore the name "functions" served to indicate that a life-activity is not anything going on *between* one thing, the organism, and another thing, the environment, but that, *as* life-activity, it is a simple event over and across that distinction (not to say separation). Anything that can be entitled to either of these names has first to be located and identified as it is incorporated, engrossed, in life-activity. Hence there was presented in an acute form the following problem: Under what conditions of life-activity and to what consequences in the latter is the distinction relevant?

The issue involved in this question coalesced, almost of itself, with the point of view independently reached in regard to knowing as inquiry with respect to its origin in the event of a hitch, blockage, or break, in the ongoing of an active situation. The coalescence worked both ways. With respect to the distinction within the course of physiological life-activity, the obvious suggestion was that the subjectmatters to which the names "organism" and "environment," respectively, apply are distinguished when some function, say digestion, is disturbed, unsettled, and it is necessary, *in order to do something about it* which will restore the normal activity (in which organs and foods work together in a single unified process) to *locate* the source of the trouble. Is there something wrong inside? Or is the source of the disturbance located in water or in food that has been taken into the system? When such a distinction is once clearly made there are those who devote themselves especially to inquiry into the structures and processes that can be *referred* distinctively to the organisms, (although they could not take place and be capable of such reference without continuous partnership in a single transaction), while others study the relations of air, climate, foods, water, etc., to the maintenance of health—that is, of unified functionings.

What happens when distinctions which are indispensable to form and use in an efficient conduct of inquiry—that is to say, one which meets its own conditions *as* inquiry—are converted into something ontological, that is to say, into something taken to exist on its own account prior to inquiry and to which inquiry must conform, is exhibited, I submit, my dear questioner, in the epistemological phase of modern philosophy; and yet the new science could not have accomplished its revolution in astronomy, physics, and physiology if it had not *in the course of its own development* of method been able, by means of such distinctions as those to which theory gave the names "subject" and "object," "mind" and "the world," etc., to slough off the vast mass of irrelevant pre-conceptions which kept ancient and medieval cosmology from attaining scientific standing.

It is not implied, however, that what has just been said covers the whole scope of the problem. There remains the question of why at a particular time the distinction between knower and the subjectmatter to be known became so conspicuous and so central as to be for two centuries or more one of *the* outstanding philosophical issues. No such problem was urgent in either ancient or medieval philosophy. The idea that most directly suggests itself as an indication of a solution of this problem is that

the rather sudden and certainly striking emergence of the "subject-object" problem is intimately connected with the cultural conditions that mark the transition of the medieval period into that age that is called *modern*. This view of the matter is, I believe, an interesting and even important hypothesis; it is one which in another connection might be followed out with advantage. It is introduced here, however, solely for whatever service it may render in understanding a position which, like that set forth in *The Theory of Inquiry*, transfers what had been taken to be ontological separations into distinctions that serve a useful, indeed necessary, function in conduct of inquiry.

Before leaving this endeavor to clarify my position through reference to well-known events in the history of philosophy, I shall mention a third matter which, unlike the two already mentioned, is still more or less actively pursued in contemporary philosophical discussion. I refer here to the extraordinary contrast that exists beyond peradventure between the subjectmatters that are known in science and those known in the course of our everyday and common living—common not only in the sense of the usual but of that which is shared by large numbers of human beings in the conduct of the affairs of their life. To avoid misunderstanding it should be observed that the word "practical" has a much fuller meaning when used to designate these affairs than it has when it is used in a narrow utilitarian way, since it includes the moral, the political, and the artistic. A simple but fairly typical example of the undeniable contrast between the subjectmatters of this common life and the knowings that are appropriate to it, and the subjectmatter and method of scientific knowing, is found in the radical unlikeness of the water we drink, wash with, sail boats upon, use to extinguish fires, etc., etc., and the H_2O of scientific subjectmatter.

It would appear dogmatic were I to say that the problem involved in this radical unlikeness of subjectmatters is insoluble if its terms are placed in an ontological context. But the differences between, say, a spiritualistic and a materialistic ontological solution remind us how far away we are from any agreed-upon solution. It hardly seems unreasonable to suggest that parties to the controversy are lined up on the basis of preferences which are external to the terms of the issue rather than on grounds which are logically related to it. When the issue pertaining to and derived from this contrast is placed and treated in the context of different types of *problems* demanding different methods of treatment and different types of subjectmatter, the problem involved assumes a very different shape from that which it has when it is taken to concern the ontological "reality." It would be irrelevant to the present issue were I to attempt to tell just what form the problem and its solution assume when they are seen and treated in the context of inquiry. It is relevant, however, to the understanding of the point of view to say that it demands statement on the ground of types of problems so different that they are capable of solution only in terms of types of subjectmatter as unlike one another as are those exemplified in the case of "*water.*" I may, however, at least point out that a thirsty man seeking water to drink in a dry land would hardly be furthered in the emergency in which he finds himself by calling upon H_2O as his subjectmatter; while, on the other hand, the physicist engaged in his type of problem and inquiry would soon be brought to a halt if he could not treat water as H_2O. For it is on account of *that* mode of treatment that water is taken out of isolation as a subject of knowledge and brought into vital and intimate connection with an indefinitely extensive range of other matters qualitatively and immediately of radically different kinds from water and from one another.

It seems pertinent at this point, my dear A___ , to refer to that aspect of my theory of knowledge to which I gave the name "instrumentalism." For it was intended to deal with the problem just mentioned on the basis of the idea or hypothesis that scientific subjectmatter grows out of and returns into the subjectmatter of the everyday kind;—the kind of subjectmatter to which *on the basis of ontological interpretation* it is totally and unqualifiedly opposed. Upon the basis of this view the metaphysical problem which so divided Berkeley from Sir Isaac Newton, and which has occupied such a prominent place in philosophy ever since the rise of new physical science, is not so much resolved as dissolved. Moreover, new construction accrues to the subjectmatter of physical science just because of its extreme unlikeness to the subjectmatters which for the sake of brevity may be called those of common sense. There is presented in this unlikeness a striking example of the view of the function of thoroughgoing abstraction mentioned shortly ago. The extreme remoteness of the subjectmatter of physical science from the subjectmatter of everyday living is precisely that which renders the former applicable to an immense variety of the occasions that present themselves in the course of everyday living. Today there is probably no case of everyday living in which physical conditions hold a place that is beyond the reach of being effectively dealt with on the ground of available *scientific* subjectmatter. A similar statement is now coming to hold regarding matters which are specifically physiological! Note, in evidence, the revolution that is taking place in matters relating to illness and health. Negative illustration, if not confirmation, may be supplied by the backward state of both knowledge and practice in matters that are distinctively human and moral. The latter in my best judgment will continue to be matter of customs and of conflict of customs until inquiry has found a method of abstraction which, because of its degree of remoteness from established customs, will bring them into a light in which their nature will be indefinitely more clearly seen than is now the case.

As I see the matter, what marks the scientific movement that began a few centuries ago and that has accomplished a veritable revolution in the methods and the conclusions of natural science are its *experimental* conduct and the fact that even the best established theories retain *hypothetical* status. Moreover, these two traits hang together. Theories as hypotheses are developed and tested through being put to use in the conducting of experimental activities which bring to the light of observation new areas of fact. Before the scientific revolution some theories were taken to be inherently settled beyond question because they dealt with Being that was eternal and immutable. During that period the word "hypothesis" meant that which *was placed under* subjectmatters so firmly as to be beyond the possibility of doubt or question. I do not know how I could better exemplify what I mean to be understood by the functional and operational character of ideational subjectmatter than by the radical change that in the development of scientific inquiry has taken place in the working position now attached to hypothesis, and to *theory* as hypothetical.

Let me say, my friend, that I have engaged in this fairly long, even if condensed, historical exposition solely for the sake of promoting understanding of my position. As I have already indicated, I did not originate the main figures that play their parts in my theory of knowing. I tried the experiment of transferring the old well-known figures from the stage of ontology to the stage of inquiry. As a consequence of this transfer, the scene as it presented itself to me was not only more coherent but indefinitely more instructive and humanly dramatic.

In any event the various factors, ancient and modern, of historical discussion and controversy were precipitated in the book whose subjectmatter is the occasion of this present exchange of views. I am aware that I have not made the kind of reply which in all probability you felt you had a right to anticipate. At the same time, while I have taken advantage of considerations that have occurred to me since the text in question was written, I do not believe that I have departed from its substantial intent and spirit. Yet I am bound to acknowledge that the occasion of precipitating historical materials into the treatise under discussion was the great variety of works on logical theory that appeared during the nineteenth century. As I look back I am led to the conclusion that the attempt conscientiously to do my full duty by these treatises is accountable for a certain cloudiness which obscures clear vision of what the book was trying to do. The force of the word "Logic," in all probability, has overshadowed for the reader the import of what in my intention was the significant expression, *The Theory of Inquiry*. For that source of misapprehension I accept full responsibility. I am, accordingly, the more grateful to you, my dear friendly critic, for affording me this opportunity for restatement, which, I venture to hope, is free from some of the encumbrances that load down the text. I shall be content if I have succeeded in this response to your request for clarification in conveying a better understanding of the *problem* that occupied me. As I reflect upon the historical course of philosophy I am unable to find its course marked by notable successes in the matter of conclusions attained. I yield to none, however, in admiring appreciation of the liberating work it has accomplished in opening new perspectives of vision through its sensitivity to problems it has laid hold of in ways which, over and over again, have loosened the hold upon us exerted by predispositions that owe their strength to conformities which became so habitual as not to be questioned, and which in all probability would still be unquestioned were it not for the debt we owe to philosophers.

Very sincerely yours,

John Dewey

INDEX

Section D

ANALYSIS OF RECENT INQUIRIES

Editorial Note:

In this section we reprint some analyses we (individually or jointly) have made of recent inquiries in the behavioral field. Our analyses reflect the procedures of inquiry that we suggested for trial earlier in this volume. Although those analyses were written within the last few years, in some respects we believe we now could improve what was said earlier. The more important of those changes are indicated in notes inserted in the chapters. The body of each chapter, however, is reprinted as it originally appeared.

R. H.
E. C. H.

I.

THE AUTHORITARIAN PERSONALITY [1]

Rollo Handy

In contrast to the work mentioned so far in this book, that done in the project now to be described[2] puts much emphasis on those factors of which the person is not consciously aware. Although the F Scale (Fascism Scale), which will be our major topic of consideration, is not labelled simply or exclusively as a measure of value, its content in many ways is similar to other scales behavioral scientists have developed under the rubric of value measurement. The book as a whole is a blending of psychoanalytic and academic psychological theory with empirical inquiry of the type often found in social psychological research.

In view of the overall aim of the volume, many decisions were made throughout the research that would be challenged by other workers. That, of course, is not said as a negative criticism, but only to point out that estimates of the achievements of the research are subject to a vast array of possible criticisms.

The authors state as a major hypothesis:

"...that the political, economic, and social convictions of an individual often form a broad and coherent pattern, as if bound together by a 'mentality' or 'spirit,' and that this pattern is an expression of deep-lying trends in his personality."[3]

Their main concern was with people who were "potentially fascistic," or whose personality structures were likely to make them especially susceptible to antidemocratic propaganda. Such people "have a great deal in common," including a cluster of "opinions, attitudes, and values."

According to the authors, opinions, attitudes, and values are "expressed more or less openly in words," and in that sense are psychologically "on the surface." However, the degree of openness with which a person responds depends upon the context, especially in the case of sensitive issues such as those concerning current ideological debates, minority groups, etc. There may well be a difference between what a person says in a fairly public setting and what he will express confidentially to friends. But even the latter, the authors say, can be observed directly using appropriate psychological techniques. In addition, however, a person may have "secret" views that he almost never will reveal to anyone else, or thoughts that he cannot admit even to himself, or ideas that are so vague or ill-formed that he cannot put them into words. The authors say: "To gain access to these deeper trends is particularly important, for precisely here may lie the individual's potential for democratic or antidemocratic thought and action in crucial situations."[4]

All the aspects of an individual's attitudes, values, etc., are part of an organized structure that may contain contradictions and inconsistencies, but in which the constituents are "related in psychologically meaningful ways." To understand such a structure, a theory of the total personality is required. The authors took Freud's views as basic in working out their theory of personality structure, and were guided by academic psychologists in their attempt to formulate the "more directly observable and measurable aspects of personality."

The authors view personality as "a more or less enduring organization of forces within the individual." These "forces" help to determine a person's behavior, and help explain what consistency it has, but behavior is not viewed as the same thing as personality: "personality lies *behind* behavior and *within* the individual." The "forces" are not responses, but "readinesses for response," and are primarily needs that vary from one person to another in terms of quality, intensity, mode of gratification, etc. Since personality "is essentially an organization of needs," and opinions, attitudes, and values depend upon needs, personality is a "*determinant*" of ideological preferences." However, the personality is not hypostatized as some ultimate determinant, and it is said to evolve in relation to the social environment.[5]

In describing their general methodology, the authors say:

"A particular methodological challenge was imposed by the conception of *levels* in the person; this made it necessary to devise techniques for surveying opinions, attitudes, and values that were on the surface, for revealing ideological trends that were more or less inhibited and reached the surface only in indirect manifestations, and for bringing to light personality forces that lay in the subject's unconscious. And since the major concern was with *patterns* of dynamically related factors—something that requires study of the total individual—it seemed that the proper approach was through intensive clinical studies. The significance and practical importance of such studies could not be gauged, however, until there was knowledge of how far it was possible to generalize from them. Thus it was necessary to perform group studies as well as individual studies, and to find ways and means for integrating the two."[6]

Groups were studied through questionnaires, and individuals through interviews and clinical tests; both approaches were carried on in close conjunction. Clinical studies of an individual's underlying wishes, fears, defenses, etc., were used to help arrive at items for the group questionnaires, and the group studies helped show what opinions, attitudes, and values were associated together and their relation to life histories and the contemporary situations of individuals.

To help identify "potentially antidemocratic" individuals, a questionnaire was filled out by many people. Among other items, the questionnaire contained antidemocratic statements with which the respondent was

---

1 Reprinted from Rollo Handy, *The Measurement of Values*, St. Louis, Warren H. Green, 1970.
2 T.W. Adorno, Else Frenkel-Brunswik, Daniel J. Levinson, and R. Nevitt Sanford, *The Authoritarian Personality*, New York, Harper & Row, 1950.
3 *Ibid.*, p. 1.
4 *Ibid.*, p. 4.

5 *Ibid.*, pp. 5-6.
6 *Ibid.*, pp. 11-12.

asked to agree or disagree. Those who showed the greatest amount of agreement with such statements (and for control purposes, also those who showed most disagreement, and some who were neutral) were given psychiatric interviews and tested clinically through the use of the Thematic Apperception Test (TAT). The questionnaire was then revised in view of the clinical findings. The notion of validity here was to find questionnaire items that correlated highly with opinions people would express in the clinical situation.

Since it was assumed that many people would not be willing to speak frankly about the ideological issues the authors were trying to get at, the scale had to be constructed in a different way than those designed to measure only surface issues. The procedure was to bring together items in the scale that, on the basis of clinical experience and the authors' theories, were presumed to indicate trends lying "relatively deep" within the personality and which constituted a disposition either to express, or to be influenced by, fascistic ideas. In the main the questionnaire items were designed to serve as rationalizations for irrational tendencies.

To illustrate, two statements on the scale were (1) "Nowadays when so many different kinds of people move around so much and mix together so freely, a person has to be especially careful to protect himself against infection and disease," and (2) "Homosexuality is a particularly rotten form of delinquency and ought to be severely punished." Individuals who agreed with one of those statements tended to agree with the other, and also tended to agree with openly antidemocratic statements.

Questionnaires were collected from over 2,000 respondents. Many were college students, but also there were prison inmates, psychiatric patients, labor union members, Kiwanis Club members, etc. In general, most subjects were drawn from the middle socioeconomic class, and the inquirers found early in their study that somewhat different instruments and procedures would be necessary for subjects having a lower socioeconomic background. The authors say that the "findings of the study may be expected to hold fairly well for non-Jewish, white, native-born, middle-class Americans," but that their population is "rather inadequate" as a basis for generalizing about the total population of the U.S.[1]

In constructing the F scale, the authors hoped to develop an instrument that would measure prejudice without mentioning any minority group by name and without appearing to have the aim that it did have. By circumventing certain defenses an individual might use when asked directly about race issues, a better measure of prejudice could result. Each item in the scale was based on an hypothesis (or hypotheses) as to the item's connection with prejudice. These hypotheses in turn resulted from earlier research. The authors did not begin with hundreds of items chosen almost randomly and then find out which ones might be associated with what they wanted to measure, but rather began with items that were theoretically linked to potential fascism.

The original F Scale was derived from the larger set of questionnaire items which the respondents answered. Nine different variables made up the basic content of the scale. Each such variable was regarded "as a more or less central trend in the person" which (according to the authors' view of personality dynamics) expresses itself in ethnocentric behavior.

Listed below are the nine variables, the brief description given of each, and a sample F Scale item:

1. *Conventionalism.* Rigid adherence to conventional, middleclass values. (A sample item is: One should avoid doing things in public which appear wrong to others, even though one knows that these things are really all right.)

2. *Authoritarian submission.* Submissive, uncritical attitude toward idealized moral authorities of the ingroup. (Every person should have a deep faith in some supernatural force higher than himself to which he gives a total allegiance and whose decisions he does not question.)

3. *Authoritarian aggression.* Tendency to be on the lookout for, and to condemn, reject, and punish people who violate conventional values. (Homosexuality is a particularly rotten form of delinquency and ought to be severely punished.)

4. *Anti-intraception.* Opposition to the subjective, the imaginative, the tender-minded. (There is too much emphasis in colleges on intellectual and theoretical topics, and not enough emphasis on practical matters and on the homely virtues of living.)

5. *Superstition and stereotypy.* The belief in mystical determinants of the individual's fate; the disposition to think in rigid categories. (Although many people may scoff, it may yet be shown that astrology can explain a lot of things.)

6. *Power and "toughness."* Preoccupation with the dominance-submission, strong-weak, leader-follower dimension; identification with power figures; overemphasis upon the conventionalized attributes of the ego; exaggerated assertion of strength and toughness. (No insult to our honor should ever go unpunished.)

7. *Destructiveness and cynicism.* Generalized hostility, vilification of the human. (No matter how they act on the surface, men are interested in women for only one reason.)

8. *Projectivity.* The disposition to believe that wild and dangerous things go on in the world; the projection outwards of unconscious emotional impulses. (The sexual orgies of the old Greeks and Romans are nursery school stuff compared to some of the goings-on in this country today, even in circles where people might least expect it.)

9. *Sex.* Exaggerated concern with sexual "goings-on." (Sex crimes, such as rape and attacks on children, deserve more than mere imprisonment; such criminals ought to be publicly whipped.)[2] A single item may represent more than one of the variables, and the different variables are represented by different numbers of items; the main concern was the overall pattern in which the variables fitted.

The authors also mention three principles that had particular significance for the development of the F Scale. The first was that an item chosen should have the maximum of indirectness. The second was that each item should have some balance between irrationality and "objective truth"; it should neither be so "wild" that hardly anyone would agree with it nor so correct that nearly everyone would agree with it. Third, each item had to "contribute to the structural unity of the scale as a whole."

The respondents indicated whether they agreed or disagreed with each item, and also to what degree, on a scale of three. When the first version of the F Scale (made up of 38 items) was administered, the mean reliability (.74) was not bad, but was "well below what is required of a truly accurate instrument." On an item analysis, some items turned out especially poor statistically because they were unclear or ambiguous, some

---

[1] *Ibid.*, p. 23.

[2] *Ibid.*, pp. 228-241.

were so "true" that nearly everyone agreed with them, and some were so "crude or openly aggressive" that nearly everyone tended to disagree. So a revision was made to increase reliability, and a reliability of .87 resulted. But the scale still contained some items that were poor statistically and a few items that needed to be dropped because they were no longer timely. Also a shorter scale was deemed desirable. So a third version was constructed. On this final revision the average of the reliability coefficients turned out to be .90 (ranging from .81 to .97), and the authors felt they now had a scale that "meets rigorous statistical standards."

In validating the F Scale, attention was given to comparing the results to those obtained in the case studies. The responses of Larry and Mack are described in some detail. Mack was a relatively high scorer on the F Scale (above the mean on all of the nine variables except *Superstition* and *Power and "toughness")*, and this seemed in general harmony with his interview materials.

As an illustration of some of the problems encountered, however, we might note that the authors expected on the basis of Mack's interview that one of his highest scores would be on *Authoritarian submission*, but his actual score was only at the group mean. The explanation given is that the items in this variable on which he scored above the mean are those expressing authoritarian submission in its purest form, while his low scores on other items in that variable resulted from the influence of his objective-scientific values. The authors say perhaps "Mack's submissive tendencies are insufficiently sublimated to permit their expression in abstract religious terms." They also note that one item which he disagreed with was an item they expected him to accept, and suggest that for some "truly submissive subjects" an item can come "too close to home," and that those subjects therefore respond contrary to their strongest feeling.

Larry scored lower than the group mean on all variables except *Authoritarian aggression*. In general, there was harmony between his responses on the F Scale and his interview materials, although there were also some surprises. For example, there was nothing in the interview material to suggest that Larry was superstitious, and yet he did agree with the astrology item. The authors suggest that perhaps "it should not be surprising to find an element of mysticism in this weak and rather passive character."[1]

The Authoritarian Personality is nearly 1000 pages long and contains an enormous amount of material not even faintly alluded to here. And although there are repeated references to the measurement of "opinions, attitudes, and values," the scales constructed do not pretend to measure values in general or even a broad range of values (in addition to the F Scale, scales were developed for Anti-Semitism, Ethnocentrism, and Politico-Economic

Conservatism). Even so, the book does illustrate some of the difficulties and problems likely to be encountered whenever one attempts to combine depth psychology with scale construction to measure something in the value realm.

Among the major questions arising in such an approach, in my opinion, are the following:

(1) Are there actually widespread behavior patterns of the type assumed when potential fascistic personalities are discussed? I am not here saying the answer is "no," but raising the question of what typologies are most useful, since so many different typologies can be imagined and in fact have been offered. The spatio-temporal setting of the studies in this volume helps to explain the emphasis on fascistic personalities, but perhaps in a different time and place that emphasis would seem misdirected.

(2) Clearly the authors have great confidence in various depth psychology theories, clinical interviews, the TAT, etc. Without going into the enormous literature critical of such tendencies, I will only note here a marked willingness on the part of some clinicians to rely on a kind of "fittingness" or "falling into place" of the various materials they deal with. They seem to show a penchant for developing "likely stories" that may have little or no scientific support, and often a kind of intuition is uncritically relied upon. Whatever the merits of psychoanalytically oriented theories, various projective tests, etc., it seems clear that many aspects of those materials have not as yet been scientifically warranted. In any event, the heavy reliance on them by the authors in assessing the validity of the F Scale obviously poses the question of the validity of the clinical instruments.

(3) *The Authoritarian Personality* was a mammoth undertaking, and some questions arise about the worth of the findings compared to the amount of effort that went into the project. Even if the various scales developed turn out to have less significance than was thought when the study first appeared, the overall project still might have merit as a guide to how similar inquiries could be improved in the future. But how to assess the merits of the output of the investigation is not an easy matter, in view of the controversies about almost all of its major aspects. In my opinion, in projects such as *The Authoritarian Personality* a disproportionate amount of effort goes into the elaboration of hypotheses and their modification on the basis of soft evidence, and it probably would be more profitable to concentrate attention on the confirmation of the guiding hypotheses right from the beginning.

Despite the negative tone of much that has been said, I see no reason to doubt that a cluster of opinions and attitudes about "antidemocratic" phenomena is found in some behavior and that this cluster has been measured to some degree by the use of the F Scale. I also think that the whole project indirectly indicates how relatively superficial many other studies of value are and how willing some inquirers are to take the most "surface" responses as indicative of values.

---

[1] The materials on validation are on pp. 269 ff; those on reliability are on pp. 242 ff.

II.

AN "EXTREMELY SIMPLE" MEASUREMENT OF VALUES [1]

Rollo Handy

Author's note: The selection reprinted below was originally preceded by a consideration of some criticisms made of alleged measurements of value, such as that only "trivial" findings can result from scientific inquiry into valuing behavior. I rejected such criticisms in general.

* * * * * *

On the other hand, and without condemning many of the value measurements that have been made, I think there often is an unwarranted and self-defeating inflation of the significance of what has been achieved. Rather simple and uncritically done work on occasion is discussed pretentiously, as though a great breakthrough had resulted. So, without agreeing at all with the underlying themes of many critics who maintain that scientific measures of value are trivial, I do take a rather skeptical attitude toward what has been achieved to date.

To illustrate, some work of Milton Rokeach will be discussed. He indicates deep dissatisfaction with conventional social psychology. The "great majority of experimental findings" on attitude change, he says, actually have little to do with attitude change, and a simpler way can be found to account for the results. Rokeach believes the "time is now perhaps ripe" to shift the main focus in social psychology away from attitude organization and to study instead value organization and change. This presupposes a "clear-cut conceptual distinction between attitude and value," although he emphasizes also that beliefs, attitudes, and values "form a functionally integrated cognitive system."[2]

Rokeach differentiates between 'value' and 'attitude' as follows:

"An attitude. . .is an organization of several beliefs focused on a specific object (physical or social, concrete or abstract) or situation, predisposing one to respond in some preferential manner. Some of these beliefs about an object or situation concern matters of fact and others concern matters of evaluation. . . . Values, on the other hand, have to do with modes of conduct and end-states of existence. To say that a person 'has a value' is to say that he has an enduring belief that a specific mode of conduct or end-state of existence is personally and socially preferable to alternative modes of conduct or end-states of existence. . . . While an attitude represents several beliefs focused on a single specific object or situation, a value is a single belief that transcendentally guides action and judgments across specific objects and situations, and beyond immediate goals to more ultimate end-states of existence. Moreover, a value, unlike an attitude, is an imperative to action, not only a belief about the preferable but also a preference for the preferable."[3]

He further distinguishes between "instrumental" and "terminal" values. In his formulation, an instrumental value is a "single belief" that a specific mode of conduct (the value) is "personally and socially preferable in all situations with respect to all objects," and a terminal value is a belief that "an end-state of existence" (the value) is "personally and socially worth striving for." (His formulation of the distinction is different from those often found.)

He then describes his current research program. His first problem was to find a way to measure value systems, and he describes his approach as "extremely simple." Initially, he took a dozen of what he called instrumental values (e.g., *broadminded, forgiving, responsible)* and a dozen terminal values (e.g., *equality, freedom, salvation)*, alphabetized each set, and asked his subjects to rank-order them in importance. After improving the items, he got data on the instrumental and terminal value rank-orderings of many groups differing in age, sex, education, etc.

In his discussion of the validity of his instrument, Rokeach makes much of the finding that the ranking of one terminal value alone, *salvation,* "highly predicts church attendance." Those who reported they were sympathetic to and participated in civil rights demonstrations tended to rank *freedom* first, and *equality* third, among 12 terminal values. Unemployed Negroes ranked *freedom* tenth and *equality* first. Students at a Calvinist college ranked both *freedom* and *equality* relatively low. Many similar results were obtained, and Rokeach is optimistic about the validity of his research.

He further says that his data on *freedom* and *equality* point to "the presence of a simple, nonetheless comprehensive, two-dimensional model for describing all the major variations among various political orientations." To illustrate the model, he compares it to the four points of a compass. The north pole represents those who value highly both *freedom* and *equality* (he lists here liberal democrats, socialists, and humanists); the south pole represents those putting a low value on both (he lists fascists, Nazis, and Ku Klux Klan members); the east is the location of those who value *freedom* highly and put a low value on *equality* (John Birch Society, conservative Republicans, followers of Ayn Rand); and on the west are those who put a high value on *equality* and a low value on *freedom* (Stalinists and Maoists).

In support of this model, Rokeach cites a word-count technique: 25,000-word samples were selected from political writings representing the four poles, and a count was made of the number of times various terminal and instrumental values were favorably mentioned. (Samples came from socialist writers such as Norman Thomas and Erich Fromm, from Hitler, Barry Goldwater, and Lenin.) The socialists mentioned *freedom* favorably 66 times and *equality* 62 times, and ranked *freedom* first and *equality* second in a group of 17 terminal values. In Hitler's *Mein Kampf, freedom* was ranked 16th and *equality* 17th among the same 17 terminal values. For Goldwater, *freedom* was first and *equality* 16th; for Lenin, *freedom*

[1] Reprinted from Rollo Handy, *The Measurement of Values*, St. Louis, Warren H. Green, 1970.

[2] Milton Rokeach, *Beliefs, Attitudes, and Values*, San Francisco, Jossey-Bass, 1968, pp. ix-xii.

[3] *Ibid.*, pp. 159-160.

was 17th and *equality* first. Rokeach says: "All in all, these data seem to fit the two-dimensional model almost perfectly."

At the end of the chapter Rokeach says:

"As I conclude this chapter I become acutely aware of at least a few questions that should be raised about the methods and findings reported here. Do the various value terms have the same meaning for different subjects? What ethical precautions are especially necessary in research on value change? Are the systematic value and attitude changes and the sleeper effects reported here genuine changes or are they artifacts of the experimental situation? Can we expect behavioral changes to follow from such value and attitude changes? Is it just as consistent for a person to move *freedom* down to *equality* as to move *equality* up to *freedom?* . . . What are the implications of our formulations and findings for education, therapy, and other areas of human concern that necessarily engage people's values?"[1]

Since the research discussed here is not presented fully by Rokeach (he promises a fuller report in a subsequent publication), it may not be fair to criticize him on the basis of the materials presently available. Yet the hubris exhibited is rather remarkable, and helps to illustrate why some critics are unimpressed by alleged measurements of value. Starting with a very simple and in many ways questionable instrument, we are quickly plunged into broad questions about implications for education, therapy, etc.

The apparent lack of self-criticism in some parts of his work is also notable. Translation problems alone might

_____
[1] *Ibid.,* p. 178.

suggest some caution in comparing word counts of the terms 'freedom' and 'equality' in the writings of Hitler and Lenin to those of American authors, let alone the notorious problems about the referents of those terms in English. Indeed, the various semantic and terminological problems are acute throughout Rokeach's research. Further, one would like evidence for the notion that a value, taken as an "enduring belief" that something is "personally and socially preferable" to its alternatives, can be gotten at through word counts, rank-ordering of a selected group of words, or similar procedures.

Many writers maintain that there can be a great difference between what is taken as personally preferable and what is socially preferable, but Rokeach insists (by definition) that a value is not a value unless both are involved. Interesting questions occur also about the sense in which values such as *freedom* and *equality* are single beliefs rather than composites of beliefs, especially since that distinction is basic for Rokeach's differentiation between 'attitude' and 'value,' which in turn is important for his proposed reform of social psychology.

Without dwelling further on possible criticisms, and without denying some merits in Rokeach's research program, his project appears to be an attempt to arrive at highly significant results about complex human behavior through a very simple approach that either bypasses many pertinent issues or glosses over them. The great danger in such approaches, in my opinion, is not that the unwary will be taken in and believe there is a strong scientific warrant for the results, but rather that those negatively disposed toward a scientific inquiry into values will interpret such attempts as a kind of *reductio ad absurdum* of any scientific approach.

III.

SOME DIFFICULTIES WITH UTILITIES[1]

Rollo Handy

In the literature on the type of formalized structures we are now considering, rational behavior is often closely related to the maximizing of utilities. The literature on utilities is extensive and diverse, and only certain aspects are taken up here.

In economic theory 'utility' was often used roughly as 'ability to satisfy a want,' or more formally, as the 'indicator of the level of want-gratification.' Jerome Rothenberg has given a helpful brief history of that notion. He says that in much traditional theory economic behavior is viewed as an attempt to maximize something; for consumers utility is posited as that which is being maximized. Rothenberg continues:

"The concept of utility here has been a useful buffer between the action of choice and the supposed psychological ground of this action. By being able to speak of maximizing utility, the economist has not had to say that individuals try to maximize gratification, or satisfaction, or pleasure, or happiness, or virtue, etc., each one of which would seem to be making an empirical commitment in the field of psychology. Utility seems philosophically neutral, while the others seem to assert something about the substantive quality of the ultimate inner goad—if indeed it is unitary."[2]

Whether this neutrality is really possible is another matter. I suggest that either implicit or explicit commitments are likely to be made. For example, Rothenberg says that economists tend to view the person as having a set of drives consisting of organic states disposing "him to activity aimed at reducing or transforming these same states in a way that leads to gratification." Drives "impose directionality on behavior," and indicate the instrumentalities through which the gratifying transformations can occur. This hardly seems neutral, since the emphasis is on organic, rather than on either environmental or biosocial, factors, and on "gratification" rather than on something else.

He goes on to say that if a person's preferences are structured so that he can give a complete transitive preference ordering of all the alternatives, we can describe his choices as if he had assigned different levels of utility to the alternatives and had selected the alternative(s) having the highest preference level of those available. Utility maximization, then, "refers almost entirely to the *structural* characteristics of preferences—namely, the presence or absence of complete preference orderings of alternatives."

In a recent article, Boulding discusses the choice process, and says that to describe such processes economists postulate a utility function in which every relevant state of the field is given an ordinal number indicating its order of preference. In a strong ordering, each state has a unique ordinal number; but in a weak ordering different states may have the same number. He continues:

"As the economist sees it, then, the problem of valuation is that of ordering a field of choice and then selecting the first on the order of preference. This is the famous principle of maximizing behavior, as it is called, which is simply a mathematical elaboration of the rather obvious principle that people always do what seems to them best at the time. It has always surprised me, as I have remarked elsewhere, that such a seemingly empty principle should be capable of such enormous mathematical elaboration."[3]

(In passing, we may note that far from being surprising that a nearly empty principle should permit enormous mathematical elaboration, it is precisely such "empty" principles that do allow such elaboration, as the history of much intellectual endeavor shows.)

Without intending any completeness at all in our survey, we have seen that there is a strong tendency to view utilities as logical constructs that somehow indicate want-gratifications through the ordering of preferences. From time to time, however, utilities are also taken as existing empirically.

Turning back now to measurement, Rothenberg mentions two traditions in utility measurement. The work of Thurstone and others in deriving an individual's preference scales and the von Neumann-Morgenstern axiomatization of utility theory under conditions of risk helped form one tradition, in which the search was for experimental techniques to measure utility functions. The second tradition stems from Paul Samuelson's "revealed preference" theory. Individual market behavior was focused on as the important observable, and preferences (the utility function) were "only logical entities useful for achieving logical closure of the system." The analytic task was construed as the postulation of the logically weakest assumptions about preference from which the properties of the observable market choices could be deduced. Rothenberg goes on:

"There has been some linkage between the two traditions in measurement technique in connection with the quantification of Von Neumann-Morgenstern utility. An increasingly behavioristic interpretation of this utility concept has led essentially to the notion that Von Neumann-Morgenstern utility is simply a construct revealed by the pattern of observed risk choices. One important feature of this construct is that it is used for predictive purposes, not solely to achieve logical closure of the system."[4]

Boulding points out that to give content to the models, we must say what the preference field is and describe the preference function. If the field is made up of a set of possible exchanges in a given system of exchange opportunities (prices), at least some properties of the

[1] Reprinted from Rollo Handy, *The Measurement of Values*, St. Louis, Warren H. Green, 1970.

[2] Jerome Rothenberg, "Values and Value Theory in Economics," in Sherman Roy Krupp, ed., *The Structure of Economic Science*, Englewood Cliffs, Prentice-Hall, 1966, p. 227.

[3] Kenneth E. Boulding, "The Emerging Superculture," in Kurt Baier and Nicholas Rescher, eds., *Values and the Future*, New York, Free Press, 1969, p. 337.

[4] Rothenberg, *op. cit.*, p. 239.

function can be "deduced" from observing behavioral differences in response to different prices; i.e., the preferences are "revealed." In theory, if we can observe the individual's behavior under different price structures, we can determine his preference function, but in practice this is so difficult that what is observed is the aggregate behavior of many individuals under different price structures, "and we deduce from this some kind of aggregate or average preference function." Boulding concludes that there is some justification for so doing if "the preference functions of different individuals are not widely dissimilar."[1]

Boulding's remarks here perplex me. He apparently is saying that the main concern is the individual's preference function, but in practice we cannot observe that, so we observe group behavior instead and derive an aggregate preference function, but that this procedure is justified only if the preference functions (which we *cannot* observe) of the individuals making up the group are fairly similar. I suggest that warranted conclusions about group preferences may be useful even when we cannot observe individual preferences or when those individual preferences are dissimilar. As an even more radical suggestion, perhaps it would be more productive to forget about utilities and just investigate what can be observed.

This brings us back to an old problem about expressed preferences. Sometimes the emphasis is on what a "rational" ordering of preferences would be, and sometimes on "actual" preferences. When "actual" preferences are involved, one approach is to measure them through Thurstone-like techniques in which the respondent says what he would prefer under specified conditions. The other approach is to study the preferences the person (or group) exhibits among alternatives "in real life." The two sets of preferences may be similar or dissimilar; which we want to measure will depend on what behavior we want to predict.

In my opinion, much of the literature on utilities and their maximization fluctuates between taking 'utility' simply as a logical construct useful for achieving closure in a formal model and taking it as designating something in behavior. This fluctuation comes about because one of the goals in this whole area of inquiry is to construct some formal models that either will describe or somehow illuminate human behavior.

To illustrate the logical closure aspect further, let us note what von Neumann and Morgenstern say about their treatment of utility:

"We have treated the concept of utility in a rather narrow and dogmatic way. We have not only assumed that it is numerical—for which a tolerably good case can be made. . .but also that it is substitutable and unrestrictedly transferable between the various players. . . . We proceeded in this way for technical reasons: The numerical utilities were needed for the theory of the zero-sum two-person game—particularly because of the role that expectation values had to play in it. The substitutability and transferability were necessary for the theory of the zero-sum *n*-person game. . . ."[2]

They add that their notion should be modified and generalized, but foresaw definite difficulties in making those improvements.

At this point, we might note that sometimes formalists show a certain irritation when they find that human behavior does not follow the paths their models indicate. Frequently, they strive to modify the model suitably, but on occasion they feel so intuitively confident in the model that behavior deviating from it is "disposed of" as irrational. In any event, I think Rothenberg is correct when he described utility as intended to be a buffer between a choice and its supposed psychological ground. The use of utilities may enable the investigator to avoid direct inquiry into those grounds, but at the price of making utility a somewhat mysterious entity, especially since there is such reluctance to accept the utility as merely an artifact of a formal model.

If it is agreed that the subject of inquiry is the maximizing of something, what that something is becomes pertinent. Especially in the context of game theory, where the game may contain a numerical measure, care must be taken not to assume that a player's utility function is identical with that measure. As Luce and Raiffa point out:

"For example, poker, when it is played for money, is a game with numerical payoffs assigned to each of the outcomes, and one way to play the game is to maximize one's expected money outcome. But there are players who enjoy the thrill of bluffing for its own sake, and they bluff with little or no regard to the expected payoff. Their utility functions cannot be identified with the game payments."[3]

In an attempt to overcome some of these problems, some writers on decision theory have been careful not to identify utility with the mathematical expectation of gain. Anatol Rapoport, for example, says that if people actually behaved according to mathematical expectation of gain in making decisions under conditions of risk, or if it were "intuitively obvious" that a rational person should make his decisions that way, then the theory of decisions under risk would reduce itself to the computation of mathematical expectations. But, he says, not only do people usually not make decisions that way, often they should not; the mathematical expectation of a person buying fire insurance is negative, but that person still may be wise to purchase it.

He goes on to note that in decision theory models utility assignments are made that are not necessarily proportional to the amount of gain expected, and also subjective probabilities (the individual's estimate of probability which may be quite different from "objective" probability) are postulated. But if the decision theory model is supposed to be descriptive, we have to know how to infer the utilities and the subjective probabilities "on the basis of which the decisions will appear consistent and predictable." He then says:

"But posing the problem in this way reveals the strong tacit assumption that behavior of individuals or of classes of individuals *is* consistent and predictable, once the underlying utilities and subjective probabilities of the alternatives are uncovered; i.e., it is assumed that such utilities and probabilities *exist*. And this may by no means be the case. There may be chance factors governing decisions, for example, chance reversals of preferences or chance fluctuations in probability estimates (depending, perhaps, on what aspect of the situation is in the focus of attention)."[4]

Scodel, Ratoosh, and Minas have described much of the

[1] Boulding, *op. cit.*, p. 337.

[2] John von Neumann and Oskar Morgenstern, *Theory of Games and Economic Behavior*, 3rd ed., Princeton University Press, 1953, p. 604.

[3] R. Duncan Luce and Howard Raiffa, *Games and Decisions*, New York, Wiley, 1957, p. 5.

[4] Anatol Rapoport, "Introduction," in Dorothy Willner, ed., *Decisions, Values and Groups*, Vol. 1, New York, Pergamon Press, 1960, p. xv.

work in this field. According to them, formalists often begin with a notion such as the maximization of expected utility, and then attempt to explain why the behavior of the experimental subjects deviates from the norms of the model. They say a "principal difficulty" in such work is that assumptions are made which involve a product of utility and subjective probability, when neither is known. If utility is taken to be a linear function of money, and subjective probability is equated to "objective" probability, it is easy to construct a theoretical model for predicting decisions, but such models "are extremely poor in making predictions about the way persons actually behave in risk-taking situations."[1]

In view of the "deviant" responses found when actual behavior is studied, the authors felt it would be useful to "examine the influence of personality variables." In their study of decision making in a dice game, they gave the subjects an IQ test (Wechsler), the Thematic Apperception Test, and the Allport-Vernon-Lindzey Study of Values test, among others. Some of their findings were: 1) In determining betting preferences, expected dollar value has negligible importance; 2) Intelligence was significantly related to variability in risk-taking, but not to the degree of risk-taking; 3) High payoff subjects scored higher on the Allport-Vernon-Lindzey *theoretical* and *aesthetic* values and lower on the *economic, social,* and *political* values than the low payoff subjects (within the college group; other subjects were not given this test); and 4) The low payoff group scored higher on need achievement as measured in the TAT than the high payoff group. They

characterize their results as "far from overwhelming," but as pointing to the importance of personality variables in risk-taking behavior.[2]

Without denigrating the findings of these authors, it may be pointed out that there are some problems with the use of projective techniques and the Allport-Vernon-Lindzey values test and that in a sense we may have taken the long road home. If we find utility somewhat elusive in the beginning, to attempt to learn more about it through the use of personality tests may not be particularly productive.

Such issues, in my opinion, are also connected with important methodological issues in behavioral science. As mentioned earlier, Rothenberg suggested that some workers were attracted to the notion of utility because it seemed "neutral," whereas an attempt to talk instead about maximizing satisfaction, or pleasure, etc., would seem "to assert something about the substantive quality of the ultimate inner goad—if indeed, it is unitary." I suggest the fundamental mistake here is to assume an ultimate and inner goad, unitary or not. The tendency to separate sharply man from his environment, mind from body, and "inner" and "outer" in behavior, although deeply entrenched in our intellectual tradition, has given rise to a great many methodogenic difficulties and problems. A transactional approach which does not assume separates that somehow have to come together and affect each other seems much more adequate, and does not have as a consequence that we must postulate either some mysterious ultimate inner goad or an equally mysterious "neutral" utility.

1 Alvin Scodel, Philburn Ratoosh, and J. Sayer Minas, "Some Personality Correlates of Decision Making Under Conditions of Risk," in Dorothy Willner, *op. cit.*, p. 37.

2 *Ibid.*, p. 48.

IV.

NOT SCHIZOPHRENIA, DR. FRIEDMAN[1]

E. C. Harwood

IN a recent essay,[2] Dr. Milton Friedman concluded that the "long-run optimum" quantity of "money"[3] will be that which results in a rate of decline in prices sufficient to offset the "nominal rate of interest." In other words, if the market rate of interest were 5 percent on loans involving virtually no risk, the quantity of "money" should be reduced as needed to insure an annual decrease in prices at the rate of 5 percent annually.

However, because of various "practicable considerations," Dr. Friedman considered this conclusion, although apparently proven to his satisfaction, to involve great risks. He suggested that holding "the absolute quantity of money constant" would be "a policy fairly close to the optimum," and then suggested that "a rise in the quantity of money at the rate of about 2 percent per year" as an "especially appealing" compromise that he believed would "stabilize the price of factor services."

Dr. Friedman added what he called "a Final Schizophrenic Note" in which he pointed out that he has heretofore advocated the policy of keeping prices stable by increasing the quantity of money at a rate "something like 4 to 5 percent per year." He asserted that he did "not want to gloss over the real contradiction between these two policies." His two "reasons" for the contradiction are: (1) that his "5-percent rule" was based on "primarily short-run considerations"; and (2) what he called a "more basic reason," that he had not earlier made the analysis presented in the latest essay.

Dr. Friedman offered as his ultimate conclusion, at the end of his "Final Schizophrenic Note," the belief that shifting to his 5-percent rule would provide such a great gain in comparison with results actually achieved in the past that it would "...dwarf the further gain from going to the 2-percent rule...." For this reason he will "continue to support the 5-percent rule as an intermediate objective greatly superior to present practice."

One of the difficulties encountered in analyzing Dr. Friedman's work is his use of words for loose or vague characterization rather than for accurate specification. For example, how long, for him is the "long run," and how short is his "short run"? Does the phrase "present practice" designate the money-credit policies of the Federal Reserve Board in the present century since it was formed in 1914 or those during the period since World War II, or those during the decade of the 1960's (which still would have been the "present" decade as of 1969 when the essay was published)? Surely, Dr. Friedman was not referring to the almost negligible increase in the "money supply" that occurred late in 1969, which presumably was not apparent when he wrote his essay.

[1] Reprinted from E.C. Harwood, *Reconstruction of Economics*, 3rd ed., Great Barrington, Mass.: American Institute for Economic Research, (1970).
[2] Milton Friedman, "The Optimum Quantity of Money," in a book of the same title including other essays, Chicago: Aldine Publishing Company, (1969).
[3] Quotations are used primarily to indicate Dr. Friedman's choices of words. In many instances loose or vague charactization rather than accurate specification seems to result.

WHAT ACTUALLY OCCURRED

In any event, the facts involved may be examined. During 1914 to 1969, the "money supply," as described by Dr. Friedman to include currency in circulation, plus demand deposits or checking accounts, and the time deposits of the Nation's commercial banks, increased from \$19.2 billion to \$412.7 billion. This increase is equal to an average annual compound rate of 5.7 percent. The index of prices paid by consumers increased from 35.4 to 131.3, an average annual compound rate of 2.4 percent; and the index of prices of commodities at wholesale increased from 36.8 to 115.1, an average annual compound rate of 2.2 percent.

The annual rates of increase in the "money supply" are not greatly different if one uses the average of the 5 years 1924 to 1928 as a base period instead of 1914. (The compound annual rate of increase for the "money supply" since then has been 4.9 percent.) And, if one used the 3 years 1947-49 as a base period, the annual rate of increase also is not greatly different (likewise 4.9 percent). To summarize, Dr. Friedman's 5-percent rule was approximately the "present practice" in each of the periods referred to (i.e., 1914 to 1969, and especially 1924-28 to 1969, or 1947-49 to 1969).

On the other hand, if Dr. Friedman's 2-percent rule had been applied and prices had remained stable, as his analysis indicates they should have, the dollar still would have its pre-World War II buying power, more or less. Assuredly, the price consequences of shifting to the 2-percent rule would greatly have exceeded the price consequences of having adhered more nearly to Dr. Friedman's 5-percent rule. Yet he has concluded that "The gain from shifting to the 5-percent rule would, I believe, dwarf the further gain from going to the 2-percent rule...." Apparently, he believes that, although the price consequences of a shift to the 5-percent rule from the actual practice of some recent period probably would have been small in comparison with the price consequences of a shift to the 2-percent rule, the "gain" would be much greater in the first instance.

For what did Dr. Friedman use "gain" as a label or name? Did he attempt to designate an increase in real wealth during the period he had in mind, or an increase in consumable products and services, or what does he refer to? In the absence of accurate specification, or scientific naming, no answer seems to be available.

This raises the question, How can one fairly and usefully appraise Dr. Friedman's prescription for the Nation's ailing money-credit system? When an economist writes his prescription not with the illegibility (for laymen) achieved by many physicians in writing their prescriptions, but with what seems a kind of poetic license in his choice of technical terms, what can be done?

FRIEDMAN vs. SAMUELSON AND KEYNES

That Dr. Friedman's is one of the superior intellects exploring obscure aspects of the "dismal science" seems apparent. Unlike some of the brilliant intellectuals who expound their views on economic matters with similarly great verbal facility, Dr. Friedman has what we should call the saving grace of common sense, to a considerable degree.

Everyone who has read much of his writing and who has heard him discussing various economic problems surely must be impressed by the breadth and depth of his scholarship as well as by what might be called extraordinary verbal facility. Almost invariably he exhibits a sometimes well-justified confidence in the usefulness of what he is sure that he clearly knows. No one can justly accuse him of obscurantism or of unwillingness to take a position when he believes that he can offer a useful policy prescription.

These aspects of Dr. Friedman's personality are mentioned not for the purpose of praising him, although we think that he well deserves much praise, but in order better to describe his methods of inquiry and compare what he has done or seems to think he has done with what modern scientists purport to do. (Unfortunately, the word "science" has become so widely misused that we use it as little as seems practicable herein.)

Dr. Samuelson, a leading disciple of Lord Keynes and having views at variance with those of Dr. Friedman, considers himself a scientist; and his presence as head of the Department of Economics at Massachusetts Institute of Technology, one of the Nation's leading scientific institutions, would seem to support that viewpoint. Dr. Friedman also regards himself as a scientist. Their opinions differ, and also do not accord with those held by the relatively few *modern* economic scientists conducting research in the money-credit field.

VARYING METHODS OF INQUIRY

The difference in methods of inquiry account in large part, we believe, for Dr. Friedman's need to add a "Schizophrenic Note" to his essay. His methods, like those applied by Keynes, Samuelson, and many other earlier as well as still living economists, are the traditional methods applied for many centuries. More than 300 years ago, those earlier methods were superseded in the physical sciences when Galileo, among others, began developing the new and more useful methods of inquiry. Less than two centuries later, the new methods were being applied successfully in the physiological sciences, including medicine. Only in the present century have adequate descriptions of the new methods and their applicability in the behavioral sciences, including economics, become available. Application of them still is in the early stages.[1]

Dr. Friedman's methods of inquiry as revealed in his essay, "The Optimum Quantity of Money," may be described as:

a. Imagining a simple society, which he labels "Hypothetical Simple Society;"

b. Assuming some (apparently believed to be) axiomatic or unvarying types of behavior;

c. Describing in words what would logically follow if certain changes were initiated;

d. Translating into mathematical terminology some of the verbal logic;

e. Presenting charts showing the hypothetical mathematical relationships;

f. Drawing conclusions that, when modified by certain "practical considerations," become a suitable basis for policy recommendations.

Although he did not say so explicitly, Dr. Friedman's confidently offered prescription clearly is based on the belief that he had successfully carried out useful inquiry. He presumably regards his use of mathematics as a step in his proof and the charts as graphic confirmation. If so, he has followed precisely the same method of inquiry as that used by Dr. Samuelson at one point in his widely used economic textbook.[2] That two able individuals both using the same *methods* arrive at conflicting results would provide reason for reconsideration to inquirers applying more modern methods of inquiry.

THE ROLE OF THE LABORATORIAN

Inasmuch as Dr. Friedman referred to earlier work,[3] it should properly be regarded as an exhibit in connection with the article under consideration. In that earlier volume, Dr. Friedman and his co-author provided detailed and extensive records of what he has called "The Key Facts As We Now Know Them." Therefore, Dr. Friedman cannot justly be accused of ignoring the role of the laboratorian in the course of scientific inquiry.[4] Nevertheless, that Dr. Friedman has failed to use the methods of modern inquiry is clear.

Modern inquirers into physical, physiological, and behavioral (including economic) problems do observe what seem, in the course of inquiry, to be the pertinent facts (consisting largely of measured changes). Modern inquirers do develop hypotheses and carefully check the internal logic as well as the logical implications of such hypotheses. At this point, however, the resemblance to Dr. Friedman's methods ends.

Modern inquirers do not assume that transpositions of

---

[1] What we consider the best report available to date on modern methods of inquiry is found in John Dewey, *Logic, The Theory of Inquiry*, New York: Holt, Rinehart and Winston, (1938). However, as Dewey himself later pointed out, aspects of his description were in error or at least could have been stated better if he had had his prolonged correspondence with Arthur Bentley before he had written the *Logic*. (See Sidney Ratner and Jules Altman, Eds., *John Dewey and Arthur F. Bentley: A Philosophical Correspondence 1932-51*, New Brunswick: Rutgers, (1964). In their articles later published in book form (John Dewey and Arthur Bentley, *Knowing and the Known*, Boston: Beacon Press, 1949), they provided much useful material for anyone seeking to understand and apply the new methods. As an aid in understanding the older methods and thereby more clearly grasping the significance and usefulness of the new, we suggest Joseph Ratner's 241-page "Introduction" in his volume *Intelligence in the Modern World*, New York: Random House, (1939). Because John Dewey's style of writing makes difficult for many readers a grasp of the new ideas presented, we suggest study of the publications listed above in the reverse of the order given.

In the present analysis of Dr. Friedman's work we can provide little more than an outline and must leave to readers the more adequate examination of the notions we summarize.

[2] See the discussion of his method of inquiry in our review of his textbook. (Reprinted as pp. 206-209 of the present volume.)

[3] M. Friedman and A. Schwartz, *A Monetary History of the United States 1867-1960*, Princeton, New Jersey: Princeton University Press for the National Bureau of Economic Research, (1963).

[4] The laboratorian is the measurer of changes. He and the theoretician or developer of hypotheses jointly participate in modern inquiry. In some fields of inquiry, the laboratorian initiates the changes that he measures. In economics, the laboratorian usually functions in a statistical laboratory and records for future reference the changes that occur in such economic aspects of life as prices, the quantity of "money" in use, wage rates, etc. On rare occasions an inquirer may be found who combines within himself both the expert statistician and the expert theoretician or developer of hypotheses. (In recent years, many economists seem to prefer to be considered model builders rather than theoreticians, but the function in inquiry is the same.) In this connection see Rollo Handy and Paul Kurtz, *A Current Appraisal of the Behavioral Sciences*, Great Barrington: Behavioral Research Council for Scientific Inquiry into the Problems of Men in Society, (1964).

verbal logic to mathematical forms have any evidential value or constitute a successive step in the proof of anything other than their own facility with mathematical transpositions. In recent years, much economic literature has been loaded, some would say overloaded, with such mathematical transpositions. A recent issue of *The American Economic Review*, for example, offered such mathematical transpositions in nearly every article. The econometricians long have followed this procedure.

Inasmuch as we have urged that economists should have a thorough grounding in the differential calculus and advanced statistics,[1] we do not wish even to appear to belittle the usefulness of mathematics in inquiry. But the viewpoint of modern inquirers is that mathematics is, in a manner of speaking, shorthand logic. The shorthand symbols used greatly facilitate various transpositions and analysis of possible relations among things (including events). The fact that verbal descriptions of what happens or may happen under certain circumstances can be formulated in mathematical symbols adds no assurance that the descriptions are sufficiently accurate to be useful. No modern inquirer would draw conclusions, much less offer policy prescriptions, merely because he could successfully express his verbal argument in the shorthand of mathematics.

THE FUNCTION OF HYPOTHESES

How does a modern inquirer use hypotheses? He uses hypotheses, including their logical implications, as signposts pointing to aspects of the problem requiring further research. Usually this requires more measurements of changes among the facts already considered, sometimes the consideration of new data not previously considered pertinent to the problem, and sometimes the discarding of what had seemed to be pertinent facts. In short, modern inquiry involves the mutual efforts of theoretician and laboratorian (even if the efforts are made by one individual). In the course of inquiry many more measurements of changes than those originally observed will have been made; hypotheses will have been formulated and tested (usually in parts rather than as a whole) by reference to the developed facts; parts or sometimes entire hypotheses will have been discarded or modified; and in the end the modern inquirer can offer a conclusion, an assertion warranted by the research done but certified only as hopefully useful, not as ultimate TRUTH or as a panacea for economic ills.[2]

Nothing thus far asserted should be interpreted as implying that the conclusions developed by outmoded methods of inquiry necessarily are wrong. Quite the contrary. The medieval methods still so generally applied in economics can, and have been, used to develop contradictory conclusions, one of which frequently was useful. Usually in such instances the practical experience and common sense of the inquirer guided him in selecting his assumptions or in rejecting conclusions that seemed to him incongruous.

However, adverse consequences may follow when the conclusions of inquiry conducted by medieval methods are used as guides to policy prescriptions. For example, Dr. Samuelson in an early edition of his textbook

advocated a rise in prices approximating 5 percent annually. In subsequent editions he successively reduced this percentage until in the latest he seems to prefer "gently rising" prices. If his original policy prescription had been followed to date, the buying power of the dollar would be only 21 cents of the 1940 dollar instead of 37 cents, a quite serious difference for holders of savings and life insurance.

DR. FRIEDMAN'S PRESCRIPTION

Consider also Dr. Friedman's policy prescription of ever-expanding money supply at the rate of 5 percent annually. In his analysis of the 1929 to 1933 contraction of business activity, the slide into the Great Depression, Dr. Friedman blamed the Federal Reserve Board, arguing that it greatly contracted the money supply. Evidently, Dr. Friedman assumed that the prosperity of the 1920's was soundly based in all important respects and that the Great Depression was primarily a consequence of improper money-credit policies in the 1930's.

An alternative hypothesis describes the prosperity of the 1920's as forced by inflating (excessive expansion of the "money supply"), which resulted in numerous economic distortions, such as the preparation of Florida subdivisions and apartment hotels far in advance of economic need and the construction of high-rise buildings in the Nation's major cities to a greater extent than could be justified by occupancy. This hypothesis includes the money-credit aspect of developments, prolonged inflating; and a logical implication of it is that the contraction of business activity from 1929 to 1932 reflected the misapplication of resources and the reorientation of economic activity required in order to remove distortions and make possible orderly economic growth. Continued application of Dr. Friedman's panacea (expanding the "money supply" continuously at the rate of 5 percent annually) might have made possible far greater distortions of the economy with even more adverse consequences. Something like this apparently has happened since World War II, during a prolonged period when Dr. Friedman's policy prescription has been closely approximated.

In a recent article,[3] Dr. Friedman has discussed the problem confronting Dr. Burns, the new Chairman of the Federal Reserve Board. In what we believe is a repetition of an erroneous analysis of the 1929 to 1932 contraction of business, Dr. Friedman asserted, "Burns takes office as the economy not only is slowing down but seems on the verge of sliding into a full-fledged and fairly severe recession—thanks to an unduly restrictive monetary policy." The implications are that the economy is "fundamentally sound," as Mr. Hoover repeatedly asserted in the early stages of the Great Depression, and that the only or at least the decisive adverse influence is the approximate leveling off instead of continued expansion of the "money supply" for several months.

CONCLUSION

Our extensive research on money-credit matters during the past 35 years has revealed aspects of the problem that Dr. Friedman apparently did not consider worth mentioning if, indeed, he was aware of them. For a quarter of a century almost continuous inflating (issue of excess purchasing media by the Nation's commercial banking system) has fostered distortions in the economy much more extensive than those of the 1920's. For example,

---

[1] See Section IX. (*Not* reprinted in the present volume.)

[2] Although the essay written by Dr. Friedman alone reveals an apparent unfamiliarity with modern methods of inquiry, in other writings for which Dr. Friedman is listed as co-author, we find evidence of greater familiarity with modern methods. Evidently, Dr. Friedman either has not read or has not fully grasped the significance of much that his co-authors have written.

[3] *Newsweek*, February 2, 1970.

the inflating at that time barely was sufficient to maintain prices generally on a plateau some 40 percent above 1914 prices; but during the period since 1940, prices have risen so far that holders of savings and life insurance have lost nearly $700 billion of real wealth. More than half of the Nation's elderly have been reduced to the poverty level by the continued depreciation of the dollar and other developments. In the meantime, construction costs, for example, have skyrocketed to 5.4 times 1940 levels. Space is not available here to describe the numerous economic distortions now existing. That they will not be corrected, thereby enabling a restoration of sound economic growth merely by applying Dr. Friedman's panacea, seems to us highly probable.

In the months and years ahead we may witness an experiment with Dr. Friedman's panacea somewhat similar to Mr. Roosevelt's experiment with Dr. Keynes' spend-for-prosperity panacea in 1934-35. We doubt that the end result will be much more desirable than was the severe depression of 1937-39.

Presumably, Dr. Friedman's labeling of his seemingly contradictory conclusions as "schizophrenic" was not done on the advice of a psychiatrist but was merely a somewhat vague means of disarming his critics. In any event, our analysis suggests that the difficulty is *not* schizophrenia, but simply the retardation of Dr. Friedman's development, at the medieval level, as a scientific inquirer.

Authors' Note: Earlier in this book we emphasized the lack of finality and completeness in inquiry; the objective is not the attainment of unalterable truth, but more and more adequate solutions to problems. Our latest work (See Section A, Chapter 5) suggests that improvements can be made in the discussions of hypotheses and the roles of the laboratorian and the theoretician as found in the preceding criticism of Dr. Friedman's procedures. We now regard "conjecture" as more useful than "hypothesis," and we emphasize the emergence of conjectures about possible connections when description is blocked, the merging of those conjectures into observations of new or modified facts, the emergence of new conjectures, the merging. . . , etc. The author directly concerned hopes that such improved description illustrates both the self-correcting aspects of the new procedures and his own capacity for development.

The following review of a recent article by Friedman includes a further discussion of conjectures in inquiry.

*   *   *   *   *

BACKWARD MARCH—UPSIDE DOWN[1]

Dr. Milton Friedman's recent article on monetary analysis has been published as an Occasional Paper by the National Bureau of Economic Research.[2] It is described as ". . .a study by the National Bureau of Economic Research based on two papers" previously published in the *Journal of Political Economy.* Thus the National Bureau has assumed the usual responsibility with reference to this publication. Presumably therefore, it has been read by at least three directors and has met with their approval or has been approved by a majority of the Board of Directors.

Inasmuch as a majority of the directors almost

certainly could not understand the publication and surely have not read it, we assume that three selected directors constituted, in the usual manner, the special committee that read and approved the manuscript. Unfortunately, along with Dr. Friedman, they apparently understand little about the early work of the National Bureau and its great contribution to economic research.

The founder of the National Bureau for Economic Research was Dr. Wesley Clair Mitchell, who directed its activities for many years. Dr. Mitchell was one of the few economists who was fortunate in learning about John Dewey's early progress in developing an adequate description of modern methods of scientific inquiry applicable to the behavioral sciences including economics. As one of Dewey's students and later as a faculty associate at Columbia University, Dr. Mitchell apparently benefited from Dewey's insights, which first appeared as various papers in the journals and later were published in book form.[3] Unfortunately, Dr. Mitchell did not live long enough to benefit from Dewey's later work with Arthur Bentley.[4] If that material had been available to Dr. Mitchell, his earlier work suggests that he would have been able to keep the National Bureau leading the advance in the development of economics as a science.

Fortunately, Wesley Mitchell had understood an important aspect of the useful procedures of inquiry later described so adequately by Dewey and Bentley. He realized that, following awareness of a problem situation the inquirer should *begin* with observation of the apparently pertinent facts and measurement of changes among them. When blocked in developing tentative descriptions of what happens under specified circumstances, the inquirer lets his imagination roam, develops conjectures (sometimes called "hypotheses") about what may happen under specified circumstances, chooses among such conjectures as a basis for further investigation of the facts or a search for new facts, new measurements of change, etc. The conjectures may be most elaborate and involve extensive mathematical transformations, but the significant aspect is that the conjectures emerge *after* study of some facts believed to be pertinent and merge back into the process of inquiry by pointing to the possible existence of new facts or the desirability of new measurements. When progress in inquiry again is blocked, the process of conjecture followed by return to observation is repeated, perhaps many times, until an adequate description of what happens under specified circumstances is developed as a solution to the problem situation warranted by the procedures of inquiry.

Mitchell was concerned primarily with that problem situation labeled the "business cycle." He began in the appropriate manner by investigating what appeared to be the pertinent facts. Under his guidance, the National Bureau of Economic Research provided much of the needed first step for modern scientific inquiry into that problem situation called the business cycle. This was no light task. Mitchell died before much progress could be made beyond the initial step, but the National Bureau did seem to be well oriented toward the more useful procedures of scientific inquiry that were being described adequately for the first time by Dewey and Bentley.

Dr. Friedman's "A Theoretical Framework for Monetary Analysis" is a major departure from the line of

---

1 By E. C. Harwood; reprinted from *Research Reports,* American Institute for Economic Research, Sept. 18, 1972.

2 Milton Friedman, *A Theoretical Framework for Monetary Analysis,* Occasional Paper Number 112 by the National Bureau of Economic Research, New York, 1971.

3 John Dewey, *Logic: The Theory of Inquiry,* Holt, Rinehart and Winston, Inc., New York, 1938.

4 John Dewey and Arthur F. Bentley, *Knowing and the Known,* The Beacon Press, Boston, 1949, and *John Dewey and Arthur F. Bentley: A Philosophical Correspondence 1932-1951,* Rutgers University Press, New Brunswick, 1964.

advance exemplified by the work of the National Bureau under Mitchell's guidance. Dr. Friedman begins by asserting "Every empirical study rests on a theoretical framework, on a set of tentative hypotheses that the evidence is designed to test or to adumbrate. It may help the readers of the series of monographs on money that Anna J. Schwartz and I have been writing to set out explicitly the general theoretical framework that underlies them."

He then expounds at some length on traditional monetary theories. That he does so with his usual great verbal facility and some mathematical competence is wholly beside the point, insofar as this review is concerned. Anyone who dreams that the traditional monetary theory including the Keynesian variations of it can be considered a useful body of consistent and coherent hypotheses (or theory, depending on one's choice of terms) should read Dr. Arthur W. Marget's remarkable two-volume work, *The Theory of Prices*,[1] especially the extensive footnotes, which must provide well over half the material on nearly 1,400 pages of fine print. At the end of such an experience one is almost sure to be either an incurable verbalist, more or less like Dr. Friedman, who seems to be seriously addicted to the further extension of elaborate hypotheses, or he will conclude that monetary economists should make a fresh beginning in the manner of modern scientific inquiry.

The crucial importance of using *each successive* hypothesis or conjecture as a guide for further observation and measurement readily can be understood. In the course of inquiry, each time that progress is blocked and a tentative description of what happens under specified circumstances remains inadequate, the inquirer in imagination conjectures (develops hypotheses) about the possibilities. Almost invariably more than one possibility may exist and sometimes many are potentially involved at each successive step. If the inquirer selects the possibility that

to him seems most plausible, then *without* returning to observation and measurement proceeds on to the next blockage and similarly selects among new conjectures, he is confronted with rapidly increasing odds against the success of his inquiry. For example, if the average number of possible hypotheses at each step is 10, his chances of selecting the correct conjecture 10 times in succession would be only 1 chance in 10 billion. Even if there were only 2 alternative possibilities at each successive point his chances of selecting all the correct or more useful conjectures in a succession of 10 would be only 1 chance in 1,024. Clearly, the procedure of developing successive hypotheses without returning to observation and measurement at each step will be a "losing game" in that the chances of thus conducting useful inquiry are negligible.[2]

The elaborate classical and neoclassical including Keynesian theories of money, which Dr. Friedman uses as a "theoretical framework," perhaps are the best examples available of "bucking the odds." Moreover, the successive conjectures included far exceed 10, depending on what portions of the various theories one chooses to use. Counting the conjectures in Dr. Friedman's "theoretical framework" might be interesting for someone who has more confidence in the usefulness of such procedures than does this reviewer. In any event, the chances that his "theoretical framework" is useful for the purposes of modern scientific inquiry are so slight as to be hardly worth mentioning.

The important point that this reviewer would make is that if Dr. Friedman's work represents the present procedures of inquiry approved by the National Bureau of Economic Research, I doubt that progress in economic research by that organization during the next few decades will be similar to that during its first few decades. How can progress be expected while executing a backward march upside down?

---

[1] Arthur W. Marget, *The Theory of Prices*, Volume I and II, Prentice-Hall, Inc., New York, 1942.

[2] As John Dewey says (using "idea" for what we call "conjecture"): "Every *idea* is an *escape* [from blocked inquiry], but escapes are saved from being *evasions* so far as they are put to use in evoking and directing observations of further factual material." *(Knowing and the Known*, p. 183.)

V.

BETRAYAL OF INTELLIGENCE [1]

E. C. Harwood

AN excellent binding and good reproduction of type and charts combine to give this book the appearance of a scientific treatise. Unfortunately, readers who therefore assume that it offers the last word in scientific economic analysis will be seriously disappointed. Still included in this fifth edition are major flaws that were among the reasons for my comment on an earlier edition, "...that such a book should have the implied stamp of approval of the Nation's leading scientific institution is a tragedy; in a sense it is a betrayal of intelligence in the modern world."

Before the reasons for such adverse criticism are described, Professor Samuelson should be commended for the marked improvements in this edition. In contrast with the earliest edition we reviewed, this volume does include some charts that show the long-term economic growth of the United States. Therefore, student readers at least have evidence that economic growth proceeded rapidly for decades before the advocates of creeping inflation developed their modified version of the Keynesian spend-for-prosperity notions.

That the charts do not include years prior to 1890 is unfortunate. From 1875 to 1890, gradual deflation of the Civil War and post-Civil War inflation was reflected in a 40-percent decline of commodity prices; yet the Nation's economic growth persisted at a rate not subsequently equaled. With that picture in front of them, even sophomores might question Professor Samuelson's apparent predilection for creeping inflation.

Professor Samuelson has altered his opinion markedly in recent years. Earlier he had said, "...such a mild steady inflation [a rise in prices of 5 percent per year] need not cause too great concern" (p. 302, second edition). In his fourth edition he suggested that creeping inflation be "...held down to, say, 2 percent per year, ..." (p. 270), and in this fifth edition he says "Price increases that could be held down *below* 2 percent per year are one thing," (p. 305, italics supplied) as though that would be negligible. To the casual reader the difference between 5 percent and 2 percent may not seem important, but from the viewpoint of anyone who would live under those conditions the difference is striking. At 5 percent per year, a dollar's worth of life insurance or funds for a retirement pension would decrease in 60 years to a little more than 5 cents worth, a loss of nearly 95 percent of one's life insurance and pension funds; but at 2 percent per year, the loss would be much less, only about 65 percent.

When one views the matter considering the amount that would be left after prolonged creeping inflation, the significance of the Professor's progress is seen to be even more striking. Now that he approves of somewhat less than 2 percent rather than 5 percent per year, he is implying, in effect, that the buyer of life insurance should be permitted to have at least 35 cents left of his dollar instead of only 5 cents. Surely this sevenfold increase in what is left for the victim of creeping inflation is a gratifying change. Professor Samuelson may yet come to believe that life insurance buyers should not have any of their savings "embezzled" by the subtle processes of inflation.

Professor Samuelson offers no purportedly scientific or economic explanation for the change from 5 to less than 2 percent. The reader of his successive editions can conclude only that Professor Samuelson has had the benefit of some secular revelation on this matter.

Although I did not find in this volume Professor Samuelson's justification for any creeping inflation, he perhaps would argue as did Professor Slichter of Harvard that "...creeping price inflation is part of the price we must pay to achieve maximum growth."[2] American economic developments from 1875 to 1890 suggest that such an assertion was not true then, and no one has provided scientifically based proof that it is true today. West Germany's experience since regaining the prewar level of output in 1950 also casts doubt on the creeping inflation theory. From 1950 to 1960 industrial production in West Germany increased 150 percent; but in Sweden the increase was only 38 percent, although the rate of creeping inflation there (measured by the rise in the cost of living) was nearly 6 times that in West Germany. The latter nation increased its industrial production 2½ times accompanied by a negligible amount of creeping inflation while Sweden was increasing its industrial production only a little more than one-third accompanied by creeping inflation at the rate of 57 percent in a decade.

Perhaps the most convincing argument against the creeping inflation theory of inducing maximum economic growth is found in the fact that inflation makes possible an excess of dollars chasing goods, which in turn provides windfall profits for many businesses including some that in the absence of inflation would incur losses. When inflation occurs, businesses that otherwise would fail or at least curtail output and release factors of production (men, capital, and natural resources) for transfer to the growing industries are enabled to remain in business with a resulting delay in the shift of resources to more rapidly

[1] *Economics, An Introductory Analysis*, by Paul A. Samuelson, Professor of Economics, Massachusetts Institute of Technology, New York: McGraw-Hill Book Company, 5th Edition, 1961. Although subsequent editions have appeared, this evaluation still is pertinent (through the eighth edition, at least). This review is reprinted from E.C. Harwood, *Reconstruction of Economics*, Great Barrington, Mass.: American Institute for Economic Research, 3rd Edition, 1970.

[2] Sumner H. Slichter, "Current Trends, Problems and Prospects in the American Economy," *The Commercial and Financial Chronicle*, February 19, 1959, p. 3.

growing industries. Change, not creeping inflation, is the price of economic growth; and experience suggests that change is inhibited and delayed by inflationary prosperity.

Dr. Samuelson's recognition of what he calls the "miracle" of West German postwar economic developments is encouraging to those who hope that his progress will continue. He describes the basis for the "miracle" as "a thoroughgoing currency reform. . . ." (p. 39), which seems an inadequate description of reforms that restored free markets as well as a redeemable currency, and, in effect, tossed into the discard the depression panacea Professor Samuelson evidently favors. Would it not be worthwhile in an economics textbook to devote more than a few lines to the experience of West Germany in recent years? Surely an economic "miracle" merits more detailed comment, especially when such consideration would reveal so much about significant aspects of American foreign and domestic economic policies.[1]

Many writers of economics textbooks have given only superficial consideration to the potential effects of a tax on site values as differentiated from a tax on the value of improvements. In a brief but clear discussion of this point (p. 597) Professor Samuelson describes how a tax on site values would fall in its entirety on those privileged to hold exclusive titles to such sites and would not burden either those who labor or those who invest in the reproducible capital of our economy. An obvious conclusion is that shifting of the tax burden from investors and earners would encourage new investment as well as production and would inhibit speculative withholding of valuable sites and resources from production. An equally obvious conclusion is that the net result could be more rapid economic growth with output more equitably distributed among those who participate in the productive process.

The potentially far-reaching consequences of taking much of site rent for public uses might well have been discussed in greater detail. The Institute of Research of Lehigh University, another distinguished school of engineering, has analyzed and reported on the potential effects of exempting improvements and taxing only land values in the city of Bethlehem, Pennsylvania.[2] Here is substantial evidence that the slum areas of a city reflect prolonged unwise apportionment of the tax burden and that the simplest remedy for "sick" urban areas would be shifting present taxes on improvements to taxes on land values. Moreover, the experience of Sydney, Australia, and several other cities indicates that even most of the landowners, surprising as it may seem, would benefit from such a shift of the tax burden. The experiences of Denmark, of New Zealand, and even of Pittsburgh, Pennsylvania, with its partial application of the principle, merit consideration by every student of economics.

ERRONEOUS ASSERTIONS

So much for the evidence of some progress by Professor Samuelson. In other respects the lack of progress is evident.

He asserts (p. 22) that, during World War II, American "civilization living standards" were "higher than ever before." Such statements frequently are made by economists enamored of the spend-for-prosperity notions, perhaps because their theories suggest that the vast monetization of Government debt should have had that result or perhaps because they are so naive as to believe that money incomes correctly reflect the standard of living. Here are the facts:

a. Production of passenger automobiles for civilian use during World War II virtually ceased. With reference to the public's huge investment in passenger automobiles, the standard of living greatly decreased as a result of wear and tear, depreciation and obsolescence, lack of replacements for vehicles scrapped, and lack of additional new vehicles to maintain the per-capita quota.

b. Construction of new residential housing decreased 85 percent and remained at a low level until after World War II. Inevitably the standard of living with reference to housing decreased during the war for reasons similar to those in a, above.

c. A comprehensive index of production of new consumer goods per capita[3] shows that a 25 percent decrease in the production of all consumer goods occurred from mid-1941 to 1945.

d. In large part because automobiles, new homes, etc., were not available, individuals hoarded about $15,000,000,000 of their wartime wages in the form of currency and many billions more in the form of idle checking accounts.[4] In addition, many billions of wartime incomes were invested in U.S. Savings Bonds.

In view of these facts, civilian standards of living could not have reached unprecedented levels during the war years. To imply otherwise may suggest to many readers that monetization of deficits, i.e., inflation, somehow offers an easy route to perpetual prosperity.

Additional erroneous assertions are scattered through this book. For example, Professor Samuelson asserts (p. 286) that the Great Depression of the 1930's was the longest "sustained" in the Nation's history. The National Bureau of Economic Research, whose research in this aspect of economic developments is far more extensive and detailed than that of any other agency, reports that the duration of the 1873-82 recession and recovery from peak to peak was 101 months compared with 93 months for the depression of the early 1930's.

On page 377 Professor Samuelson refers to "classical views that there can never be unemployment." This representation of the classical views also is erroneous. The classical economists argued that unemployment would be extensive if some elements of labor refused to accept lower wages whenever that was indicated in order that they might be employed.

Also, on page 406 Professor Samuelson says, "From the early 1870s to the middle 1890s, depressions were deep and prolonged, booms were short-lived and relatively anemic, the price level was declining." That the price level was declining is correct, but the National Bureau's record shows that, from the time the gold standard was resumed in 1879 until 1895 there were 4 recessions having an average duration of less than 20 months, almost exactly the average for more than 100 years. None of the 4 was "deep and prolonged," and during this period the Nation enjoyed its most rapid and consistent growth as measured by the expansion of industrial production.

When he attempts to discuss "money," Professor

---

[1] See Melchior Palyi, *Managed Money at the Crossroads*, University of Notre Dame Press, Notre Dame, Indiana, 1958, p. 100 *et seq.*

[2] Eli Schwartz and James E. Wert, *An Analysis of the Potential Effects of a Movement Toward a Land Value Based Property Tax*, a project of the Institute of Research of Lehigh University, published by Economic Education League, Albany, N. Y., 1958.

[3] A chart of this index usually is published monthly in the weekly publication of the American Institute for Economic Research, *Research Reports*.

[4] *Current Economic Trends*, American Institute for Economic Research, 1960, p. 6.

Samuelson gives his readers inadequate information. For example, what is meant by the words on a $10 bill, "The United States of America will pay to the bearer on demand Ten Dollars"? I could find no evidence in the Professor's discussion that he knows of this promise or its significance, in spite of his attributing West Germany's "miracle" to "currency reforms," a principal feature of which has been a sound currency now redeemable in gold on demand. Surely, differentiating between dollars (1/38 of an ounce of gold) and promises to pay dollars is elementary in any attempt to describe a money-credit system. The foreign central bankers who have demanded that such promises be kept in recent years, with a resulting loss to the United States of more than $5,000,000,000 in gold, have a clear understanding of the difference between promises to deliver something and the thing promised. Should not American students be equally well informed?

Apparently in an attempt to justify increasing Government debt, Professor Samuelson asserts (p. 399), "If there were no public debt. . .(1) charitable institutions would have to be supported by public and private contributions more than by interest on endowments, (2) social security and annuities would have to take the place of *rentier* interest, and (3) service charges by banks would have to be increasingly relied upon instead of government bond interest." Evidently he is not aware that virtually all the private colleges in the United States, until a few decades ago, depended largely on endowment funds invested primarily in other than government bonds. Moreover, in those days, when there were almost no U.S. Government bonds in existence, most banks not only had no service charges but also paid interest even on checking account balances in excess of specified minimum amounts.

Perhaps in an effort to add what he considers to be the weight of recognized authority to his assertions, Professor Samuelson repeatedly says that most economists agree with various views he offers (pp. 9, 241, 242, 256, 298, 299, 364, 375, 380, and 829). For example, he asserts (page 241) that the basic Keynesian analysis is ". . .increasingly accepted by economists of all schools of thought. . .," and on the next page he says of his so-called synthesis of Keynesian and older economics, "The result might be called 'neoclassical economics' and is accepted in its broad outlines by all but a few extreme left-wing and right-wing writers." The Keynesian analysis assuredly is not accepted by members of the Economists' National Committee on Monetary Policy. This group of experts in the money-credit field cannot properly be classified as "left-wing" or "right-wing" inasmuch as they are primarily economic scientists. They constitute a substantial number (75) of those who specialize in this field.

SWEDEN'S ECONOMY

Another interesting point is the Professor's reference to Sweden. (Sweden has for some years been regarded by the Keynesian state planners and government interventionists as a nearly ideal country because of its, at first, seemingly successful application of semi-socialistic and spend-for-prosperity notions.) The reference is, "A great economic statistician, Simon Kuznets of Harvard, has recently shown that the leading Western nations have for decades been averaging rapid rates of growth of output per head. How rapid a growth? About 10 percent per decade for France and England; about 16 percent for Canada and the United States. And almost 30 percent per

decade for Sweden!" (p. 116).

Persuading students to believe that Sweden is now exceeding, or recently has far outpaced, other nations of Europe and the United States in economic growth seems an inexcusable falsification of the record to this reviewer. Sweden's economy once was growing at the rapid rate indicated, but that was before the semi-socialist planners and spend-for-prosperity theorists gained a dominating influence in Sweden's government during the fourth decade of the present century. During the 1950's Sweden's industrial production per capita increased about one-quarter; during the same period, in the rest of Western Europe the increase was nearly twice as great and in West Germany about four times as great.[1] Figures now are available for all of the 1960's and the data for Sweden are about 20 percent better than those for West Germany. However, there already is adequate documentation that the rate of economic growth in Sweden since World War II was much less than the comparable rates in much of Europe, including the nations that suffered extensive war damage during World War II.

KEYNESIAN "THEORY"

In part 2, Dr. Samuelson presents the familiar Keynesian notions with numerous charts and formulas. The subject matter is presented much as a chemist or a physicist would write about an accepted theory in his field. There the resemblance ends, however. What Professor Samuelson offers is not a scientific theory but a set of hypotheses for which proof has not been provided between the covers of his book or elsewhere. Unwary students may at first assume that what is called the "theory" of income determination is like Einstein's theory of relativity in that adequate testing of the factual implications of the original hypothesis has elevated it to the rank of a warranted assertion or accepted theory.

In attempting to convince students that the Keynesian notions are sound, Professor Samuelson reveals what I assume is his understanding of modern scientific method. After describing the Keynesian hypothesis concerning income determination, Professor Samuelson says (p. 262), "An arithmetic example may help verify this important matter." He then offers a table not of recorded economic changes but of changes that he has imagined and that merely summarize in figures his earlier argument.

Then, on the next page, he asserts, "Now we can use Fig. 5 to confirm what has just been shown by the arithmetic of Table 1." Students at an institution like M.I.T. are accustomed to the idea that verification involves proof of some kind, that what has "been shown" has been demonstrated or proved, and that what has been "confirmed" has been "established firmly" or put "beyond doubt," to use phrases from the Oxford dictionary. And that is what Professor Samuelson seems to believe he has accomplished. Has that been done?

The first statement of the Keynesian notions is in words. The second summary statement, the table, is in the symbols of mathematics or shorthand logic. The third, and the final alleged confirmation, is a chart presenting the imagined relations in pictorial form. The Professor, although he seems not to realize it, is saying something like this: "Here is my story about the Keynesian revelation; next, I verify it by writing it in shorthand; finally, I prove it beyond doubt by drawing a picture of

---

[1] These data are derived from the monthly and the 1967/68 supplement to *International Financial Statistics* published by the International Monetary Fund.

it." One wonders how students at M.I.T. could be induced to regard seriously such anachronistic nonsense. Aristotle convinced some of his disciples 2,000 years ago that such procedures provided useful knowledge, but that was long before modern methods of scientific inquiry had exposed the futility of such dialectical quests for certainty.

The Keynesians generally have followed the outmoded procedure of judging the usefulness of a theory by its plausibility instead of by checking its implications against measured economic changes. In the realm of science, theory is controlled by the facts. When scientists find facts at variance with theory, that theory is discarded; but many Keynesian economists do not even bother to seek the measurements of changes implied by their theory. In this respect, Professor Samuelson simply is following the too-long established precedent in his field.

THE "NEOCLASSICAL SYNTHESIS"

Professor Samuelson claims (Preface p. vii, p. 403, and elsewhere) that he has achieved or is in the process of achieving a "neoclassical synthesis" that will join in fruitful wedlock classical economics and that portion of the Keynesian ideas deemed by Samuelson to be worthy of the union. If what von Mises or Hayek, as examples of economists in the classical tradition, have written about the Keynesian ideas may be taken at face value, either would be decidedly reluctant to see his brainchild a "groom" at the "wedding" Professor Samuelson plans.

Moreover, the present writer's position is that such a "wedding," whether of the "shotgun" variety or otherwise, would not be fruitful for the simple reason that both bride and groom, that is, the Keynesian notions and much of classical economics, are "dead ducks." My reasons for so believing have been discussed in detail in other chapters. Here there is room for only a summary explanation.

The methods of conducting inquiries applied by the Keynesians and to a substantial extent by the classical economists were the older, now obsolete methods. Briefly, those methods included Aristotelian logic, introspection, what may be called secular revelation (a process at which Lord Keynes was especially adept), and the quest for certainty so long persisted in also by philosophers. Such methods give great weight to the internal logical consistency and general plausibility of an hypothesis but accord little weight to the desirability of testing its logical implications against measurements of economic changes before offering the hypothesis as a warranted assertion applicable to the problems of men.

Anyone who will observe its consequence in several fields can see that a revolution in methods of inquiry is well underway in the behavioral sciences, including economics. This revolution is comparable to the Galilean revolution of three centuries ago in the physical sciences and to the similar revolution in the physiological sciences marked by the advent of graduate schools of medicine more than a hundred years ago.

Evidently Professor Samuelson is determined to continue as one of the last of the alchemist-economists, using as his model Lord Keynes (whom Professor Samuelson on page 241 describes as "a many-sided genius"). As everyone who recalls the discussions in economic journals during the 1930's is well aware, Lord Keynes' method of escape from every blind alley in which his economist critics nearly cornered him was the simple process of abandoning successive positions and dashing down other blind alleys. The verbal skill that facilitated his Houdini-like "escapes" was widely accepted as proof of his "brilliance" by those to whom the scintillating flash of words seemed more significant than the humdrum facts preferred by others who have rejected perpetual-motion theories and alchemists' dreams. However, following in Lord Keynes' footsteps may not be practicable. Times have changed; the revolution in methods of inquiry proceeds with increasing speed; and an emulator of Lord Keynes may discover, as the alchemist professors did long ago, that the market for outmoded textbooks can rather suddenly disappear.

An alternative would be to learn as rapidly as possible and apply modern methods of conducting scientific inquiries in the behavioral field. This choice could in time make Professor Samuelson an eminent associate for the distinguished scientists on the faculty at M.I.T. instead of the anachronistic pseudo-scientist that he now seems in the light of our present understanding of scientific method.

In spite of its flaws, we have reviewed this book because it is reported to be the most widely used economics textbook today. Many college students are being indoctrinated with the Keynesian notions although much evidence indicates that application of these notions has brought Sweden to the brink of disaster, all but ruined France prior to the fiscal and other economic reforms effected by De Gaulle, and has greatly endangered the future of the United States, to mention only a few of the consequences. Resolute discarding of such notions made a vital contribution to the "miracle" of West German economic growth. In the light of these developments, the importance of teaching American youth scientifically warranted assertions instead of the doctrines offered by Professor Samuelson seems obvious.

VI.

MEDIEVAL SCHOLASTICISM vs. MODERN SCIENCE[1]

E. C. Harwood

IN this comprehensive volume (nearly 900 pages) the distinguished author has presented the results of a lifetime of study, a work that deserves careful analysis. Its principal value, in my opinion, is not found in the conclusions, pleasing as they may be to proponents of economic freedom, nor in the criticism of "welfare" economics, valid as such criticism may be; rather do I find that the book's principal merit is its frank statement of the author's method, his assumption as to what constitutes knowledge in the economic field, and his procedures based on that assumption. Dr. von Mises' treatise illustrates the main weaknesses of economics as it is written about and widely taught today. As evidence of the urgent need for reconstruction in economics, this book perhaps is without an equal.

Lest these comments be misinterpreted as disparagement of the author or belittlement of his efforts, I assure readers that such is not my intent. I consider this volume one of the outstanding and representative works in the field. Dr. von Mises was professor of economics at the University of Vienna for a full quarter century. Subsequently, for 16 years prior to World War II, he was Professor of International Economic Relations at the Graduate Institute of International Studies in Geneva. More recently he has been a visiting professor at the Graduate School of Business Administration, New York University. Thus for more than a half century, Dr. von Mises has taught and written in his chosen field. His writings indicate that his scholarly ability and the breadth and depth of his work in the field are equaled in few living men.

In this brief review, there is not room for an adequate discussion of the author's comments on recent economic developments. With much of his criticism of specific policies and proposals, particularly those that constitute a revival of "mercantilism" as practiced in France during Colbert's time nearly three centuries ago, I agree; and with his conclusion that many of the proposals for economic planning are one-way routes to socialism or communism, I likewise agree. However, much of my criticism of these recent developments would be on grounds that underlie, as it were, von Mises' criticisms; in short, I doubt the validity and usefulness of much that he criticizes on the same basic grounds on which I question his own work.

We begin with a question, the answer to which is basic to economics as well as to all other fields of inquiry. What is knowledge? The only satisfactory answer that I have found is, to use Dewey and Bentley's phraseology, the knowing and the known.[2] In short, the only referent

(thing referred to) that I can find for the word symbol "knowledge" is the integrated knowing and the known.

I am not prepared at this writing (any more than Dewey and Bentley apparently were in 1949) to urge that the symbol "knowledge" can now be safely used to specify (or scientifically name) anything. But I do feel reasonably sure that, if the symbol "knowledge" is found satisfactory for scientific discourse, whatever it refers to will have this characteristic: the knower's "knowledge" will enable him to predict and control, predict what will occur under certain circumstances and in the light of that prediction to control (or adjust) in some degree either man's behavior or the external environment, or both. Facilitation of prediction and control is an essential characteristic of "knowledge" in what seems to be the emerging modern scientific usage of this word symbol.

We turn now to von Mises' usage of economic "knowledge." The clearest summary statement seems to be that on page 858, which is as follows:

"What assigns economics its peculiar and unique position in the orbit both of pure knowledge and of the practical utilization of knowledge is the fact that its particular theorems are not open to any verification or falsification on the ground of experience. . . . It can never, as has been pointed out, prove or disprove any particular theorem. . . . The ultimate yardstick of an economic theorem's correctness or incorrectness is solely reason unaided by experience."

Additional light on von Mises' answer to the question, What is (economic) knowledge, is provided by the following:

"It is impossible for the human mind to conceive logical relations at variance with the logical structure of our mind." (page 25)

"For the comprehension of action [including economic behavior] there is but one scheme of interpretation and analysis available, namely, that provided by the cognition and analysis of our own purposeful behavior." That is, introspection. (page 26)

"The human mind is not a tabula rasa on which the external events write their own history. It is equipped with a set of tools for grasping reality. Man acquired these tools, i.e., the logical structure of his mind, in the course of his evolution from an amoeba to his present state. But these tools are logically prior to any experience." (page 35)

"However, the sciences of human action differ radically from the natural sciences. All authors eager to construct an epistemological system of the sciences of human action according to the pattern of the natural sciences err lamentably.

"The real thing which is the subject matter of praxeology, human action, stems from the same source as human reasoning. . . . The theorems attained by correct praxeological reasoning are not only perfectly certain and incontestable, like the correct mathematical theorems. They refer, moreover, with the full rigidity of their apodictic certainty and incontestability to the reality of

[1] *Human Action, A Treatise on Economics,* by Ludwig von Mises, New Haven: Yale University Press, 1949. This review is reprinted from E.C. Harwood, *Reconstruction of Economics,* 3rd Ed., Great Barrington, Mass.: American Institute for Economic Research, 1970.

[2] John Dewey and Arthur F. Bentley, *Knowing and the Known,* Boston: Beacon Press, 1949, particularly pages 296 and 297 (page 176 of the present volume), from which the following is quoted: "*Knowings:* Organic phases of transactionally observed behaviors. Here considered is a familiar central range of namings-knowings. *Knowns:* Environmental phases of transactionally observed behaviors."

action as it appears in life and history. Praxeology conveys exact and precise knowledge of real things." (page 39)

"Economics is not, as ignorant positivists repeat again and again, backward because it is not 'quantitative.' It is not quantitative and does not measure because there are no constants. Statistical figures referring to economic events are historical data." (page 56)

"Such problems do not allow any treatment other than that of understanding." (page 57)

"All that is needed for the deduction of all praxeological theorems is knowledge of the essence of human action. . . .We must bethink ourselves and reflect upon the structure of human action. Like logic and mathematics, praxeological knowledge is in us; it does not come from without." (page 64)

"The fundamental logical relations and the categories of thought and action are the ultimate source of all human knowledge." (page 86)

I have quoted at length in order to minimize the risk that presenting material out of its context might misrepresent the author's views. How does Dr. von Mises' answer to the question, What is knowledge, compare with the answer that seems to be emerging from the latest studies of man's knowing behavior? Where does his understanding of knowing and method fit in the historical succession of man's procedures of knowing? Evidently, he has not abandoned the Greek ideal, the quest for certainty; on the contrary, he is convinced that he has succeeded where so many others have failed and in spite of the fact that modern men seeking knowledge no longer consider his objective a reasonable goal.

A LEAP BACKWARD

Dr. von Mises denies not once but several times that his theories can ever be disproved by the facts. This point of view represents a leap backward to Platonic idealism or one of its subsequent offspring in various disguises. Theories thus derived are medieval scholasticism, albeit on a par with much that is taught as economic knowledge today.

There is even ground for alleging that some aspects of his methods are even farther out of date and have their roots millions of years ago. What else are his assertions about "conception" and "understanding" if not an acceptance of revelation as a road to knowledge?

Dr. von Mises' conception of the mind and its function in his search for knowledge may be compared with: "Reason pure of all influence from prior habit is fiction."[1] Also of interest in this connection is the following: "Many who think themselves scientifically emancipated and who freely advertise the soul for a superstition, perpetuate a false notion of what knows, that is, of a separate knower. . .by dismissing psychology as irrelevant to knowledge and logic, they think to conceal the psychological monster they have conjured up."[2]

Like the Greeks, Dr. von Mises disparages change: "Praxeology is not concerned with the changing content of acting, but with its pure form and its categorial structure." (page 47) No one who appreciates the long struggle of man toward more adequate knowing would criticize Aristotle greatly for his adoption of a similar viewpoint 2,000 years ago, but after all that *was* 2,000

years ago; surely economists can do better than seek light on their subject from a beacon that was extinguished by the Galilean revolution in the 17th century. In this connection, Dr. Ratner again is helpful: "Modern scientists, however, began by taking precisely the world of change as their subject for scientific study, and to help them on their way, they introduced the method of experimentation which is no less and no other than a method whereby the natural changes going on can be further increased and complicated in manifold ways by changes deliberately made. From the Greek point of view (and in this case, *not* excepting any Greek), this is confounding confusion, science gone insane. But as events have fully demonstrated, it is science really come to its senses, and intelligence come into its own."[3]

As for von Mises' assertion that economists must rely on "cognition and analysis of our own purposeful behavior," this is the thoroughly discredited mode of knowing by introspection. Moreover, how can even the method of introspection be used if the knowledge praxeology provides is "a priori," is "not subject to verification or falsification on the ground of experience and facts?" If we find neither experience nor facts when we "analyze our own purposeful behavior," do we find anything at all?

Von Mises differentiates the natural sciences from what he calls the a priori sciences including praxeology, which he considers the basis for his economics. This, too, is an outmoded distinction, although many, perhaps most, economists agree with von Mises' views. As for such differentiation, Dr. Ratner comments as follows: "Why is it that in the technical fields of science, the revolution in *method* initiated by Galileo has already been substantially completed, has, in our time, carried through its last fundamental reform, whereas in other fields, including fields as intellectual as philosophy and logic, the revolution is just about now seriously getting under way? The easy answer is to invoke a distinction between 'natural' sciences and 'social' sciences. . .The 'distinction' simply repeats, as an explanation, the fact to be explained. . .there is a difference in the development of scientific investigation of the natural and the social *because* the former is 'natural' and the latter 'social.'

"The backwardness of philosophy, logic and all social inquiries does not explain the *forwardness* of the natural sciences. It simply exposes and emphasizes the need for an explanation. . .Let it be granted, for the sake of argument, that the natural sciences are *now* beyond the reach of influence or connection with social institutions, forces and all that goes with the latter. It is an undeniable fact of modern history—let alone of all human history—that they were not *always* there. Hence the more you conceive the social to be retarding or inherently inimical to the development of science, the more must the 'natural' sciences have been able to overcome in reaching their present estate. In so far as the 'natural' sciences are *now* distinguished and distinguishable from the 'social' sciences it is a *distinction* they have achieved; it is a *result*, not a gift ('something given' or a 'datum'); it is a *consequence*, not a cause. The invocation of the 'distinction' between 'natural' and 'social' *subject-matters* to explain the differences between 'natural' and 'social' *sciences* doesn't even explain the differences away. It just leaves them precisely where and as it *finds* them.

"A philosophy or logic of science cannot, without

[1] John Dewey, *Human Nature and Conduct*, New York: Modern Library, 1922, p. 31.
[2] *Ibid.*, pp. 176 and 177.

[3] Joseph Ratner, *Intelligence in the Modern World, John Dewey's Philosophy*, New York: Modern Library, 1939, p. 52. (Page 35 of the present volume.)

being foolish, take refuge in a 'distinction' in subject-matter to explain the advance of the natural sciences in modern times. And the more the 'distinction' is asserted to be *in rerum natura* as a ground for explanation, the greater the folly of the philosophy or logic becomes. "[1]

MODERN INFLUENCE

But the careful reader of von Mises' treatise will encounter at least one specific example where the influence of modern scientific method seemingly has overcome the author's best intentions. On page 547 et seq., he discusses inflation and business cycles, and on page 798 in criticizing various theories of the business cycle, he emphasizes certain facts of economic life and their relationships. He points out that both production and prices increase during a boom, a development that obviously would be impossible ordinarily without credit expansion. To summarize, he asserts in effect that any theory failing to include the part that credit expansion must play is discredited by the facts.

Now it seems obvious that prices are quite precise measures of the ratios at which some goods exchange for purchasing media or for other goods. Therefore, price rises are measures of change in such exchange relationships. Moreover, any assertion that production has increased implies more or less accurate measurement of the change alleged to have occurred. So also with the expansion of bank credit; it can be known only through measurement stated in terms of the additional amounts of purchasing media involved. Here are the significant aspects, for science, of what is widely called experiment, namely, measurement of change and study of the relationships between or among the changes.[2]

And if we turn to another passage, we find what von

Mises thinks of such proceedings. "Those economists who want to substitute 'quantitative economics' for what they call 'qualitative economics' are utterly mistaken. There are, in the field of economics, no constant relations, and consequently no measurement is possible." (page 55) In such disparaging terms does the author dispose of the statistical laboratorians who provided the test that he insists all theories of the business cycle, including his own, must meet.

I think it to von Mises' credit that he cannot resist the temptation to be a modern economic scientist. Far from considering this an inconsistency that should make him the butt of ridicule, I regard this particular inconsistency as one of his outstanding achievements. If only his fellow economists could similarly break the bonds that shackle them with the past, the science of economics probably would advance much farther, much faster.

VON MISES' CONTRIBUTION

Finally, I repeat that Dr. von Mises' treatise seems to me an outstanding contribution. He has boldly attempted to explain the assumptions and preconceptions on which his work is based, a task that few economists have had the ability or perhaps the courage to undertake. Unfortunately, he has but lightly touched on many semantic problems, particularly that of specification (scientific naming or use of word symbols) but few economists have troubled seriously about definitions in recent years. A few decades ago, the situation was different; nearly every textbook began with attempts to define the terms to be used. But, like many philosophers, the economists have all but given up this aspect of their task, apparently in the hopeless conviction that semantic confusion is a small price to pay for the retention of old and familiar methods.

Therefore, I believe that Dr. von Mises may have contributed far more than he had previously realized to the needed reconstruction in economics. The first task in reconstruction always is the demolition and removal of the structure that must be replaced. That task, I believe he has facilitated.

---

[1] *Ibid.*, pp. 69, 70, and 71. (p. 40 of the present volume.)

[2] True, the scope of the experiment is much larger than that of the chemist's experiment in a test tube, but it is on an exceedingly small scale in comparison with the scope of an astronomer's measurements. In the aspects significant for science, and therefore for knowing in the modern sense, this is the essence of experiment.

VII.

WOULD GOVERNMENT SUPPORT BE A

"BOOBY TRAP" FOR BEHAVIORAL SCIENTISTS?[1]

E. C. Harwood & Rollo Handy

THE book[2] reviewed here is by The Behavioral and Social Sciences Survey Committee, a group appointed jointly by the National Academy of Sciences and the Social Science Research Council. The number of influential persons associated with the project (see listing of names on pp. 275-279 and pp. vi-viii of the book) suggests that the general point of view expressed in the Report is shared by many contemporary behavioral scientists, possibly a majority of them.

In the first half of the book the authors discuss the field of behavioral inquiry, briefly describe many of the relevant disciplines, review the research methods currently in use, and discuss the social import of behavioral science findings. They recommend the following: the further development, with Congressional support, of a system of social indicators analogous to the President's Council of Economic Advisers' economic indicators; the private (non-governmental) development of an annual Social Report to the nation; the formation of a commission to devise a national data system for behavioral research purposes; the establishment of a federal government group to work on ways of protecting the anonymity of the individuals studied; the creation of graduate schools of applied behavioral science; and an increase in federal support funds for basic and applied behavioral research from between 12 per cent and 18 per cent per year over the next decade.

The second half of the book contains much useful, detailed information on the following: students and degrees granted in behavioral science fields; PhD–granting departments; the role of behavioral scientists in professional schools; behavioral research institutes; non-university behavioral research; federal and private support of behavioral research; and the situation in countries other than the U.S.A. The factual data are primarily for the 1966-67 academic year; numerous projections are made for 1976-77.

I

Many sections of the Report will not withstand critical inquiry. *At times* the procedures adopted in the Report are unscientific and superficial, the language used is loose, the relevant evidence is only partially mentioned, and controversial issues are slurred over. Some illustrations follow:

(1) The authors say: "The fact that social prediction will always be contingent upon subsequent events, and hence will always lack complete accuracy, means only that some estimate of the degree of uncertainty must enter into a responsible prediction" (p. 21). But *all* scientific prediction is "contingent upon subsequent events" (*e.g.*, a predicted eclipse of the moon will occur only if something untoward doesn't happen in the meantime); social prediction is no different from physical prediction in that respect.

(2) In discussing political science, the authors note the development of the "behavioralist" movement after World War II. They go on to say: "At first, this new approach was resisted by some who held to more classical political theory. Fortunately, the tensions that arose have largely disappeared, and now there is a recognized division of labor between the more classically oriented political theorists and the contemporary quantitatively oriented empiricists" (pp. 38-39). The use of modern scientific methods urged by the behavioralists is so fundamentally opposed to the nonscientific procedures commonly used by the classical political theorists that one wonders how a productive division of labor could be established between the two groups.

(3) In discussing the ways in which Game Theory can "illuminate" many types of group behavior, the authors note that in this nation no political party has been able to maintain a stable level of support much over the 50 per cent level for more than a short time, and that in European parliamentary coalitions usually only a small majority is maintained. The authors then say: "Unless we are to attribute these observed facts to coincidence, they must have an explanation somewhere deep in the machinery of democracy. Game Theory shows how an explanation of the 'minimum-size principle' can be derived rigorously from simple assumptions. The basic idea is that minimum-size majorities have all the power they need to govern; the price they must pay (in terms of concessions and compromises on issues) to attract or to retain additional adherents will be greater, the theorem shows, than anything the core group can hope to gain from the additional strength" (pp. 79-80). The reader is not told how the Game Theory explanation "illuminates" the "deep" workings of democracies. Nor is the reader given any evidence that Game Theory yields a more useful description of the occurrence of "minimum-size majorities" than a description simply in terms of the group-in-power's unwillingness to make more concessions than necessary to retain power. To translate the latter description into Game Theory notions (even if they "can be derived rigorously from simple assumptions") doesn't improve whatever scientific warrant the description may have.

(4) The authors cite a study of 80 cultures, based on the Human Relations Area Files, showing that in cultures "in which male children were subjected to various sorts of physical stress during the first two years

[1] Reprinted from *The American Journal of Economics and Sociology*, Vol. 30, No. 2, April, 1971.
[2] *The Behavioral and Social Sciences: Outlook and Needs.* By The Behavioral and Social Sciences Survey Committee. (Englewood Cliffs, N.J.: Prentice-Hall, 1969), 320 + xv pp., $7.95.

of their lives, the adult males averaged 2.7 inches taller than the adult males in those cultures in which the male infants were not so stressed, even though the racial backgrounds of the cultures were matched as carefully as possible. These somewhat surprising results are being checked by contemporary studies in Africa, for they appear to have implications for childrearing that should not be overlooked. Apparently an appropriate amount of physical stimulation may be a good thing in infancy" (p. 111). As it stands, such a statement (whatever may be in the original research report) strongly suggests a *post hoc, ergo propter hoc* procedure, and seems to assume, without giving evidence, that increasing height is an unqualified "good thing."

(5) In illustrating the benefits students may expect from studying behavioral science materials, the authors say that a student will learn "that mental illness is a product of traumatic relationships between individuals—parent and child, husband and wife, worker and supervisor, and so forth—with perhaps a genetic component as well in some kinds of illness" (p. 262). The emphasis on *individuals* ignores the socio-cultural aspects that many workers believe to be involved, and the passage quoted suggests that traumatic relationships always or often lead to mental illness.

(6) On the last page of the main text, the authors say: "the behavioral and social sciences are *potentially* some of the most revolutionary intellectual enterprises ever conceived by the mind of man. This is true basically because their findings call into question traditional assumptions about the nature of human nature, about the structure of society, and the unfolding of social processes. They challenge the inevitability of business cycles, the instructional and rehabilitative value of punishment, and the superiority of white skin. Psychology has already had a powerful impact on child-rearing and on adults' views of their own sexuality. Economics has shaken traditional faith in the unregulated market and weakened resistance to planned and directed economies" (p. 272). The illustrations given in the latter part of the quotation are vague. No evidence is mentioned that psychologists or economists in fact have had the influences stated. Possibly those workers were as much influenced by the social changes vaguely referred to as they influenced those developments. And even if psychologists and economists had the impact suggested, we are given no evidence that their views are scientifically sound.

II

In addition to the criticisms indicated above, naive and sometimes inconsistent remarks are found in the book about the relation of behavioral scientists to the government. The authors give several reasons for urging that the proposed annual Social Report to the nation be "tried out on a private basis," including their fears that a Report sponsored by the government would be caught up in partisan issues and be less objective than a privately sponsored Report (pp. 106-07). Yet they also suggest that the Report might be taken over by government eventually, and in other parts of the book show little awareness of the problems posed by direct governmental sponsorship of behavioral research. The blithe ignoring of such issues seems especially inappropriate for behavioral scientists, who presumably should have shed their political innocence and should be especially vigilant in defending their freedom of inquiry.

The authors are favorably impressed by the work of the Council of Economic Advisers and suggest the establishment at some future time of a "permanent council of social advisers" (p. 109). Yet as the Council of Economic Advisers now functions, the policy advice it gives can hardly deviate from what the President deems politically expedient. Such an official advisory group is simply not in a position to urge publicly the elimination of unsound economic policies that are strongly supported by the President and the Congress. A group of scientists could report privately to the President without encountering such difficulties. Or a scientific group could be responsible to the general public. But in our political system official public status for an advisory group is a strong guarantee that its policy recommendations will harmonize with the views of the group in power.

Although the authors emphasize how controversial issues in behavioral inquiry can be, they give no indication that such matters have a bearing on increased federal funding. Scientifically warranted assertions and policy recommendations based on those assertions may be so unpalatable to politicians that they will not provide the financial support desired or needed. The lure of large amounts of federal money may lead behavioral scientists into a situation in which they cannot function as scientists but can function only as special pleaders for the politicians in power.

Already apparent is the fact that some existing governments recognize how important the work of behavioral scientists will become. For example: the Communist party in the Soviet Union has taken great pains to control the work of behavioral scientists with a view of ensuring that they serve the interests of those in power; and in the United States during recent years each political party when in office has used some behavioral scientists in ways evidently intended to further the retention of power.

Many people today are so impressed with the benign aspects of democratic or republican forms of government that they forget the lessons of history. The first democratic government in Europe following the French Revolution, which was inspired in part by the success of the American Revolution, beheaded Lavoisier, the father of modern chemistry. On the other hand, much early scientific work in the 17th and 18th centuries was made possible because benevolent despots in various European countries chose to defy some religious leaders and protect a few scientists as well as support their inquiries. More recently, the economic advisers of an American President apparently have endorsed economic action so unsound that, in the words of a distinguished Harvard professor, it should "make every economist blush."

III

In short, an important lesson to be learned from the experiences of history is that scientists should not expect to be assured of unrestricted freedom of inquiry and discussion as the servants of the government, any form of government, nor by any vested interests having special privileges or positions of power that those interests desire to defend and perpetuate. Especially should behavioral scientists be wary of becoming the tool of agencies that may inhibit full freedom of inquiry and discussion, because, of the three major fields of science—physical, physiological, and behavioral—the last deals almost continuously with controversial matters of consequence to one or another vested interest.

In recent years, the Behavioral Research Council[1] has suggested a code for behavioral scientists analogous to the Hippocratic Oath and the legal code of ethics. It is:

My primary and overriding moral commitment or obligation is to serve as a behavioral scientist for the purpose of seeking solutions for the problems of men in society and publicly informing my fellow citizens as to the results of such scientific research. This implies:

(1) Relying in such inquiries on the methods of modern sciences in their evolutionary development.

(2) Endeavoring continually to improve my own ability as a scientist to develop warranted 'if-then'

conclusions or assertions by applying scientific methods and by subordinating any personal biases in order to assure objectivity in my work and findings.

(3) Avoiding all conflicts of interest (such as might result from employment by special interests, etc.) that might inhibit scientific work or bias me in any way tending to pervert scientific inquiry.

(4) Differentiating clearly in all writings and public statements so that those to whom I communicate will understand whether I am speaking or writing in my role as a scientist within my field of competence or am simply urging in my role as a citizen or in some other specified role a course of action that I personally prefer.

(5) Criticizing as unscientific, without fear or favor, all purportedly scientific reports within my field of competence that (in the absence of such criticism) could be expected seriously to mislead my fellow citizens, whom I have chosen to serve.

Does anyone imagine that men who were conscientiously following such a code would choose to be dependent on funds from any government?

Note: We now (1972) should amend the last sentence above to read, "Does anyone imagine that men who were conscientiously following such a code would choose to be dependent on funds from any government unless provided a status independent of political pressures comparable with that of the United States Supreme Court?" Hopefully, this also reflects a further advance in our development.

[1] During the 1950's, George A. Lundberg, Stuart C. Dodd, and E.C. Harwood held conferences with leading behavioral scientists in Claremont, California; Seattle, and New York for the purpose of forming a new organization expected to facilitate cooperation among behavioral scientists in the various fields. Development of common methods, of technical terminology (to the extent practicable) applicable in all the fields of inquiry, and of the cross-fertilization that might be expected to result were to be aims of the new organization. The Behavioral Research Council was formally organized in 1960 at Claremont, California.

The first research project undertaken was a survey of progress in all of the behavioral sciences. The results of this research were published in *A Current Appraisal of the Behavioral Sciences* in 1964. Because this publication has been widely acclaimed by reviewers in many of the scientific journals as the most comprehensive and useful publication of its kind ever published, the Behavioral Research Council is undertaking a revision of the first edition with anticipation of a completed second edition in 1971.

CONCLUSION

IN the course of this volume a good many specific issues have been considered, including some that are highly technical and complex. At this stage the reader may find it useful to refresh his memory by reviewing the short Orientation chapter (p. 3) in order to recall the general direction of the advance we believe has been made in describing useful procedures of inquiry.

We harbor no illusion that the procedures we tentatively recommend for trial in inquiry will find widespread and immediate acceptance. The cultural lag will not be easily overcome. Yet the price to be paid for the comfort of adhering to outmoded "ways of knowing" is so high that we believe further progress in civilization may well depend upon replacing those outmoded procedures. The problems of men-in-society have become so acute that they may form a more serious threat to human survival and welfare than were the scourges of famine and plague in the past.

There is, of course, no dearth of enthusiastically supported pseudo-solutions to our pressing social problems. Unfortunately, diametrically opposed policies are often supported by "experts," each of whom is confident of having achieved the "truth." Man's long experience with a variety of medicine men has apparently not been sufficient to turn us away from a reliance on such procedures, disappointing and destructive as they often have been.

Some 35 years ago, in the final paragraph of his *Logic*, John Dewey described the consequences of adhering to traditional views about "knowledge":

"Theories of knowledge that constitute what are now called epistemologies have arisen because knowledge and obtaining knowledge have not been conceived in terms of the operations by which, in the continuum of experiential inquiry, stable beliefs are progressively obtained and utilized. Because they are not constructed upon the ground of operations and conceived in terms of their actual procedures and consequences, they are necessarily formed in terms of preconceptions derived from various sources, mainly cosmological in ancient and mainly psychological (directly or indirectly) in modern theory. Logic thus loses its autonomy, a fact which signifies more than that a formal theory has been crippled. The loss signifies that logic as the generalized account of the means by which sound beliefs on any subject are attained and tested has parted company with the actual practices by means of which such beliefs are established. Failure to institute a logic based inclusively and exclusively upon the operations of inquiry has enormous cultural consequences. It encourages obscurantism; it promotes acceptance of beliefs formed before methods of inquiry had reached their present estate; and it tends to relegate scientific (that is, competent) methods of inquiry to a specialized technical field. Since scientific methods simply exhibit free intelligence operating in the best manner available at a given time, the cultural waste, confusion and distortion that results from the failure to use these methods, in all fields of connection with all problems, is incalculable. These considerations reinforce the claim of logical theory, as the theory of inquiry, to assume and to hold a position of primary human importance."[1]

To which we add only the following: Having given up the "quest for certainty," we do *not* assert that applying modern procedures of inquiry will surely solve the most pressing problems of men-in-society. However, the past successes of those procedures when applied, along with the repeated failures of other procedures, offers great hope that present social problems may be solved as usefully as physical and physiological problems have been solved.

---

1 John Dewey, *Logic: The Theory of Inquiry*, New York, Holt, Rinehart, and Winston, 1938, pp. 534-535.

APPENDIX: TRIAL NAMES[1]

Rollo Handy & E. C. Harwood

A. PRELIMINARY COMMENTS

Language problems frequently impede communication in behavioral scientists' discussions of their inquiries and of the methods applied in such inquiries. This report presents a glossary of some important terms in order to diagnose some of the inconsistencies, incoherencies, or other inadequacies of language and to suggest trial names that may prove useful to behavioral scientists.[2] Unfortunately, misunderstandings easily occur, even in the initial stages of discussion; consequently, aspects of the problem will be discussed before the trial names are suggested.

Many attempts have been made to improve naming in the behavioral sciences, and an extensive literature is concerned with definitions. In this report, no detailed attempt is made to compare our procedures with others. We begin with the procedures developed by Dewey and Bentley. In order to avoid misunderstanding, we emphasize that we are not attempting to develop or prescribe any final group of names. As Dewey and Bentley say:

"The scientific method neither presupposes nor implies any set, rigid, theoretical position. We are too well aware of the futility of efforts to achieve greater dependability of communication and consequent mutual understanding by methods of imposition. In advancing fields of research, inquirers proceed by doing all they can to make clear to themselves and to others the points of view and the hypotheses by means of which their work is carried on." (Page 89; all citations to *Knowing and the Known* are to the pages as numbered in the reprinting of that book in the present volume.)

They further say of their procedure:

"It demands that statements be made as descriptions of events in terms of durations in time and areas in space. It excludes assertions of fixity and attempts to impose them. It installs openness and flexibility in the very process of knowing. . . .We wish the tests of openness and flexibility to be applied to our work; any attempts to impose fixity would be a denial—a rupture—of the very method we employ." (*K&K*, p. 89).

Our intention has been to continue the Dewey-Bentley line of advance, if it is an advance, without assuming that it necessarily is the only or even the best way to proceed. If improvement in efficiency of communication results, some progress will have been made. If instead our work impedes communication, it should be superseded by something more useful.

"Trial" is used here then, to indicate that we do not seek to fix permanently, or even standardize for a long time, the terminology suggested. Under some circumstances, standardization of terminology may have little or no scientific use. The standardization of names in alchemy or astrology, for example, would be pointless for scientific purposes (except in the sense that if all astrologers agreed, refutation of their views might be easier). As scientific inquirers proceed, new similiarities and differences will be discovered in the subject matter of inquiry; consequently, a fixed terminology probably would be a barrier to progress.

"Name" is used here in the Dewey-Bentley manner (See *K&K*, pp. 132-133), although we realize that others use that word differently. Names here are *not* regarded as things separate from, and intermediate between, the organism and its environment. Rather the focus is on naming *behavior*; on an organism-environmental transaction. Conventionally, a sharp separation has been made between a word and its so-called "meaning," but here we attempt to keep the whole naming process in view. For us, the import of "H_2O" as a scientific name is understood in relation to current scientific practices; "H_2O" is a shorthand label for certain aspects of a subject matter of inquiry, including the relations among those aspects, as observed by scientists. To concentrate on "H_2O" as a set of marks or sounds radically separated from the thing named, as some epistemologists do, is considered an undesirable separation of things that, from the viewpoint of our purpose here, usually are found together. Specifically, separation of the word, its so-called "meaning," and the word user, frequently results in hypostatization and seemingly insoluble problems of the locus and status of "meanings" and of "knowledge."

In the present context naming is the aspect of knowing with which we are concerned. Naming behavior, as Dewey and Bentley say, "selects, discriminates, identifies, locates, orders, arranges, systematizes." (*K&K*, p. 133.)

Naming can be made "firmer," be more consistently useful, without restricting future revisions. For crude everyday purposes, naming a whale a fish may be useful; but to name it a mammal marks an improvement from the viewpoint of scientific usefulness. Revisions as to what "atom" is used to designate or name also have provided improved naming.

Our procedures in preparing this report are transactional. "Transaction" here designates or is a name for

[1] Reprinted, with revisions, from Alfred de Grazia, Rollo Handy, E.C. Harwood, and Paul Kurtz, eds., *The Behavioral Sciences: Essays in Honor of George A. Lundberg*, Great Barrington, Mass., Behavioral Research Council, 1968.

[2] This report relies heavily on the work of John Dewey and Arthur Bentley. See especially their *Knowing and the Known*, Boston, Beacon Press, 1949; reprinted in full in Rollo Handy and E.C. Harwood, *Useful Procedures of Inquiry*, Great Barrington, Mass., Behavioral Research Council, 1973; and Sidney Ratner and Jules Altman, eds., *John Dewey and Arthur F. Bentley: A Philosophical Correspondence, 1932-1951*, New Brunswick, N.J., Rutgers University Press, 1964.

This report also makes use of the survey of the behavioral sciences by Rollo Handy and Paul Kurtz, first edition, *A Current Appraisal of the Behavioral Sciences*, Great Barrington, Mass., Behavioral Research Council, 1964. A revised edition is now in preparation.

the full ongoing process in a field where all aspects and phases of the field as well as the inquirer himself are in common process. A transactional report is differentiated from self-actional reports (in which independent actors, powers, minds, etc., are assumed to function) and from interactional reports (in which presumptively independent things are found in causal interconnection). "Borrower can not borrow without lender to lend, nor lender lend without borrower to borrow, the loan being a transaction that is identifiable only in the wider transaction of the full legal-commercial system in which it is present as occurrence." *(K&K, p. 130.)*

The work and accomplishments of scientists have been described in many different ways, and no attempt is made here to settle all controversies or to endorse dogmatically any one view. Perhaps most can agree, however, that an important part of the scientist's job is the increasingly more useful description of things, including their connections and relations, that are differentiated in the cosmos.

Some authors attempt to distinguish sharply between "description" and "explanation." "Description" is used here to *include* what many refer to as "explanation," rather than in a way that contrasts a "mere" or "bare" description with a scientific "explanation." Obviously scientists seek to improve the crude descriptions of common sense, but their improved reports on their subject matter (i.e., what some label "explanations") are also descriptions in the broad sense. For example, a stick partially submerged in water appears to be bent, and a crude description may go no further than to so state. But if a more adequate description is given, in terms of light refraction, human processes of perception, human language habits, etc., then we have what is sometimes called an "explanation." The explanation of the bent appearance consists in a full description of the whole transactional process, which enables us to predict what normal human observers will see, given certain circumstances.

"Warranted assertion" is used here rather than "true statement" (or "true proposition"). "Warranted assertion" seems an appropriate name for the outcome of successful scientific inquiry. The term helps to remind us that the assertion involved is warranted by the procedures of inquiry and is subject to modification or rejection by further inquiry. It also helps to exorcise the ghost that scientists have as their business the discovering of final and fixed generalizations.

As inquiry proceeds, modification of naming is to be expected. The differentiation of water from the rest of the cosmos is useful for daily life, but adopting the scientific name "H_2O" marked an improvement in that further prediction and control was facilitated. Perhaps the development of physics and chemistry will some day result in the further alteration of the naming for what in everyday life is called water.

We deny emphatically that there is any kind of intrinsic or necessary relation between the marks and sounds used in naming and what is named. In that sense, naming is wholly conventional; whether "water," "aqua," or "gkim" is used to refer to a certain liquid makes no difference. (This is not to deny, of course, that specific words are part of particular languages, and identifying "water" as a noun in the English language affords many clues as to how the word will be used by English speakers.) On the other hand, some names are much more useful than others. "H_2O", for example, as used in current physical science, is quite different from "water" taken as designating one of the assumed four primordial elements. Although the whole notational system now used

for chemical elements and their combinations is in an important sense descriptive, once the system is chosen, naming within it is determined in major respects by the system. "H_2O" as shorthand for water is not capriciously chosen but rather is the outcome of painstaking and carefully controlled inquiry. In general, then, although there is no ultimately right naming, and although all naming is conventional, scientific naming is neither capricious nor arbitrary.

Sometimes those who object that naming is too simple a process to be of much importance in scientific inquiry take a much different view of the naming process than that offered here. If strong emphasis is put on naming in relation to assertions warranted by testing, then some of those objections, at least, seem to be met. To have labels for differentiated aspects of the cosmos that have been thoroughly tested is one thing. To elaborate a terminology that stands either for aspects that have not been usefully differentiated, or for supposed aspects inconsistent with well-established "if—then" statements, is quite another matter. Perhaps both "phlogiston" and "caloric" had considerable merit as names consistently usable for various processes that at one time were assumed to occur in heat phenomena. Their deficiencies, from the present point of view, were precisely that they did not name differentiated aspects of the cosmos as found by scientific inquiry.

When those terms became entrenched in scientific discourse, however, they were not easily evicted; they were part of a semantic vested interest. Much the same almost certainly applies to many behavioral science terms now in wide and frequent use. Sometimes suggested changes in naming are rejected on the ground that new specifications (scientific namings) omit important connotations the term had in ordinary discourse or in earlier science. Here again the importance of testing can hardly be overemphasized. Rejection of "phlogiston" doubtless omitted what was once dear to many people, yet scientific progress apparently benefited from those omissions.

"Specification" is used here to refer to the naming that has been found useful in science. Specification is a different process than some of the processes frequently named "definition." "Definition" has been used to refer to such diverse things that confusion often results. As Dewey and Bentley say:

"The one word 'definition' is expected to cover acts and products, words and things, accurate descriptions and tentative descriptions, mathematical equivalences and exact formulations, ostensive definitions, sensations and perceptions in logical report, 'ultimates,' and finally even 'indefinables.' No one word, anywhere in careful technical research, should be required to handle so many tasks." *(K&K, p. 148.)*

Broadly speaking, "definition" often is used to apply to almost any procedure for saying what the so-called "meaning" of a term is. Much of the difficulty with "definition" seems to be just its linkage with "meaning." But leaving that problem aside, a considerable variety of procedures have been used in attempts somehow to designate what a term stands for or has been applied to, and many of those procedures are highly dubious from a scientific point of view.

In this report, "specification" is used as a name for scientific naming; i.e., the efficient (especially useful) kind of designation found in modern scientific inquiry.

B. FURTHER DEVELOPMENT OF SOME
BASIC NAMES

In striving for agreement on some firm, coherent, and consistent naming, proceeding initially along roughly evolutionary lines may be helpful. "Cosmos" was selected to name the sum total of the things we can see, smell, taste, hear, and feel (often aided by instruments), including connections among those things, so that we can talk about the sum total of things without repeatedly having to describe them in detail. "Cosmos" is applied to the universe as a whole system, including the speaking-naming thing who uses that name. Moreover, "cosmos" is the name for all that is included in man's knowing behavior from the most distant past discussed in scientifically warranted assertions to the probable future insofar as it is known by scientifically warranted predictions.

Next we differentiate among the vast number of things in the cosmos and select the living things; for these we choose the name "organism." Note that selecting for naming does not imply detaching the physical thing from the cosmos. Everything named remains a part of cosmos with innumerable relations to other parts.

Among the organisms, we further differentiate for the purpose of the present discussion and select for naming ourselves, our ancestors, and our progeny; these we name "man."

We then observe the transactions of man with other aspects and phases of cosmos and note the transactions named "eating," "breathing," etc. Among those numerous transactions, we differentiate further and select for naming the transactions typical of man but found infrequently or not at all in other organisms.

This type of behavior involves processes of a kind such that something stands for or is assumed to refer to something else. Such processes we name "sign behavior," or simply "sign." Note that "sign" is not the name of the thing that stands for something else; "sign," as used here, is the name of the transaction as a whole; i.e., "sign" is the short name for "sign process." For example, the word "cup" is not taken as the sign for the vessel we drink coffee from; rather the word, the container, and the word user all are regarded as aspects or phases (sometimes both) of the full situation. Sign process is the type of transaction that distinguishes some behavioral from physiological processes, a knowing behavior transaction from a transaction such as eating, digesting, seeing, etc. (But no absolute or ultimate separation is suggested; sign processes always include physiological processes and may affect those processes, as when the reading of a telegram containing bad news affects respiration.)

Sign process in evolutionary development has progressed through the following still-existing stages:

a. The signaling or perceptive-manipulative stage of sign in transactions such as beckoning, whistling, frowning, etc.

b. The naming stage as used generally in speaking and writing.

c. The symboling stage as used in symbolic logic and mathematics.

Focusing our attention now on the naming stage of sign process, we choose to name it "designating." Designating always is behavior, an organism-environmental transaction typical primarily, if not exclusively, of man in the cosmos. Designating includes:

1. The earliest stage of designating or naming in the evolutionary scale, which we shall name "cueing."

"By Cue is to be understood the most primitive language-behavior. . . .Cue, as primitive naming, is so close to the situation of its origin that at times it enters almost as if a signal itself. Face-to-face perceptive situations are characteristic of its type of locus. It may include cry, expletive, or other single-word sentences, or any onomatopoeic utterance; and in fully developed language it may appear as an interjection, exclamation, abbreviated utterance, or other casually practical communicative convenience." *(K&K,* p. 136.)

2. A more advanced level of designating or naming in the evolutionary scale, which we shall name "characterizing." This name applies to the everyday use of words; usage that is reasonably adequate for many practical purposes of life.

3. For the, at present, farthest advanced level of designation we use "specifying." This name applies to the highly developed naming behavior best exhibited in modern scientific inquiry.

For the purpose of economizing words in discourse, we need a general name for the bits and pieces of cosmos differentiated and named. For this general name we choose "fact." Fact is the name for cosmos in course of being known through naming by man (with man included among the aspects of cosmos) in a statement sufficiently developed to exhibit temporal and spatial localizations. Fact includes all namings-named durationally and extensionally spread; it is not limited to what is differentiated and named by any one man at any moment or in his life time.

Frequently we need to discuss a limited range of fact where our attention is focused for the time being. For this we choose the name "situation." This is the blanket name for those facts localized in time and space for our immediate attention.

Within a situation we frequently have occasion to refer to durational changes among facts. For these we choose the name "events."

Finally, in discussing events we usually have occasion to refer to aspects of the fact involved that are least vague or more firmly determined and more accurately specified. For those we choose the name "object." Object is an aspect of the subject matter of inquiry insofar as it has reached an orderly and settled form, at least for the time being.

Further tentative comments on sign process may be helpful. The transition from sign process at the perceptive-manipulative stage (here designated "signaling") to the initial naming stage (designated "cueing") is a change from the simplest attention-getting procedures, by evolutionary stages, to a somewhat more complex sign process that begins to describe things and events. No clear line of demarcation is found. Some perceptive-manipulative signalings and primitive word cues are descriptive as well as simple alerting behavior.

The transition from cueing to characterizing also reflects evolutionary development with increasing complexity of process, including formal grammar, etc. The further transition from characterizing to specifying in the manner of modern science reflects the further evolutionary development of sign process, a still more complicated procedure.

At first thought the stage we have here designated "symboling" may seem to be a marked departure from, or to reflect a break in, the evolutionary development of sign process. However, mathematical symboling, at least as frequently used in scientific inquiry, may be considered shorthand specifying. Each symbol replaces one or more words. A single mathematical equation may replace a long and involved sentence, even a paragraph, or a longer description in words.

Sometimes symboling is considered to be different from naming, and even Dewey and Bentley speak of it as an "advance of sign beyond naming, accompanied by disappearance of specific reference such as naming develops." *(K&K*, p. 178.) Mathematical inquiry seems in some respects to differ in kind from the designation used in empirical inquiry, yet the mathematical symbols used in scientific inquiry designate something quite specific; equivalences or other relations, for example. For the purposes of empirical inquiry, aspects of the formal mathematical structure are used to facilitate summarizing and focusing attention on relations among things.

Thus sign process in its evolutionary progress to date may be described as the efforts of man to communicate: first by simple perceptive-manipulative processes; then by verbal processes of increasing complexity, until this increasing complexity of verbal procedure became so much of a barrier to further progress that a shorthand system was devised in order to facilitate further communication. This shorthand system has been most extensively developed in mathematical symboling.

C. LIST OF TRIAL NAMES

Many of the names below were taken from Ch. 11 of Dewey and Bentley's *Knowing and the Known*, while others were used in *A Current Appraisal of the Behavioral Sciences*. The importance of the names does not stem from their sources, but rather from their aid in facilitating communication. The names below are provisionally claimed to be important in the sense that we found them useful in trying to communicate more successfully among ourselves. (In some instances names are listed because we found them to be barriers to mutual understanding.) However, other names overlooked by us may prove to be even more useful than those we here discuss, and some of those presently regarded as useful may prove to be grossly misleading on further inquiry.

A final suggestion to the reader: The prevalence of interactional and self-actional theoretical assumptions may make transactional procedures unfamiliar at first sight. With reference to nomenclature, what seems obvious in self-actional or interactional terms frequently is deficient from a transactional point of view.[1] What may seem odd, peculiar, or overly simple—judged in terms of an acceptance of other procedures—becomes useful, appropriate, and sometimes necessary, given transactional procedures.

For example, Dewey and Bentley have been severely criticized for neglecting what the critics regard as obvious and necessary for all work in the field: distinguishing radically between psychology and logic. Their reply follows:

"We may assure all such critics that from early youth we have been aware of an *academic* and

---

[1] The prevalence of nontransactional behavior in inquiry reflects linguistic habits not easily changed. For example, although the authors of *A Current Appraisal of the Behavioral Sciences* adopted a transactional method, in the first edition they sometimes inadvertently separated "internal"—"external," "individual"—"social," "organism"—"environment," and a word from its so-called "meaning," with resulting incoherence. The discussion in the glossary section of the present report suggests the dangers of fusing "biological" and "physiological," and helps to point out the lack of clarity in some of the uses of "operational" and "specification."

pedagogical distinction of logical from psychological. We certainly make no attempt to deny it, and we do not disregard it. Quite the contrary. Facing this distinction in the presence of actual life processes and behaviors of human beings, we deny any rigid *factual* difference such as the academic treatment implies. . . .We have as strong an objection to the assumption of a science of psychology severed from a logic and yet held basic to that logic, as we have to a logic severed from a psychology and proclaimed as if it existed in a realm of its own where it regards itself as basic to the psychology. *We regard knowings and reasonings and mathematical and scientific adventurings even up to their highest abstractions, as activities of men—as veritably men's behaviors*—and we regard the study of these particular knowing behaviors as lying within the general field of behavioral inquiry. . . ." *(K&K*, p. 180; emphasis in last sentence not in original.)

Note: In the entries below, some quotations are taken from *Knowing and the Known*, Ch. 11. Unless otherwise indicated, we agree with the material quoted.

ACCURATE: Dewey and Bentley suggest this adjective to "characterize degrees of achievement" in the range of specification. However, "degrees of achievement" seems to imply some standards of comparison; standards that we do not have. We suggest that names in the range of scientific specification may be more or less accurate in the sense of more or less painstakingly chosen and applied. Perhaps Dewey and Bentley were naming the same characteristics of naming behavior by their phrase "degrees of achievement." We suggest that "accurate" be used as a short name for "to date found most useful scientifically or by scientists." See PRECISE.

ACTION, ACTIVITY: These words are used here only to characterize loosely durational-extensional subject matters of inquiry. The words suggest self-actional or interactional assumptions in which actions are the doings of independent selves, minds, etc., separated from the full organism-environmental transaction; procedures that are rejected here for inquiry into knowings-known. See INTERACTION; SELF– ACTION; BEHAVIOR.

ACTOR: A confusing although widely used word. "Actor" often is used in ways that unfortunately separate the doer too sharply from the complex behavioral transaction. "Actor" here is used only in the sense of "Trans-actor," the human aspect of a behavioral situation.

APPLICATION: In the terminology adopted here, a name is said to be *applied* to the thing named. Use of "application" helps to avoid the connotation of some intrinsic or necessary relation between the thing named and the marks or sounds used in naming.

ASPECT: The name for any differentiated part of a full transaction, without special durational stress. (For the latter see PHASE.) The aspects are not taken as independent "reals." In a borrower-lender transaction, the borrower, the lender, and what is lent are among the aspects of the transaction. Those aspects are inseparable in that there is no borrowing without lending, and vice-versa.

BEHAVIOR: The name here covers all the adjustmental processes of organism-in-environment. This differs from other uses that limit "behavior" to the muscular and glandular actions of organisms in "purposive" processes, or to the "external" rather than "internal" processes of the organism. "Behavior" here is always used transactionally, never as of the organism alone, but instead as of the organism-environmental process. (This is not to deny that *provisional* separation of organism and environment, within a transactional framework, can be useful in inquiry.)

BEHAVIORISM: Although many conflicting behaviorist procedures of inquiry can be found, a common feature is the rejection of traditional mentalistic and nonscientific procedures. We agree that the latter should be rejected. However, care should be taken to distinguish our transactional procedures from many types of behaviorism, because some behaviorists regard behavior as occurring strictly within the organism or regard behavior as physiological. Our rejection of traditional presuppositions should not be understood as implying exclusion of physiological processes; we include them as aspects of sign behavior. (See SIGN BEHAVIOR; TRANSACTION.)

BIOLOGICAL: The name given here to those processes in living organisms that are not currently explorable by the techniques of the physical sciences alone. Biological inquiry covers inquiry into both physiological and sign-behavior. No ultimate separation between the physical and biological "realms" is assumed, nor do we assume that present physical and physiological techniques of inquiry will remain unchanged. Perhaps future inquiry will make our present divis.ons of subject matters unsuitable. See PHYSICAL.

CHARACTERIZING: This name is applied to the everyday use of words that is reasonably adequate for many practical purposes. Characterizing is more advanced than cueing, but less advanced than specifying.

CIRCULARITY: In self-actional and interactional procedures, circularity may constitute a grievous fault. In explicitly transactional inquiry, some circularity is to be expected. For example, the description of useful procedures of inquiry is based on the observation of past successful inquiries; that description in turn may help to improve future inquiries; which in turn may lead to an improved description of procedures; etc. Some critics of Dewey and Bentley regard the type of circularity found in *Knowing and the Known* as a major flaw, but they apparently fail to grasp the significance of the Dewey-Bentley procedures.

COHERENCE: The word is applied by us not to the internal consistency of a set of symbols, but to the connection found in scientific inquiry to obtain between or among objects. Not logical connection, then, but the kind of "hanging together" that occurs in observed regularities, is what is named.

CONCEPT, CONCEPTION: "Concept" is used in so many ways, especially in mentalistic and hypostatized forms, and in ways separating the sign from the sign-user, that its total avoidance is here recommended. "Conception" is frequently construed as a "mentalistic entity," but sometimes as a synonym for a point of view provisionally held and to be inquired into. Even in the latter instance, the word may have mentalistic connotations. We are convinced that it is not useful because it so often is a semantic trap for the unwary.

CONJECTURE: When description is blocked in inquiry, the inquirer imagines what may be happening; "conjecture" designates such a tentative notion about possible connections among facts. In view of the other applications found for "hypothesis," we suggest "conjecture" as a replacement for that name. See HYPOTHESIS, THEORY.

CONNECTION: In naming-knowing transactions, the general name for the linkages among the aspects of a process, as found through inquiry. In an observed regularity, the things involved in the regularity are said to be connected. "Connection" covers the relations some-

times referred to as "causal," "statistical," "probabilistic," "structural-functional," etc.

CONSCIOUSNESS: Not used by us unless as a synonym for "awareness."

CONSISTENCY: Discourse found to be free of contradictory and of contrary assertions is characterized as consistent.

CONTEXT: Here used transactionally to refer to the mutually related circumstances and conditions under which things (objects and events) are observed.

COSMOS: Names the sum total of things we can see, smell, taste, hear, and feel (often aided by instruments), including connections among those things. "Cosmos" is applied to the universe as a whole system, including the speaking-naming thing who uses the name "cosmos." Observable durations extend across cultures, backward into the historical-geological record, and forward into indefinite futures as subject matters of inquiry. Not to be construed as something underlying knowing-knowns yet itself unknowable.

CUE: The earliest stage of designating or naming in the evolutionary scale. Primitive naming, here called "cueing," is close to signaling, and no clear line of demarcation between them is found. The differentiation is made on the basis that organized language occurs in cueing. Some psychologists apply "cue" to what we name "signal," and vice-versa. If such psychological use develops firmly, our use will be superseded.

DEFINITION: Often used in a broad sense to cover any procedure for indicating the "meaning" of a term, including: the stipulation of the application of a term in technical contexts (as when "ohm" is chosen as the name for a unit of electrical resistance); descriptions of the uses a term has in everyday speech; equations relating a single symbol and a combination of symbols for which the single symbol is an abbreviation (as in symbolic logic); what is here called "specifying"; as well as many other procedures. Also used to refer to a description of the "nature" or "essence" of a thing. In view of the many widely varying procedures to which "definition" has been applied, we avoid the term here. See SPECIFYING.

DESCRIPTION: Expansion of naming or designating in order to communicate about things (including situations, events, objects, and relations) on which attention is focused.

DESIGNATING: Always considered here transactionally as behavior. Includes cueing, characterizing, and specifying. When naming and named are viewed in common process, "designating" refers to the naming aspect of the transaction. Designating is the knowing-naming *aspect* of fact.

ENTITY: Its use often presupposes self-actional or interactional procedures, and especially some independent-of-all-else kind of existence. Not used here. See THING.

ENVIRONMENT: *Not* considered here as something surrounding, and fully separable from organisms; but as one aspect of organism-environmental transactions. The apparently plausible separation of organism from environment breaks down when one attempts to locate and consistently describe the exact demarcation between organism and environment. For some purposes of inquiry, focusing attention primarily on either the organic or the environmental aspect of the whole transaction may be useful.

EPISTEMOLOGICAL: To the extent the use of "epistemological" supposes that knowers and knowns are fully separable, the word is incompatible with transactional procedures and is not used here.

EVENT: The name chosen here for durational changes among facts upon which attention is focused for purposes of inquiry.

EXACT: See PRECISE, ACCURATE.

EXCITATION: To be used in reference to physiological organism-environmental processes when differentiation between such physiological stimulation and sign-behavioral stimulation is desired. See STIMULUS.

EXISTENCE: The known-named aspect of fact. Physical, physiological, and behavioral subject matters are regarded here as equally existing. However, "existence" should not be considered as referring to any "reality" supposedly supporting the known but itself unknowable.

EXPERIENCE: "This word has two radically opposed uses in current discussion. These overlap and shift so as to cause continual confusion and unintentional misrepresentation. One stands for short extensive-durational process, an extreme form of which is identification of an isolated sensory event or 'sensation' as an ultimate unit of inquiry. The other covers the entire spatially extensive, temporally durational application; and here it is a counterpart for the word 'cosmos'." "Experience" sometimes is used to name something considered to be primarily localized in the organism ("he experienced delight") or to what includes much beyond the organism ("the experience of the nation at war"); to relatively short durational-extensional processes ("he experienced a twinge") and to relatively vast processes ("the experience of the race"). "The word 'experience' should be dropped entirely from discussion unless held strictly to a single definite use: that, namely, of calling attention to the fact that *Existence* has organism and environment as its aspects, and cannot be identified with either as an independent isolate." See BEHAVIOR.

FACT: The cosmos in course of being known through naming by organisms, themselves being always among its aspects. Fact is the general name for bits and pieces of cosmos as known through naming, in a statement sufficiently developed to exhibit temporal and spatial localizations. Man is included among the aspects of cosmos. "It is knowings-knowns, durationally and extensionally spread; not what is known to and named by any one organism in any passing moment, nor to any one organism in its lifetime. Fact is under way among organisms advancing in a cosmos, itself under advance as known. The word 'fact,' etymologically from *factum*, something done, with its temporal implications, is much better fitted for the broad use here suggested than for either of its extreme and less common, though more pretentious applications: on the one hand for an independent 'real'; on the other for a 'mentally' endorsed report."

FIELD: "On physical analogies this word should have important application in behavioral inquiry. The physicist's uses, however, are still undergoing reconstructions, and the definite correspondence needed for behavioral application can not be established. Too many current projects for the use of the word have been parasitic. Thorough transactional studies of behaviors on their own account are needed to establish behavioral field in its own right." "Field" here names a cluster of connected facts as found in inquiry. We do not use "field" as the name for a presumed separate environment in which independent facts are found; "field" names the entire complex process of mutually connected things and their relations on which attention is focused, and includes the observer in the transaction.

FIRM: Namings are firm to the extent that they are found to be useful for consistent and coherent communication about things, including events. Firmness, thus demonstrated, involves no implication of finality or of immunity to being superseded as scientific inquiry advances.

HUMAN: The word used to differentiate ourselves, our ancestors, and our progeny from the remainder of the cosmos. No ultimate division of the cosmos into man, other organisms, and physical objects is intended. Nor, obviously, do we intend by our naming to deny man's evolutionary development from other organisms, or the myriad connections man has with other aspects of the cosmos.

HYPOTHESIS: In the literature on methodology, "hypothesis" sometimes is applied to any conjecture about possible connections among facts, but sometimes is restricted to relatively exact formulations that may emerge in an advanced stage of inquiry. Sometimes "hypothesis" is embedded in the terminology of traditional logic and epistemology, as when a hypothesis is said to be a proposition not known to be true or false initially, but from which consequences are deduced; if sufficient deductions are confirmed, the "hypothesis" is said to become a "truth." To avoid confusion, we suggest replacing "hypothesis" by "conjecture." See CONJECTURE, THEORY.

IDEA, IDEAL: "Underlying differences of employment are so many and wide that, where these words are used, it should be made clear whether they are used behaviorally or as names of presumed existences taken to be strictly mental." "Idea" may be serviceable as referring to a notion about things.

INDIVIDUAL: "Abandonment of this word and of all substitutes for it seems essential wherever a positive *general theory* is undertaken or planned. Minor specialized studies in individualized phrasing should expressly name the limits of the application of the word, and beyond that should hold themselves firmly within such limits." In the transactional framework here adopted, "behavior" covers both so-called "individual" and "social" behavior, which are aspects of behavioral transactions. See BEHAVIOR.

INQUIRY: "A strictly transactional name. It is an equivalent of knowing, but preferable as a name because of its freedom from 'mentalistic' associations." Scientific inquiry is the attempt to develop ever more accurate descriptions (including what are often called "explanations") of the things and their relations that are differentiated in cosmos, in order to facilitate prediction and control (or adjustive behavior thereto). Statements about the observed regularities, measurements of change, etc., are formulated as warranted assertions.

INTER: "This prefix has two sets of applications (see Oxford Dictionary). One is for 'between,' 'in-between,' or 'between the parts of.' The other is for 'mutually,' 'reciprocally.'" (E.g., this prefix sometimes is applied to the relation "in-between," as when mind and body are said to interact in the pineal gland, or that a tennis ball is intermediate in size between a golf ball and a soft ball. Sometimes "inter" is used for mutually reciprocal relations, as in the interaction of borrower and lender. "The result of this shifting use as it enters philosophy, logic, and psychology, no matter how inadvertent, is ambiguity and undependability." The habit of mingling without clarification the two sets of implications is easily acquired; we use "inter" for instances in which the "in-between" sense is dominant, and the prefix "trans" is used where mutually reciprocal influence is included.

INTERACTION: "This word, because of its prefix, is

undoubtedly the source of much of the more serious difficulty in discussion at the present time." Some authors use "interaction" in the way "transaction" is used here. We restrict "interaction" to instances in which presumptively independent things are balanced against each other in causal interconnection, as in Newtonian mechanics. For inquiry into knowing-knowns, such an interactional procedure is rejected. See TRANSACTION.

KNOWINGS: Organic aspects of transactionally observed behaviors. Here considered in the familiar central range of namings-knowings.

KNOWLEDGE: "In current employment this word is too wide and vague to be a *name* of anything in particular. The butterfly 'knows' how to mate, presumably without learning; the dog 'knows' its master through learning; man 'knows' through learning how to do an immense number of things in the way of arts or abilities; he also 'knows' physics, and 'knows' mathematics; he knows *that*, *what*, and *how*. It should require only a moderate acquaintance with philosophical literature to observe that the vagueness and ambiguity of the word 'knowledge' accounts for a large number of the traditional 'problems' called *the problem of knowledge*. The issues that must be faced before firm use is gained are: Does the word 'knowledge' indicate something the organism possesses or produces? Or does it indicate something the organism confronts or with which it comes into contact? Can either of these viewpoints be coherently maintained? If not, what change in preliminary description must be sought?" See WARRANTED ASSERTION.

KNOWNS: "Known" refers to one aspect of transactionally observed behaviors, i.e., to what is named. "In the case of namings-knowings the range of the knowns is that of existence within fact or cosmos, not in a limitation to the recognized affirmations of the moment, but in process of advance in long durations."

LANGUAGE: Here viewed transactionally as behavior of men (with the possibility open that inquiry may show that other organisms also exhibit language behavior). Word-users here are not split from word-meanings, nor word-meanings from words.

MANIPULATION: See PERCEPTION-MANIPULATION.

MATHEMATICS: Here regarded as a behavior developing out of naming activities and specializing in symboling, or shorthand naming. See SYMBOLING.

MATTER, MATERIAL: See PHYSICAL. If the word "mental" is dropped, the word "material" (in the sense of matter as opposed to mind) falls out also.

MEANING: Not used here, because of confusion engendered by past and current uses. Transactional procedures of inquiry reject the split between bodies-devoid-of-meaning and disembodied meanings.

MENTAL: Not used here. Its use typically reflects the hypostatization of one aspect of sign behavior.

NAME, NAMING, NAMED: Naming is here regarded as a form of knowing. Names are *not* considered here as third things separate from and intermediate between the organism and its environment. Naming transactions are language behavior in its central ranges. Naming behavior states, selects, identifies, orders, systematizes, etc. We at times use "designating" as a synonym for "naming."

OBJECT: Within fact, and within its existential phase, object is that which has been most firmly specified, and is thus distinguished from situation and event. Object is an aspect of situation inquired into insofar as useful description or firm naming of that aspect has been achieved.

OBJECTIVE: Used here only in the sense of "impartial" or "unbiased."

OBSERVATION: Used here transactionally, rather than as a separated "activity" of the observer. Observation and reports upon it are regarded as tentative and hypothetical. Observation is not limited to "sense-perception" in the narrow sense; i.e., to a "simple" sensory quality or some other supposed "content" of such short time-span as to have no or few connections. Observation refers to what is accessible and attainable publicly. Both knowings and electrons, for example, are taken as being as observable as trees or chairs.

OPERATION: "The word 'operation' as applied to behavior in recent methodological discussions should be thoroughly overhauled and given the full transactional status that such words as 'process' and 'activity' require. The military use of the word is suggestive of the way to deal with it."

OPERATIONISM: This has become a confusing word, and sometimes seems to be merely an invocation of scientific virtue. "Operational definition" sometimes refers to defining phrases having an "if—then" form ("x is water soluble"="if x is immersed in water, then it dissolves"); sometimes to the insistence that the criteria of application of a word be expressed in terms of experimental procedures; and sometimes to a statement of the observable objects and events that are covered in the use of a word. On some occasions, "operational definition" apparently is used to refer to something similar to, if not identical with, what we call "specification" or scientific naming. See SPECIFYING.

ORGANISM: Used here to differentiate living things from other things in the cosmos, but *not* to detach organisms from their many connections with other aspects of cosmos. Organisms are selected for separate naming for methodological purposes, not as constituting something separated from the rest of cosmos.

PERCEPT: In the transactional framework, a percept is regarded as an aspect of signaling behavior, not as a hypostatized independent something.

PERCEPTION-MANIPULATION: Although perception and manipulation are regarded as radically different in some procedures of inquiry, transactionally viewed they have a common behavioral status. They occur jointly and inseparably in the range of what is here called signal behavior.

PHASE: Used for an aspect of cosmos when attention is focused on the duration of a time sequence, as when referring to the various phases of the manufacture and distribution of products.

PHENOMENON: Used here for provisional identification of situations. Not to be construed as "subjective," nor as a mere appearance of an underlying reality.

PHYSICAL: At present, we find three major divisions of subject matter of inquiry: physical, physiological, and sign-behavioral. These divisions are made on the basis of present techniques of inquiry, not on the basis of assumed essential or ontological differences. See BIOLOGICAL.

PHYSIOLOGICAL: "That portion of biological inquiry which forms the second outstanding division of the subjectmatter of all inquiry as at present in process; differentiated from the physical by the techniques of inquiry employed more significantly than by mention of its specialized organic locus." See BEHAVIORISM.

PRECISE: Dewey and Bentley use "exact" as an adjective to describe symbols, and "accurate" to describe specifying. We question the usefulness of differentiating between specifying and symboling other than to point out that the latter seems to be shorthand for the former. Because symbols are often used in connection with relatively precise measurements for the purposes of scientific inquiry, we suggest that "precise" may be more useful than "exact" as an adjective characterizing any symbolizing. Symbols are precise to the extent that they are shorthand names for precise measurements or what could be precise measurements. See ACCURATE.

PROCESS: To be used aspectually or phasally as naming a series of related events.

PROPOSITION: Used sometimes in the context of logic to name the states-of-affairs to which statements (or assertions, or sentences) refer. Thus "The dog is black" and "Der Hund ist schwarz" are said to express the same proposition. Generally such procedures make sharp distinctions among words, word-users, and "meanings," or among namers, nameds, and names. Such separations are here rejected, and along with them go many related distinctions. We regard the talkings (including namings, thinkings, reasonings, etc.) of man as human behavior rather than as third things somehow occurring between men and what they talk about, and we believe that proceeding in this manner not only avoids many needless mysteries but aids scientific inquiry into such talkings.

QUEST FOR CERTAINTY: In prescientific inquiry, the attempt to discover an eternal and immutable "reality" that can be known with complete certainty. We do not assert the absolute nonexistence of such "reality," but point out the failure to find it and the barrier such a notion has been to scientific progress. In somewhat disguised forms, the quest for certainty crops up in purportedly scientific investigations, as in attempts to find a certain and indubitable base upon which inquiry rests.

REACTION: In physiological stimulation (as contrasted with sign-behavioral stimulation), "excitation" and "reaction" are coupled as aspects of the stimulation transaction. See EXCITATION, STIMULUS.

REAL: Used sparingly as a synonym for "genuine," in opposition to "sham" or "counterfeit."

REALITY: "As commonly used, it may rank as the most metaphysical of all words in the most obnoxious sense of metaphysics, since it is supposed to name something which lies underneath and behind all knowing, and yet, as Reality, something incapable of being known in fact and as fact."

RESPONSE: In signaling behavior, as differentiated from physiological stimulation, "stimulus" and "response" are coupled as aspects of the stimulation transaction.

SCIENCE, SCIENTIFIC: "Our use of this word is to designate the most advanced stage of specification of our times—the 'best knowledge' by the tests of employment and indicated growth."

SELF: Within the framework here adopted, "self" names one aspect of organism-environmental transactions, rather than an hypostatized "entity."

SELF-ACTION: "Used to indicate various primitive treatments of the known, prior in historical development to interactional and transactional treatments." That is, used to refer to frameworks in which presumptively independent actors, minds, selves, etc., are viewed as causing events (as, for example, when gods are said to cause meteorological phenomena, or minds to create new ideas). "Rarely found today except in philosophical, logical, epistemological, and a few limited psychological regions of inquiry."

SIGN: The name applied here to organism-environmental transactions in which the organism involved in a situation accepts one thing as a reference or pointing to some other thing. "Sign" here is *not* the name of the thing that is taken as referring to something else; rather "sign" names the whole transaction. The evolutionary stages of "sign" are here named "signal," "name," and "symbol."

SIGNAL: Used here to refer to the perceptive-manipulative stage of sign process in transactions such as beckoning, whistling, frowning, etc. No clear line of demarcation between signaling and cueing is found; some perceptive-manipulative signalings are not only alerting behaviors, but also may begin to describe aspects of cosmos.

SIGN-BEHAVIOR: Sign-behavior refers to that range of biological inquiry in which the processes studied are not currently explorable by physical or physiological techniques alone. Human behavior here covers both so-called "social" and "individual" behavior. No ultimate or ontological separation of physical, physiological, and sign-behavior is assumed; the distinction made here concerns the techniques of inquiry found useful for various types of subject matters. See PHYSICAL, PHYSIOLOGICAL.

SIGN-PROCESS: Synonym for SIGN.

SITUATION: Used here as a blanket name for a limited range of fact, localized in time and space, upon which attention is focused. "In our transactional development, the word is not used in the sense of environment; if so used, it should not be allowed to introduce transactional implications tacitly."

SOCIAL: See INDIVIDUAL.

SPACE-TIME: Space and time are here used transactionally and behaviorally, rather than as fixed, given frames (formal, absolute, or Newtonian) or physical somethings. Bentley's words suggest our present approach: "The behaviors are present events conveying pasts into futures. They cannot be reduced to successions of instants nor to successions of locations. They themselves span extension and duration. The pasts and the futures are rather phases of behavior than its control."[1]

SPECIFYING: Used here to refer to the naming that has been found useful in science. "The most highly perfected naming behavior. Best exhibited in modern science. Requires freedom from the defectively realistic application of the form of syllogism commonly known as Aristotelian." Should *not* be mistaken as a synonym for "definition," at least in many senses of the latter word.

STIMULUS: Used in various ways in current inquiry, sometimes designating an object or group of objects in the environment, sometimes something in the organism (events in the receptors, for example), and sometimes something located elsewhere. The near chaos connected with this word strongly suggests the need for a transactional procedure. "Stimulating" may be a preferable term, inasmuch as it suggests a transactional process.

SUBJECT: Used here in the sense of "topics," as in "subject matter being inquired into," rather than in any sense postulating a radical separation of subject and object.

SUBJECTIVE: The usual subjective-objective dichotomy is rejected here, and what commonly are called "subject" and "object" are regarded as aspects of relevant transactions. However, inasmuch as some inquiries in philosophy and psychology still use procedures based on "subjective" analysis or introspection, we emphasize our objection to whatever is not publicly observable. Subjectivism, understood as a procedure of inquiry

[1] Arthur F. Bentley, *Inquiry Into Inquiries* (Sidney Ratner, ed.) Boston, Beacon Press, 1954, p. 222.

attempting to obtain scientifically useful "knowledge" from what is not publicly accessible, is rejected here.

SUBJECT MATTER: "Whatever is before inquiry where inquiry has the range of namings-named. The main divisions in present-day research are into physical, physiological, and behavioral."

SUBSTANCE: No word of this type has a place in the present system of naming.

SYMBOL: A shorthand naming component of symboling behavior. As used here, not to be hypostatized, but viewed transactionally and comparable with "name" and "signal."

SYMBOLING: Symboling, in scientific inquiry, is a shorthand means of specifying or scientifically naming. In the development of pure mathematics structures, consistency within the symbol system is of primary importance. In such instances the symbols do not directly designate specific things and events but rather designate potential relations. (E.g., "2" does not name the type of thing that "dog" does.) However, when mathematics is used in scientific inquiry, the mathematical symbols are applied to the subject matter; then the symbols become shorthand specifications or abbreviated names.

SYSTEM: Used here as a blanket name to refer to sets or assemblages of things associated together and viewed as a whole. Systems may be self-actional, interactional, or transactional. Typically used here in the transactional sense of "full-system," in which the components or aspects are not viewed as separate things except provisionally and for special purposes other than a full report on the whole situation.

TERM: "This word has today accurate use as a name only in mathematical formulation where, even permitting it several different applications, no confusion results. The phrase 'in terms of' is often convenient and, simply used, is harmless. In the older syllogism term long retained a surface appearance of exactness which it lost when the language-existence issues involved became too prominent. For the most part in current writing it seems to be used loosely for 'word carefully employed.' It is, however, frequently entangled in the difficulties of concept. Given sufficient agreement among workers, term could perhaps be safely used for the range of specification, and this without complications arising from its mathematical uses."

THEORY: Widely used in many differing applications; i.e., as conjecture, notion, hypothesis, final outcome of inquiry, etc. We suggest that "theory" be used to designate the description of what happens under specified circumstances. So used, a theory is highly warranted by the evidence presently available (e.g., the theory of evolution), but is subject to future correction, modification, or abandonment. See DESCRIPTION, WARRANTED ASSERTION.

THING: Used here as the general name for whatever is named. Things include both objects and events; any and every aspect of cosmos.

TIME: See SPACE-TIME.

TRANS: This prefix is used to indicate mutually reciprocal relations. See INTER.

TRANSACTION: Refers here to the full ongoing process in a field. In knowing-naming transactions, the connections among aspects of the field and the inquirer himself are in common process. To be distinguished from "interaction" and "self-action." See INTERACTION and SELF-ACTION.

TRUE, TRUTH: The many conflicting uses of these words incline us not to use them. In their senses of "can be relied upon," "in accordance with states-of-affairs," and "conformable to fact," they name what we call "warranted assertions." However, the connotation of permanence, fixity, and immutability suggests the quest for certainty. See WARRANTED ASSERTION.

VAGUE: This term refers to various types of inaccuracy and imprecision. Probably "vagueness" could profitably be replaced by other words indicating just what type of inaccuracy or imprecision is involved.

WARRANTED ASSERTION: Used here to refer to those assertions best certified by scientific inquiry. Such assertions are open to future correction, modification, and rejection; no finality is attributed to them. See INQUIRY.

WORD: As used here, there is no supposed separation of "meaning" from a physical vehicle somehow carrying that "meaning." Words are viewed transactionally as an aspect of knowing behavior; the subject matter is inquired into whole, as it comes, not as bifurcated.

INDEX